fourth edition

Comfort Heating

Billy C. Langley

PRENTICE HALL CAREER & TECHNOLOGY
Englewood Cliffs, New Jersey 07632

Library of Congress Cataloging-in-Publication Data

Langley, Billy C.
Comfort heating / Billy C. Langley. — 4th ed.
p. cm.
Includes index.
ISBN 0-13-151879-8
1. Heating—Equipment and supplies. 2. Heating—Equipment and supplies—Maintenance and repair. I. Title
TH7223.L28 1994 93-8220
CIP

Acquisitions Editor: Ed Francis
Editorial/production supervision and interior design:
Tally Morgan, WordCrafters Editorial Services, Inc.
Cover design: Marianne Frasco
Buyer: Ilene Sanford
Editorial assistant: Gloria Schaffer

Prentice-Hall, Inc.
A Paramount Communications Company
Englewood Cliffs, New Jersey 07632

Printed in the United States of America
10 9 8 7 6 5 4 3 2 1

ISBN 0-13-151879-8

Prentice-Hall International (UK) Limited, *London*
Prentice-Hall of Australia Pty. Limited, *Sydney*
Prentice-Hall Canada Inc., *Toronto*
Prentice-Hall Hispanoamericana, S.A., *Mexico*
Prentice-Hall of India Private Limited, *New Delhi*
Prentice-Hall of Japan, Inc., *Tokyo*
Simon & Schuster Asia Pte. Ltd., *Singapore*
Editora Prentice-Hall do Brasil, Ltda., *Rio de Janeiro*

contents

preface

TO THE INSTRUCTOR

Comfort Heating, fourth edition, is designed to provide the necessary facets of theory and practice for anyone interested in the subject of comfort heating with gas, oil, electricity, coal, or solar energy. The book is intended to be used as a curriculum guide, a textbook, or a course for independent study. The practical fundamentals are covered, as are recommended service and installation procedures. *Comfort Heating* serves as a comprehensive textbook for the second-year student and a valuable reference for the experienced technician.

Comfort Heating is the accumulation of more than 35 years of field experience in the areas of research, service, installation, business, and teaching. It is written from the dual viewpoints of education and field experience using practical and everyday terminology and language, thus preparing the reader for the working environment. Some of the newer types of heating equipment, including geothermal heat pump systems, and components are included.

In each area of the book, the author presents the necessary theory for both the reader and the service technician. Examples are used to help reinforce the presented theory. Operating, installation, service, and maintenance instructions from the various manufacturers who provided these instructions are incorporated wherever practical.

APPROACH, PHILOSOPHY, AND UNIQUE FEATURES

While the material in *Comfort Heating* is based on both theory and experience, emphasis is constantly placed on the practical facets of the heating phase of the heating, ventilating, air conditioning, and refrigeration industry to aid the reader in making efficient and correct diagnosis and repairs to the equipment.

Whenever practical, troubleshooting procedures are presented to the reader. The operating, maintenance, installation, troubleshooting, and service procedures are presented using both theory and field experience. The newer technologies involving heating equipment and service procedures are presented at appropriate points in the text. This information is presented in a manner that does not interfere with the reader's study of the material. The newer types of heating equipment are included to aid the reader in understanding how rapidly this industry is changing.

As a component is presented, its purpose, function, and operation are completely covered and discussed. This type of presentation aids the reader in understanding the purpose and operation of that component in the overall system without the need to search through the complete book page by page, thus breaking his or her concentration.

The performance-based objectives provided at the beginning of each chapter give the reader an indication of what minimum knowledge he or she can expect to receive from that portion of the text.

Each chapter presents a specific area of the comfort heating industry. Thus, the reader can refer to the appropriate chapter for specific information. Each chapter ends with a set of Review Questions, indicating the minimum material that the reader should understand from that chapter. Each chapter concludes with a summary. This summary may be used as a quick review for class-administered exams or for a brief review for reinforcement of the theories covered.

The text begins with an entire chapter devoted to a brief history of heating. Safety procedures and precautions are integrated into the text at the appropriate places. This will cause the reader to be more safety conscious while performing the maintenance and service procedures on the equipment.

The Review Questions at the end of each chapter provide an opportunity for student self-evaluation before taking an instructor-administered test. The Review Questions provide immediate feedback on the key concepts presented in that chapter. An extensive glossary provides the reader with a single place for reference to the technical terms used in the heating phase of this industry.

SUPPLEMENTS PACKAGE

There is an instructor's manual which is designed to aid the instructor in grading exams that are taken from the Review Questions at the end of each chapter.

CHAPTER ORGANIZATION

Chapter 1 is a brief history of heating, acquainting the reader with the first methods used for comfort heating. Chapter 2 is a presentation of the sources of heat used in comfort heating equipment and the combustion process used to obtain the heat from these sources.

Chapter 3 covers the orifices and burners used in comfort heating equipment. Their use and operation in the combustion process is also presented. Both natural-draft and forced-draft burners are covered. The different types of flames and what causes them are also discussed. Chapter 4 is a presentation of oil burners and their operation. The proper treatment of fuel oil and the necessary piping systems are presented. Gas-oil burners are also presented to aid the persons that may be working on these types of systems.

Chapter 5 presents the different types of heat exchangers used in comfort heating equipment to remove the heat from the heating medium and deliver it to the circulating air for the building. Chapter 6 discusses the venting systems that are used to remove the vent gases to outdoors for safe operation of the heating equipment. Power venting and gas burner combustion testing procedures are discussed in this chapter, as are automatic flue dampers and their operation.

Chapter 7 presents the methods used for electric heating. The different types of electric heating equipment, the required air flow, and how to calculate the amount of electricity that should be used for a given application are also presented. Heat pump systems are covered in Chapter 8. The conventional systems as well as the geothermal types are presented for study. The installation, maintenance, service, and operating procedures are discussed in plain, everyday language.

Chapter 9 presents refrigerant recovery, recycling, and reclaim. The Clean Air Act and what it means to the refrigeration and air conditioning industry; the reasons for the phaseout of CFC and HCFC refrigerants; the laws as they pertain to the service technician and the procedures used for recovering and recycling refrigerant from a system.

Chapter 10 presents comfort heating boilers and their use, operation, and maintenance. Both hot water and steam systems are discussed. Steam generation theory is also presented. Comfort heating furnaces are covered in Chapter 11. The ventilation requirements for gas furnaces are presented so that the student will be aware of the amount of air required for the proper operation of a gas furnace. Electric heating is also presented. Btu comparison methods are presented so that the reader will be able to select the correct equipment for the job at hand. High efficiency furnaces and their operation are discussed along with their special requirements.

Chapter 12 discusses infrared heating and how it is used in special applications for comfort heating. The Roberts-Gordon CO-RAY-VAC unit is covered in detail. Humidification and its benefits are presented in Chapter 13. How relative humidity is measured and the effects that it has on both the occupants and the structure are covered. Humidifier sizing and operation are also presented to the reader.

Chapter 14 is devoted to electric and electronic ignition systems used on comfort heating equipment. The operation, maintenance, and service of these systems are covered in detail.

Basic heating controls are presented in Chapter 15. The operation of these controls and how each one contributes to the safe and economical operation of the furnace is covered. Adjustment, maintenance, and troubleshooting techniques are covered in this chapter.

Chapter 16 presents the methods used to calculate the heat loss of a residential structure. Also included in this chapter are the procedures for estimating the operating cost of a furnace and how to compare the costs to other pieces of equipment or determine the approximate operating cost of the unit after it is installed.

Chapter 17 presents troubleshooting charts that can be used to troubleshoot electric heating units, gas heating units, oil units, and heat pump systems in all their various cycles. The room thermostat and other heating system controls are discussed. Chapter 18 is the glossary of terms used in the comfort heating industry.

TO THE STUDENT

Almost every student has asked at one time or another why he or she must learn a given topic, especially when the basic fundamentals are being studied. In most cases the student does not want to learn the basics, just the repair procedures. However, this is not a good way to learn anything because understanding of the basics of a subject must be known before effective troubleshooting is possible. The more a person knows about what is being done the easier the job will be. Without knowledge of the basics, unfamiliar problems will be extremely difficult, if not impossible, to diagnose.

This is especially true in the heating field. A wrong diagnosis and repair could result in property damage or even the death of someone due to faulty equipment operation. Even after completion of the formal studies it is a good idea to review the basics periodically so that you will not forget them. Often we do not review the basics and sometimes even try to bypass their requirements, resulting in a hazardous situation. Also, the more that you know the easier it is to diagnose and repair the equipment properly. The more effective you are at installation, service, maintenance, troubleshooting, and repair the faster you will get promotions and pay raises. Remember, someone can take away your material things but they cannot take away your knowledge. With knowledge you can get more material things. But without it little is possible.

Effective learning requires several steps. First it is necessary to take an active part in your education. Getting involved requires self-discipline and changing of habits. You must attend class regularly. Take an active part in the class activities. Ask questions about information that is not clear to you. When you are in the laboratory make use of the equipment to do the assigned activities. If you stand back and let someone else do all the laboratory exercises, you will not get the reinforcement that is needed to fully understand the material studied. You should also read the material for the next lesson before the next class period. In this way you can be prepared to ask questions about any material that is not clear to you. Also, when you ask questions someone else may also bring up a question that you had not thought of. This is all part of the learning process. Use it!

It is also best to set aside a particular period of time that you will study. Study at this same time on a regular basis, preferably every day, even if you just review the material. Many times something is revealed through review that was missed in the original study session. If a term is used that you are not familiar with, look it up in the Glossary and remember what it means for future use. Without knowing the terminology, a full understanding of the material is not possible. Learn the terminology. Every industry has a special terminology: learn the terminology for the heating industry so that you can speak intelligently about it and know what is being discussed.

Study the examples in the textbook. They are included to help you understand the material. When they are ignored, the learning process is slowed down. The figures used in the book will also help you understand the material. Use them! When a figure is referred to, study it along with the text for a maximum learning experience and retention of the material.

Sometimes it is helpful to read the Review Questions at the end of the chapter before studying the material. Do not try to memorize them. When you are reading through the text and find material that is related to the Review Questions you will probably remember it. When you have finished studying the chapter be sure to answer all of the Review Questions; this will allow you to review material that you do not understand before class time. Then the instructor can help you with any problems that you may have. The Review Questions indicate the minimum amount of knowledge that you should get from that chapter.

Take class notes: seldom do two instructors consider the same topic equally important. Frequently the instructor will put more emphasis on the areas that can be covered in his particular laboratory. This is perfectly satisfactory because learning several things well is usually better than learning a little bit about a lot of things. The laboratory is used to reinforce theories learned in the classroom and the textbook. The class notes are an additional source of review for the exam and an excellent reference for later use.

Active learning requires a great deal of effort on your part, but it will be well worth the time and effort in the long run. Your understanding of each area covered in the textbook will permit easier learning of other material later in life.

Be sure to ask questions about the areas that you do not fully understand. Remember, there are no dumb questions—only dumb answers. Continue to study and success will surely be yours.

Billy C. Langley

1

a brief history of heating

The objectives of this chapter are:

- To acquaint the reader with the evolution of heating from primitive to present times.
- To cause the reader to better appreciate the advancements made in the heating industry.

The originators of home heating were probably Stone Age people. With his primitive ways, Stone Age man had no means of producing heat for himself. He had to depend on nature to provide a means of keeping warm or a method of cooking food. This came in the form of lightning which started dried leaves or wood burning. Lightning, therefore, is akin to the modern electric ignitors used on modern heating units.

Archeologists have lamps that have been traced back thousands of years. Some of these lamps were made like bowls while others were made from human skulls. These crude lamps were used by prehistoric artists in old caves where they made etchings on the cave walls. They produced light to work by and enough heat to warm the artists' hands during the cold season. The fuel for the lamps was animal fat, with wicks made from dried moss, twisted into ropelike strands. By slightly stretching the imagination, this could be called the first oil burner.

The first central heating units were put to use by the ancient Romans and Chinese. Not wanting to get their magnificent castles dirty with soot, the Romans built a furnace pit, or hypocaust, beneath the buildings. The heat would penetrate the castles through a 12- or 14-inch floor and thus heat their homes with the first human-made radiant heating systems.

The Chinese approached the matter in a more advanced way. They built ovens with air passages beneath their homes and installed warm air pipes to the rooms to be heated. In the bedrooms, the beds were placed directly over the outlets. In this manner, the Chinese heated their homes with the first central warm air systems.

Chimneys came into being in the twelfth century. With the chimney came the fireplace, and the fuel was burned within the area to be heated, providing better use of the heat-giving source. They were, however, very expensive at first and only a few buildings were equipped with them. The chimney and fireplace removed smoke, smells, and a good deal of heat from the home. When burning, they created such a draft that special furniture was designed to protect the occupants.

Ben Thompson helped alleviate the draft problem somewhat by building a restrictor, or throat, in the chimney. The restriction slowed down the burning and allowed more of the heat to be put into the room. Ben Franklin improved on the restrictor by adding a path for the smoke that made it go through the restrictor, then down behind the fireplace, below the hearth, then out the chimney, thereby making greater use of the fuel.

The early common fuels for fireplace use were wood, peat, and charcoal. Shepherds burned dried manure and in some cases dried bones, called a bone-fire, which has been changed in modern language to bonfire. Coal was recognized by Theophrastus as a fuel. However, it was usually used as a semiprecious jewel to ward off evil spirits. People were executed in England for burning coal. In the fourteenth century, the forests in England were diminishing so rapidly that Parliament asked the king to change the law governing the burning of coal and allow its use as a fuel. However, it was Queen Elizabeth I who actually passed the act to conserve the forest and allow coal to be put to use.

Louis Savot, a Paris physician, designed and installed the first heat-circulating fireplace. It was installed in Louvre Palace around 1600 A.D. The bottom and back were made of metal. Air circulated through the fireplace by entering at the bottom, below the hearth, rose to the top by convection, and exited under the mantle, very much resembling heat exchangers in modern forced-air furnaces. The same principle is used in today's fireplaces.

The slower-burning coal fires allowed the grate to be used. The coal to be burned was placed on the grate, allowing greater aeration of the fuel and more complete burning. Also, the ashes would fall beneath the grate and could be removed more easily. Outlets were placed in the first floor ceiling and allowed the heat to enter the upstairs rooms and, after a fashion, heat them. Thus, they brought central heating one step closer. Because of the grate, the Franklin stove and its descendents caused the fireplace to become more decorative than useful.

Gas, the most popular of the modern fuels, was first used in the western world by balloonists. Pilatre de Rozier, the leading balloonist of France at the time, attempted to cross the English Channel in a combination airship that included a hot-air fire balloon connected with a gas balloon. In June 1785, high over the French coast, the airship exploded. All passengers were sent to a fiery death. Because of this incident, gas became known as a mysterious and powerful source of energy. Humans did not devise a means for controlling gas until many years later.

Fuel oil began to be recognized around 1860 as a plentiful source of energy for economical heating. In 1861, Werner, a mechanic, developed the first oil burner. In his invention, oil was trickled over preheated plates, where it turned into a vapor and could be readily ignited. This method is not too different from modern pot-type oil burners. In 1863, the first pressure-type spray oil burner was introduced by Brydges Adams.

Fuel oil, however, was not advancing alone. Coal was also making great strides in comfort heating. Living room stoves were being moved to the basement. The home was heated with ducts, pipes, and radiators. All of these devices, together with the advancements which made it cleaner to use coal, helped bring about practical central heating in the nineteenth century. The man of the household shook the grate and stoked the furnace in the morning and again in the evening. However, the pipes, ducts, and registers were an eyesore in the decor of the home. Because the design of the ducts was difficult and the home could not be evenly heated, steam boilers and radiators became more popular. However, the noise accompanying these early radiators made them undesirable.

Through highly sophisticated engineering, completely automatic oil heat emerged in about 1920. These units were controlled by a thermostat. There was no mess as with coal. Oil heat was accompanied by carefree operation, and the equipment was very durable. The heating equipment had automatic safety controls, which was a strong factor in oil replacing coal as a heating fuel. By about 1928 there were more than one-half million oil burners in use.

In the 1950s, pipelines brought natural gas to most of the larger cities and to the homes of most users. This convenience soon made natural gas the most popular home heating fuel. About this same time public utilities operating with the security of government franchises forced their way into the comfort heating markets. Each of these utilities claimed to be the most economical and dependable.

In the 1960s, electric utilities emerged on the heating market. At this time sophisticated graphs and charts were used to disguise the operating costs, which were more expensive than most people cared to pay.

Both gas and electricity claimed to be the only truly modern fuel. However, when we search back in history we find that the Chinese used gas for heating almost 3,000 years ago. Also, electric heat was being installed and put to use as far back as 1889 in Minneapolis, Minnesota.

2

heat sources and combustion

The objectives of this chapter are:

- To introduce to you the different sources of heat for comfort heating.
- To provide you with the basic characteristics of heating fuels.
- To acquaint you with the distribution systems used for different sources.
- To demonstrate to you the need for knowledge of proper combustion.
- To bring to your attention the requirements of a safe, economical heating system.
- To acquaint you with fuel oil combustion.
- To provide you with the proper oil burner combustion exchanger design.
- To acquaint you with the role of excess air in combustion.

HEAT SOURCES

Strictly speaking, fuel is any substance that releases heat when mixed with the proper amount of oxygen. Only those materials that ignite at relatively low temperatures, burn rapidly, and are easily obtained in large quantities at relatively low prices are considered good fuels.

The value of a fuel is derived from the amount of heat released when it is burned and the heat of combustion is measured. This value is obtained when a given amount of a fuel is burned under controlled conditions. The apparatus used for this purpose is called a *calorimeter.* The released heat is absorbed by a definite volume of water and the rise in temperature of the water is measured. The common ratings are given in British thermal units per pound (Btu/lb) or cubic foot (ft^3) of the fuel burned.

When a fuel contains hydrogen, the heat given off during combustion will depend on the state of the water vapor (H_2O) formed when the hydrogen is burned.

A heating value known as the higher, or gross, heating value is obtained when this water vapor is condensed and the latent heat of condensation is salvaged. If this water vapor is not condensed, the latent heat of vaporization is lost and this is known as the lower, or net, heating value.

The heating values of solid fuels are given in Btu/lb. Liquid fuels may be rated in Btu/lb or Btu/gal of the fuel. The Btu rating of gaseous fuels is given per ft^3. The gas industry uses the standard conditions of a temperature of 60° F, 30 inches of mercury pressure with a saturated condition with water vapor to determine these values.

The function of a heating system depends on some source of energy for proper operation. There are four basic sources of energy used to accomplish this tremendous job. They are (1) *solids,* (2) *liquids,* (3) *gases,* and (4) *electricity.*

Even with modern advancements and techniques, *coal* is still the most popular solid fuel used for heating. Coal is a mixture of carbonaceous material, minerals, and water. Coal was formed from vegetation that had gone through the decaying processes over a long period of time. This decayed matter was then covered with layer upon layer of earth and rock. The heat caused by the decaying vegetation and the pressure of the layers of earth caused the composition to turn into several ranks of coal.

Coal is obtained by one of two methods: open-pit or deep mining. When the open-pit method is used, the overlying earth is removed with earth-moving equipment and then the coal is recovered. When the deep mining method is used, a tunnel is sunk into the coal shaft proper. The coal is then removed by using explosives and loading it into cars or on conveyors that take it to the surface and then through the refining process.

Coal is divided into four classifications in decreasing order of rank. The classifications are anthracite, bituminous, subbituminous, and lignite. The higher ranking coals are classified according to their carbon content when they are dry. The lower ranking coals are classified according to their calorific value.

The two most popular coals used for heating are anthracite, or hard, and bituminous, or soft. Anthracite coal will release from 13,000 to 14,000 Btu/lb when burned with 9.6 pounds of air. Bituminous coal will release from 12,000 to 15,000 Btu/lb when burned with 10.3 pounds of air. Dry air at about 70° F has a volume of 13.3 ft^3/lb.

Coal is burned by the use of several methods. The oldest method used is hand-firing. In this method, coal is thrown onto the grates through the opening where the primary air is drawn. The fuel bed consists of a layer of ashes lying directly on the grates, a hot zone where the combustion occurs, a cooler, or distillation, zone where the gases are driven from the coal, and then a layer of green coal (Fig. 2-1). It is necessary to admit secondary air over the fire to obtain complete combustion for two reasons. First, hot carbon and carbon dioxide react to form combustible carbon monoxide. Second, there is distillation of gases from the distillation zone.

When hand-firing was used, one method was to push the hot coals to the back of the fire box and leave the green coal toward the front. The secondary air enter-

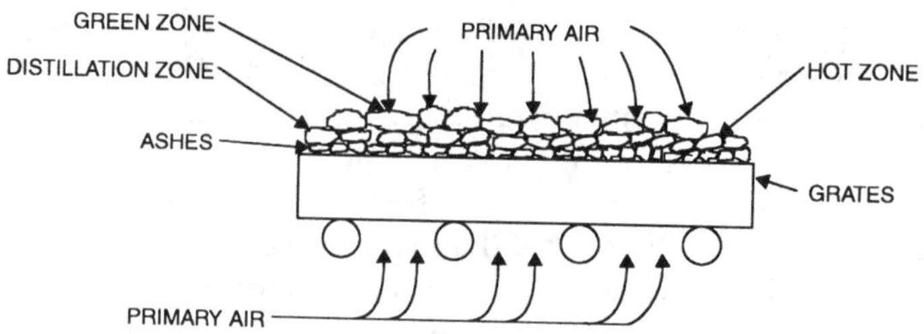
GREEN ZONE
DISTILLATION ZONE
ASHES
PRIMARY AIR
HOT ZONE
GRATES
PRIMARY AIR

FIGURE 2–1 Coal bed.

ing through the door would carry the gases over the hot coals, causing them to burn more completely than if the green coal were placed on the top.

Mechanical firing was the next method to appear for the use of coal as a heating fuel. This method used the principle of feeding coal and air simultaneously into the fire box. The coal is pushed into the fire box by means of a rotating worm or screw. The hotter part of the fire was either pushed to the rear or to the top of the fire box, depending on whether the front feed or the bottom feed was used.

Coal is not as popular as a heating fuel as it was at the turn of the century. Some of the factors contributing to its lack of use are inconvenience, storage problems, and pollution.

Fuel oil is the most common liquid used in the heating industry today. It is a petroleum product and is comprised of a mixture of liquid hydrocarbons produced as a byproduct of the refining of petroleum. Petroleum has been known for thousands of years. Seepages of oil and gas around the Caspian Sea and Black Sea were known and used for heating and cooking before the birth of Christ. The Chinese drilled for oil long before the Christian era. They used percussion bits, bamboo piping, and much human labor. The Chinese discovered gas and oil while drilling for salt.

Fuel oil is a stiff competitor of natural gas in the home heating industry. Fuel oil is graded and classified according to the range of distillation. The grades range from 1 to 6, omitting number 3. Grades 1, 2, and 4 are used for heating, with 1 and 2 the most popular for domestic use. Number 4 is generally used in light industrial furnaces. The lower graded numbers are more expensive than the higher numbered oils because of impurities, such as asphalt. Grades of 5 and 6 are too thick to be used in domestic equipment and require preheating to ensure a steady flow to the burner. The specifications governing fuel oils set forth by the U.S. Department of Commerce conform to American Society for Testing Materials (ASTM) specifications for fuel oils.

The proper combustion of fuel oils can be obtained only when the oil is properly atomized and mixed with air. The heat emitted by burning fuel oil ranges from 137,000 to 151,000 Btu/gal, depending on the grade. The heat content of number 1 is 137,000 Btu/gal and of number 2, the most popular domestic fuel oil, is 140,000 Btu/gal. (See Table 2–1.) The flash points of fuel oils will vary considerably because of the refining methods.

TABLE 2–1 Btu ratings of fuel oil

Grade	Heat/Gal
No. 1	132,900–137,000
No. 2	135,800–141,800
No. 4	143,100–148,100
No. 5	146,800–150,000
No. 6	151,300–155,900

The storage of fuel oil is a contributing factor in the efficient operation of the system. In cold climates, the oil should be stored inside or in some type of heating device, such as steam or hot water pipes around the tank or electric immersion heaters used to preheat it before entering the combustion area. The fuel oil may be stored outside in milder climates without it becoming excessively thick, thereby reducing the burner efficiency.

The use of fuel oils for domestic purposes is a result of the convenience and cleanliness of oil compared to coal. Despite the competition of natural gas, fuel oil is still a leader in home heating equipment, especially in the northeastern sections of the United States.

There are two major classifications of gaseous fuels used for comfort heating. The most familiar is *natural gas*. The second one is known as *LP gas*. The principal components of natural gas are methane, about 85%; ethane, about 12%; the other 3% is made up of propane and butane. They have no carbon monoxide, oxygen, olefins, or acetylene in their composition; however, some have large quantities of carbon dioxide, nitrogen, and hydrogen sulphide.

Natural gases are the lightest of all petroleum products. They are usually found where oil is found, but in some cases they are found elsewhere. Theorists have long argued about the exact origin of natural gas. However, most agree that natural gas was formed during the decomposition of plant and animal remains that were buried in prehistoric times. Because these plants and animals lived during the same period as those that are presently found as fossils, natural gas is sometimes called a *fossil fuel*.

Both natural gas and petroleum are mixtures of hydrocarbons. Both are considered fossil fuels and are composed of various chemicals obtained from the hydrogen and carbon contained in the prehistoric plants and animals.

The gas industry may be broken down into the various areas of exploration, production, transmission, and distribution.

The exploration section of the industry performs the function that its name implies. The people who are employed in exploration simply explore new areas, determine the location of the gas or petroleum field, and make the necessary reports, purchases, and other essential duties prior to the actual drilling.

When all the exploration functions are completed, the production department accomplishes the actual drilling of the well. The gas, or crude oil, is brought to the earth's surface and is blocked at this point. The well remains in this state until the gas or oil is needed.

When the need arises, the transmission department receives the gas from the well at pressures from 500 to 3,000 pounds per square inch gauge (psig). Even with these high pressures, the resistance of the pipe, and the distance covered, booster pumps are used to transfer the gas from the well to the refinery. At the refinery, the gas passes through a drying process that removes moisture, propane, and butane. During this process, most of the odor is also removed from the raw gas and an odorant is added to aid in leak detection.

After the refining processes are completed, the distribution department receives the gas through measuring gates where the number of ft^3 of gas is recorded. The gas is then passed through a series of regulators that reduce the pressure in steps. The steps are necessary to prevent the regulators from freezing and becoming inoperative.

The gas pressure is reduced to correspond with the requirements of one of two distribution systems, either the intermediate or the low pressure system. In most cases, the low pressure system is taken from the intermediate system. The intermediate distribution system maintains pressures from 18 to 20 psig while the low pressure system has a pressure of approximately 8 ounces. The low pressure system is generally employed where cast iron pipe is used for distribution. Cast iron pipe does not seem to hold the gas as well at higher pressures as does the copper or polyethylene pipe used in the intermediate systems.

When the intermediate system is used, the gas pressure is reduced from between 18 and 20 psig to 4 ½ ounces where it enters the house. Both systems use gas meters at this point, but the low pressure system does not require a regulator. Even though the 8 ounces of pressure in the low pressure distribution system is higher than that required in the meter loop (the meter loop is all components from the main line through the meter and regulator, if used), no regulator is needed as a result of the pressure drop. This pressure drop occurs because gas does not flow in a straight line but rolls instead, reducing the flow of gas about 10 ft^3 for each turn. This rolling action also brings about the need for straightening vanes in the line directly ahead of the gas meter. If these vanes were left out, the meter would not measure the flow of gas correctly.

On leaving the meter loop, the gas flows into the house piping. Most of the appliances used are manufactured to operate on 4 ½ ounces of gas pressure; however, natural gas furnaces are built to operate on 3 ½ inches of water column of gas pressure in the furnace manifold. This requires an additional regulator at the furnace.

The measurement of small pressures requires a manometer, a U-tube, or a manifold pressure gauge. A pressure indicated by water column is a very small amount. A pressure of one psi will support a column of water 2.31 ft high, or 27.7 inches. A pressure as low as 0.05 psi will support a column of water 1.39 inches high (27.7×0.05).

When the pressure is measured inside a pipe that is slightly higher than the atmospheric pressure, the water column in the U-tube will be depressed in one leg and pushed up in the other (Figure 2–2).

Natural gas has a specific gravity of 0.65, an ignition temperature of 1,100° F, and a burning temperature of 3,500° F. One cubic foot of natural gas will emit from 900 to 1,400 Btu/ft^3, with the greater amount of natural gas used as a heating fuel emitting approximately 1,100 Btu/ft^3. The Btu content of natural gas will vary from area to area. The local gas company should be consulted when the exact Btu content is desired.

Natural gas is made up of 55 to 98% methane (CH_4), 0.1 to 14% ethane (C_3H_8), and 0.5% carbon dioxide (CO_2). It requires 15 ft^3 of air per ft^3 of gas for proper combustion. It is lighter than air. Because methane and ethane have such low boiling

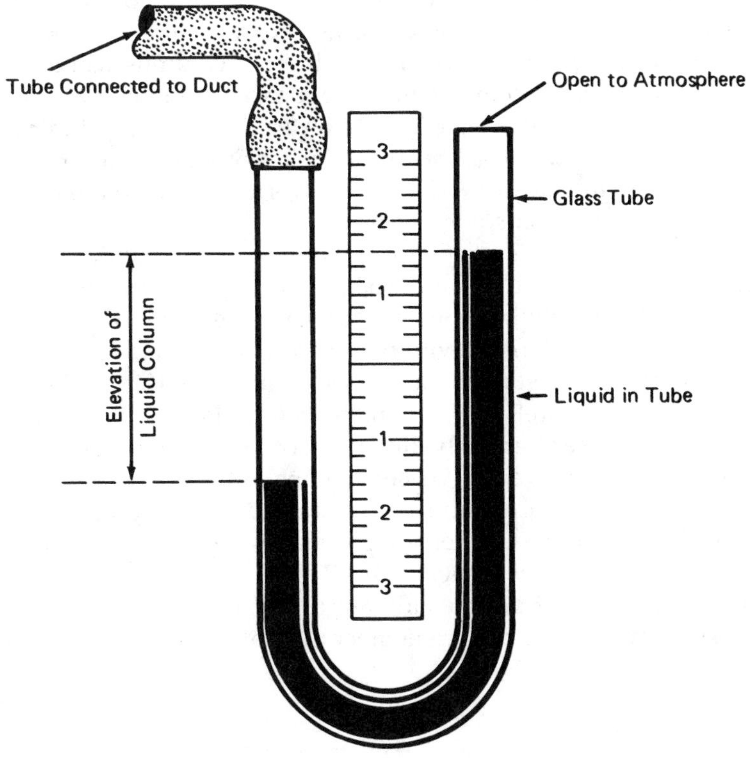
Tube Connected to Duct
Open to Atmosphere
Glass Tube
Elevation of Liquid Column
Liquid in Tube
3
2
1
1
2
3

FIGURE 2–2 U-tube manometer.

points, methane -258.7° F and ethane -127.5° F, natural gas remains a gas under the pressures and temperatures encountered during its distribution. Because of the varying amounts of methane and ethane, the boiling point of natural gas will vary according to the mixture.

The second classification of gaseous fuels is *liquefied petroleum (LP)* gas. Liquefied petroleum is both butane and propane, and in some cases a mixture of the two. These two fuels are refined natural gases and were developed for use in rural areas. They are transported by truck and stored in containers specifically made for LP gas installations.

Liquefied petroleum is a liquid until the vapor above it is drawn off. When these fuels are extracted from raw gas at the refinery, it is in the liquid state, under pressure, and remains in this state during storage and transportation. Only after the pressure is reduced does liquefied petroleum become a gaseous fuel.

LP gas has at least one definite advantage in that it is stored in the liquid state and thus the heat content is concentrated. This concentration of heat makes it economically feasible to provide service anywhere that portable cylinders may be used. The fact that 1 gallon of liquid propane turns to 36.31 ft^3 of gas when it is evaporated illustrates its feasibility. The height of the water column in the U-tube is 3.2 inches, which corresponds to a pressure of 0.116 psi inside the pipe to which the tube is connected. If there was no pressure difference between the inside and outside of the pipe, the water level in both columns would stand at exactly the same level. The pressures which are measured in gas furnace manifolds and air ducts are always measured with a water gauge, which indicates the pressure in inches of water column. The use of inches of water column when measuring low pressures eliminates the need for converting measurements made with a water gauge to pounds per square inch.

When LP gas is stored in the container, it is in both the liquid and gaseous state (Figure 2–3). To make the action of LP gas more easily understood, let's review briefly the boiling point and pressures of water (Figure 2–4). When the pressure cooker is first filled with water with no heat applied and the top remaining off, there is no pressure on the surface of the water [Figure 2–4(a)]. However, when the top is put securely in place and heat is applied, as in Figure 2–4(b), the pressure will begin to rise after the boiling point of the water is reached. As more heat is applied, more pressure will be created above the water by the evaporating liquid. When a constant temperature is maintained, a corresponding pressure will also be maintained. Likewise, when the pressure is reduced, the boiling point is reduced.

If we apply this principle to LP gas in a storage tank, we can see that as vapor is withdrawn from the tank more liquid will evaporate to replace that which was withdrawn. We must remember that each liquid has its own boiling point and pressure.

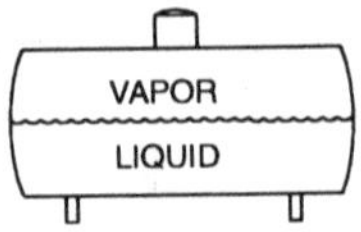
VAPOR
LIQUID

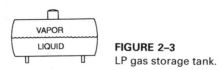

FIGURE 2–3
LP gas storage tank.

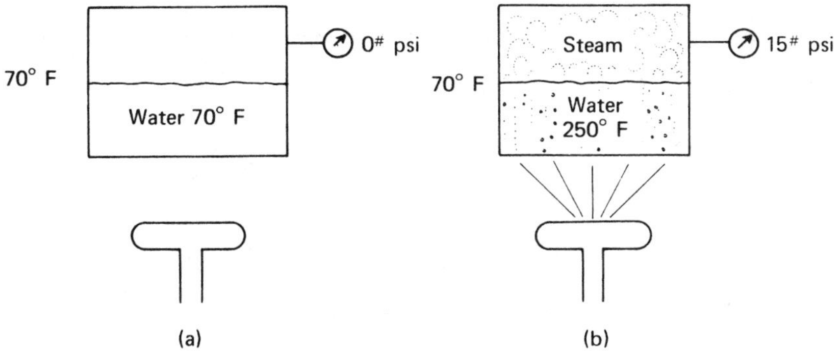
0# psi
70° F
Water 70° F
(a)
Steam
15# psi
70° F
Water
250° F
(b)

FIGURE 2–4 Pressure cooker.

Butane (C_4H_{10}), like propane (C_3H_8) and natural gas, is placed in the hydrocarbon series of gaseous fuels, because they are composed of hydrogen and carbon. Butane has a boiling point of 31.1° F, a specific gravity of 2, and a heating value of 3,267 Btu/ft^3 of vapor. It requires 30.97 ft^3 of air per ft^3 of vapor for proper combustion. At sea level, it has a gauge pressure of 36.9 pounds at 100° F. Butane expands to 31.75 ft^3 of vapor per gallon of liquid. It is heavier than air. The ignition temperature is approximately 1,100° F and the burning temperature is 3,300° F.

Propane has a boiling point of -43.8° F, a specific gravity of 1.52, and a heating value of 2,521 Btu/ft^3 of vapor. It requires 23.82 ft^3 of air per ft^3 of vapor for proper combustion. At sea level, it has a gauge pressure of 175.3 pounds at 100° F. Propane expands to 36.3 ft^3 of vapor per gallon of liquid. It also is heavier than air. It has an ignition temperature of 1,100° F and a burning temperature of 2,975° F.

When we study the physical properties of LP gases, we can see that each has both good and bad properties. These properties should be given a great deal of consideration when determining which fuel is to be used for any given application. The two characteristics deserving the most consideration are the Btu content and the vapor pressure. See Table 2–2 for a comparison of vapor pressures. When considering these fuels, the pressure is the major limiting factor, especially in colder climates. As we study the table, we can see that when the temperature of liquid butane reaches 30° F or lower, there is no pressure in the tank. Therefore, butane would not be suitable as a fuel at these lower temperatures without some source of heat for the storage tank. This source of heat may be steam or hot water pipes around the tank, electrical heaters around the tank, or even having the tank buried in the ground. However, these all add to the initial cost of the equipment.

If we look at the pressures of propane, we see that they are suitable throughout a wide range of temperatures. Therefore, from the pressure standpoint, propane would be the ideal fuel. On the other hand, the lower Btu rating makes it less desirable than butane. To overcome this dilemma, the two fuels may be mixed to obtain some of the desirable characteristics of each gas. An example of this may be a mixture of 60% butane and 40% propane. At a temperature of 30° F, it will have a vapor pressure of approximately 24 psig and a heat content of 2,950 Btu/ft^3. Since

TABLE 2–2 LP gas vapor pressures

TEMP. °F	PROPANE	BUTANE	TEMP. °F	PROPANE	BUTANE
0	38.2 psig		70	124 psig	16.9 psig
10	46		80	142.8	22.9
20	55.5		90	164	29.8
30	66.3		100	187	37.5
40	78	3 psig	110	212	46.1
50	91.8	6.9	120	240	56.1
60	107.1	11.6	130	272	66.1

these fuels are usually mixed before delivery to the local distributor, it is difficult to know exactly what the tank pressure and Btu/ft^3 are. As long as there is enough pressure in the storage tank to allow 11 inches water column of pressure to enter the house piping, there is little or nothing a service technician can do.

Before getting involved too deeply in work involving LP gases, the state and local authorities should be consulted. Some states maintain strict control over the personnel working with these fuels.

The use of *electricity* is becoming more important and more frequently used in comfort heating. Electricity is not a new phenomenon; humans have known of its existence for centuries. The early applications of the heat-producing ability of electricity were limited. It was used only in a few specialized areas, mainly industrial processes and as portable heaters to help supplement inefficient heating systems. Today, electric heating is adaptable to almost any type of construction and in any climate, and at a cost that most people can afford.

Electrical power is generated at the utility company's generating station. As it leaves the generating station, it passes through a bank of transformers to increase the voltage. The voltage is increased to aid in the transmission of the electricity. From the transformer bank, the electricity is distributed to user substations. The voltage is reduced at the substation by another bank of transformers to voltages that can be used by commercial manufacturing plants. The voltage is again reduced by the building's current transformer and is carried through the meter loop to the disconnect switch. From the disconnect switch, the electricity is distributed through the house wiring to the various appliances and electric heating units (Figure 2–5).

The major methods of using electricity for comfort heating are resistance heating, heat pump, and a combination of the two. There are three types of resistance elements used for electric heating: (1) the open wire, (2) the open ribbon, and (3) the tubular cased wire (Figure 2–6). An electric resistance heating element may be defined as an assembly consisting of a resistance wire, insulated supports, and terminals for connecting the electrical supply wire to the resistance wire. Resistance heating will convert electrical energy to heat energy at the rate of 3,410

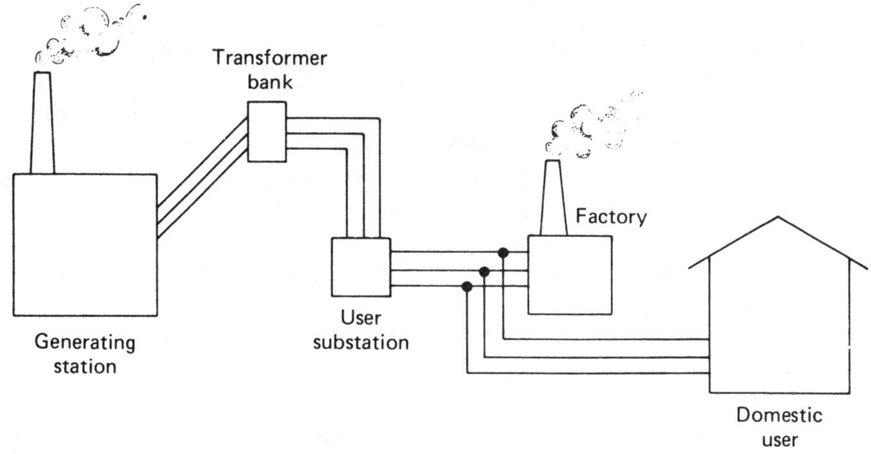
Transformer
bank
Factory
User
substation
Generating
station
Domestic
user

FIGURE 2–5 Electrical distribution.

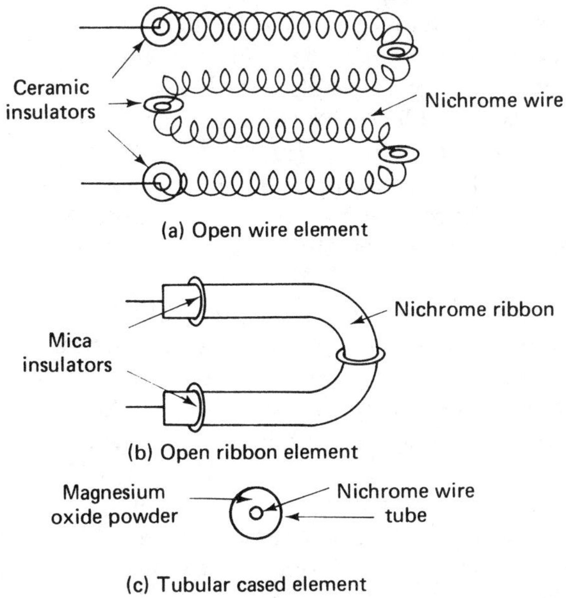
Ceramic
insulators
Nichrome wire
(a) Open wire element
Nichrome ribbon
Mica
insulators
(b) Open ribbon element
Magnesium
oxide powder
Nichrome wire
tube
(c) Tubular cased element

FIGURE 2–6
Resistance elements.

Btu/kW (1,000 watts). Theoretically, electrical heating is 100% efficient; that is, for each Btu input to the heating equipment, 1 Btu in usable heat is recovered.

The open wire heating elements are usually made of nichrome wire, which is wire made from nickel and chromium, but without iron, wound in a spring-like shape and mounted in ceramic insulators to prevent electrical shorting to the metal frame. The open wire elements have a longer life than the others because they operate cooler due to releasing all the heat directly into the air stream. They also have a lower air pressure drop as compared to the other types.

The open ribbon elements are also made of nichrome wire and are insulated in the same manner as the open wire elements. The flat design allows more intimate contact between the wire and the air because it has more surface area. Thereby, its efficiency is increased over the open wire element. However, due to the increase in manufacturing costs, the open wire element is more popular.

The tubular cased heating element is the same type as those used in cook stoves. The nichrome wire is placed inside a tube and insulated from it by magnesium oxide powder. Thus, tubular cased heating elements do not require external insulation as do the other two resistance elements. They are also less efficient because of the energy loss caused by the extra material that the heat must pass through before reaching the moving air. They are, however, safer to use because of the interior insulation used in their manufacturing process. The tubular cased elements have a shorter life than either of the other two elements because of their higher operating temperature. The control of these elements is more difficult than the others because of the extra material involved. Also, the air must be circulated sufficiently long to ensure the proper cooling of these elements.

The *heat pump* is the most efficient method of electric heating. A heat pump is a refrigeration unit that reverses the flow of refrigerant according to the different seasons of the year. By use of a series of valves, it cools the home in the summer and provides heat for it in the winter. Electricity is used to drive the compressor, and if the system design conditions—usually 72° F and 50% relative humidity (RH) indoor conditions—are maintained, a heat pump will release three to four times as much heat as could be obtained from resistance elements. Therefore, with an input of 1 kW (3,410 Btu), approximately three times the input or 10,230 Btu are released by the heat pump.

A heat pump is less efficient at lower outdoor temperatures because the evaporator temperature must be lower than the ambient temperature in order to absorb enough heat to evaporate the refrigerant. This lower temperature is accompanied by a lower suction pressure that reduces the compressor capacity. Also, at the lower temperatures, there is not enough heat absorbed to replace the heat loss through the walls of the home.

A heat pump is classified by its heat source. These heat sources are air to air, water to air, water to water, ground to air, and solar assist. The first of each of these combinations refers to the heat source for winter operation. However, the air-to-air unit is the most common due to the installation costs.

When these units are installed in cold climates, additional heat is usually required to heat the home satisfactorily. Because of the reduced efficiency, resistance elements are used in conjunction with these systems.

If the cost of equipment per hour of operation is considered, a heat pump is more economical to own than a cooling system with a separate heating unit. The reason for this is that when one part, either heating or cooling, is used, the other part is not operating. Therefore, the cost per hour of operation is more.

GAS COMBUSTION

The available energy contained in a fuel is converted to heat energy by a process known as *combustion*. Combustion may be defined as the chemical action of a substance with oxygen resulting in the evolution of heat and some light.

There are three basic requirements for combustion: sufficiently high temperatures, oxygen, and fuel (Figure 2–7). When the air-fuel mixture is admitted to the combustion chamber, some means must be provided to bring the mixture to its flash point. This is usually done by a pilot light or an electronic ignitor. If, for any reason, the temperature of the mixture is reduced below its flash point, the flame will automatically be extinguished. For example, if the temperature of a mixture of natural gas and air is reduced below its flash point of 1,100° F, there will be no flame.

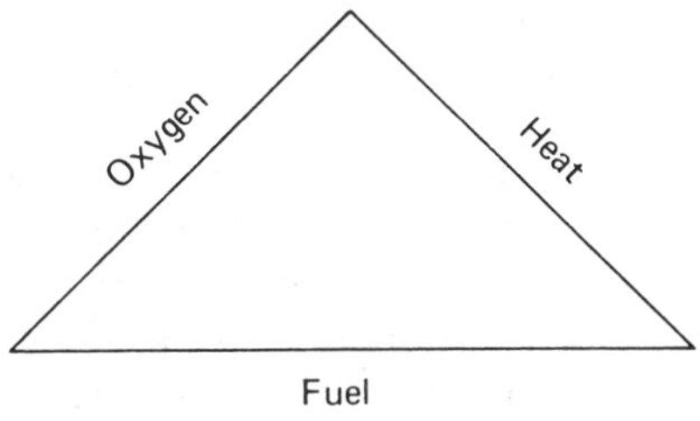
Oxygen
Heat
Fuel

FIGURE 2–7
Basic combustion requirements.

Also, an ample supply of properly distributed oxygen must be provided. The oxygen requirements governing the combustion process will vary with each different fuel. They also depend on whether the fuel and air are properly mixed in the correct proportions.

The third requirement for combustion is the fuel. Properties of fuel were dealt with earlier in this chapter. The physical properties of each fuel must be considered when determining its requirements for combustion. All of the basic requirements for combustion must be met or there will be no combustion.

An important factor to keep in mind when making adjustments involving gaseous fuels is the limits of flammability, which are stated in percentages of the gas in air of the mixture that would allow combustion to take place. To simplify this, if there is too much gas in the air, the mixture will be too rich to burn. If there is too little gas in the air, the mixture will be too lean to burn. The upper and lower limits are shown for the more common gaseous fuels in Table 2–3.

Complete combustion can be obtained only when all of the combustible elements are oxidized by all the oxygen with which they will combine. The products

TABLE 2–3 Fuel gas limits of flammability

Gas	Upper Limit	Lower Limit
Methane	14	5.3
Ethane	12.5	3.2
Natural	14	3
Propane	9.5	2.4
Butane	8.5	1.9
Manufactured	29	4

of combustion are harmless when all of the fuel is completely burned. These products are carbon dioxide (CO_2) and water vapor (H_2O).

The rate of combustion, or burning, depends on three factors:

1. The rate of reaction of the substance with the oxygen.
2. The rate at which the oxygen is supplied.
3. The temperature due to the surrounding conditions.

All the oxygen supplied to the flame is not generally used. This is commonly called *excess oxygen,* or *excess air.* This excess oxygen is expressed as a percentage, usually 50%, of the air required for the complete combustion of a fuel. An example of this is that natural gas requires 10 ft^3 of air for each ft^3 of gas. When 50% excess air is added to this figure, the quantity of air supplied is calculated to be 15 ft^3 of air to each ft^3 of natural gas. There are several factors governing the excess air requirements. They are the uniformity of air distribution and mixing; the direction of gas flow from the burner; and the height and temperature of the combustion area. Excess air constitutes a loss and should be kept to a minimum. However, it cannot usually be less than 25 to 30% of the air required for complete combustion.

Excess air has both good and bad effects in the combustion area operation. It is added as a safety factor in case the 10 ft^3 of required air is reduced for some reason, such as dirty burners, improper primary air adjustments, or a lack in the supply of primary air. The adverse effect is that the nitrogen in the air does not change chemically and tends to reduce the burning temperature and the flue gas temperature, thereby reducing the efficiency of the heating equipment. The air supplied for combustion contains about 79% nitrogen and 21% oxygen.

The products of combustion produced when 1 ft^3 of natural gas is completely burned are 8 ft^3 of nitrogen, 1 ft^3 of carbon dioxide, and 2 ft^3 of water vapor (Figure 2–8). These products are harmless to human beings. In fact, carbon dioxide is the ingredient added to water that makes soft drinks fizz.

The byproducts of incomplete combustion are carbon monoxide—a deadly product; aldehydes—a colorless, inflammable, volatile liquid with a strong pun-

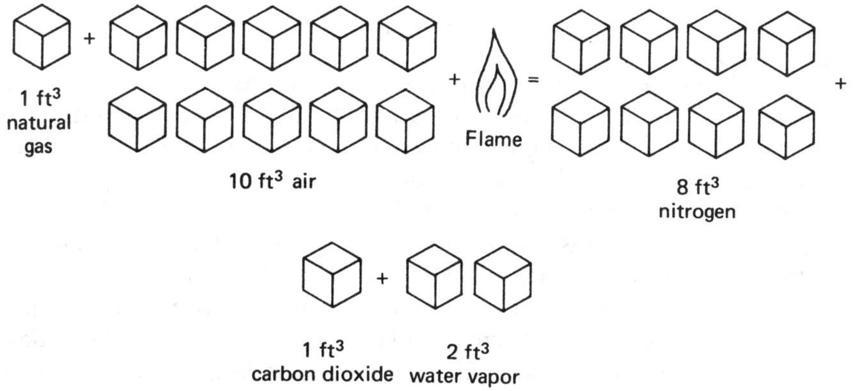
1 ft3
natural
gas
10 ft3 air
Flame
8 ft3
nitrogen
1 ft3
carbon dioxide
2 ft3
water vapor

FIGURE 2–8 Elements of combustion.

gent odor and an irritant to the eyes, nose, and throat; ketones—used as paint removers; oxygen acids; glycols; and phenols. These byproducts are harmful and must be guarded against by the proper cleaning and adjustment of heating equipment.

Probably the most important step in maintaining good combustion is the proper adjustment of the ratio of primary air to secondary air. This can be accomplished, to any degree of efficiency, only by flue gas analysis, covered in detail in Chapter 6.

When considering the combustion requirements for any given installation, the air supply to the equipment must be calculated. The air supply is governed by size of the equipment by Btu rating; type of fuel; size of equipment room; building construction tightness; exhaust operation; and the city code.

Heating equipment regulated by experienced personnel will produce clean, economical, efficient combustion. However, when the designing of the equipment and the fundamentals of combustion are ignored, a potential hazard exists.

OIL COMBUSTION

Even though some of the information in this section was presented in the preceding section, it is worth reading because it develops the foundation from which every dependable oil service technician should work.

Fuel Oil

Number 2 distillate fuel oil (domestic heating oil) is a product of the refining of crude oil, which is formed underground over millions of years through decomposition of marine organisms, fish, and vegetation. This organic matter eventually becomes liquid or gas concentrated underground in pockets or pools. All petroleum products, including natural gas, gasoline, kerosene, and Number 2 fuel oil, are chemical compounds that make up crude oil, and they all contain carbon and hydrogen.

The process of separating these various components can be complex and is commonly referred to as *refining*. Eventually, one of the products of the refining process is No. 2 fuel oil, which has characteristics suitable for use as a fuel in residential oil burners. The designation "No. 2" is used as a specification guide that defines some physical characteristics such as flash point, ash, and viscosity.

When fuel oil is burned, the chemical energy that is stored in the oil is released in another form of energy: heat. But to create this conversion of energy, an external source of heat must be applied to the oil droplets to start the reaction. The electric spark generated by the electrodes of an oil burner provides the initial heat. The heat from the electrodes causes the oil droplets to become oil vapors and eventually burn continuously. This burning then heats the surrounding oil droplets, causing them to burn. This process continues until all or most of the droplets are vaporizing and burning. If the conditions for combustion are ideal, all oil droplets will burn completely and cleanly within the combustion zone. Combustion is the process of burning.

Combustion

Combustion, as we normally think of it, is generally described as the rapid oxidation of any material which is classified as combustible matter. The term *oxidation* simply means the adding of oxygen in a chemical reaction. *Combustible matter* means any substance which combines readily and rapidly with oxygen under certain favorable conditions. Since fuel oil primarily consists of carbon (85%) and hydrogen (15%), combustion of fuel oil, according to our previous definition, is the rapid combining of the carbon and hydrogen with oxygen.

As you know, the oxygen needed for combustion comes from the air provided by the burner blower. Approximately 21% of the air is oxygen, while the remainder (79%) is nitrogen. Therefore, to supply the oxygen needed for combustion, a great deal of nitrogen goes along for a free ride. This will become an important factor in later discussions of proper oil burner adjustment.

What we see and feel from combustion—flames, smoke, heat—is a result of chemical reactions. Since we cannot see carbon, hydrogen, or oxygen atoms (the smallest units to combine), we symbolize the reactions with formulas that describe the process. For example,

carbon + air (oxygen + nitrogen) forms carbon dioxide + nitrogen + heat

hydrogen + air (oxygen + nitrogen) forms water vapor + nitrogen + heat.

These reactions can be written using symbols in the following manner:

$$C + O_2 + N_2 \rightarrow CO_2 + N_2 + \text{heat} \qquad (2\text{–}1)$$

$$2H_2 + O_2 + N_2 \rightarrow 2H_2O + N_2 + \text{heat.} \qquad (2\text{–}2)$$

Both chemical reactions produce entirely new products, and each reaction gives off heat; however, in each reaction the nitrogen (N_2) has not changed, indicating that nitrogen does not participate in the reaction. If pure oxygen rather than air were used in these reactions, the final products of each reaction would be heat plus 100% H_2O, respectively. It is known that the flue gas does not contain 100% CO_2, and now it is obvious why. Because of the large amounts of nitrogen in the air, the bulk of the flue gas is made up of unreacted nitrogen.

If exactly the right amount of air (no excess air) were supplied for complete combustion of the carbon and hydrogen in the fuel oil, the products of combustion would be as indicated in Figure 2–9. However, with the actual oil burner equipment, it is not possible to get a perfect mixture in which all the carbon and hydrogen are supplied with the exactly correct quantity of oxygen. To ensure that all the carbon and hydrogen come into contact with enough oxygen to burn completely, excess air must be supplied. The excess air is simply air over and above the theoretical requirement for the combustion of the fuel oil. With excess air needed for combustion, reaction (2-1) becomes

$$C + O_2 + N_2 \rightarrow CO_2 + N_2 + O_2 + \text{heat.} \qquad (2\text{–}3)$$

Note that the only difference between reaction (2–3) and reaction (2–1) is that O_2 (oxygen) is a product of the reaction. This O_2 is the oxygen in the excess air that does not combine with carbon to make carbon dioxide. In essence, extra O_2 is pro-

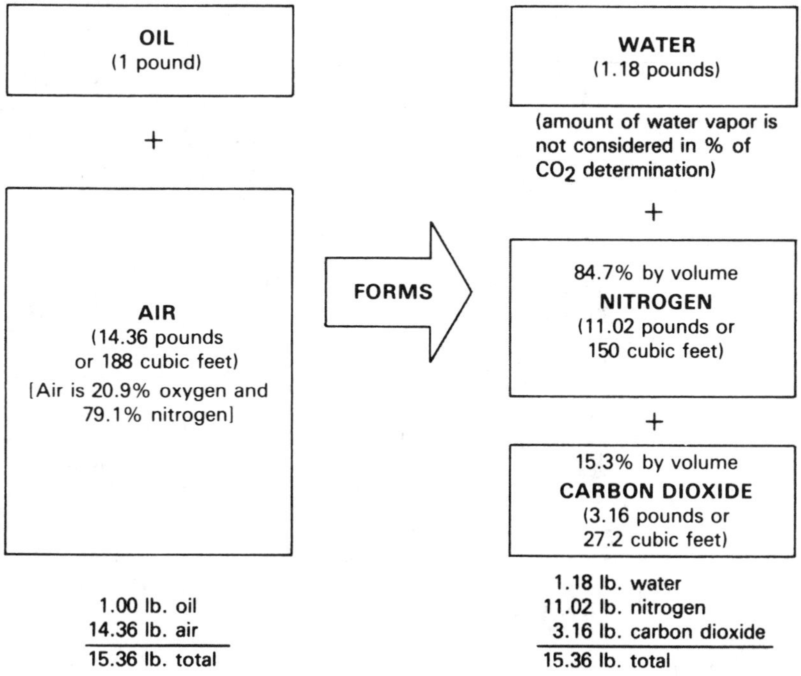
OIL
(1 pound)
+
AIR
(14.36 pounds
or 188 cubic feet)
[Air is 20.9% oxygen and
79.1% nitrogen]
FORMS
WATER
(1.18 pounds)
(amount of water vapor is
not considered in % of
CO2 determination)
+
84.7% by volume
NITROGEN
(11.02 pounds or
150 cubic feet)
+
15.3% by volume
CARBON DIOXIDE
(3.16 pounds or
27.2 cubic feet)
1.00 lb. oil
14.36 lb. air
15.36 lb. total
1.18 lb. water
11.02 lb. nitrogen
3.16 lb. carbon dioxide
15.36 lb. total

FIGURE 2–9
Amount by weight and volume of combustion products when 1 lb of fuel oil is burned (0% excess air). *(Courtesy of R. W. Beckett Corp.).*

vided, as a component of excess air, to guarantee that all the carbon and hydrogen come in contact with the oxygen and burn.

This excess nitrogen does not react during the combustion process but enters the heating unit at room temperature and reduces the temperature of the combustion gases so that less heat is available to be transferred to the distribution medium. As a result, excess air is a source of heat loss. By introducing 50% excess air, the situation shown in Figure 2–10 is created. Compare this with Figure 2–9 and note that:

- The amount (weight) of H_2O, CO_2, and N_2 formed in Figure 2–10 is the same as that in Figure 2–9.
- Percent by volume of CO_2 and N_2 in Figure 2–10 is less than that formed in Figure 2—9.
- Oxygen (as part of excess air) is a product in Figure 2–10 but not in Figure 2–9.

In Figure 2–10, since 20.9% of the excess air is oxygen, 7.1% of all the combustion gases is oxygen. This can be determined by multiplying the percent excess

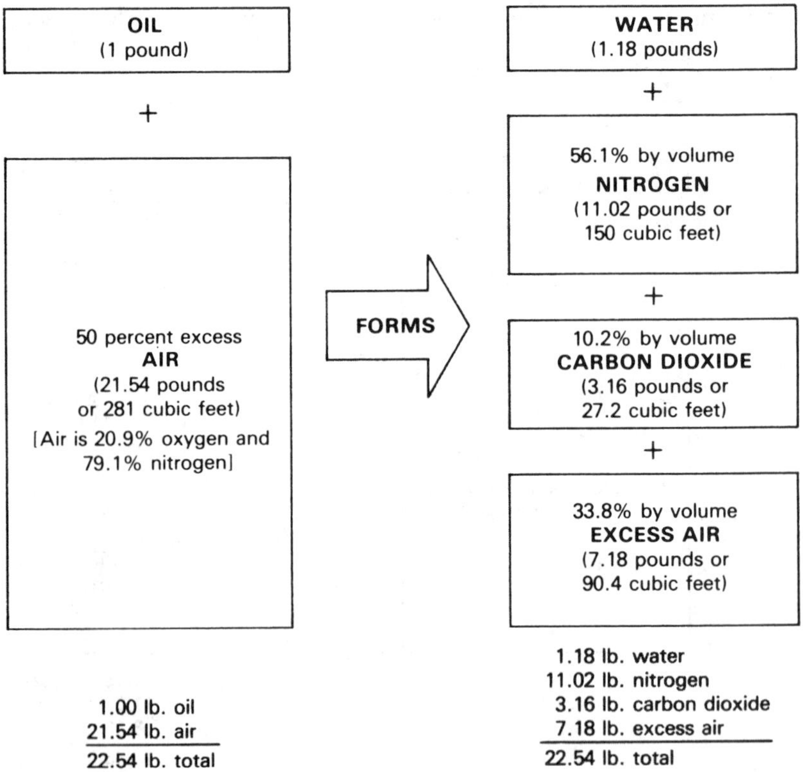
OIL
(1 pound)
+
50 percent excess
AIR
(21.54 pounds
or 281 cubic feet)
[Air is 20.9% oxygen and
79.1% nitrogen]
FORMS
WATER
(1.18 pounds)
+
56.1% by volume
NITROGEN
(11.02 pounds or
150 cubic feet)
+
10.2% by volume
CARBON DIOXIDE
(3.16 pounds or
27.2 cubic feet)
+
33.8% by volume
EXCESS AIR
(7.18 pounds or
90.4 cubic feet)
1.00 lb. oil
21.54 lb. air
22.54 lb. total
1.18 lb. water
11.02 lb. nitrogen
3.16 lb. carbon dioxide
7.18 lb. excess air
22.54 lb. total

FIGURE 2–10 Amount by weight and volume of combustion products when 1 lb of fuel oil is burned (50% excess air). *(Courtesy of R. W. Beckett Corp.).*

air (38.8%) by that portion of excess air which is oxygen (0.209). This gives approximately 7.1% oxygen.

Note that in Figure 2–10 the percentage of CO_2 or O_2 changed from Figure 2–9 is a result of excess air; therefore, the percent CO_2 or O_2 in the flue can be used as a measure of excess air or vice versa, as a general rule.

- The greater the CO_2, the less excess air.
- The greater the O_2, the more excess air.

Figure 2–11 displays the relationship between CO_2 and excess air.

Oil Burner Combustion Chamber

The function of the combustion chamber is to surround the flame and to radiate heat back into the flame to aid in combustion. The combustion chamber design and construction help determine whether the fuel will be burned efficiently. The chamber must be made of the correct material, properly sized for the nozzle firing rate, shaped correctly, and of the proper height.

The chamber should be designed and built to provide the maximum space required to burn the oil needed to fire the heating plant and to meet its load. Unburned droplets of oil should not touch the chamber surface, especially a cold surface. A cold surface will reduce combustion temperatures and cause soot and carbon formation. The hotter the area around the burning zone, the easier the

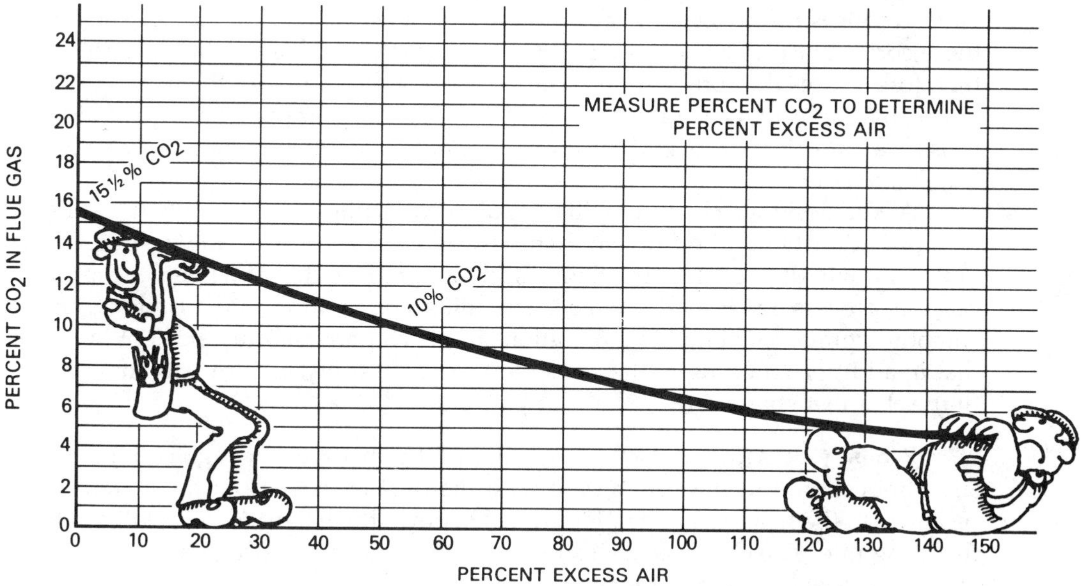
MEASURE PERCENT CO2 TO DETERMINE PERCENT EXCESS AIR
15½% CO2
10% CO2
PERCENT CO2 IN FLUE GAS
0
2
4
6
8
10
12
14
16
18
20
22
24
0
10
20
30
40
50
60
70
80
90
100
110
120
130
140
150
PERCENT EXCESS AIR

FIGURE 2–11 Relationship between excess air and % CO_2. (*Courtesy of R. W. Beckett Corp.*).

droplets will vaporize and ignite, and the hotter the flame will be. If the chamber is too small, the oil will not have enough time to complete combustion before it strikes the colder walls.

When the chamber is too large, there will be areas in the chamber which the flame will not fill. This causes cooler chamber surfaces and reduces the reflected heat from the chamber walls. As a result, the fuel droplets will not evaporate as rapidly in the cooler chamber and will be more difficult to burn completely. More air will be required to burn smoke free, and the result will be low CO_2 (high O_2) and lowered efficiency.

Floor size. The size of the combustion chamber is measured in square inches of floor space. The ideal size for a home is about 80 to 90 square inches per gallon of fuel oil. If the burner is functioning well, and the chamber is made from quick heating refractory material and is properly designed, it is possible in most cases to use this formula up to 1.50 gph. For residential use, the chamber should not exceed 95 square inches per gallon for a high pressure burner.

When the combustion chamber is accurately sized to the heating plant capacity, using 80 or 90 square inches per gallon, it is extremely important that the nozzle spray pattern and spray angle conform to characteristics of the burner air pattern and that the oil pressure at the nozzle not exceed 100 psi.

Shape. The majority of combustion chambers are square, rectangular, or round. Curved surfaces generally produce more complete mixing of the oil and air and eliminate the pockets of air in the corners of square or rectangular chambers, which reduces the reflected heat from the chamber walls to the flame. The air in these corners also does not usually become a part of the combustion process and therefore dilutes the combustion products as they flow through the heating plant. This is particularly true of the corners of the chamber that the oil is sprayed in because the flame is narrow and the oil has not heated up to maximum temperature. See Figure 2–12.

A well designed chamber will have all corners filled in. Generally this should be more extensive in the front so that the wall will be closer to the desired flame pattern. It will then confine the flame, and more reflected heat will enter the combustion process in its early stages. This aids combustion and provides much smoother ignition. On making alterations in the chamber, it must be kept in mind that in addition to filling in the corners, the nozzle spray pattern and angle must fit the chamber shape.

Walls. It is important that the walls of the chambers be high enough to assist combustion, but not to interfere with the heat transfer from the combustion products to the heat exchanger. Figure 2–13 shows the height to be used based on the firing rate. The chamber wall should be 2 to 2 ¼ times as high above the nozzle as it is from the floor to the nozzle.

If the base of a heating unit has a tendency to overheat, the walls should be 2 ½ to 3 times the height from the floor to the nozzle. This is sometimes a problem in gravity-type air duct systems or boilers that have been converted from coal to

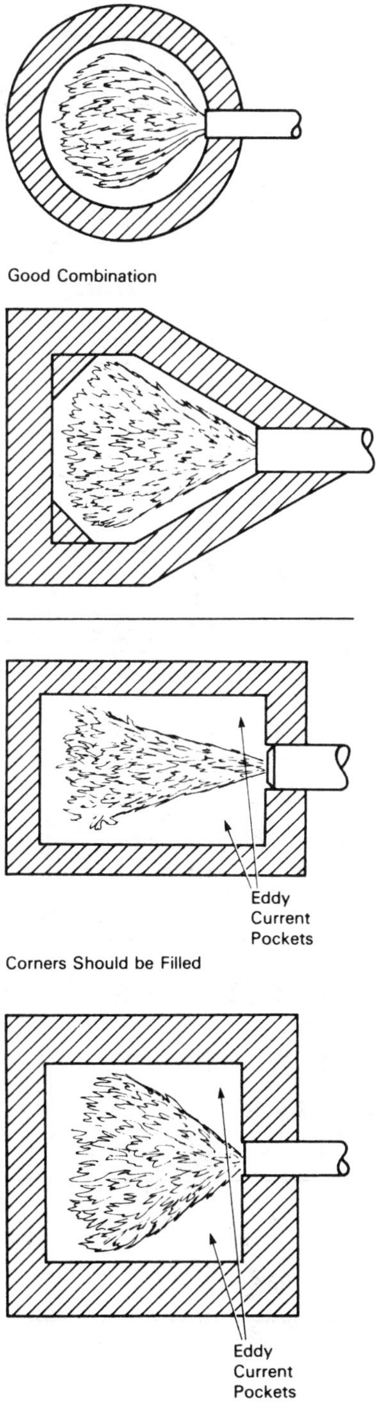

Good Combination
Eddy
Current
Pockets
Corners Should be Filled
Eddy
Current
Pockets

FIGURE 2–12
Combustion chamber design. *(Courtesy of R. W. Beckett Corp.).*

	Oil Consumption G.P.H.	Sq. Inch Area Combustion Chamber	Square Combustion Chamber Inches	Dia. Round Combustion Chamber Inches	Rectangular Combustion Chamber Inches Inches	HEIGHT FROM NOZZLE TO FLOOR INCHES			
						Conventional Burner Wth. x Lgth.	Conventional Burner Single Nozzle	Sunflower Flame Burner Single Nozzle	Sunflower Flame Burner Twin Nozzle
80 Sq. In. Per Gal.	.75	60	8 x 8	9	.....	5	x	5	x
	.85	68	8.5 x 8.5	9	.....	5	x	5	x
	1.00	80	9 x 9	10 1/8	.....	5	x	5	x
	1.25	100	10 x 10	11 ¼	.....	5	x	5	x
	1.35	108	10½ x 10½	11¾	.....	5	x	5	x
	1.50	120	11 x 11	12 3/8	10 x 12	5	x	6	x
	1.65	132	11½ x 11½	13	10 x 13	5	x	6	x
	2.00	160	12 5/8 x 12 5/8	14¼	6	x	7	x	
	2.50	200	14¼ x 14¼	16	12 x 16½	6.5	x	7.5	x
	3.00	240	15½ x 15½	17½	13 x 18½	7	5	8	6.5
90 Sq. In. Per Gal.	3.50	315	17¾ x 17¾	20	15 x 21	7.5	6	8.5	7
	4.00	360	19 x 19	21½	16 x 22½	8	6	9	7
	4.50	405	20 x 20		17 x 23½	8.5	6.5	9.5	7.5
	5.00	450	21¼ x 21¼		18 x 25	9	6.5	10	8
100 Sq. In. Per Gal.	5.50	550	23½ x 23½	Round Combustion Chambers usually not used in these sizes	20 x 27½	9.5	7	10.5	8
	6.00	600	24½ x 24½		21 x 28½	10	7	11	8.5
	6.50	650	25½ x 25½		22 x 29½	10.5	7.5	11.5	9
	7.00	700	26½ x 26½		23 x 30½	11	7.5	12	9.5
	7.50	750	27¼ x 27¼		24 x 31	11.5	7.5	12.5	10
	8.00	800	28¼ x 28¼		25 x 32	12	8	13	10
	8.50	850	29¼ x 29¼		25 x 34	12.5	8.5	13.5	10.5
	9.00	900	30 x 30		25 x 36	13	8.5	14	11
	9.50	950	31 x 31		26 x 36½	13.5	9	14.5	11.5
	10.00	1000	31¾ x 31¾		26 x 38½	14	9	15	12
	11.00	1100	33¼ x 33¼		28 x 29½	14.5	9.5	15.5	12.5
	12.00	1200	34½ x 34½		28 x 43	15	10	16	13
	13.00	1300	36 x 36		29 x 45	15.5	10.5	16.5	14
	14.00	1400	37½ x 37½		31 x 45	16	11	17	14.5
	15.00	1500	38¾ x 38¾		32 x 47	16.5	11.5	17.5	15
	16.00	1600	40 x 40		33 x 48½	17	12	18	15
	17.00	1700	41¼ x 41¼		34 x 50	17.5	12.5	18.5	15.5
	18.00	1800	42½ x 42½		35 x 51½	18	13	19	16

FIGURE 2–13 Combustion chamber sizing data. (*Courtesy of R. W. Beckett Corp.*).

oil. Be sure to use insulation between the furnace and the chamber wall all the way to the top of the wall.

Space between the chamber wall and the heating plant should be filled with an insulating material, such as mica pellets, except for wet leg or wet base boilers. A poor grade of backfill shortens the life of the chamber, reduces the efficiency at which the oil burns, and increases combustion noise.

Burner setting. The chamber must be installed so that the oil can burn clean without impinging on the floor and causing carbon to form. Figure 2–14

1 Firing Rate (GPH)	2 Length (L)	3 Width (W)	4 Dimension (C)	5 Suggested Height (H)	6 Minimum Dia. Vertical Cyl.
0.50	8	7	4	8	8
0.65	8	7	4.5	9	8
0.75	9	8	4.5	9	9
.085	9	8	4.5	9	9
1.00	10	9	5	10	10
1.10	10	9	5	10	10
1.25	11	10	5	10	11
1.35	12	10	5	10	11
1.50	12	11	5.5	11	12
1.65	12	11	5.5	11	13
1.75	14	11	5.5	11	13
2.00	15	12	5.5	11	14
2.25	16	12	6	12	15
2.50	17	13	6	12	16
2.75	18	14	6	12	18

NOTES:

1. Flame lengths are approximately as shown in column (2). Often, tested boilers or furnaces will operate well with chambers shorter than the lengths shown in column (2).
2. As a general practice any of these dimensions can be exceeded without much effect on combustion.
3. Chambers in the form of horizontal cylinders should be at least as large in diameter as the dimension in column (3). Horizontal stainless steel cylindrical chambers should be 1 to 4 inches larger in diameter than the figures in column (3) and should be used only on wet base boilers with non-retention burners.
4. Wing walls are not recommended. Corbels are not necessary although they might be of benefit to good heat distribution in certain boiler or furnace designs, especially with non-retention burners.

FIGURE 2–14 Recommended minimum inside dimensions of refractory-type combustion chambers. *(Courtesy of R. W. Beckett Corp.)*.

shows the recommended inside dimensions. The burner and cone should be installed ¼ inch back from the inside chamber wall. It is recommended that the refractory fiber be installed around the outside diameter of the burner end cone and air tube. If insulating material is not available, and the chamber opening exceeds 4 ⅜ inches, the burner end cone setback must be increased. See Figure 2–14(a).

Soft fiber refractory. Refractories of low specific heat and low conductivity (insulating) will rise in temperature more rapidly from a cold start and maintain a higher temperature during steady operation of an oil burner. This will help produce more complete combustion and increase the heat transfer by radiation to the heat transfer of the heat exchanger.

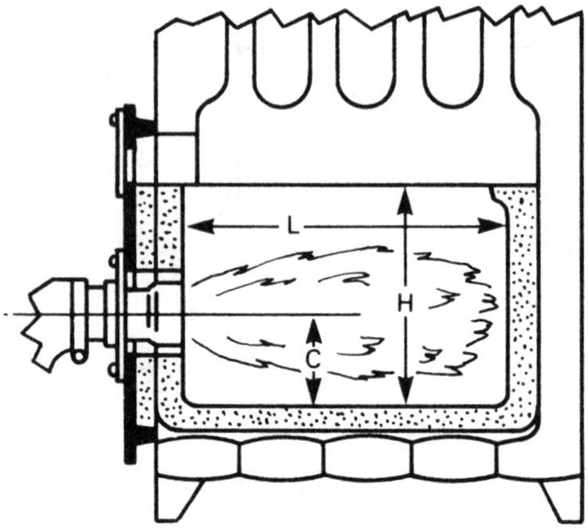
L
H
C

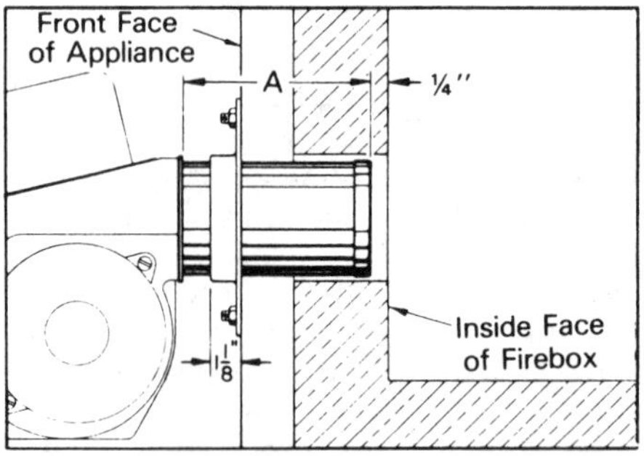
Front Face of Appliance
A
¼"
Inside Face of Firebox
1⅛"

"A" = Usable air tube length.

The burner head should be ¼" back from the inside wall of the combustion chamber. Under no circumstances should the burner head extend into the combustion chamber. If chamber opening is in excess of 4 ³/⁸", additional set back may be required.

FIGURE 2–14a
Air tube insertion. (*Courtesy of R. W. Beckett Corp.*).

Tests by the National Bureau of Standards comparing a hard brick to a precast soft chamber in the same boiler determined that losses by radiation, conduction, convection, and incomplete combustion were 13.4% for the brick and 8.6% for the precast chamber. The difference was equal to 8,300 Btu per hour in favor of the precast chamber. This amounts to a possible savings of 6%. Examples of soft refractory chambers are shown in Figure 2–15.

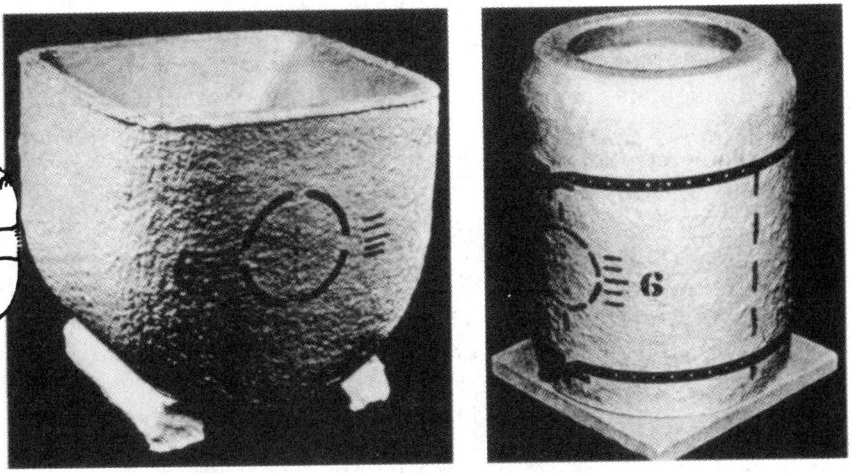
6

FIGURE 2–15 Soft fiber refractory combustion chambers. *(Courtesy of R. W. Beckett Corp.).*

Although it is possible to obtain a relatively good fire without a chamber, a properly sized and shaped combustion chamber will substantially improve combustion, provide a hotter flame, and reduce the amount of soot accumulation associated with start-up and shutdown. Large fire commercial burners are frequently fired without a chamber, but with small residential burners the chamber becomes extremely important. The modern materials for chamber construction become red hot within 20 seconds after starting the fire, causing heat to be reflected back into the oil spray, speeding up the conversion of the liquid oil to vapor, and making the flame smaller but hotter.

In general, combustion temperatures of high speed flame-retention burners will be 100° F to 200° F higher than non-flame-retention burners, even though the same oil rate, same air-fuel ratio, and the same chamber are used.

When flame-retention burners are fired without a chamber in a wet base boiler, the flame envelope becomes larger and the flame temperature is lowered by as much as 200° F to 400° F. Without the chamber, the products of incomplete combustion are increased and so is the undesired production of carbon monoxide. Consequently, although some boiler manufacturers may employ flame-retention burners without combustion chambers, the efficiency could be improved in most instances by using them.

In some applications economics dictate the installation of a flame-retention burner without the chamber. For example, if a customer has an obsolete rotary wall-flame burner in his home and is unable to afford the replacement of the boiler, a common solution would be to remove the rotary burner, seal the hearth with refractory cement, and install a flame-retention burner fired through the door. This type of installation would be far less costly than the more desirable boiler-burner replacement which must eventually follow, but will permit the homeowner an interim improvement.

While there has been a lot said about the improved combustion achieved through the use of a chamber, there are also some other benefits to be considered. Chambers act as sound absorbers, and this feature is highly desirable since some flame-retention burners have more intense flame noise than the older burners they are replacing. Another benefit obtained from combustion chambers is the protection of the dry base boiler or furnace, which could not withstand prolonged exposure to intense heat or the rapid heating and cooling of the metal.

When the correct firing rate to match the heat load has been determined, the proper size combustion chamber should be selected to match that firing rate. This will result in maximum efficiency operation. The relation between the size of an existing chamber and the determination of the correct firing rate to fit that chamber is important and should be considered whenever the firing rate is altered.

Role of Excess Air in Combustion

Excess air must be supplied to ensure adequate mixing of the fuel and oxygen. However, excess air is one of the major causes of low efficiencies. To see how this occurs, consider that excess air.

- Dilutes combustion gases.
- Absorbs heat.
- Drops the overall temperature of combustion gases.

The dilution of combustion gases occurs simply because of the presence of additional gas in the form of excess air. The excess air absorbs heat in the combustion zone and reduces the flame temperature. This, in turn, reduces the transfer of heat to the heat exchanger since a significant amount of heat is transferred by radiation. Moreover, as excess air is introduced, the overall temperature of the combustion gases drops because the heat from these combustion gases is used to raise the temperature of the excess air. See Figure 2–16. Think of this process as being similar to adding refrigerated cream to a cup of coffee, as shown in Figure 2–17. The cup of coffee is originally 160° F (high temperature) and occupies a small volume (half a cup). Adding cream at 40° F increases the volume (almost a full cup) and lowers the overall temperature to 120° F (mild temperature). Note that the temperature of the mixed coffee and cream is higher than the temperature of the cream alone and lower than the temperature of the coffee alone. Heat from the coffee went

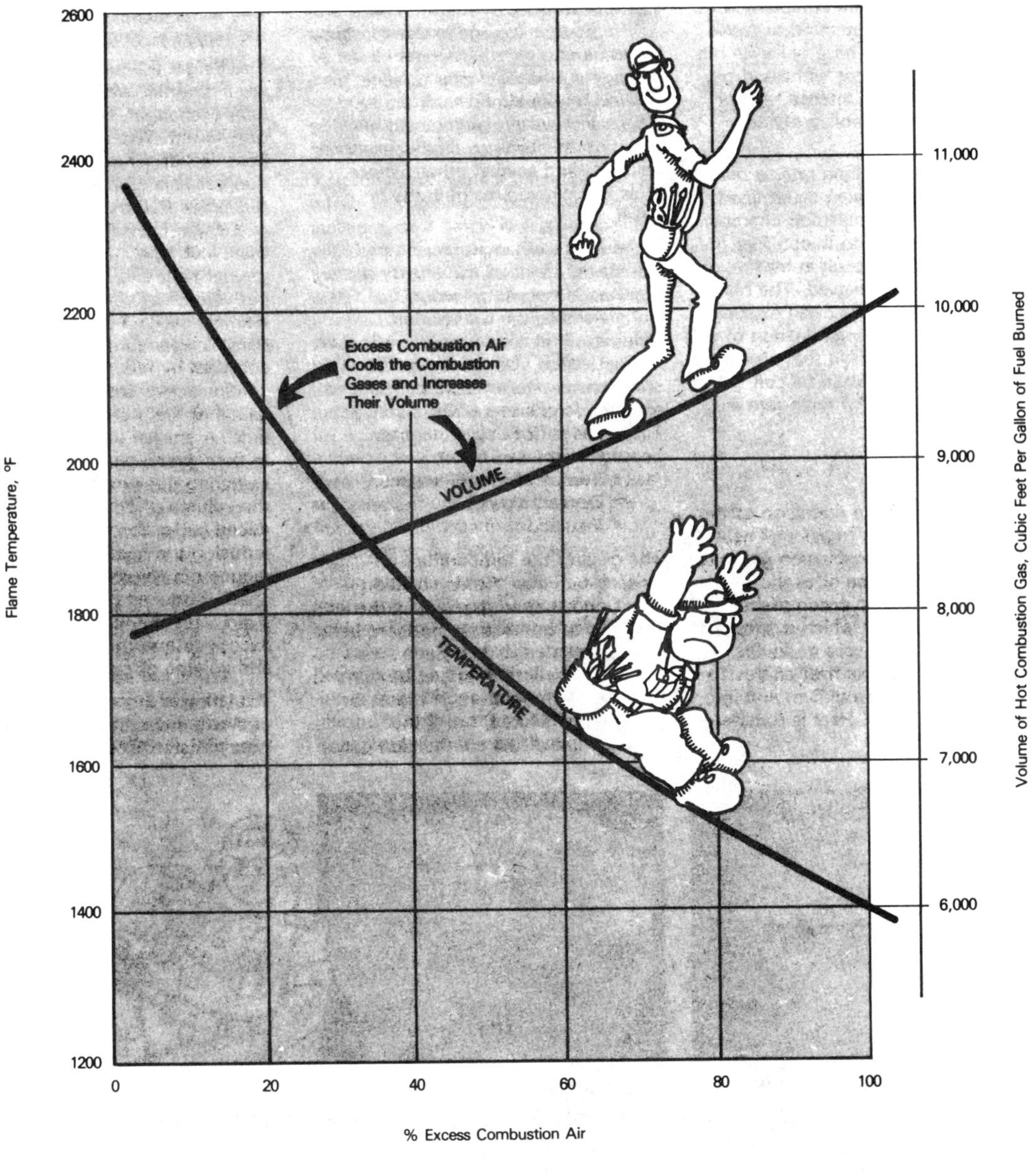
2600
2400
2200
2000
1800
1600
1400
1200
Flame Temperature, °F
11,000
10,000
9,000
8,000
7,000
6,000
Volume of Hot Combustion Gas, Cubic Feet Per Gallon of Fuel Burned
Excess Combustion Air Cools the Combustion Gases and Increases Their Volume
VOLUME
TEMPERATURE
0
20
40
60
80
100
% Excess Combustion Air

FIGURE 2–16 Approximate relationship of % excess air with flame temperature and volume of gases. *(Courtesy of R. W. Beckett Corp.).*

Adding excess air to flame is like adding cream to a cup of coffee.

40° Cream
160° Coffee
160° Coffee
120° Coffee and Cream

• Small Volume
• High Temperature

• Larger Volume
• Lower Temperature

FIGURE 2–17 Representation of the effect of excess air on combustion gas temperature. *(Courtesy of R. W. Beckett Corp.).*

into heating the cream and the overall temperature dropped or, in other words, the cream absorbed some heat from the coffee.

Also, by looking at Figure 2–18 we can see that the coffee example illustrates the effect of excess air (shown as water) in diluting the gas (coffee) and the resulting reduction in the CO_2 percent.

Bear in mind that this temperature reduction and dilution take place in the combustion zone, not in the flue or stack. It is important to note that the effect of excess air on the temperature of the flue gas is different—with more excess air the flue gas temperature tends to rise. This happens because the volume of the combustion gas per unit of fuel burned is so that the gases pass over the heat exchanger surfaces more rapidly, reducing the contact time. This reduces the heat transfer rate to the heat exchanger.

To review, remember that excess air causes the following.

- Lower flame temperature.
- Lower combustion gas temperature.
- Higher flue stack gas temperature.
- Poorer heat exchange to the distribution medium.

All of these changes reduce the efficiency of the heating system. So minimizing excess air is essential in the proper adjustment of oil burners. However, as discussed in the next section, simply reducing the excess air without concern for other factors could lead to a great deal of trouble.

Excess Air-Smoke Relationship

During the combustion of oil, some smoke is usually generated since some of the oil droplets do not contact enough oxygen to complete the reaction which forms carbon dioxide. The smoke consists of small particles of mainly unburned carbon. Some of these particles stick to the heat exchanger surfaces, acting as insulation,

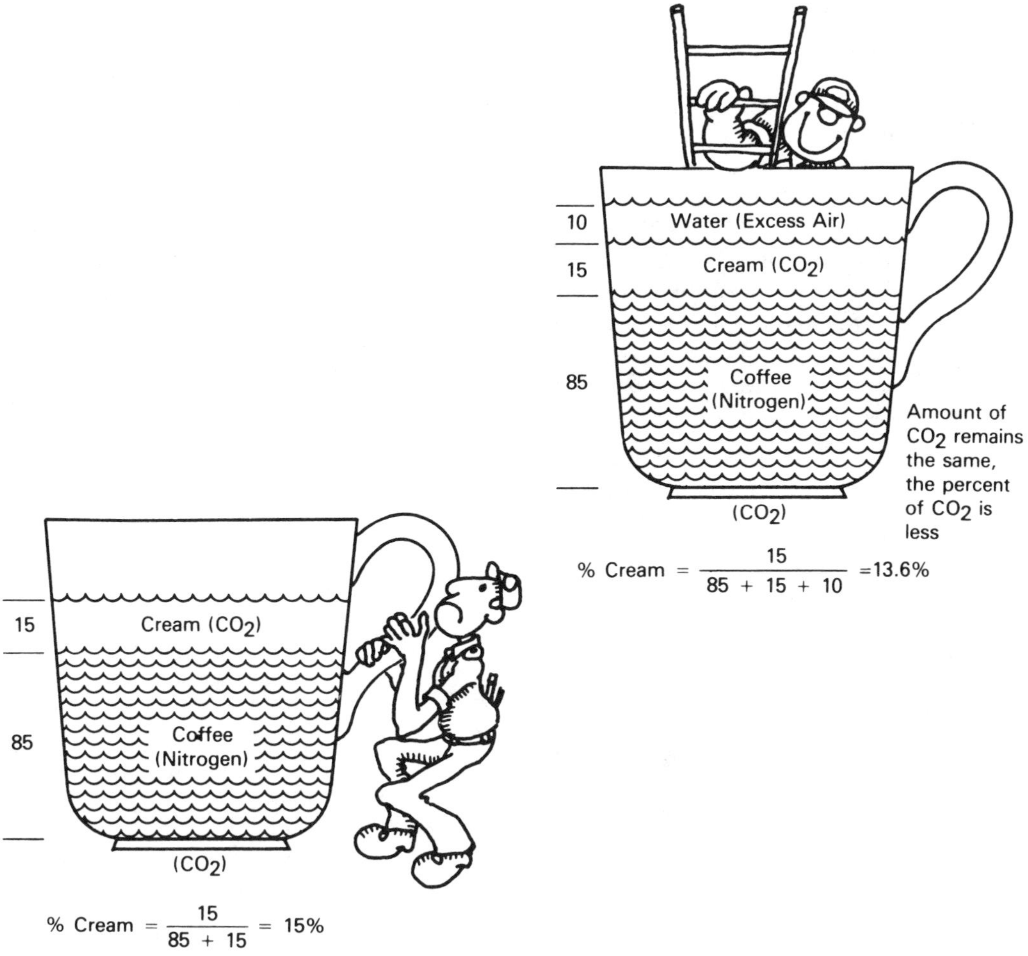
10
Water (Excess Air)
15
Cream (CO_2)
85
Coffee
(Nitrogen)
Amount of CO_2 remains the same, the percent of CO_2 is less
(CO_2)
% Cream = $\frac{15}{85 + 15 + 10}$ = 13.6%
15
Cream (CO_2)
85
Coffee
(Nitrogen)
(CO_2)
% Cream = $\frac{15}{85 + 15}$ = 15%

FIGURE 2–18 The effect of excess air on CO_2. (*Courtesy of R. W. Beckett Corp.).*

and can eventually clog up the flue passages while others are exhausted through the stack and add to the pollution of the air.

The excess air-smoke relationship is all important to the proper adjustment of oil burners. As discussed earlier, there must be sufficient excess air to provide good mixing of the combustion air and fuel oil. Without this excess air, incomplete combustion occurs and smoke is formed. Thus, to minimize smoke, more excess air is generally added. Unfortunately, as the amount of excess air is increased, the transfer of heat to the heat exchange medium (hot water, warm air, or steam) is reduced. A delicate balance must be achieved between smoke generation (caused by insufficient excess air) and reduced heat transfer due to a reduced combustion gas temperature and an increased volume of combustion products (caused by unnecessary excess air). Figure 2–19 illustrates the typical relationship between smoke and excess air. Notice that smoke and efficiency increase as the excess air is decreased. The exact shape of this curve varies from unit to unit. Knowing this, the curve can provide a general idea of where the burner air should be adjusted. The highest efficiency occurs when a proper balance is obtained by a trade-off between smoke and excess air.

Effects of air leaks. After discussing what happens inside the heating plant, it is much easier to understand why air leaks cause lost efficiency. Air which leaks into the combustion gases before they pass through the heat exchanger acts like excess air. The air leaks dilute the combustion gases, cooling them and increasing their volume so that they pass through the heat exchanger more quickly. However, an air leak is even worse than excess air because an air leak cannot reduce the amount of smoke formed in the combustion zone.

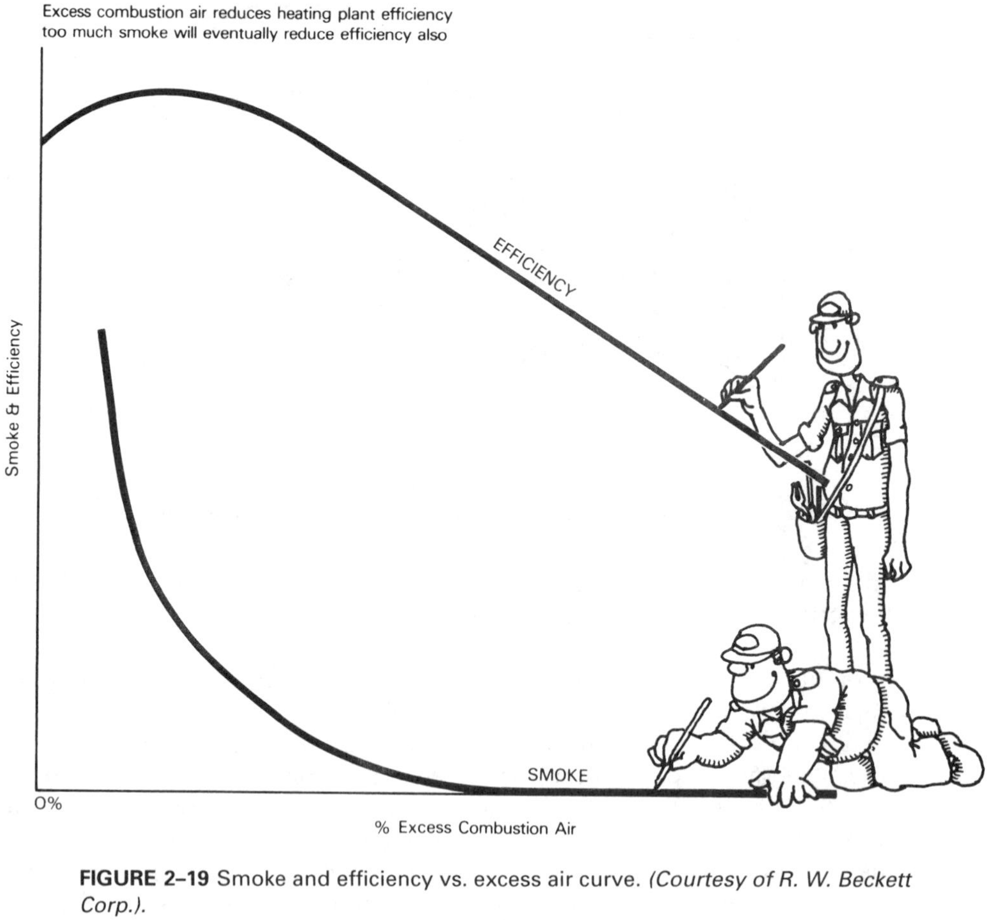
Excess combustion air reduces heating plant efficiency
too much smoke will eventually reduce efficiency also
EFFICIENCY
SMOKE
Smoke & Efficiency
0%
% Excess Combustion Air

FIGURE 2–19 Smoke and efficiency vs. excess air curve. *(Courtesy of R. W. Beckett Corp.).*

REVIEW QUESTIONS

1. Define a fuel.
2. How are fuels rated?
3. When the moisture is condensed and saved from flue gas products, what is obtained?
4. How are gaseous fuels rated?
5. Give the standard conditions used by the gas industry when determining heating values.
6. How is fuel oil graded and classified?
7. What are the most popular grades of fuel oil for domestic burner use?
8. Of what is LP gas comprised?
9. Which type of gas is the heaviest?
10. What is natural gas sometimes called?
11. Natural gas enters a residence at what pressure?
12. Why is an additional regulator needed at a natural gas furnace?
13. What is the Btu content of most natural gas?
14. How is LP gas stored?
15. How many cubic feet of vapor does a gallon of propane produce when it is evaporated?
16. Does butane or propane have a higher heat content?
17. If an LP storage tank was installed above ground with no source of heating, which gas must be used in an ambient temperature of 25° F or below?
18. How much heat is released when 1 watt of electricity is used by an electric heating element?
19. How much heat does a heat pump deliver when compared to resistance heating elements?
20. What is the process that changes fuel to heat?
21. What is the purpose of a pilot or an ignitor on a gas furnace?
22. Name the three basic requirements for combustion.
23. What should be kept in mind when adjusting gas burners?
24. What is the usual percentage of excess air supplied to a gas furnace?
25. What is the purpose of excess air to a gas furnace?
26. How much carbon dioxide is produced when 1 ft^3 of natural gas is burned?
27. What two elements are found in crude oil?
28. With ideal conditions, what will happen to the fuel oil during the combustion process?
29. What percentage of air is oxygen?
30. What happens to the nitrogen in air during the combustion process?
31. What will cause the formation of carbon in a combustion zone?
32. How will the oil burner flame appear in a properly designed combustion chamber?
33. Where is excess air introduced into the combustion process?
34. What is extremely important in the proper adjustment of oil burners?
35. With a decrease in excess air to an oil burner, what happens to the smoke and efficiency?

3

gas orifices, burners, and flames

The objectives of this chapter are:

- To introduce you to the path through which the gas flows on entering a gas heating unit.
- To familiarize you with the purpose and operation of the orifices used in gas heating equipment.
- To acquaint you with the requirements, types, and operation of the gas burners used in a gas heating system.
- To cause you to become aware of the various types of flames encountered in gas heating.
- To further acquaint you with the principles of combustion.
- To acquaint you with the different burner adjustments to obtain the proper flame.

After the gaseous fuel has left the meter loop, or tank piping, and has been distributed through the house piping, it comes to the heating equipment. As the gas progresses through the various controls, such as the gas pressure regulator—not needed on LP gas—and the main gas valve, it enters the equipment gas manifold. The gas pressure in the manifold must be kept as near as possible to 3 ½ inches water column for natural gas and 11 inches water column for LP gases. The gas then proceeds out of the manifold through the orifice and into the main burner. From the main burner it goes into the combustion chamber of the heating equipment, where it is ignited by the pilot burner or electronic ignitor. The combustion chamber of the heating equipment is where the heat is given up to the air or water that provides heat for our homes.

ORIFICES

An *orifice* (Figure 3–1) may be defined as the opening through which gas is admitted to the main gas burner. It is mounted on the gas manifolds by means of pipe threads and projects into the burner (Figure 3–2). It is normally referred to as the orifice spud.

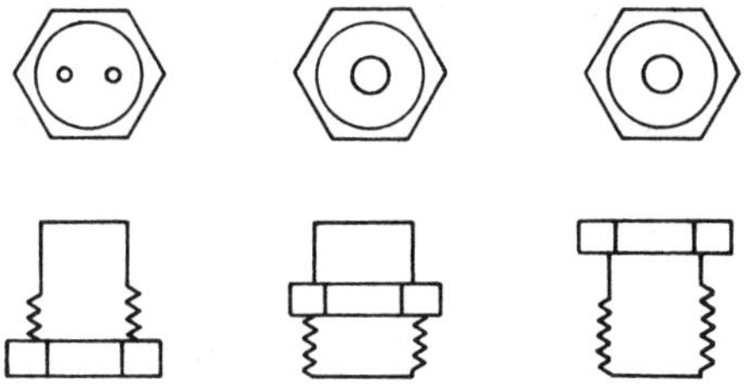

FIGURE 3–1
Basic types of orifice spuds.

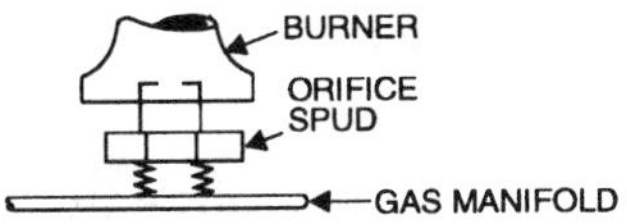
BURNER
ORIFICE SPUD
GAS MANIFOLD

FIGURE 3–2
Orifice location.

For a burner to operate satisfactorily, it must be supplied with the proper Btu input. This is the function of the orifice. The size of the orifice, the gas manifold pressure, and the gas density determine the rate of gas flow to the burner. The orifice and burner must be matched in Btu ratings. If the type of gas is changed, say from natural to LP, the orifice size must also be changed. When the orifice is oversized, enough primary air cannot be drawn into the burner. Thus, incomplete combustion results and a yellow, inefficient flame is produced.

An orifice that is sized too small will also have adverse effects on the burner. There will be delayed ignition accompanied by a loud boom when the gas is ignited. Also, the burner will not operate to its full capacity, resulting in poor heating.

The orifice must direct the gas stream exactly down the center of the burner (Figure 3–3). The velocity of gas down the burner tube causes the primary air to be

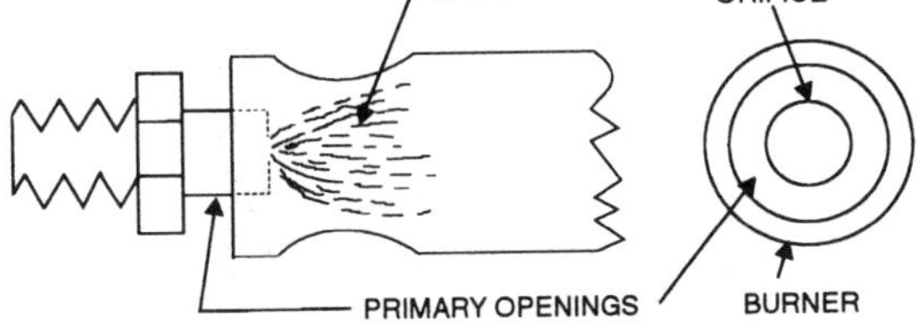
GAS STREAM
ORIFICE
PRIMARY OPENINGS
BURNER

FIGURE 3–3
Orifice in relation to burner.

drawn in through the burner face and mixed with the gas. If the gas velocity is reduced, insufficient primary air will be the result. Therefore, the orifice must be drilled straight and in the exact center of the spud. This may be accomplished easily by drilling the hole from the rear (Figure 3–4). The *V* shape will ensure the desired hole position.

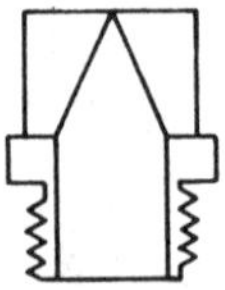

FIGURE 3–4
Orifice cutaway.

The preceding information should establish that the orifice is a precision piece of equipment and should be treated as such when service is required. To clean an orifice, a soft instrument, such as a broom straw, wire brush bristle, etc., must be used. Care must be taken not to change the shape or size of the hole.

A table of orifice sizes can be used to ensure the proper orifice for the job (see Table 3–1).

MAIN GAS BURNERS

A *gas burner* is defined as a device that provides for the mixing of gas and air in the proper ratio to ensure satisfactory combustion.

The first burners were used for lighting. The carbon in the flame was a result of incomplete combustion that produced the light. This flame was not good for heating because of the lower temperature produced.

The blow pipe was the first burner to produce a blue flame and, therefore, any great heat intensity. The blue flame was produced by blowing primary air into the base of the flame (Figure 3–5). The modern blow pipe accomplishes this same thing

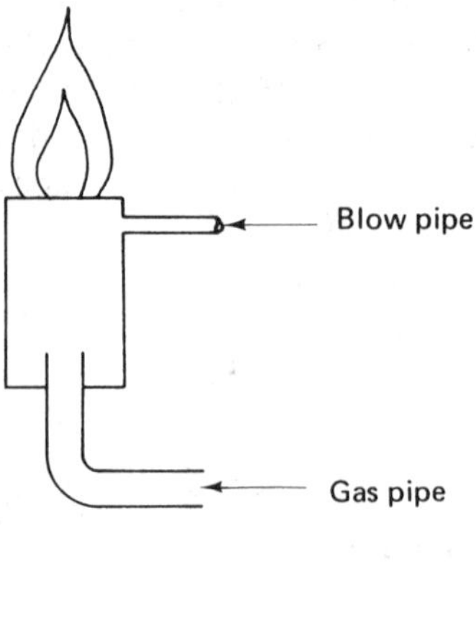
Blow pipe
Gas pipe

FIGURE 3–5
Basic blow pipe.

TABLE 3–1 Orifice capacities for natural gas, 1,000 Btu/ft^3, manifold pressure 3 1/2 in. water column

Wire Gauge Drill Size	Rate ft^3/hr	Rate Btu/hr
70	1.34	2,340
68	1.65	1,650
66	1.80	1,870
64	2.22	2,250
62	2.45	2,540
60	2.75	2,750
58	3.50	3,050
56	3.69	3,695
54	5.13	5,125
52	6.92	6,925
50	8.35	8,350
48	9.87	9,875
46	11.25	11,250
44	12.62	12,625
42	15.00	15,000
40	16.55	16,550
38	17.70	17,700
36	19.50	19,500
34	21.05	21,050
32	23.70	23,075
30	28.50	28,500
28	34.12	34,125
26	37.25	37,250
24	38.75	39,750
22	42.50	42,500
20	44.75	44,750

by blowing compressed air in the same end that the gas enters. The gas and air is thus mixed in the pipe as in the burners in modern–day furnaces.

Modern furnaces make use of one of four different types of main gas burners. These burners are (1) *drilled port;* (2) *slotted port;* (3) *ribbon;* and (4) *inshot* (Figure 3–6). The inshot is probably the most popular. Burners are made of either cast iron or stamped steel, depending on the type and purpose. Usually the drilled port burner is made of cast iron. The others are made by forming steel into the desired shape. The type of burner used depends on the particular equipment design. Each manufacturer will use the burner that best lends itself to the requirements of the equipment.

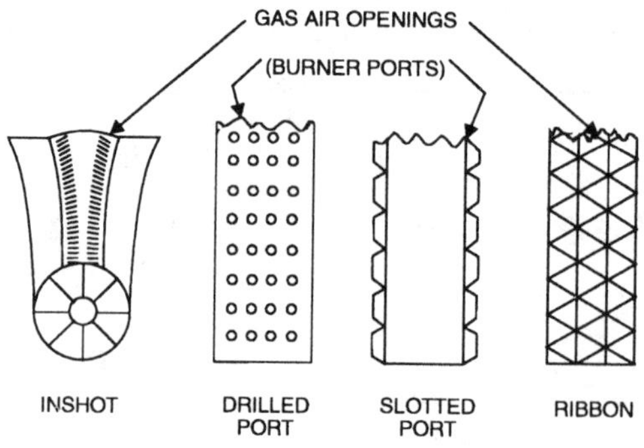
GAS AIR OPENINGS
(BURNER PORTS)
INSHOT
DRILLED PORT
SLOTTED PORT
RIBBON

FIGURE 3–6
Types of burners.

Burner Design

The very first part of the burner that is put to use is the *burner face* (Figure 3–7). The face of the burner is called the air mixer. It is through this opening that the primary air is admitted to the burner.

The *burner mixing tube* (Figure 3–8) is the next part of the burner with which the gas and air mixture comes into contact. The mixing tube provides the necessary space for the proper mixing of the gas and air and extends the full length of the burner. The venturi type of mixing tube is much more successful than a straight piece of pipe. The venturi reduces the turbulence of the mixture and allows a more even distribution to the burner ports.

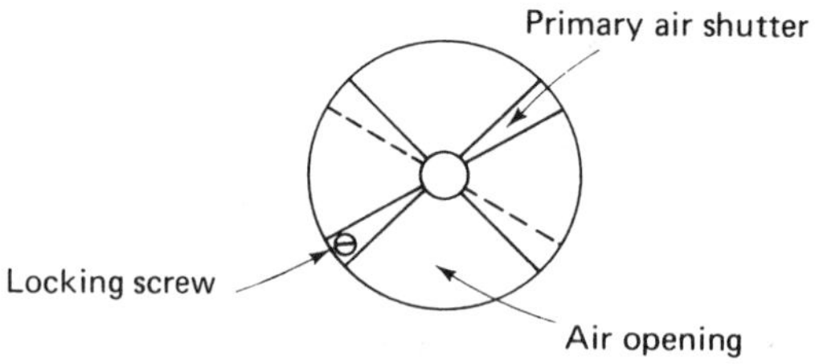
Primary air shutter
Locking screw
Air opening

FIGURE 3–7
Burner face.

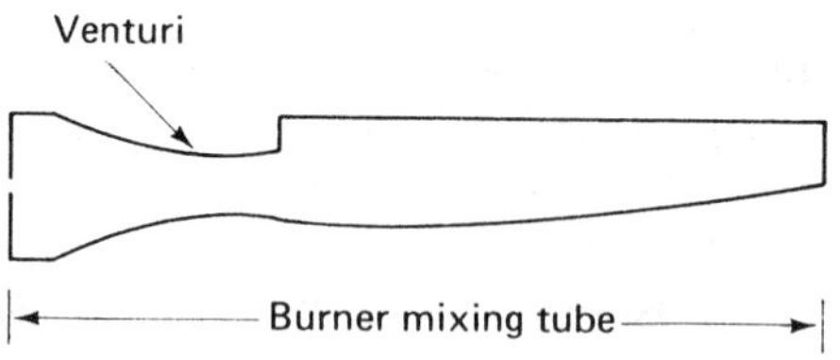
Venturi
Burner mixing tube

FIGURE 3–8
Side view of burner.

The *burner head* is the last part of the burner involved in the mixing and distribution of the gas and air mixture. The burner head is the top part of the burner (Figure 3–9). It is here that the ports are placed that determine the type of burner (drilled port, slotted port, ribbon, or inshot). The proper design of the burner head

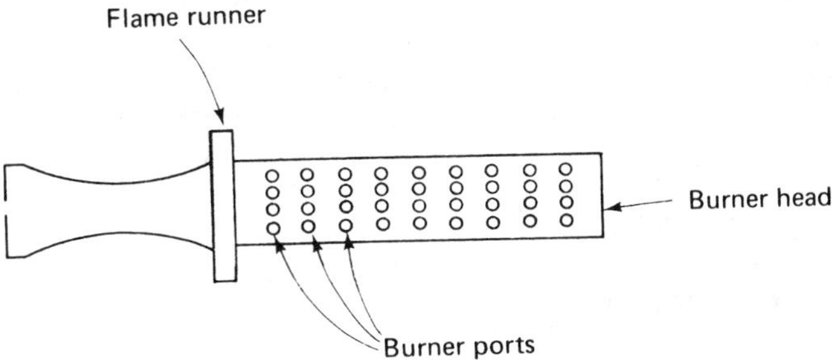
Flame runner
Burner head
Burner ports

FIGURE 3–9 Burner head.

is necessary if complete combustion of the gas is to be obtained. The burner ports are openings that release the gas and air mixture into the combustion chamber.

There should not be more than two rows of ports on a burner head without providing sufficient room for the secondary air to circulate over the burner head. This circulation of air cools the burner and helps prevent burner burn–out. For this reason it is important to keep the rust and scale formations removed from the burners.

The control of *primary air* is accomplished on the face of the main burner. There are several methods used for this purpose. The adjustable shutter, however, is the most common method used (Figure 3–10). The adjustment is made by enlarging or reducing the size of the opening through which the air is admitted. The threaded shutter is less common and is used only on cast iron burners (Figure 3–10). The deflector type, or adjustable baffle (again see Figure 3–10) is gaining acceptance in the industry. This burner has larger air openings than necessary. This

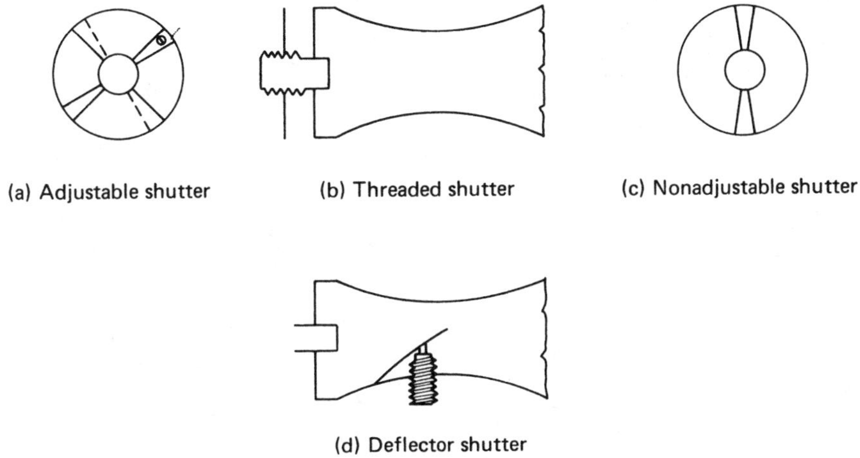
(a) Adjustable shutter
(b) Threaded shutter
(c) Nonadjustable shutter
(d) Deflector shutter

FIGURE 3–10 Primary air adjustment methods.

is an advantage because it will not clog up as fast as the other types. The primary air adjustment is made by pushing the deflector into the gas stream. This added restriction reduces the gas velocity and reduces the intake of primary air. Some burners have fixed openings and no adjustment can be made. These openings are sized according to the burner Btu rating, and the proper flame is automatically obtained when the correct orifice spud is used.

FORCED DRAFT BURNERS

The burners discussed previously have been of the atmospheric type. With the advent of the roof mount and the higher efficiency units, the forced draft burner is gaining in popularity. A small blower is used to force the air into and through the combustion area and out the vent system. See Figure 3–11. The air is forced into the combustion area under a slightly positive pressure. Both the combustion area and the vent system are under this slightly positive pressure. The combustion area is sealed from atmospheric conditions.

The amount of air supplied and the inside air pressure are controlled by adjusting the air flow. By forcing a fixed amount of air into the combustion chamber, preset conditions are maintained for the combustion area. The combustion air adjustments must be made according to the equipment manufacturer's specifications for the furnace in question. The blower supplies all of the air needed for proper combustion and the force required to move the products of combustion through the vent system and into the atmosphere. Anytime that a leak occurs anywhere in

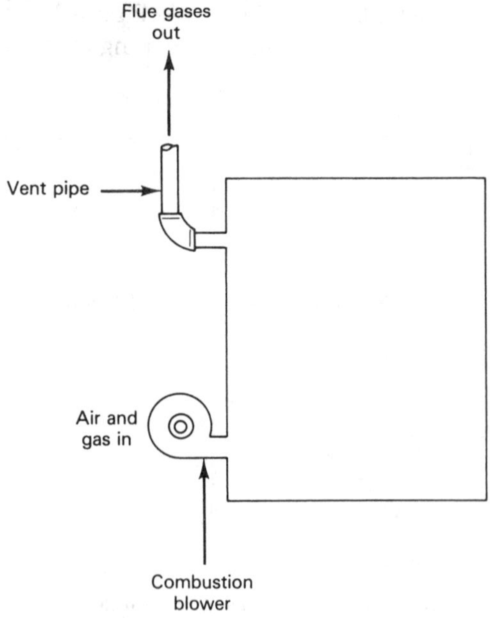
Flue gases out
Vent pipe
Air and gas in
Combustion blower

FIGURE 3–11
Forced draft heat exchanger. *(Courtesy of Billy C. Langley,* High Efficiency Gas Furnace Troubleshooting Handbook, *©1991, p. 4; reprinted by permission of Prentice-Hall, Inc., Englewood Cliffs, NJ).*

the system, the combustion process will be upset and proper, if any, operation will not be possible.

Forced draft is ideal for outdoor heating equipment, where a fixed condition is desired for combustion control, where unusual conditions prevail, or where vent pipe lengths are restricted. The main gas valve cannot open on these units until a definite pressure is obtained inside the combustion area.

INDUCED DRAFT BURNERS

Another type of burner that has gained in popularity due to the roof mount and the higher efficiency units is the induced draft burner. The induced draft burner, as its name implies, draws the combustion air into the combustion area, through the heat exchanger, and discharges it into the vent system. See Figure 3–12. In this type of system, the combustion area and the heat exchanger are under a slightly negative pressure. The vent system may have either a slightly positive or atmospheric pressure inside it. Any adjustments that are made to the induced draft burner must be made in accordance with the manufacturer's specifications for the furnace in question. The blower supplies all of the air needed for proper combustion and the force necessary to force the products of combustion out of the equipment and into the atmosphere. Anytime an air leak occurs anywhere in the system, the combustion process will be upset and proper, if any, operation will not be possible.

POWER GAS BURNERS

Another type of burner that is being used is known as the *power gas burner* (Figure 3–13). These burners are small, highly efficient, and suited for use with natural gas. Power burners have a very wide range of capacities. The maximum ca-

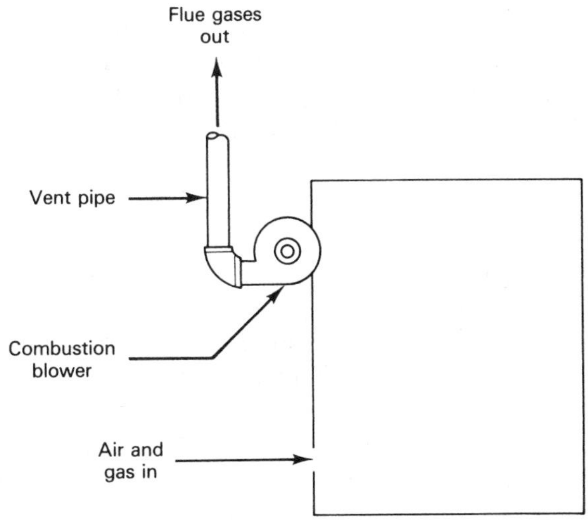
Flue gases out
Vent pipe
Combustion blower
Air and gas in

FIGURE 3–12
Induced draft heat exchanger. *(Courtesy of Billy C. Langley,* High Efficiency Gas Furnace Troubleshooting Handbook, *©1991, p. 5; reprinted by permission of Prentice-Hall, Inc., Englewood Cliffs, NJ).*

FIGURE 3–13 Power gas burner.

pacity is 500,000 Btu and the minimum is 120,000 Btu. They are applicable to any heating application where oil or an equivalent capacity is presently being used or as an integral component of a new boiler–burner system. They can also be directly installed and fired through existing doors of cast iron or steel firebox boilers without requiring further brickwork or modification of the boiler base.

Power burners furnish 100% of the combustion air through the burner for firing rates up to their rated capacities. These burners have locked air shutters, but with a reduced input start by means of a slow opening gas valve. Power burners are easy to install and to adjust due to the use of an orifice to meter the gas and a dial to adjust the air shutter to match the gas orifice used. An automatic electric pilot ignition is used that incorporates a 6,000–volt (V) transformer to provide the necessary spark to ignite the pilot gas.

In operation, these burners are more noisy than the atmospheric–type burner. Power burners are applicable to small boilers and to residential and industrial furnaces. They are also used as antipollution after–burners. These burners are automatic and are equipped with the latest types of safety and operating controls.

PILOT BURNERS

A *pilot burner* is defined as a small burner used to ignite the gas at the main burner ports as a source of heat for the pilot safety device, and for flame rectification in electronic ignition systems. (See Chapter 13.) Approximately 50% of all heating equipment failures can be attributed to this burner.

The automatic pilot is probably the most important safety device used on modern gas heating equipment. During normal operation, the pilot flame provides the heat necessary to actuate the pilot safety device. Also, the pilot is located in the combustion chamber in proper relation to the main burners so that the flame will ignite the gas admitted when the thermostat demands heating. Should something happen to cause the pilot to become unsafe, the pilot safety will prevent any unburned gas from escaping into the combustion chamber.

There are basically two types of pilot burners: (1) the *millivolt* and (2) *the bimetal.*

The *millivolt* type can be divided into two types: the *thermocouple* and the *thermopile.* The thermocouple is an electricity–producing device made of dissimilar metals. One end of the metal is heated and is called the *hot junction,* while the other end remains cool and is called the *cold junction* (Figure 3–14). The output of a thermocouple is approximately 30 millivolts (mV) when heated to a temperature of about 3,200° F. The thermopile is a series of thermocouples wired so that the output voltage will be approximately 750 mV (Figure 3–15). The greater the numbers of thermocouples in series, the higher will be the output voltage.

The *bimetal pilot* is named so because the sensing element is made of two dissimilar metals welded together so that they become one piece (Figure 3–16). Because the two metals have different expansion rates, they bend when heated. As the bimetal element is heated by the pilot flame, a movable contact is pushed toward a stationary contact. When these contacts touch, an electrical circuit is completed through them. This control is usually employed to operate in the control, or low voltage, circuit (Figure 3–17).

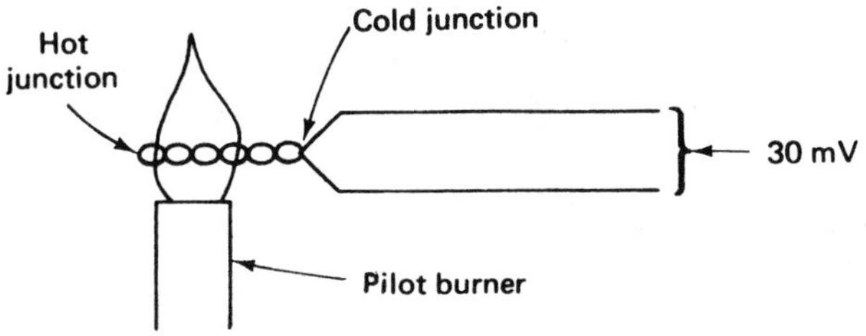
Hot junction
Cold junction
30 mV
Pilot burner

FIGURE 3–14
Thermocouple principle.

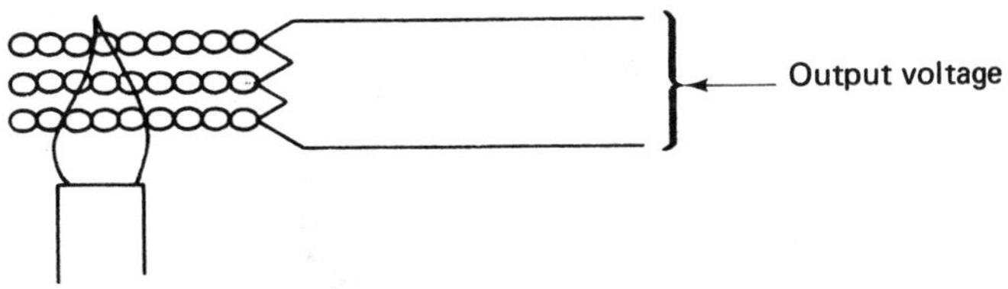
Output voltage

FIGURE 3–15 Thermopile principle.

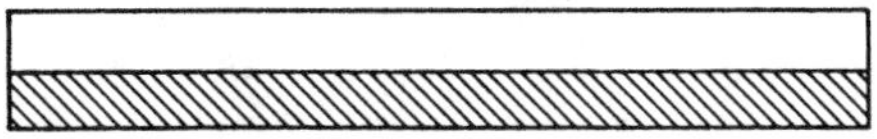

FIGURE 3–16
Operation of bimetal element.

ELECTRIC IGNITION

Several types of electric ignitors are in use on modern heating equipment. The main types are the glow coil and the direct spark ignitor. Some state and local laws require the use of automatic ignitors and have done away with the standard standing pilot on new equipment.

Glow Coil Ignitors

These ignitors are used to light or relight pilot burners in the case of flame failure (Figure 3–18). These units are assembled in a draft protected holder which provides means for either horizontal or vertical mounting to a pilot burner. Glow coil type ignitors are connected to an electrical power source of 2.5 volts. The coil is heated because of electrical resistance. After the pilot is lighted, the electrical power is interrupted to the glow coil. Continuous electrical power to the glow coil will cause it to burn out and become inoperative. Their use is limited to the lighting of pilot burners only, and they should not be used for main burner ignition.

Direct Spark Ignitors

These systems (Figure 3–19) may be used as direct replacement pilot lighters on equipment that has been certified by an approved testing agency with the pilot lighter as part of the original equipment. Where automatic pilot relighters are not certified as an original component of the unit, their application must be limited to

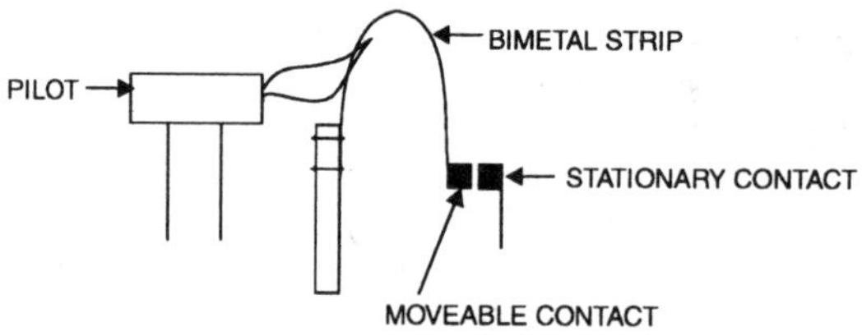
BIMETAL STRIP
PILOT
STATIONARY CONTACT
MOVEABLE CONTACT

FIGURE 3–17
Operation of bimetal element.

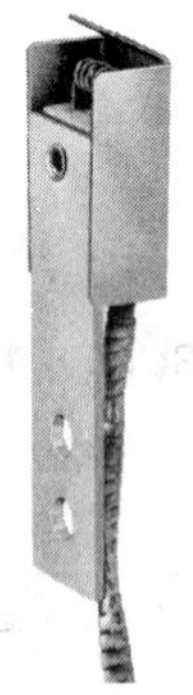

FIGURE 3–18
Glow coil ignitor. (*Courtesy of ITT General Controls.*)

rooftop heating units, space heaters, open bay heaters, or on installations where the unburned gases are quickly vented.

These units have a spark frequency of approximately 100 sparks per minute. They may be used on either 24 volts or 115 volts. They are satisfactory for use in a temperature range of -40° F to +160° F.

FLAME TYPES

There are basically two types of flames—the *yellow* and the *blue* flame. There are different variations of these two flames that are produced by changes in the primary air supplied to gaseous fuels. Primary air is the air that is mixed with the gas before ignition. The counterpart of primary air is secondary air, which is mixed with the flame after ignition (Figure 3–20).

The following is a list of flames and their indications that are encountered in heating systems using gas as a fuel. The first is the yellow flame, and we will progress to the blue flame.

Yellow Flame

This flame has a small, blue–colored area at the bottom of the flame (Figure 3–21). The outer portion, or outer envelope, is completely yellow and is usually smoking. This smoke is unburned carbon from the fuel. The yellow, or luminous, portion of the flame is caused by the slow burning of the carbon that is being burnt. Soot is the outcome of the yellow flame, which produces a lower temperature than does the blue flame. A yellow flame is an indication of insufficient primary air. Incomplete combustion is also indicated by a yellow flame—a hazardous condition that cannot be allowed.

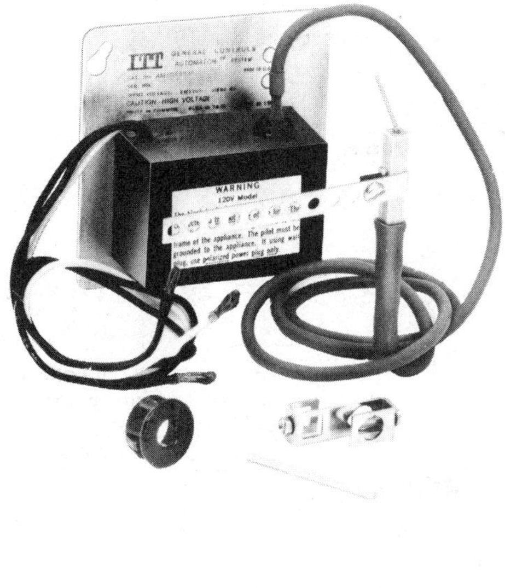
ITT
WARNING
120V Model

FIGURE 3–19
Direct spark ignitor. (*Courtesy of ITT General Controls*).

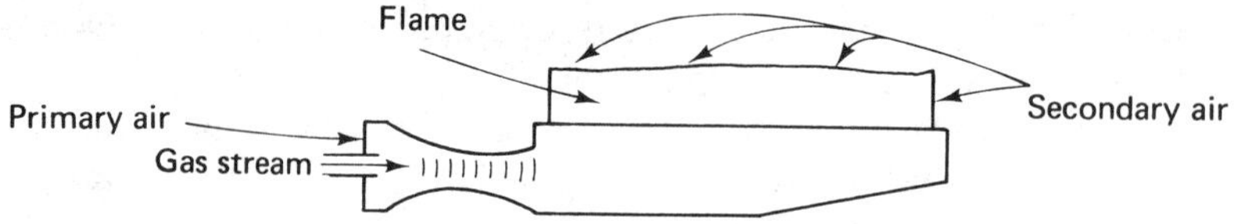
Flame
Primary air
Gas stream
Secondary air

FIGURE 3–20 Primary air and secondary air supporting combustion.

Yellow–Tipped Flame

This condition occurs when not quite enough primary air is admitted to the burner. The yellow tips will be on the upper portion of the outer mantle (Figure 3–22). This flame is also undesirable because the unburned carbon will be deposited in the furnace flues and restrictors, thus eventually restricting the passage of the products of combustion.

Orange Color

Some orange or red color in a flame, usually in the form of streaks, should not cause any concern. These streaks are caused by dust particles in the air and cannot be completely eliminated (Figure 3–23). However, when there is a great amount of red or orange in the flame, the furnace location should be changed or some means provided for filtering the combustion air to the equipment.

Soft, Lazy Flame

This condition appears only when just enough primary air is admitted to the burner to cause the yellow tips to disappear. The inner cone and the outer envelope will not

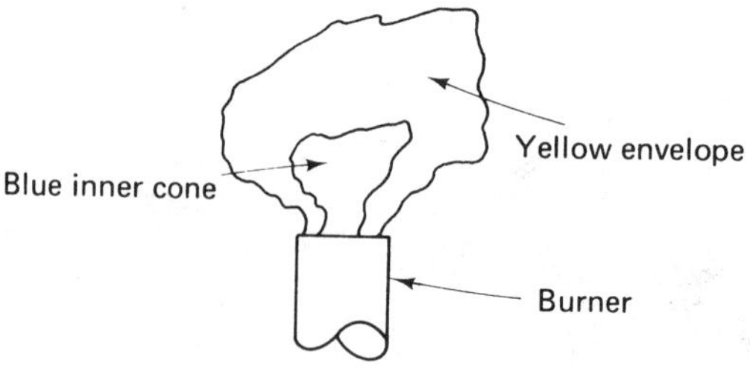
Yellow envelope
Blue inner cone
Burner

FIGURE 3–21
Yellow-tipped flame.

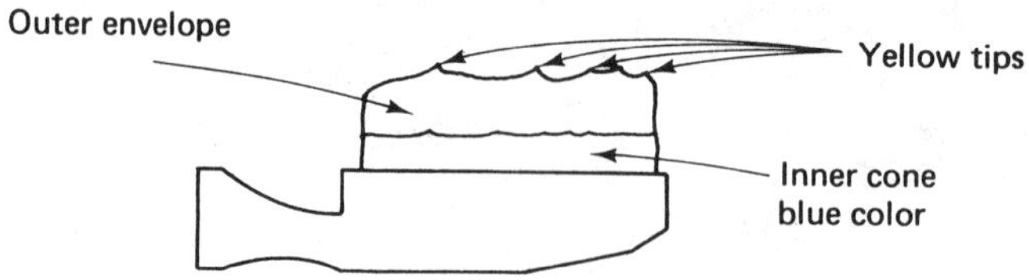
Outer envelope
Yellow tips
Inner cone
blue color

FIGURE 3–22 Yellow-tipped flame.

be as clearly defined as they are in the correct blue flame (Figure 3–24). This type of flame is best for high–low fire operation. It is not suitable, however, wherever there may be a shortage of secondary air. This flame will burn blue in the open air; however, when it touches a cooler surface, soot will be deposited on that surface.

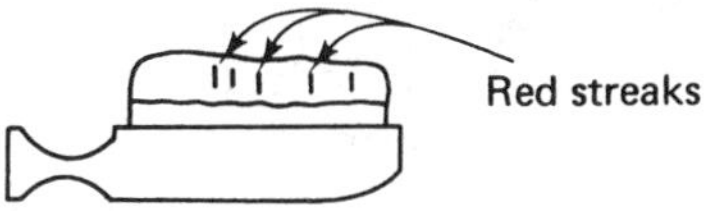
Red streaks

FIGURE 3–23
Orange streaks in flame.

Sharp, Blue Flame

When the proper ratio of gas to air is maintained, there will be a sharp, blue flame (Figure 3–25). Both the outer envelope and the inner cone will be pointed and the sides will be straight. The flame will be resting on the burner ports and there will not be any noticeable blowing noise. The flame will ignite smoothly on demand from the thermostat. Also, it will burn with a nonluminous flame. This is the most desirable flame for heating purposes on standard heating units.

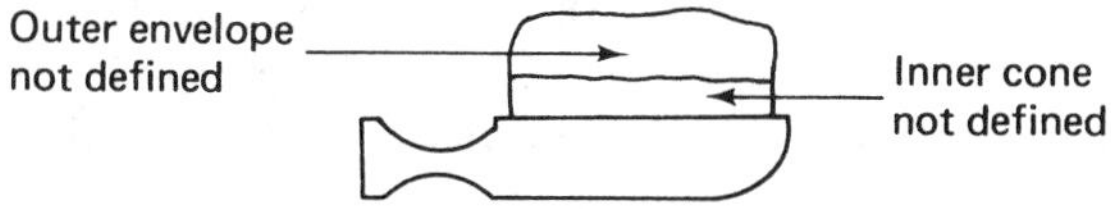
Outer envelope
not defined
Inner cone
not defined

FIGURE 3–24
Soft, lazy flame.

Lifting Flame

When too much primary air is admitted to the burner, the flame will actually lift off the burner ports. This flame is undesirable for many reasons. When the flame is raised from the ports, there is a possibility that intermediate products of combustion will escape to the atmosphere (Figure 3–26). The flame will be small and will be accompanied by a blowing noise, much like that made by a blow torch. There will be rough ignition and, if enough secondary air is available, ignition may be impossible. Intermediate products of combustion are the same as the products of incomplete combustion, which must be avoided.

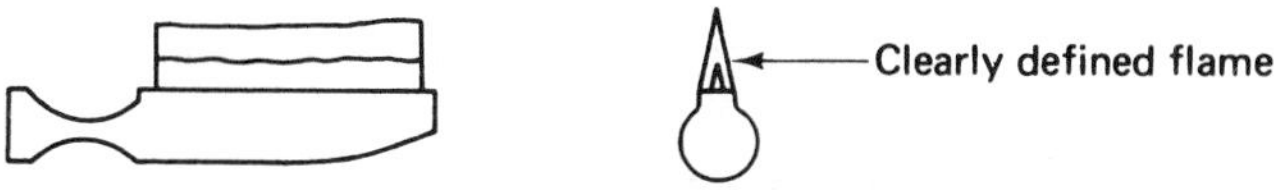
Clearly defined flame

FIGURE 3–25
Sharp flame.

Floating Flame

A floating flame is an indication that there is a lack of secondary air to the flame. In severe cases, the flame will leave the burner and have the appearance of a floating cloud (Figure 3–27). Again, the intermediate products of combustion are apt to

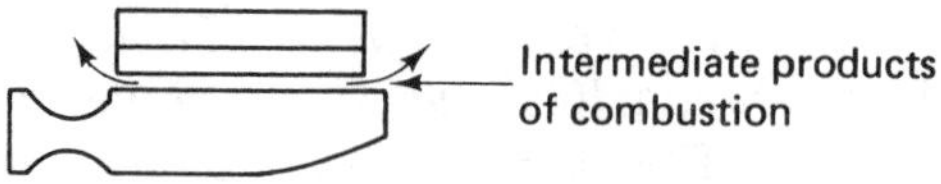
Intermediate products
of combustion

FIGURE 3–26 Lifting flame.

escape from between the burner and the flame. The flame is actually floating from place to place wherever sufficient secondary air for combustion may be found.

This flame is sometimes hard to detect because, as the door to the equipment room is opened, sufficient air will be admitted to allow the flame to rest on the burner properly. However, the obnoxious odor of aldehydes may still be detected. Another possible cause of this flame is an improperly operating vent system. If the products of combustion cannot escape, fresh air cannot reach the flame. This is a hazardous situation that must be eliminated regardless of the cause.

COMBUSTION TROUBLESHOOTING

Heating service technicians are called on to make fast, accurate analyses and corrections of heating equipment in order to provide the customer with safe, economical, and adequate heat. The major situations dealing with the flame that they will be called on to analyze are (1) delayed ignition, (2) roll–out ignition, (3) flashback, (4) resonance, (5) yellow flame, (6) floating main burner flame lifting off the burners, and (7) main burner flame too large. The following paragraphs will discuss these problems and make some recommendations for their solution.

Delayed Ignition

Delayed ignition is caused by improper or poor flame travel to the main burner, or by poor flame distribution over the burner itself, or by too small orifices. When this situation occurs, it can usually be detected by a noisy ignition—that is, a light explosion, or puff, on ignition of the main burner.

The possible causes and correction of delayed ignition are:

1. *Distorted burner or distorted carry-over wing shots.* Both the carry-over wing slots and the main burner slots should be uniformly shaped and of the proper size (Figure 3–28). If not, the distortion could cause the adjacent burner to have faulty ignition accompanied by an accumulation of gas, which causes a noisy and dangerous ignition.

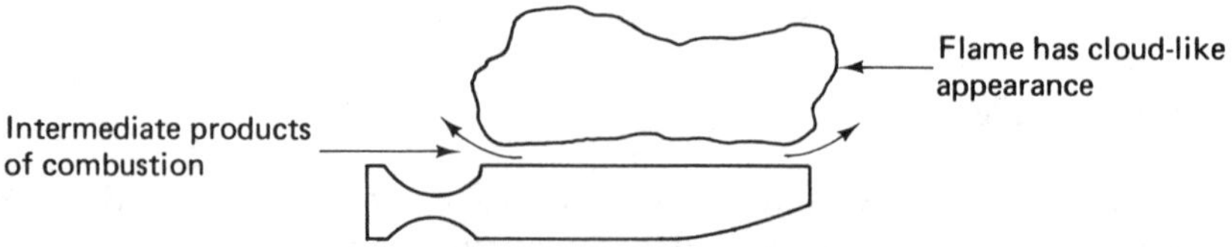
Flame has cloud-like
appearance
Intermediate products
of combustion

FIGURE 3–27 Floating flame.

2. *Misaligned carry-over wings.* The purpose of the carry-over wing is to direct the gas for ignition from one main burner to the next. The carry-over wing on each burner must be aligned with the adjacent burner so that the flame path is no more than 1/16 inch above or 1/8 inch below the adjacent burner. A delayed ignition could result if these limits are not met (Figure 3–29). This problem can usually be solved by loosening the clamp aligning the burner and retightening the clamp.

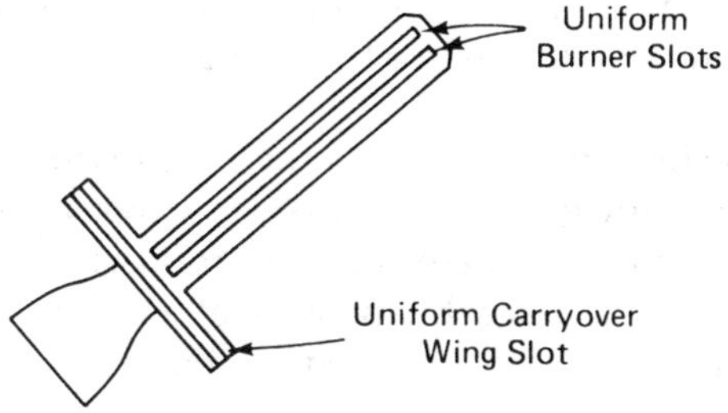
Uniform
Burner Slots
Uniform Carryover
Wing Slot

FIGURE 3–28
Carry-over wing and burner slots.

3. *Painted-over or rusted carry-over slots.* Occasionally, paint will get into the carry-over wings during their manufacture. Also, the burners and heat exchanger will rust and this rust will accumulate in the ports and slots, preventing proper flame travel during the ignition period. This poor flame travel results in delayed ignition. The solution to this problem is to clean the ports and slots of paint and rust. If rust is the problem, the owner should be instructed to turn off the pilot during summer operation to reduce the possibility of rust recurring if condensation forms on the cooler surfaces of the furnace.

4. *Low manifold gas pressure.* If the supply gas pressure to the furnace should be too low, the flow of ignition gas may be slow from one burner to the next, possibly resulting in delayed ignition of some of the burners. This possibility exists on any installation, especially during cold weather when the demand for gas is high. All furnaces are equipped with a pressure tap on the main gas valve or the manifold to measure the gas pressure. Use either a water manometer or a gas manifold pressure gauge (Figure 3–30). The manifold gas pressure for natural gas is 3 1/2 to 4 inches of water column. Propane equipment requires an 11–inch water column. These pressures should be taken with all other gas appliances on the same gas main in operation.

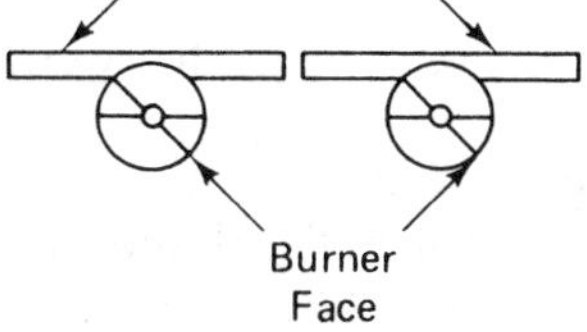
Carryover
Wings
Burner
Face

FIGURE 3–29
Carry-over wing adjustment.

5. *Orifices too small.* The only solution to this problem is to install the proper size orifices. The ones that are in the furnace may be increased in size, or new ones may be installed. Before changing the orifice size, make certain that the manifold gas pressure is properly adjusted. Then calculate the size needed and make the necessary correction. This condition is more likely to exist on new installations or units that have been changed from LP gas to natural gas.

6. *Main gas valve equipped with step opening regulator.* When delayed ignition occurs on a furnace equipped with a step opening gas valve, the remedies are:

a. Carefully check and correct any of the previously mentioned possibilities.
b. Next, measure the first step opening of the gas pressure with a manometer or manifold pressure gauge. If the first step pressure opening has less than 2 inches water column, there are two options: (1) The first step pressure may be adjusted to at least 2 inches water column. (2) Replace the step opening regulator with a standard regulator.

Roll-Out Ignition

Roll-out is usually indicated by a puff or whish sound on burner ignition. This situation is usually caused by soot or some other obstruction in the heat exchanger causing a restriction. The solution is to remove the soot or obstruction from the heat exchanger.

Flashback

Flashback occurs when the flow of the gas–air mixture entering the burner is slower than the velocity of the flame travel. This results in a popping sound and sometimes a flame burning at the orifice. It is usually caused by a low manifold gas pressure, a change in the gas–air mixture, or a sudden down draft on the flame pattern. The following are possible causes of flashback:

1. Extremely hard flame.
2. Distorted burner, or carry-over wing slots.
3. Low manifold gas pressure.
4. Defective burner orifice.
5. Misalignment of the burner and orifice.
6. Erratic operation of the main gas valve.
7. Unstable gas supply pressure.
8. Rust or soot in the burner.
9. Wrong type of LP gas, or gas–air supply mixture.

Extremely hard flame. In this situation, close off on the primary air shutter until a small yellow tip appears on the flame. A yellow tip of up to 3 inches can be tolerated by most furnaces when they are hot.

Distorted burner or carry-over wing slots. When this situation occurs, check the burner ports and carry-over wing slots for damage. Correct any damage by repairing or replacing the burner.

Low manifold gas pressure. To correct this condition, check the manifold pressure with a manometer or manifold pressure gauge after the burners have been fired for at least five minutes. Correct any pressures that do not meet the recommendations for the type of gas used.

Defective burner orifice. To determine whether or not the burner orifice is bad, turn off the main gas valve, remove the burners, and then remove each individual orifice from the manifold. Check the outside of the orifice spuds for dents or misshaping of the orifice hole. Check the inside of the orifices for foreign matter such as dirt, grease, pipe dope, etc. Remove any foreign material, being careful not to damage the orifice. Replace the orifices in the manifold and check for gas leaks with a soap and water solution or other liquid-type leak detector. Never use a match for leak testing gas piping.

Misaligned burner orifices. If flashback is persistent on any one burner, first check the orifices for damage and foreign matter. If the problem still exists, completely remove the manifold assembly from the furnace, and apply water pressure to the assembly. This procedure will indicate any orifice misalignment by a crooked spray of water from a misaligned orifice. (See Figure 3–31.) Make any necessary repairs to correct the problem. In some cases, repair may require the replacement of the entire assembly.

Unstable gas supply pressure. This condition is more likely to occur on LP gas furnaces. This condition can be identified by use of a manifold pressure gauge or a manometer connected to the gas line.

This situation is usually caused by a chattering regulator or the main gas line exposed to extreme cold temperatures. It is sometimes necessary to install two–stage gas pressure regulators to eliminate this problem.

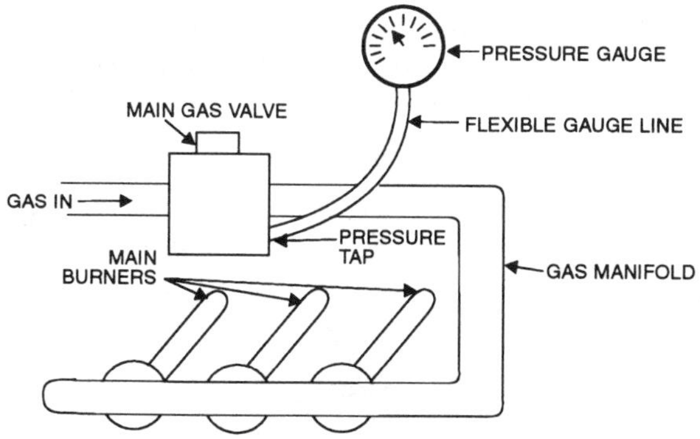
PRESSURE GAUGE
MAIN GAS VALVE
FLEXIBLE GAUGE LINE
GAS IN
PRESSURE TAP
MAIN BURNERS
GAS MANIFOLD

FIGURE 3–30
Manifold pressure connection.

Rust or soot in the burner. When this situation occurs, the main burners must be removed and all foreign matter removed from inside them. The cause of rusting should be determined and corrected. Sometimes rusting is caused by condensation if the pilot burner is left on during summer operation. If so, the user should be instructed to turn off the pilot during the cooling season.

Sooting of a furnace may be caused by low gas pressure and/or an improper setting of the primary air adjustment. Either of these conditions is generally accompanied by a large yellow flame. Precautions must be taken to prevent sooting of a furnace heat exchanger.

Wrong type LP gas, or gas mixture. The furnace name plate must be checked to be certain that the furnace is designed for the type of gas being supplied to it. Mixtures of butane and propane gases used in furnaces not approved for liquefied petroleum gas could be the cause of poor burner operation. They could result in flashback, sooting, and burner damage because of overheating. A quick check for this situation is to check the relief valve setting. Tanks equipped with relief valve settings of 100 to 150 psi will accommodate gas mixtures. Relief valve settings of 200 to 250 psi are used on tanks equipped for pure propane.

Resonance

Resonance can be identified by a loud rumbling noise or a pure tone buzz or hum. Resonance is most common when butane is used as the fuel. This situation may be caused by excessive primary air being supplied to the main burner or by a defective main burner spud. If excessive primary air is the cause, adjust the primary air shutter until a slight yellow tip appears on the flame. Lock the shutter in this position. If defective spuds are suspected, remove the spuds and check for nicks or dents on the outer edge of the orifice. Also, check for dirt or other foreign material inside the orifice. Clean or replace as necessary.

Yellow Flame

A large yellow, or luminous, flame is a good indication that almost all of the oxygen required for combustion is being taken from the secondary air. Complete combustion will occur as long as enough oxygen is supplied and nothing interferes with the flame. If the flame should touch a cooler surface, however, soot and toxic products will be released. Therefore, large yellow flames must be avoided.

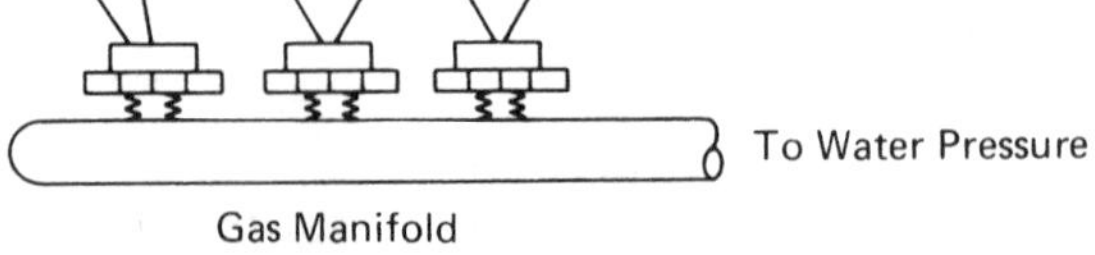
Misaligned Orifice
Good Orifices
To Water Pressure
Gas Manifold

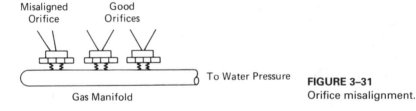

FIGURE 3–31
Orifice misalignment.

There are five possible causes of large yellow flames:

1. Primary air shutter closed off too much; to correct, adjust the primary air shutter.
2. Partially clogged main burner ports or orifices; to correct, remove the spud and inspect for damage.
3. Misaligned burner spuds; to correct, check alignment and correct as necessary.
4. Soot or foreign material inside of the heat exchanger; to correct, clean the heat exchanger to allow proper combustion.
5. Poor venting, down draft, or improper combustion air supply; to correct, eliminate any poor vent piping. It may be necessary to install an outside air duct to provide the required combustion air.

Floating Main Burner Flame

This situation is not very common, but may occur under the following conditions:

1. The heat exchanger is blocked with soot; to correct, clean the soot from the heat exchanger and correct the cause of poor combustion.
2. Air blowing into the heat exchanger; to correct, check for a leaking or cracked heat exchanger that may allow circulating air into the heat exchanger.
3. Negative interior pressure in the furnace room; to correct, recheck the combustion air requirements. Also, check for exhaust fans in the furnace room. Correct as necessary.

Main Burner Flame Too Large

The possible causes and the corrective action to be taken are:

1. Excessive gas manifold pressure; to correct, check pressure with a manifold pressure gauge or manometer and reduce to proper pressure.
2. Defective gas pressure regulator; to correct; replace the gas valve or pressure regulator.
3. Orifice size too large; to correct, replace the orifice with proper size. Do not solder orifices closed and redrill, because the solder may melt out and cause more problems.

REVIEW QUESTIONS

1. A manifold gas pressure of 11 inches water column is required for what fuel?
2. Describe the function of a burner orifice.
3. As the gas leaves the orifice, where is it directed?

4. What causes air to be drawn into a gas burner?
5. In what component is the gas and air mixed in a gas furnace?
6. Where is the primary air to a gas burner adjusted?
7. What effect does a reduction in gas velocity have on combustion air?
8. What will a leak cause when a forced draft burner is used?
9. On a forced draft burner, when can the main gas valve open?
10. In what condition is the pressure in the combustion area when an induced draft burner is used?
11. What component causes about half of the heating equipment failures?
12. Name the different types of junctions associated with a thermocouple.
13. Of what is a bimetal made?
14. For what are glow coil ignitors used?
15. Can direct spark ignition (DSI) systems be used on any furnace?
16. At what point does primary air enter the combustion process?
17. What causes the smoke in a yellow flame?
18. What causes orange streaks in a flame?
19. What causes a sharp, blue flame?
20. What are intermediate products of combustion?
21. What is slight explosion, or puff, on main burner ignition known as?
22. What will a restricted heat exchanger cause?
23. Of what is a popping sound in the main burner an indication?
24. Of what is persistent flashback on a single burner an indication?
25. What could a negative pressure in the furnace room cause?

4

fuel oil burners

The objectives of this chapter are:

- To introduce you to the different types of fuel oil burners.
- To acquaint you with fuel oil nozzles.
- To provide you with the operating principles of the most popular types of fuel oil burners used in comfort heating systems.
- To familiarize you with the ignition principles used on fuel oil burners.
- To acquaint you with the necessity of proper analysis and treatment of fuel oil.
- To introduce you to the piping systems involved in fuel oil heating systems.
- To familiarize you with the operating principles of combination gas-oil burners.

Fuel oils, like natural and LP gases, are excellent heating fuels. Among their many differences, the lighting and burning of fuel oil requires special equipment. Fuel oil in its liquid form will not burn; it must be either atomized or vaporized. From this we can define a fuel oil burner as a mechanical device that prepares oil for combustion. The actual burning of the fuel takes place in the firebox, with the atomizing method being the most popular. The atomizing type of burner prepares the oil for burning by breaking it up into a foglike vapor. This is accomplished in three ways: (1) by forcing the oil under pressure through a nozzle—air or steam may be forced through with the oil or the oil may be forced alone; (2) by allowing the oil to flow out the end of small tubes that are rapidly rotated on a distributor; or (3) by forcing the oil off the edge of a rapidly rotating cup that may be mounted either vertically or horizontally.

FUEL OIL BURNERS

A typical *high pressure* oil burner is comprised of an electric motor, blower, and fuel unit. The motor is mounted on the motor shaft and the fuel unit is directly connected to the end of the motor shaft. The fuel unit consists of a strainer, a pump, and a pressure regulating valve.

In operation, when the thermostat demands heat for the structure, the fuel oil is drawn from the storage tank. The oil is drawn through a strainer where the solid particles are removed, passed through the fuel pump, and forced to the pressure regulating valve. The oil hesitates here until the pressure has reached 100 psig. When this pressure is reached, the pressure regulating valve opens and allows the fuel oil to proceed to the burner nozzle where it is atomized. After the fuel leaves the nozzle, it is mixed with the proper amount of air and is ignited by a high voltage transformer—approximately 10,000 V and 25 milliamperes (mA) (Figure 4–1).

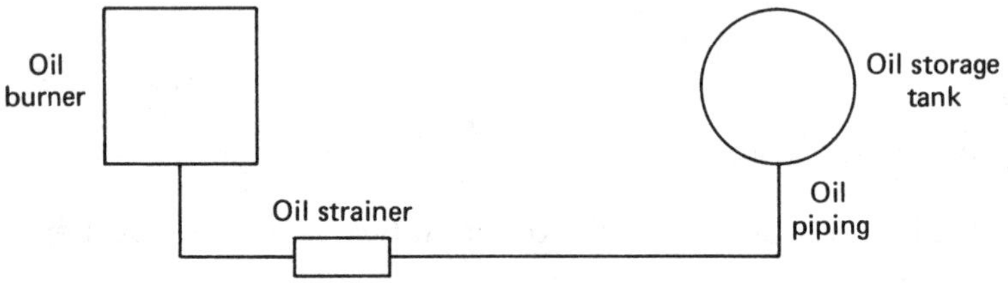
Oil
burner
Oil storage
tank
Oil
piping
Oil strainer

FIGURE 4–1 Basic fuel oil system.

When sufficient heat has been supplied to the structure, the thermostat will signal the burner motor to stop. As the motor reduces its speed, the oil pressure from the fuel pump also is reduced. As the oil pressure is reduced below 90 psig, the pressure regulating valve stops the flow of oil to the nozzle and the flame is extinguished because of the lack of fuel. The oil burner will remain at rest until the structure requires additional heat, at which time the sequence is started again.

The *low pressure* fuel oil burner is similar in appearance to the high pressure burner (Figure 4–2). It also operates much the same as the high pressure burner

FIGURE 4–2
A low-pressure oil burner. (*Courtesy The Carlin Co.*)

except for the method used to atomize the oil. In the low pressure burner, the oil is atomized by compressed air, similar to a point spray gun. The primary air and the fuel oil are forced through the nozzle at the same time with 1 to 15 pounds of pressure. The secondary air is supplied in the same manner as the air used by the high pressure burner (Figure 4–3).

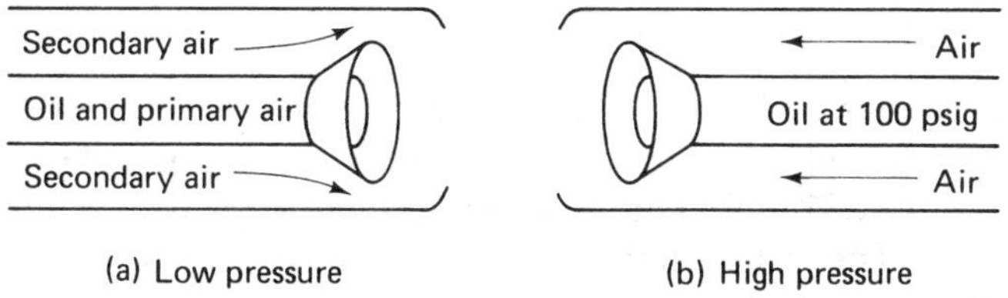
Secondary air
Oil and primary air
Secondary air
(a) Low pressure
Air
Oil at 100 psig
Air
(b) High pressure

FIGURE 4–3
Oil burner feed principle.

The *rotary-type* oil burner incorporates a completely different principle from either the high or low pressure burners. The rotary burner employs a cup to atomize the oil. The oil is forced to the edge of the cup in small droplets because of the high speed of the cup. As the droplets leave the edge of the cup, they are dispersed into a current of rapidly moving air provided by a centrifugal blower (Figure 4–4). These droplets are further atomized by this air, and the foglike fuel is projected parallel to the axis of the cup. The atomized oil is then ignited in much the same manner as any other type of oil burner. In some burners the cup may be replaced with small tubes extending to the face of the rotating member. The oil is forced to the ends of the tubes where centrifugal force distributes the oil into the air stream.

Nozzles

To understand how a nozzle fits into the overall performance of an oil burner, we shall first review the steps of efficient combustion.

Like all combustible matter, the oil must first be vaporized (that is, converted to a vapor or a gas). This requirement must be fulfilled before combustion can come about. This vaporization is usually accomplished by the application of heat.

The oil vapor must then be thoroughly mixed with air so that the necessary oxygen will be present for combustion. It is then necessary to raise the temperature of this mixture above the ignition point.

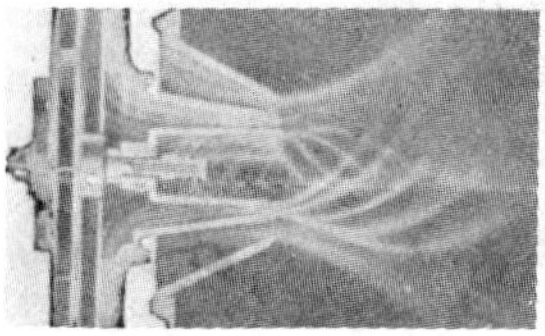

FIGURE 4–4
Rotary oil burner principle. (*Courtesy Ray Burner Co.*)

A continuous supply of air and fuel must be provided for continuous combustion. The products of combustion are then removed from the combustion chamber.

These same steps are necessary for all types of oil burners—the low pressure gun type, the high pressure gun type, as well as the rotary burners.

One of the functions of a nozzle is to atomize the fuel, or break it up into tiny droplets that can be vaporized in a short period of time. The atomizing nozzle performs three basic and vital functions for an oil burner.

1. **Atomizing.** It speeds up the vaporization process by breaking up the oil into tiny droplets. There are about 55 billion droplets per gallon of oil at a pressure of 100 psig. The exposed surface of 1 gallon of oil is expanded to approximately 690,000 in.^{2} of burning surface. The individual droplet sizes range from 0.002 to 0.010 in. The smaller droplets are necessary for the fast, quiet ignition and to establish a flame front close to the burner head. The larger droplets take longer to burn and help fill the combustion chamber with flame.
2. **Metering.** A nozzle is so designed and dimensioned that it will deliver a fixed amount of atomized fuel to the combustion chamber within a plus or minus range of 5% of its rated capacity. This means that nozzles must be available in many flow rates to satisfy a wide range of industry needs. Under 5 gallons per hour (gph), for example, 21 different flow rates and 6 different spray angles are considered as standard.
3. **Patterning.** A nozzle is also expected to deliver atomized oil to the combustion chamber in a uniform spray pattern and a spray angle best suited to the requirements of a specific installation.

Now that the operation of a nozzle is known, a cut-away showing the functional parts of a typical nozzle is shown in Figure 4–5. The flow rate, spray angle, and pattern are directly related to the design of the tangential slots, swirl chamber, and orifice.

In operation, a source of energy is needed to break up the oil into small droplets. Therefore, pressure is supplied to the nozzle from the oil pump, usually

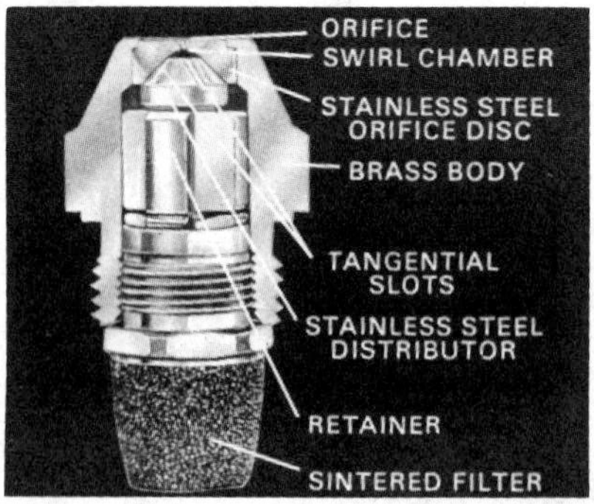
ORIFICE
SWIRL CHAMBER
STAINLESS STEEL ORIFICE DISC
BRASS BODY
TANGENTIAL SLOTS
STAINLESS STEEL DISTRIBUTOR
RETAINER
SINTERED FILTER

FIGURE 4–5
Typical high pressure nozzle. (*Courtesy Delavan, Inc.*)

at 100 psig for high pressure nozzles or 1 to 15 psig for low pressure nozzles (Figure 4–6). However, pressure energy alone is not enough to do the job. It must be converted to velocity energy. This is accomplished by directing the pressurized fuel through a set of slots. These slots are cut in the distributor at an angle, or tangentially, to produce a high velocity rotation within the swirl chamber. At this point, about half of the pressure energy is converted to velocity energy.

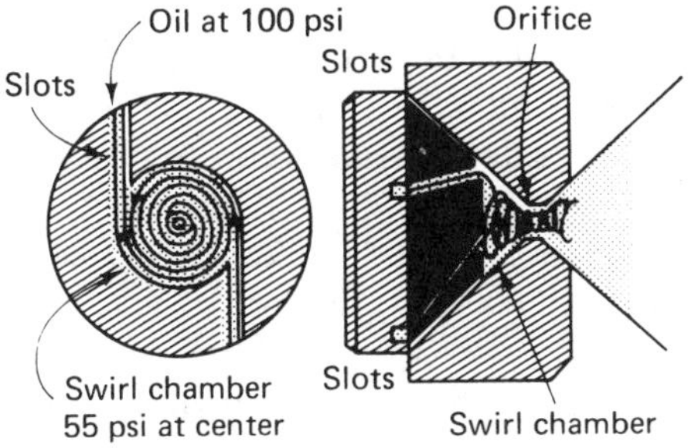
Oil at 100 psi
Orifice
Slots
Slots
Swirl chamber
55 psi at center
Slots
Swirl chamber

FIGURE 4–6
How a nozzle works. (*Courtesy Delavan, Inc.*)

As the oil swirls, centrifugal force is exerted against the sides of the chamber, driving the oil against the orifice walls, leaving a void or core of air in the center. The oil then moves forward out of the orifice in the form of a hollow tube. The tube then becomes a cone-shaped film of oil as it emerges from the orifice, ultimately stretching to a point where it ruptures and throws off droplets of liquid.

Normally, 100 psig is satisfactory for the fixed pressure supplied to a high pressure nozzle, and all manufacturers of this type of nozzle calibrate their nozzle for that pressure.

It is interesting to observe the sprays for a nozzle at various pressures (Figure 4–7). At low pressure, the cone-shaped film is long and the droplets breaking off it are large and irregular [Figure 4–7(a)]. As the pressure is increased, the spray angle

Spray at 10 psi pressure. (a) **Spray at 100 psi pressure.** (b) **Spray at 300 psi pressure.** (c)

FIGURE 4–7 Spray angles at various pressures. (*Courtesy Delavan, Inc.*)

becomes better defined. Once a stable pattern is formed, any increase in pressure does not affect the basic spray angle, measured directly in front of the orifice.

At higher pressures, however, note that beyond the area of the basic spray angle, the amount of droplets does make a slight change in direction, inward. This change at this point is because the droplets have a lower velocity due to air resistance, and the air drawn into the spray, by the spray itself, moves the droplets inward. This is the same phenomenon that causes a shower curtain to be drawn into the spray.

Pressure has another predictable effect on nozzle performance. An increase in pressure causes a corresponding increase in the flow rate of a nozzle, assuming all other factors remain equal. This relationship between pressure and flow rate is best shown in Table 4–1.

An increase in pressure also reduces the droplet size in the spray. For example, an increase from 100 to 300 psig reduces the average droplet diameter about 28%.

If the pressure is too low, the burner may be underfired. Efficiency may also drop sharply because droplet size is larger and the spray pattern may be changed. If the pressure is not carefully checked, the marking on the nozzle becomes meaningless.

Regardless of the number of spray patterns offered by manufacturers, nozzles can be grouped into two basic classifications for the purpose of this discussion (Figure 4–8).

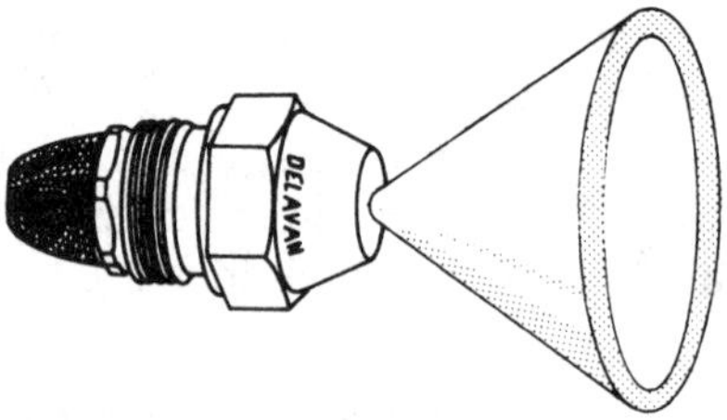
DELAVAN

(a) Hollow cone spray pattern

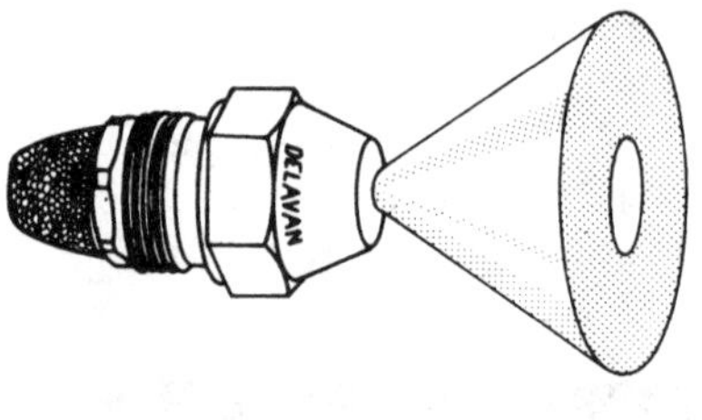
DELAVAN

(b) All-purpose spray pattern

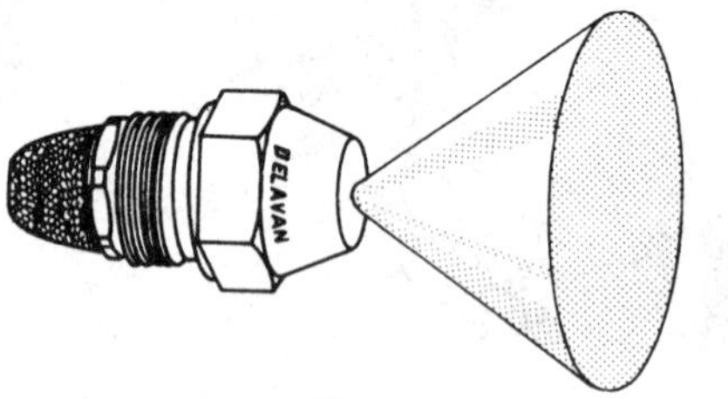
DELAVAN

(c) Solid cone spray pattern

FIGURE 4–8 Spray patterns. (*Courtesy Delavan, Inc.*)

TABLE 4–1 Effects of pressure on nozzle flow rate

Nozzle Rating at 100 psi	Nozzle Flow Rates in Gallons per Hour (Approx.)					
	80 psi	120 psi	140 psi	160 psi	200 psi	300 psi
.50	0.45	0.55	0.59	0.63	0.70	0.86
.65	0.58	0.71	0.77	0.82	0.92	1.12
.75	0.67	0.82	0.89	0.95	1.05	1.30
.85	0.76	0.93	1.00	1.08	1.20	1.47
.90	0.81	0.99	1.07	1.14	1.27	1.56
1.00	0.89	1.10	1.18	1.27	1.41	1.73
1.10	0.99	1.21	1.30	1.39	1.55	1.90
1.20	1.07	1.31	1.41	1.51	1.70	2.08
1.25	1.12	1.37	1.48	1.58	1.76	2.16
1.35	1.21	1.48	1.60	1.71	1.91	2.34
1.50	1.34	1.64	1.78	1.90	2.12	2.60
1.65	1.48	1.81	1.95	2.09	2.33	2.86
1.75	1.57	1.92	2.07	2.22	2.48	3.03
2.00	1.79	2.19	2.37	2.53	2.82	3.48
2.25	2.01	2.47	2.66	2.85	3.18	3.90
2.50	2.24	2.74	2.96	3.16	3.54	4.33
2.75	2.44	3.00	3.24	3.48	3.90	4.75
3.00	2.69	3.29	3.55	3.80	4.25	5.20
3.25	2.90	3.56	3.83	4.10	4.60	5.63
3.50	3.10	3.82	4.13	4.42	4.95	6.06
4.00	3.55	4.37	4.70	5.05	5.65	6.92
4.50	4.00	4.92	5.30	5.70	6.35	7.80
5.00	4.45	5.46	5.90	6.30	7.05	8.65
5.50	4.90	6.00	6.50	6.95	7.75	9.52
6.00	5.35	6.56	7.10	7.60	8.50	10.4
6.50	5.80	7.10	7.65	8.20	9.20	11.2
7.00	6.22	7.65	8.25	8.85	9.90	12.1
7.50	6.65	8.20	8.85	9.50	10.6	13.0
8.00	7.10	8.75	9.43	10.1	11.3	13.8
8.50	7.55	9.30	10.0	10.7	12.0	14.7
9.00	8.00	9.85	10.6	11.4	12.7	15.6
9.50	8.45	10.4	11.2	12.0	13.4	16.4
10.00	8.90	10.9	11.8	12.6	14.1	17.3
11.00	9.80	12.0	13.0	13.9	15.5	19.0
12.00	10.7	13.1	14.1	15.1	17.0	20.8
13.00	11.6	14.2	15.3	16.4	18.4	22.5
14.00	12.4	15.3	16.5	17.7	19.8	24.2
15.00	13.3	16.4	17.7	19.0	21.2	26.0
16.00	14.2	17.5	18.9	20.2	22.6	27.7
17.00	15.1	18.6	20.0	21.5	24.0	29.4
18.00	16.0	19.7	21.2	22.8	25.4	31.2
19.00	16.9	20.8	22.4	24.0	26.8	33.0
20.00	17.8	21.9	23.6	25.3	28.3	34.6
22.00	19.6	24.0	26.0	27.8	31.0	38.0
24.00	21.4	26.2	28.3	30.3	34.0	41.5

TABLE 4–1 Con't Effects of pressure on nozzle flow rate (*Courtesy Delavan, Inc.*)

26.00	23.2	28.4	30.6	32.8	36.8	45.0
28.00	25.0	30.6	33.0	35.4	39.6	48.5
30.00	26.7	32.8	35.4	38.0	42.4	52.0
32.00	28.4	35.0	37.8	40.5	45.2	55.5
35.00	31.2	38.2	41.3	44.0	49.5	60.5
40.00	35.6	43.8	47.0	50.5	56.5	69.0
45.00	40.0	49.0	53.0	57.0	63.5	78.0
50.00	44.5	54.5	59.0	63.0	70.5	86.5

In the hollow cone, as the name implies, the greatest concentration of droplets is at the outer edge of the spray, with little or no droplet distribution in the center. In general, the hollow-cone spray can be recommended for use in the smaller burners, particularly those firing 1 gph and under, regardless of air pattern. Hollow-cone nozzles have a more stable spray angle and pattern under adverse conditions than solid-cone nozzles of the same flow rate. This is an important advantage in fractional gallonage nozzles where high velocity fuel may cause a reduction in spray angle and an increase in droplet size.

In the solid cone, by definition, the distribution of droplets is fairly uniform throughout the pattern (Figure 4–9). This type of spray pattern is recommended where the air pattern of the burner is heavy in the center, or where long fires are required. It is also recommended for smoother ignition in most burners firing above 3 or 4 gph. An interesting characteristic of solid-cone patterns is that the cone tends to become more and more hollow as the flow rate increases, particularly above 8 gph.

SINTERED
FILTER
STANDARD
STRAINER

FIGURE 4–9
Filters and strainers. (*Courtesy Delavan, Inc.*)

The all-purpose nozzle is neither a true hollow-cone nor a true solid-cone nozzle. At the lower rates, it tends to be more hollow than solid, and as the flow rate increases, the pattern becomes more like a solid cone (Figure 4–10). It can be used in place of the solid- or the hollow-cone nozzles between .4 and .8 gph, and is suitable for most burners, regardless of the air pattern.

Nozzle Filter

The nozzle filter, or strainer, is designed to prevent dirt or other foreign matter from getting into the nozzle and clogging its passages. Nozzle manufacturers

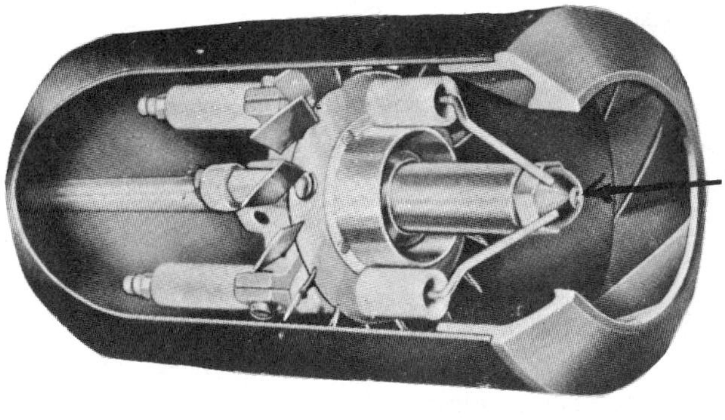
Nozzle

FIGURE 4–10 Nozzle location. (*Courtesy Monarch Manufacturing Works, Inc.*)

specify the type and size based on protecting the smallest passage in the nozzle, the slots. In all nozzle sizes, the strainer or filter openings are not larger than one-half the size of the smallest passage. For example, a 200-mesh screen has openings of 0.0029 inch and is used on .50-gph nozzles in which the smallest passage is 0.006 inch (Figure 4–9).

Porous bronze filters may have the same size openings as strainer screens, but the extra depth provides better filtering. They are also made in several densities.

The nozzle is the very last part of the burner with which the fuel oil comes into contact. It is located on the top of the nozzle tube which is installed inside the combustion chamber (Figure 4–10).

BURNER OPERATION

The foregoing discussion has dealt mainly with high pressure burners and nozzles. However, for a more complete discussion of the operation and servicing of fuel oil burners, the Ray rotary burner will be used as an example. First, we will discuss the principles of operation of oil burners, and a more complete component description will follow.

The *rotary oil burner* is a horizontal atomizing unit consisting of only one moving part, a hollow steel shaft, on which is mounted the atomizing cup, the fan, and the motor rotor. These parts comprise the shaft assembly, which rotates on two ball bearings. The fuel oil is introduced by means of a stationary fuel tube extending through the center of the hollow main shaft of the burner into the atomizing cup (Figure 4–11).

The atomizing cup, in the smaller sizes, is made of brass. In the larger sizes, the atomizing cups are made of a special metal alloy selected to meet the requirements of furnace temperatures. All Ray atomizing cups are of the long taper design that permits the oil to form a film on the inside of the cups before it reaches the rim and results in an even discharge. In the Ray rotary oil burner, the oil is introduced into the tail end of the fuel tube, passes through the tube, and flows out of the fuel

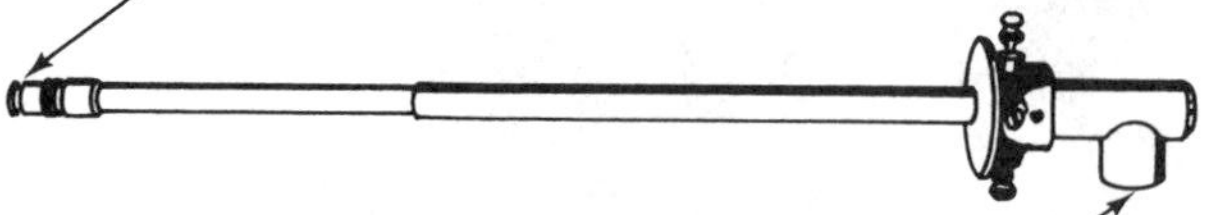
Atomizing cup end
Oil in

FIGURE 4–11
Fuel tube. (*Courtesy Ray Burner Co.*)

tip into the back portion of the atomizing cup. The fuel tube and tip are stationary, and the oil passages are of such size that high pressure is not required to obtain the rated oil capacity. The atomizing cup is carefully machined on the inside to a gradual smooth taper and rotates at high speed in a counterclockwise direction as viewed from the tail end. The oil, which flows into the rear of the atomizing cup, is forced by centrifugal action to form a thin uniform film on the inside of the atomizing cup and gradually moves forward, absorbing heat from the fire, and flows off the sharp edge of the cup in a thin film directly into a high velocity air stream. The primary air is discharged from the angular vane nozzle in a rotational direction opposite to that of the oil. The impact of the high velocity air on the film of oil instantly breaks the latter into a mist of fine particles that are readily evaporated. When intimately mixed with the proper amount of air, the vapor is quickly ignited and burns with a dense, turbulent flame. Such a flame gives a maximum of heat without smoke or soot and with a minimum of refractory deterioration.

The fan, which is mounted on the hollow shaft, supplies a relatively small amount of the total air necessary for combustion, its principal purpose being to atomize the oil and to control the shape of the flame. The remainder of the air necessary for complete combustion enters the combustion chamber as secondary air.

Nozzles

The nozzle on the Ray horizontal rotary burner is an air nozzle that serves a dual purpose: (1) to provide the primary air at the proper velocity, and (2) to shape the contour of the flame.

The nozzles are furnished in four sizes: extra large, large, small, and extra small. The size refers to the bore of the nozzle controlling the maximum air quantity and affecting its velocity. The actual velocity, however, is controlled by the primary air butterfly setting.

Since the delivery rate of the fan is reduced with a smaller nozzle, the firing capacity of the burner is automatically reduced when the smaller sizes are employed.

The larger size nozzles, in general, are used in cases where the burner is required to operate at its maximum capacity, whereas the smaller sizes are used when reduced burner capacities are required. However, there are some instances when the fixed rule does not apply, such as abnormal draft conditions and extreme grades of oil.

The nozzles for Ray rotary burners are also furnished in a number of angulations. The term *angulation* refers to the position of the vanes within the nozzles that serve to produce a flame of some definite width. The angulation of these vanes ranges from 50 to 90° in steps of 5° and 10°.

The 50° nozzles produce a short, wide flame, while the 90° nozzles produce a long, narrow flame. The shape of the flame may also be influenced by the shape of the combustion chamber and by the direction of the flow of secondary air through the combustion chamber (Figure 4–12).

The standard nozzles furnished with the various sizes of burners are indicated in Table 4–2. These were selected as standard because these sizes serve best in the majority of cases and when recommended firebox sizes can be used.

When recommended firebox dimensions cannot be followed, the selection of the nozzles should be altered accordingly. However, when the firebox is short and wide, it is necessary to provide additional height from the center line of the burner to the firebox floor to accommodate the wide nozzles properly.

There are times when the combustion of the oil will not be stable with the large size nozzles. This can be corrected by installing a nozzle of smaller size.

When any such changes are made, it is best to select a nozzle that is one size wider. For example, when a large 65° nozzle is to be replaced with a smaller nozzle, a 60° nozzle should be selected if the same flame length is to be maintained. This results from the change in air velocity. Conversely, when changing from a small to a large nozzle, the large nozzle should be one size narrower. For example, use a large 70° nozzle when changing from a small 65° nozzle.

The amount of primary air used through the nozzle should be just enough to produce good atomization and sufficient enough to prevent the oil from breaking through the air stream.

It is good practice to select a nozzle size just large enough to furnish the amount of air required for good atomization at full fire rate and without pulsation, with the butterfly valve nearly full open.

When the oil is breaking through the air stream, the flame will be smoky near the front of the firebox, which will usually result in carbon forming on the refractories at this point.

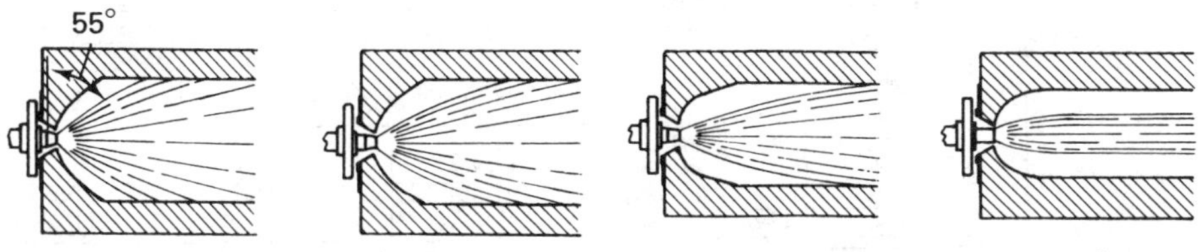
55°

(a) Ray 55° nozzle (b) Ray 65° nozzle (c) Ray 75° nozzle (d) Ray 90° nozzle

FIGURE 4–12 Comparison of flame contour produced by Ray angular vane nozzles. (*Courtesy Ray Burner Co.*)

TABLE 4–2 General nozzle specifications

BURNER SIZE	NOZZLE SIZE	OIL 550	GAS 547	GAS 550	COMBINATION 547	COMBINATION 550	F.D. SPECS. 131-134 141-144	F.D. SPECS. 101-104	VANE ANGLES*	NOZZLE DIA. (Inches)
RCR	Standard	9363-5 ●	X	X	X	X	X	X	65°	1-13/64
000	Extra Small	4228-XS-5 ●	X	X	X	X	X	X	65° – 70°	1-3/16
	Small	4228-S	X	X	X	X	X	X	"	1-1/4
0	Extra Small	4228-XS	4228-XS	X	4228-XS	X	X	X	55° – 90°	1-3/16
	Small	4228-S	4228-S	X	4228-S	X	X	X	"	1-1/4
	Large	4228-L-5 ●	4228-L	X	4228-L-15 ●	X	X	X	"	1-3/8
	Extra Large	4228-XL	4228-XL-15 ●	X	4228-XL	X	X	X	"	1-1/2
1	Extra Small	4232-XS	4232-XS	X	4232-XS	X	4232-XS	4232-XS	55° – 90°	1-7/16
	Small	4232-S	4232-S	9506-2-25 ●	4232-S	9506-2-25 ●	4232-S	4232-S	"	1-9/16
	Large	4232-L-5 ●	4232-L	X	4232-L-15 ●	X	4232-L-5 ●	4232-L-6 ●	"	1-11/16
	Extra Large	4232-XL	4232-XL-15 ●	X	4232-XL	X	4232-XL	4232-XL	"	1-13/16
2	Extra Small	4237-XS	4237-XS	9506-3-25	4237-XS	9506-3-25	4237-XS	4237-XS	55° – 90°	1-1/2
	Small	4237-S	4237-S	9506-1-25 ●	4237-S	9506-1-25 ●	4237-S	4237-S	"	1-5/8
	Large	4237-L-5 ●	4237-L	X	4237-L-15 ●	X	4237-L-5 ●	4237-L-6 ●	"	1-3/4
	Extra Large	4237-XL	4237-XL-15 ●	X	4237-XL	X	4237-XL	4237-XL	"	1-7/8
3	Extra Small	4241-XS	4241-XS	9427-2-25	4241-XS	9427-2-25	4241-XS	4241-XS	55° – 90°	1-3/4
	Small	4241-S	4241-S	9427-1-25 ●	4241-S	9427-1-25 ●	4241-S	4241-S	"	1-7/8
	Large	4241-L-5 ●	4241-L	9427-3-25	4241-L-15 ●	9427-3-25	4241-L-4 ●	4241-L-6 ●	"	2
	Extra Large	4241-XL	4241-XL-15 ●	X	4241-XL	X	4241-XL	4241-XL	"	2-1/8
3-30		4241-S-4 ●	X	9427-1-24 ●	X	9427-1-24 ●	4241-S-4 ●	4241-S-4 ●	"	1-7/8
5	Extra Small	4245-XS	4245-XS	9537-2-25	4245-XS	9537-2-25	4245-XS	4245-XS	55° – 90°	2-3/16
	Small	4245-S	4245-S	9537-1-25 ●	4245-S	9537-1-25 ●	4245-S-5 ●	4245-S-5 ●	"	2-5/16
	Large	4245-L-5 ●	4245-L	X	4245-L-15 ●	X	4245-L	4245-L	"	2-7/16
	Extra Large	4245-XL	4245-XL-15 ●	X	4245-XL	X	4245-XL	4245-XL	"	2-9/16
5-45		4245-XS-4 ●	X	9537-2-24 ●	X	9537-2-24 ●	4245-XS-4 ●	4245-XS-4 ●	"	2-3/16
6	Extra Small	4249-XS	4249-XS	9534-2-25	4249-XS	9534-2-25	4249-XS	4249-XS	50° – 90°	2-3/4
	Small	4249-S	4249-S	9534-1-25 ●	4249-S	9534-1-25 ●	4249-S-5 ●	4249-S-5 ●	"	2-7/8
	Large	4249-L-5 ●	4249-L	X	4249-L-15 ●	X	4249-L	4249-L	"	3
	Extra Large	4249-XL	4249-XL-15 ●	X	4249-XL	X	4249-XL	4249-XL	"	3-1/8
6-60		4249-XS-4 ●	X	9534-2-24 ●	X	9534-2-24 ●	4249-XS-4 ●	4249-XS-4 ●	"	2-3/4
7	Extra Small	4253-XS	4253-XS	9564-2-25	4253-XS	9564-2-25	4253-XS	4253-XS	50° – 90°	3-3/8
	Small	4253-S	4253-S	9564-1-25 ●	4253-S	9564-1-25 ●	4253-S-5 ●	4253-S-5 ●	"	3-1/2
	Large	4253-L-6 ●	4253-L	X	4253-L-16 ●	X	4253-L	4253-L	"	3-5/8
	Extra Large	4253-XL	4253-XL-16 ●	X	4253-XL	X	4253-XL	4253-XL	"	3-3/4
7-85		4253-XS-4 ●	X	9564-2-24 ●	X	9564-2-24 ●	4253-XS-4 ●	4253-XS-4 ●	"	3-3/8
8	Extra Small	4257-XS	4257-XS	9565-2-25	4257-XS	9565-2-25	4257-XS	4257-XS	50° – 90°	3-7/8
	Small	4257-S	4257-S	9565-1-25 ●	4257-S	9565-1-25 ●	4257-S-5 ●	4257-S-5 ●	"	4
	Large	4257-L-6 ●	4257-L	X	4257-L-16 ●	X	4257-L	4257-L	"	4-1/8
	Extra Large	4257-XL	4257-XL-16 ●	X	4257-XL	X	4257-XL	4257-XL	"	4-1/4
8-130		4257-XS-4 ●	X	9565-2-24 ●	X	9565-2-24 ●	X	X	"	3-7/8
9	Extra Small	4261-XS	4261-XS	X	4261-XS	X	4261-XS	4261-XS	50° – 90°	4-1/4
	Small	4261-S	4261-S	X	4261-S	X	4261-S-5 ●	4261-S-5 ●	"	4-3/8
	Large	4261-L-6 ●	4261-L	X	4261-L-16 ●	X	4261-L	4261-L	"	4-1/2
	Extra Large	4261-XL	4261-XL-16 ●	X	4261-XL	X	4261-XL	4261-XL	"	4-5/8
10	Extra Small	4265-XS	4265-XS	X	4265-XS	X	4265-XS	4265-XS	50° – 90°	5-1/8
	Small	4265-S	4265-S	X	4265-S	X	4265-S	4265-S	"	5-1/4
	Large	4265-L-6 ●	4265-L	X	4265-L-16 ●	X	4265-L-5 ●	4265-L-5 ●	"	5-3/8
	Extra Large	4265-XL	4265-XL-16 ●	X	4265-XL	X	4265-XL	4265-XL	"	5-1/2
12	Extra Small	4265-S	4265-S	X	4265-S	X	X	X	50° – 90°	5-1/4
	Small	4265-L	4265-L	X	4265-L	X	X	X	"	5-3/8
	Large	4265-XL-6 ●	4265-XL-16 ●	X	4265-XL-16 ●	X	X	X	"	5-1/2
	Extra Large	4265-XXL	4265-XXL	X	4265-XXL	X	X	X	"	5-5/8

VANE ANGLE CODE

90°	80°	75°	70°	65°	60°	55°	50°	
-0	-2	-3	-4	-5	-6	-7	-8	Model 550 Oil Burners
-10	-12	-13	-14	-15	-16	-17	-18	Model 547 Gas and Combination Burners
-20	-22	-23	-24	-25	-26	-27	-28	Model 550 Gas and Combination Burners

● Standard nozzle on burner unless specified otherwise.

* Available in 5° increments.

X (crossed box) = Not Available

All nozzles interchangeable with older model burners.

(*Courtesy Ray Burner Co.*)

An excess of primary air usually has a tendency to cause the flame to burn away from the nozzle. This frequently results in flame pulsations.

The best possible firing results are obtained when the primary air delivery through the nozzle has exactly the proper volume and velocity.

In cleaning the nozzle, all carbon or oil residue should be removed with a knife or a scraper. When this is done, the nozzle should be thoroughly cleaned with kerosene or solvent, making certain that the passages between the vanes are not obstructed.

The vanes should be checked to be certain that they are not bent, loose, or broken. Bent vanes should be straightened, and when the vanes are broken, the nozzle should be replaced.

Atomizing Cup

The purpose of the atomizing cup, in conjunction with the air blast from the nozzle, is to atomize the fuel oil.

The inside of the atomizing cup is tapered slightly. This taper, coupled with the speed at which the cup rotates, causes the oil to move forward to the edge of the cup in a thin film.

To promote good atomization, it is important that the inside of the cup be smooth and that the cup be kept clean.

It is equally important that the edge of the cup be trimmed and belled properly and that there be no ragged edges. Any of these imperfections will cause the oil film to be irregular and will, therefore, promote poor atomization.

It is also important that the outside of the cup be kept clean so that the air delivery from the nozzle will not be obstructed or deflected from its normal course.

The atomizing cup can be cleaned or polished, both internally and externally, without removing it. Ordinarily the cup can be cleaned by wiping it with a cloth or by using kerosene or solvent and a cloth.

In certain cases, when burning oil of low viscosity such as Number 2 oil, the thin edge of the cup may be belled out slightly using a heavy, round tool such as a piece of shafting. This will help stabilize a flame that otherwise may be difficult to control.

When replacing atomizing cups (Figure 4–13), the following is the proper method of trimming and dressing. First, consult the proper data sheet for exact measurements. If the cup is too long, dress it down with a file held to the spinning cup. Then taper the inside of the lip of the cup as shown and polish it with emery cloth. It is important, for perfect atomization, that this lip be absolutely smooth and correctly rounded on the inside, but sharp on the outer edge.

Fan Housing Cover

The fan housing cover encloses the fan housing and provides a mounting for the nozzle. It is equipped with a baffle and vanes that direct the air from the discharge side of the fan to the nozzle. It is important that the space between the baffle and

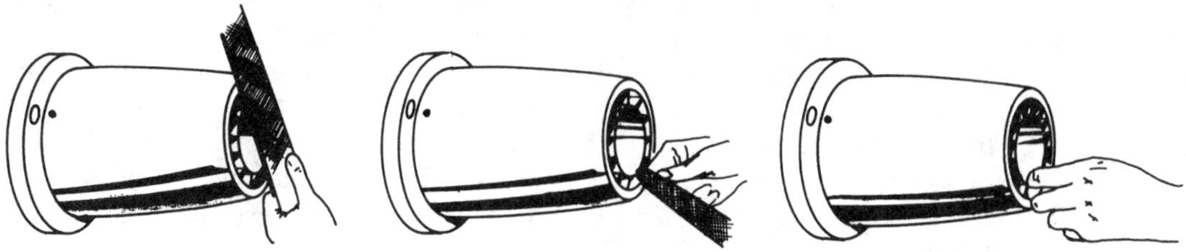

FIGURE 4–13 Dressing atomizing cups. (*Courtesy Ray Burner Co.*)

the cover be kept clean so that the flow of air from the fan will not be retarded or reduced.

To simplify locating the cover when it is to be replaced, before it is removed mark it with a center punch or a crayon to serve as a guide. Unless the cover is properly positioned, it will not be possible to position the nozzle properly.

Before replacing the cover, be certain that it is thoroughly clean and that all of the rivets holding the baffle are tight. To ensure proper alignment, the machine surfaces of the housing and the cover must be clean.

Fan

The fan delivers the primary air to the burner nozzle. Since the amount of air delivered by the fan materially affects the firing capacity of the burner, it is important that the fan, the fan housing, the air inlet housing, and the primary air butterfly be kept clean.

The fan is properly balanced at the factory to ensure freedom from vibration. Any fouling of the fan may result in an unbalanced condition, which will be reflected in the operation of the burner. The parts should be inspected at least once a season. The fan housing can be inspected by removing the nozzle and the fan housing cover, or by removing the atomizing cup. In the event these parts are fouled and the fan or fan housing requires cleaning, the fan should be removed.

Air Inlet Housing and Primary Air Butterfly

The air inlet housing serves as a duct to supply the air to the inlet of the fan. The primary air butterfly serves as a throttle to regulate the amount of air drawn into the fan and consequently also regulates the amount of air furnished to the nozzle (Figure 4–14).

It is important that these parts be kept clean. The air inlet housing ordinarily requires cleaning only once every season, except when the air is fouled with dust or lint. It is good practice to wipe the butterfly once each month to ensure positive air regulation and delivery and thereby maintain more uniform firing conditions.

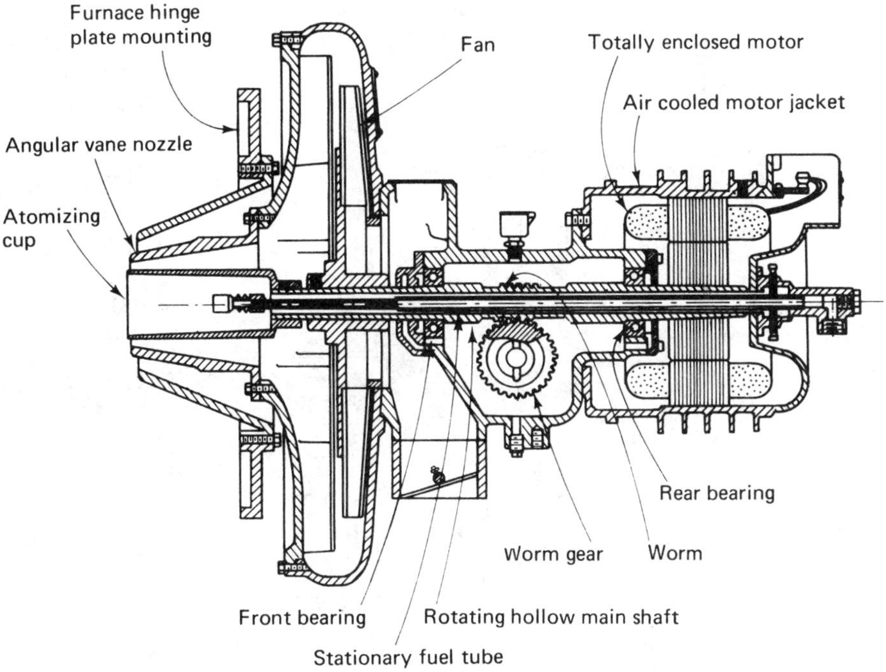
Furnace hinge plate mounting
Fan
Totally enclosed motor
Air cooled motor jacket
Angular vane nozzle
Atomizing cup
Rear bearing
Worm gear
Worm
Front bearing
Rotating hollow main shaft
Stationary fuel tube

FIGURE 4–14 Ray rotary burner components. (*Courtesy Ray Burner Co.*)

This should be done only when the burner is idle; otherwise the cloth may be drawn into the fan, clogging the intake.

Worm Gear and Worm Gear Shaft

The worm gear and the worm gear shaft drive the fuel oil pump and are driven by the worm gear on the hollow main shaft. The worm gear also distributes the lubricating oil, lubricating the bearings on the shaft (Figure 4–15).

The worm gear shaft is provided with a driving pin that meshes with the slots on the hub of the worm gear from which it is driven.

These parts require no adjustment, and as long as the burner is properly lubricated, the life of the worm gear will be unlimited. A lack of proper lubrication, however, will cause the worm gear to overheat and will materially shorten its life.

If the worm gear is damaged or the driving pin in the worm gear shaft is broken, it is recommended that the pump on the burner, or the two-stage pumps in the reservoir, be checked to be certain that they operate freely before either of these parts is replaced and the burner put into operation again. If this is not done, the new parts may be broken or damaged.

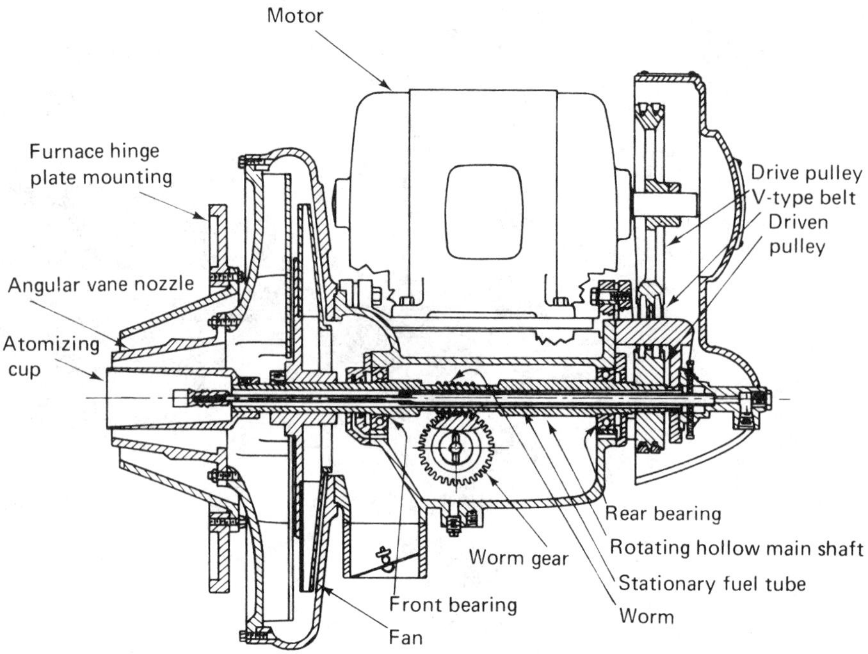
Motor
Furnace hinge plate mounting
Angular vane nozzle
Atomizing cup
Drive pulley
V-type belt
Driven pulley
Rear bearing
Rotating hollow main shaft
Stationary fuel tube
Worm
Worm gear
Front bearing
Fan

FIGURE 4–15 Ray rotary burner components. (*Courtesy Ray Burner Co.*)

Gear Housing

The gear housing encloses the bearings of the worm gear and the worm gear shaft. It also serves as a reservoir for the lubricating oil. The worm gear cover is provided with an oil gauge cup, used to check the amount of lube oil in the housing.

When the burner is idle, the gauge cup should be filled to within 1/8 inch of the top of the cup. Motor oil of number 30 weight (W) should be used for lubrication and should be added as often as necessary. When the surrounding temperature is low, it may be advantageous to use number 20 S.A.E. oil. The oil in the gear housing should be drained once each season, and the housing refilled with new oil.

Fuel Feed Assembly

The fuel feed assembly consists of a tailpiece, fuel tube, and tip assembly (Figure 4–16). Some assemblies include a vibration dampening assembly that screws on to the end of the tip. This assembly permits the transportation of the fuel oil from the burner piping to the atomizing cup through the hollow main shaft, without the use of stuffing boxes or special seals.

It is important that the fuel feed assembly be properly centered in the hollow main shaft and that both the tube and the inside of the shaft be clean so that fuel tube rattling will not occur. Rattling of the fuel tube will cause the burner to be very

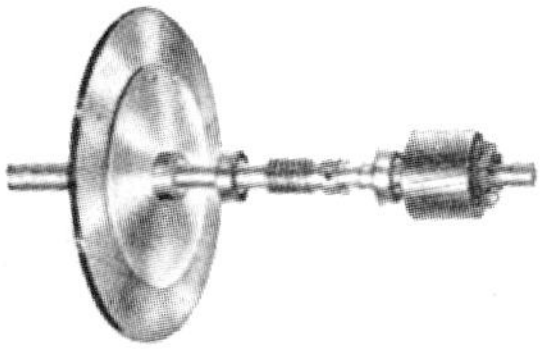

FIGURE 4–16
Fuel feed assembly. (*Courtesy Ray Burner Co.*)

noisy in its operation and may, after prolonged rattling, cause the fuel tube to break. For centering the fuel tube in the hollow shaft, four adjusting screws with locknuts are provided in the tailpiece.

To check the fuel tube alignment look at it from the front or atomizing cup end of the burner. The fuel tube should be centered in the hollow shaft as accurately as possible. A fairly accurate check of its location can be made by moving the fuel tube tip with the ends of the fingers, first from side to side, and then up and down, noting the motion required to contact the shaft at each point from the free-standing position.

Prior to making any adjustments, the tailpiece mounting screws should be checked to be sure that the tailpiece is properly secured, as these will affect the alignment of the tube.

If the fuel tube should rattle when it is found to be centered, the fuel feed assembly should be removed to determine

1. If the hollow shaft is clean.
2. If the fuel tube is clean externally.
3. If the fuel tube is bent.

With all of these in order and with the fuel tube centered, no rattling should occur.

The fuel tube should be cleaned thoroughly and then checked to be sure that it is smooth and absolutely straight. The straightedge should be placed on all four sides of the fuel tube in turn.

The fuel tube is made of steel and can be straightened through bending, but extreme care should be exercised.

The hollow shaft should then be checked for cleanliness. If necessary, it should then be cleaned by drawing a piece of cloth saturated with kerosene or solvent through it, in much the same manner as cleaning a gun barrel. At no time should a fuel feed assembly be installed without first checking the hollow shaft for cleanliness.

Fuel Tip

The fuel tip on the end of the fuel tube assembly distributes the fuel oil on the inner surface of the cup and prevents the oil from running back along the tube into the hollow shaft. (Refer to Figure 4–14.)

The orifice or discharge hole in the fuel tip should be at the top in every case to eliminate the possibility of the fuel draining from the fuel tube when the burner is idle.

Except for replacements or for relocation of the orifice, there is no occasion for removing the fuel tip, because it can be withdrawn through the hollow shaft with the feed assembly.

The fuel tip baffle of the nozzle type should be installed with the nozzle at the bottom and the tip orifice at the top, as stated previously.

Motor Jacket Cover

The motor jacket cover encloses the motor housing and provides a mounting for the tailpiece (Figure 4–17). Incorporated in the motor jacket cover is a terminal box for the connections of the motor, oil valve, etc. Since the motor jacket cover provides a support for these various items, it is important that it be properly positioned and properly secured. The motor jacket cover is secured to the motor jacket with four screws. Prior to removing the motor jacket cover, the fuel feed assembly should be removed.

Motor Jacket

The motor jacket serves as a housing for the motor parts. It is secured to the gear housing with four screws and is provided with set screws or a clamping screw to secure the stator in position. A drain hole is provided to drain off any lubricating oil which might leak from the bearing cover. In replacing the motor jacket, it is important that the hole be placed at the bottom. To ensure proper stator alignment, the bore of the motor jacket must be clean and smooth. This will simplify installation of the stator.

Hollow Shaft and Bearings

The hollow shaft is the main burner shaft and is equipped with two ball bearings (Figure 4–18). The shaft has a worm gear to drive the worm gear in the pump. The shaft also supports the rotor, the fan, and the atomizing cup. This assembly revolves as a single unit.

The ball bearings are located at each end of the gear housing, enclosed by a bearing cover with gaskets. They are fitted to the shaft with only a light tap fit and not a tight pressed fit. The fitting of the bearings in the housing is a light push fit and should permit assembly by pressing in with the two thumbs.

When the shaft and bearings are fitted on the assembly into the gear housing, there should be just a slight amount of end play, usually about 0.010 inch. The forward bearing assembly includes a thrust spring that forces the shaft assembly to the rear.

In replacing the bearing cover, the drain groove should always be located at the bottom, and the bearing covers should also clear the shaft to eliminate binding

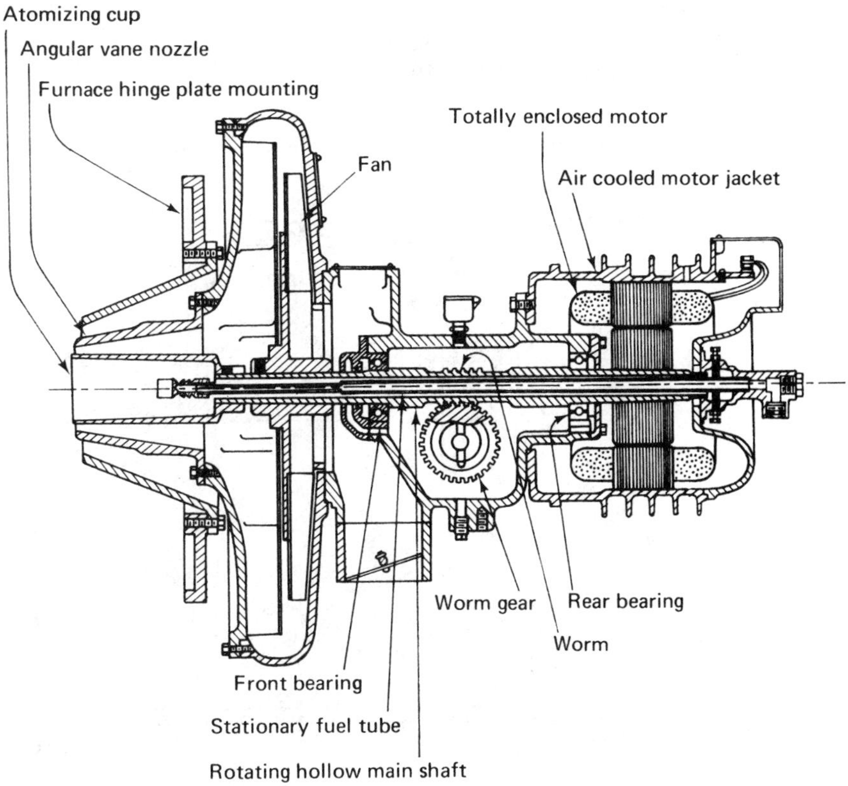
Atomizing cup
Angular vane nozzle
Furnace hinge plate mounting
Totally enclosed motor
Fan
Air cooled motor jacket
Worm gear
Rear bearing
Worm
Front bearing
Stationary fuel tube
Rotating hollow main shaft

FIGURE 4–17 Ray burner rotary burner components. (*Courtesy Ray Burner Co.*)

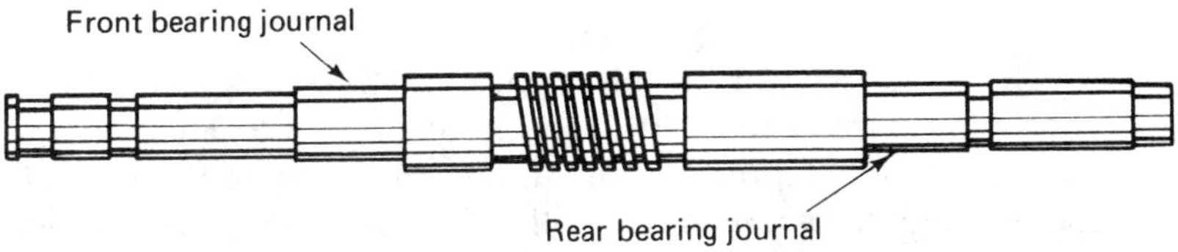
Front bearing journal
Rear bearing journal

FIGURE 4–18 Main burner shaft. (Courtesy Ray Burner Co.)

or wear. The front bearing cover is provided with a breather part that serves as a vacuum break to eliminate oil being drawn from the front bearing by the vacuum created by the fan. It is important that this port remain open and that it be cleaned when any repairs are made.

Ray Viscosity Valve

The Ray viscosity valve, in conjunction with the two-state pump and reservoir and modulating system, automatically adjusts the fuel oil rate to match the load requirements. With a single initial adjustment, this valve automatically compensates for any fluctuations in viscosity and meters the oil at a prescribed and modulated rate of flow (Figure 4–19).

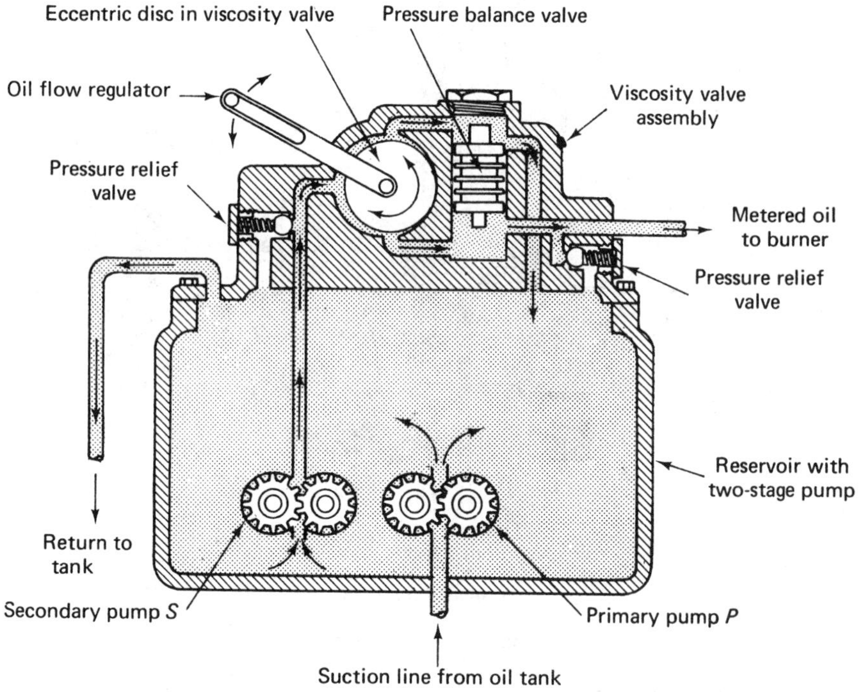
Eccentric disc in viscosity valve
Pressure balance valve
Oil flow regulator
Viscosity valve assembly
Pressure relief valve
Metered oil to burner
Pressure relief valve
Reservoir with two-stage pump
Return to tank
Secondary pump S
Primary pump P
Suction line from oil tank

FIGURE 4–19 Ray viscosity valve. (*Courtesy Ray Burner Co.*)

The primary pump *P* delivers oil from the storage tank to the reservoir in excess of the pumping rate of the secondary pump *S*. The non-variable flow of oil from the secondary pump *S* enters the inlet port of the viscosity valve. The oil passes through the valve in two paths, the flow through each of which is controlled by positioning the eccentric disc in the viscosity valve assembly. Since the flow is viscous, the rate through each side remains constant regardless of oil viscosity, and the pressure automatically varies to compensate for changes in flow resistance.

The oil from each path enters opposite ends of the pressure balancing valve in which the free-floating piston proportions the area of the outlet ports from the cylinder, maintaining equalized pressures at both ends. This assures a viscous flow rate to the burner unaffected by changes in pressure in the return line to the tank or in the line to the burner nozzle. The metered oil from one path is delivered to the burner and the remainder, from the other path, back to the reservoir.

Two pressure relief valves, one at the pump discharge and one at the outlet to the burner nozzle, operate only to relieve excessive pressure when operating against a closed valve or in case of incorrect adjustment. For an adjusted position of the eccentric disc, therefore, the flow rate to the burner remains constant, independent of oil temperature changes or line pressure variations.

When adjusting the viscosity valve, it should be noted that the end of the metering shaft has a slot for screwdriver setting. There may also be a position-indicating notch in the end of the shaft or a pin indicator through the shaft. As viewed from the shaft extension side, the notch or pin must operate within the arc represented: approximately one o'clock to five o'clock (Figure 4–20).

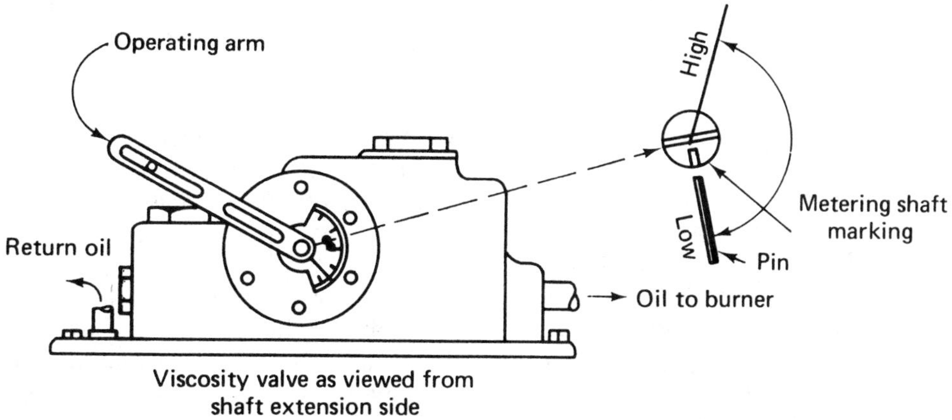
Operating arm
Return oil
High
Low
Metering shaft marking
Pin
Oil to burner
Viscosity valve as viewed from shaft extension side

FIGURE 4–20 Adjusting Ray viscosity valve.(*Courtesy Ray Burner Co.*)

Single-Stage Pump

The single pump used on manual or semiautomatic burners is the rotary gear type. It serves a dual purpose: (1) to draw the oil from the tank; and (2) to deliver the oil at some predetermined pressure to the oil regulating valve (Figure 4–21).

The pump is designed to operate at low speed (200 to 300 rpm) to eliminate wear and to ensure long life. When properly primed and in good condition, it should generate a vacuum of approximately 27 inches of mercury at sea level.

Ordinarily, the pump requires no service except for renewing the mechanical seal. If the pump does not generate sufficient vacuum after a long period of service, this condition can usually be corrected by removing one or two thin gaskets between the pump body and cover.

Care should be exercised when gaskets are removed to ensure that the adjoining gaskets are not broken or torn. When it is necessary to remove any gaskets or to make any repairs to the pump requiring the removal of the pump cover, it is recommended after assembly that the burner shaft be rotated manually to ensure that the pump operates freely before applying power to the burner motor.

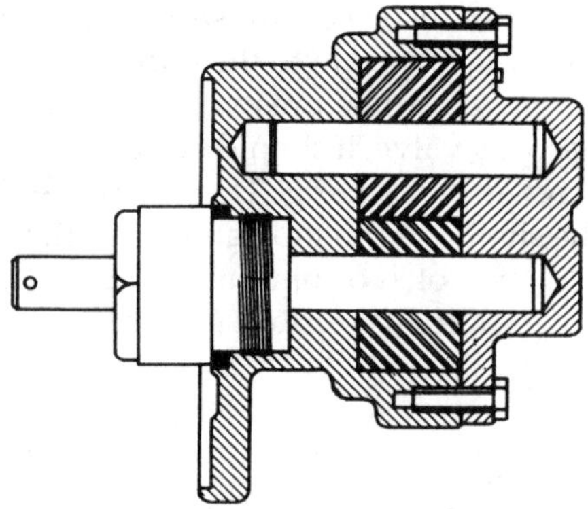

FIGURE 4–21
Single-stage pump with mechanical seal. (*Courtesy Ray Burner Co.*)

If the pump becomes badly worn after years of service, it should be replaced. Replacing the gears and shafts in an old body is not recommended, because if the gears and shafts are badly worn, it is certain that some wear has occurred in the pump body or the pump cover. Consequently, the new gears and shafts will not correct the fault.

If the pump is removed from the burner or a new pump is to be installed, it is very important that the pump be properly aligned with the worm gear shaft. Pump-aligning tools are furnished by the Ray Oil Burner Company for this purpose, and these should be used in making such repairs or replacements.

There may be some cases in which the heavier grades of Bunker fuel oils are used or, when changing from light to heavy fuel oils, in which the pump is fitted too close. An indication of this form of trouble will be the breaking of the driving pin in the worm gear, the breaking of the shear pins in the pump coupling, or damage to the fiber worm gear. To correct it, an extra gasket may be added under the pump cover. Usually the addition of one gasket of from 0.001- to 0.0015-inch thickness will correct the trouble.

Two-Stage Pump

The two-stage pump is of the same design as the single pump, except that the two pumps are contained in a single housing with pump covers at each end and are driven by a single shaft (Figure 4–22). The primary pump has greater capacity than the secondary pump. It serves only as a suction pump and is used only to draw the fuel oil from the storage tank and to discharge it into the reservoir. The secondary pump serves as a pressure pump, drawing its oil from the reservoir and discharging it under pressure into the valve assembly.

The only test required for the secondary pump is a pressure test. If the secondary pump will generate and maintain the desired operating pressure, it is in proper working condition. The general instructions and suggestions covering the single pump will also apply to the two-stage pump.

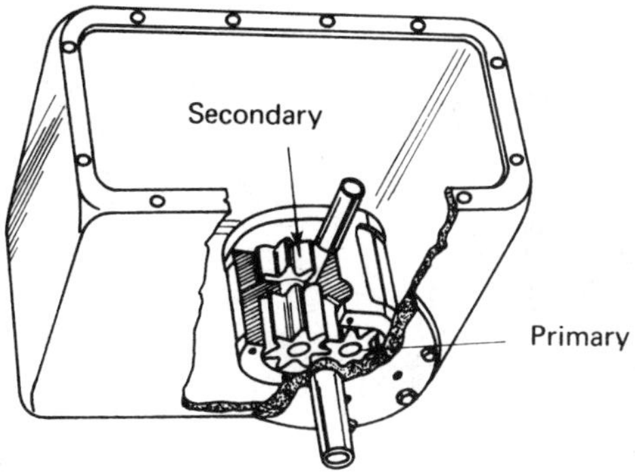
Secondary
Primary

FIGURE 4–22
Two-stage pump. (*Courtesy Ray Burner Co.*)

Reservoir

The reservoir serves as a housing for the two-stage pump to which the viscosity valve assembly or the reservoir cover and relief valve assembly are connected. The oil from the primary pump is discharged into the reservoir (Figure 4–23).

The secondary pump draws its oil from the reservoir at a point about one inch from the bottom of the reservoir. The amount of oil in excess of that amount consumed by the burner that is delivered by the secondary pump is discharged back into the reservoir. When the reservoir is filled and overflowing, this oil is discharged through the hinge and the return line to the storage tank. Thus, no oil will be discharged to the tank unless the reservoir is filled and overflowing.

There is a tendency for some small particles of dirt to filter through the strainer basket, and since these particles are heavier than the oil, some of them settle to the bottom of the reservoir. Should it be allowed to accumulate for long periods of time, it can be drawn into the secondary pump. To eliminate this form of trouble, the reservoir should be drained and flushed out occasionally. The reservoir is provided with a clean-out hole and cover, through which the cleaning of the reservoir is simplified, or with plugged holes in the bottom, through which dirt may be flushed out. It is suggested that the reservoir be cleaned or flushed out once each season to ensure trouble-free burner operation.

Pump Seals

A mechanical seal is used; this is an adjustable screw-type packing gland that uses a redesigned stuffing box in place of the spring-loaded type. Because of the direction in which the shaft rotates, the threads on the gland nut are left-handed (Figure 4–24).

To replace the mechanical seal, the following steps should be followed (Figure 4–25). All threads on the hub are left-handed. Hold the sleeve with pipe pliers on the recess to avoid marring the ground surfaces. Unscrew the seal retainer (left-hand thread). The stationary seat and O ring will come out with the retainer. Push the stationary seal out and insert the new seat into the O ring. Do not scratch the

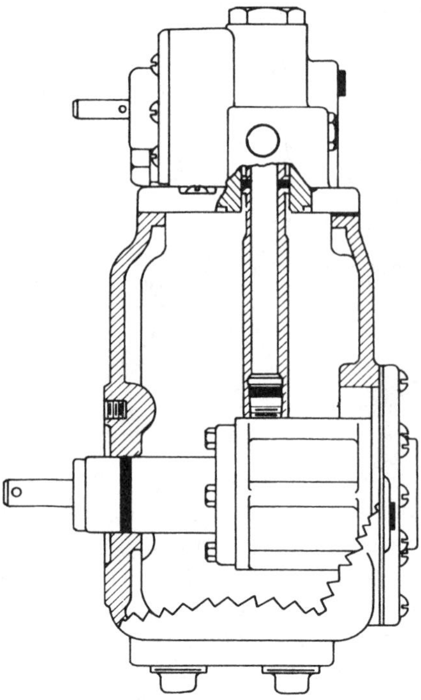

FIGURE 4–23
Two-stage pump and reservoir. (*Courtesy Ray Burner Co.*)

seat face and keep it clean. Unscrew the sleeve (left-hand) with pipe pliers. Grasp the brass body of the rotating seal and slide it off the end of the pump shaft. Slide the drive washer off the shaft. Inspect the pressed-on truarc ring. It should be at a depth of ½ inch from the near edge of the O ring groove. Normally, the drive washer and truarc ring do not need replacing.

If the truarc ring is broken, install a new one by wedging a screwdriver into the truarc slot just barely enough to install the ring on the shaft end. Overstressing will weaken it and allow it to slip on the shaft. If this ring slips or is out of position,

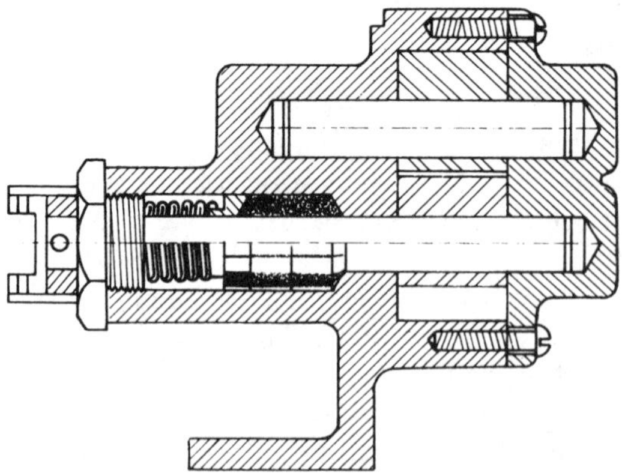

FIGURE 4–24
Single-stage pump with packing gland. (*Courtesy Ray Burner Co.*)

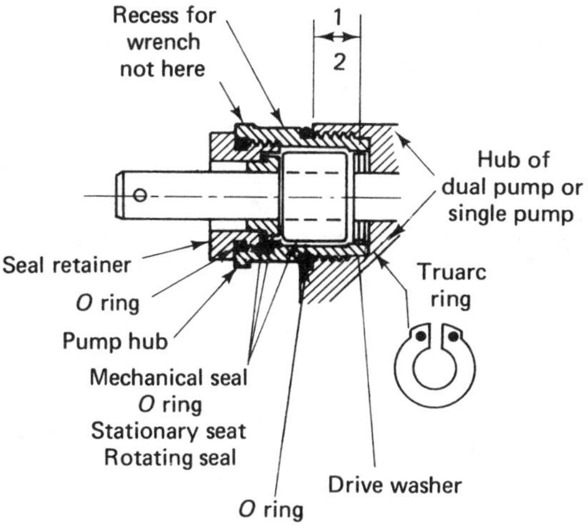
Recess for wrench not here
1
2
Hub of dual pump or single pump
Seal retainer
O ring
Pump hub
Mechanical seal
O ring
Stationary seat
Rotating seal
Truarc ring
Drive washer
O ring

FIGURE 4–25
Seal replacement. (*Courtesy Ray Burner Co.*)

the correct pressure will not be maintained between the two polished sealing surfaces and leakage may result.

Remove the screwdriver and press the truarc ring down to the ½-inch dimensions as shown, using a piece of ½-inch iron pipe with a good square end. Install the driver washer with the 45° bent finger inserted in the slot in the truarc ring.

Oil the shaft lightly, and slide the new rotating seal onto the shaft, aligning the two slots on the two fingers of the drive washer. The polished carbon face is positioned outward to run against the stationary polished steel face. Keep the carbon clean and unscratched.

Reinstall the sleeve or hub. Install the retainer with the new stationary seat, positioning the O ring between the stationary seat and the retainer and the intermediate O ring between the retainer and the sleeve. The outer O ring seals the complete assembly into the pump hub or body. Remember that seal faces must be clean and unscratched for proper operation.

Three O rings of different sizes are involved. They must all be carefully examined and, if defective in any way, must be replaced.

Solenoid Oil Valve

The solenoid oil valves are used to open or shut off the oil supply to the burner. When the solenoid coil is energized, the oil valve will open; when de-energized, the valve will be closed (Figure 4–26).

Solenoid oil valves are designed for some specific operating pressures, and these pressures should not be exceeded if positive valve action is to be expected.

Oil valve failure can usually be traced to one of the following causes:

1. Improper power characteristics.
2. Valve parts fouled with dirt or foreign matter.

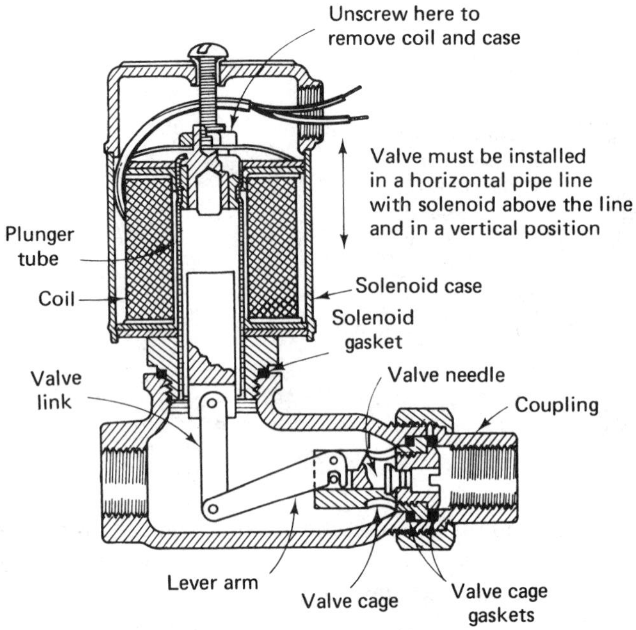
Unscrew here to remove coil and case
Valve must be installed in a horizontal pipe line with solenoid above the line and in a vertical position
Plunger tube
Coil
Solenoid case
Solenoid gasket
Valve link
Valve needle
Coupling
Lever arm
Valve cage
Valve cage gaskets

FIGURE 4–26
Solenoid valve. (*Courtesy Ray Burner Co.*)

3. Valve parts worn.
4. Valve mechanism damaged by dropping, bumping, or rough handling.

In making repairs to solenoid oil valves, follow the manufacturer's instructions.

Electric Oil Heater

The purpose of the electric oil heater is solely to maintain an initial supply of oil at a suitable temperature for starting purposes, and not for heating and maintaining the oil that is normally consumed by the burner at some required temperature (Figure 4–27).

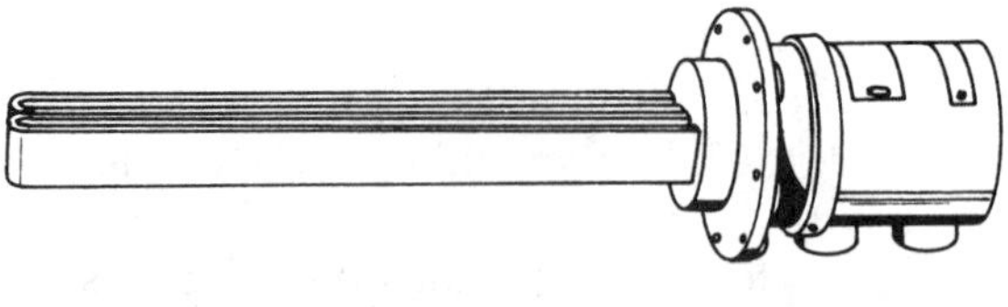

FIGURE 4–27
Electric oil heater. (*Courtesy Ray Burner Co.*)

The oil heater should be adjusted to maintain an oil temperature as low as practical. Excessive temperatures may cause gasification or carbonizing and may cause unsatisfactory starting conditions.

To ensure the best possible results from the heater, the housing and the heater element should be cleaned at least once a season. When the fuel oil contains excessive amounts of sludge, sediment, or water, these parts should be cleaned more frequently.

The heater is usually equipped with an auxiliary switch head that provides a control circuit with adjustable contacts, closed in the hot position. This circuit is usually wired into the burner circuit to allow starting only after a predetermined temperature is reached.

Nozzle Protector

The nozzle protector provides a port for the burner nozzle, protects the nozzle, and provides some secondary air for combustion, which also cools the nozzle.

When making the installation or later when making repairs, it is important that the nozzle protector be installed concentric with the nozzle so that the secondary air will be admitted uniformly. Unless this is done, the flame may become distorted.

It is also important that the plastic refractory be installed flush with the nozzle protector and completely surround it so that the damage to this part may be avoided (see Table 4–3).

Ignition Bracket Assembly

The ignition bracket assembly consists of a bracket, ignition transformer, solenoid gas valve, ignitor, fire safety switch, twist lock receptacle, terminal block, and burner latch. In mounting the ignition bracket assembly, care should be exercised to ensure that the igniters are centered in the igniter protectors (Figure 4–28).

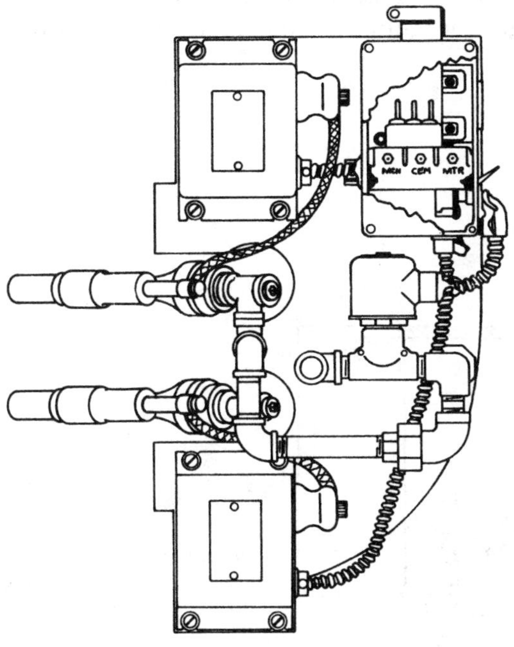

FIGURE 4–28
Igniter bracket assembly. (*Courtesy Ray Burner Co.*)

The wiring to the ignition bracket should be made with flexible conduit to permit removing the ignition bracket assembly as a unit for making inspection or repairs to the igniters, etc.

Fire Safety Switch

As the name implies, the fire safety switch is a device that operates when overheating or external fire causes the safety string to burn. In some cases, a wire and fusible link is used to replace the string. With either of these, a fusible link rated at 160°F should be used. The fire safety switch is a spring-loaded, normally open contact switch. It must be held closed with the safety string.

Igniter Assembly

Igniter assemblies of two types were originally furnished with Ray burners. Both were known as the gas-electric type, employing an electric spark at 5,000 V to ignite the gas. With these igniters, the ignition remains on constantly during the ignition period.

The *raw gas igniters* were specified and used when manufactured, mixed, or natural gas was used for ignition purposes (Figure 4–29). The *bottled gas igniters* are now standard and specified for use with bottled gases, propane, butane, and similar gases, and for natural gas (Figure 4–30). When bottle gases are used, a vapor switch of the reverse action, manual reset type should be used on the gas line to ensure that the burner will not attempt to start when the bottled gas supply falls below a safe operating pressure.

Vapor switches are also used as an added safety device on installations employing any other type of gas for ignition, and are recommended wherever complete protection is desired.

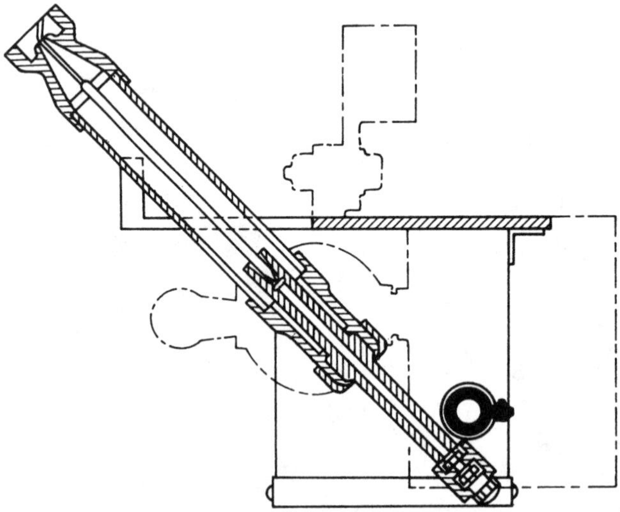

FIGURE 4–29
Raw gas igniter. (*Courtesy Ray Burner Co.*)

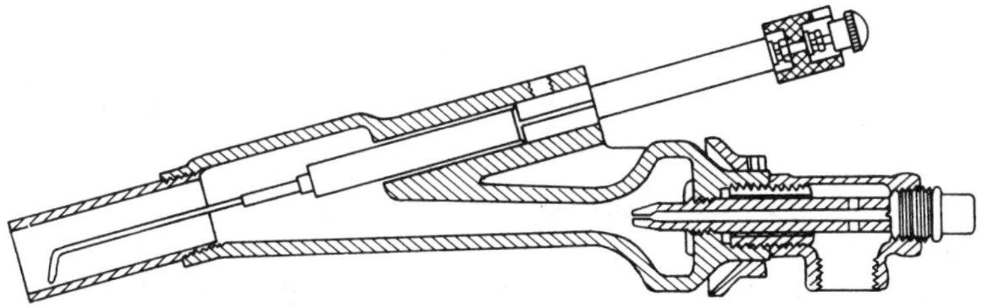

FIGURE 4–30 Bottled gas igniter. (*Courtesy Ray Burner Co.*)

Ignition Transformers

The ignition transformer, commonly known as a *step-up transformer,* is used to step up the voltage to 5,000 V in order to jump the gap at the end of the ignition electrode. Ignition transformers are trouble free and require only that the primary voltage correspond to that of the transformer, that the electrical connections be clean and tight, and that the secondary terminals be kept clean.

If the secondary terminal post becomes fouled with soot, dust, or foreign matter, the high tension spark can easily short circuit to the transformer case, resulting in no spark at the ignition electrode. A transformer in good condition should cause the spark to jump a gap of not less than 1/4 inch.

Oil Strainer

The oil strainer removes any dirt or foreign matter that might be injurious to the pump on a pump-type burner and removes the smaller particles of dirt, etc., that might cause the fouling of the viscosity valve or metering valve.

So that these components are properly protected, the mesh of the strainer must be of proper size. With the lighter grades of oil, and also with the heavier grades of oil, when the firing rates are small, the strainer mesh should be 60 or even 80 mesh. With the larger firing rates up to 60 gph, the basket should not exceed 30 mesh. Strainers coarser than 20 mesh should not be used on automatic burners. If cleaning is required too frequently, a larger strainer or two strainers in parallel should be used.

It is important that the strainer be kept clean so that the flow of oil to the pump or to the burner will be normal. There is, of course, no fixed period for cleaning the strainer basket. With some fuel oils cleaning may not be required for several weeks, while with other oils it may be necessary to clean the strainer baskets daily.

When frequent cleaning is necessary, it may be desirable to clean the tank and the lines or in some cases to treat the fuel oil with a sludge solvent. When cleaning the strainer, check the gasket each time to be certain that it is intact.

When the strainer is installed in a gravity feed or pressure feed, line leaks at the gasket are easily detected. However, when the strainer is in the suction line, a leaky gasket is not easy to detect and can be responsible for faulty oil pressure or lack of pump capacity.

Relief Valve

The relief valve is a spring-loaded valve consisting of a valve seat, valve, spring, and adjusting stem or screw that is used for regulating the oil pressure from the pump to the metering valve (Figure 4–31). The oil enters from the underside of the valve, forcing the valve from its seat. The oil pressure is regulated by the amount of pressure exerted on the valve by the spring. Consequently, the greater the spring compression, the higher the oil pressure.

The adjusting spring or screw increases or decreases the spring tension. Turning the adjusting stem or screw clockwise will increase the oil pressure; counterclockwise will decrease it. Once the oil pressure has been established at the proper value, there should be no changes required.

The oil pressure is a function of spring tension on the valve plus the resistance of oil that is discharged through the valve. With the same spring tension on the valve, the oil pressure will be somewhat greater at the higher discharge rates. Therefore, with the burner operating with no flame, the oil pressure will be greater than when the flame is established and at its normal size.

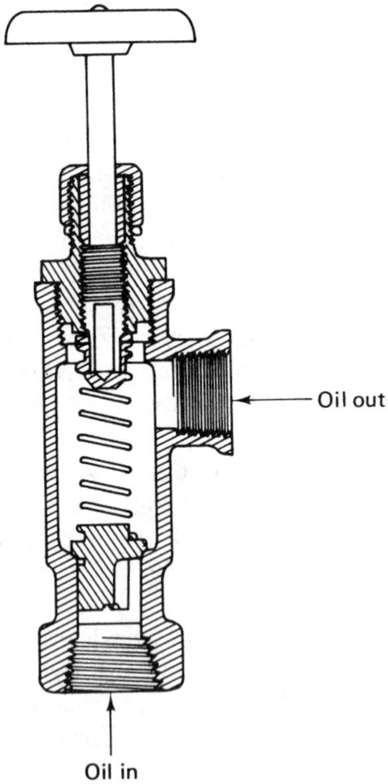
Oil out
Oil in

FIGURE 4–31
Relief valve. (*Courtesy Ray Burner Co.*)

Except for wear on the seat of the valve, or in cases of a weak or broken spring, the relief valve requires no attention. If the oil pressure is not stable or proper, the trouble can usually be traced to faulty oil lines, a fouled strainer basket, or a faulty oil pump. Therefore, if trouble is experienced, these various items should be checked before any changes or repairs are made to the relief valve.

FUEL OIL TREATMENT

The treatment of fuel oil is a complicated operation. There are many types of cure-alls on the market for this purpose. Unfortunately, cure-alls for this purpose are no better than cure-alls for other types of equipment.

Fuel oil treatment should be done by persons with the necessary facilities. To treat oil properly requires extensive laboratory tests. At any rate, some of the problems that are encountered with fuel oil cannot be immediately overcome.

To indicate the complexity of fuel oil treatment, let us explore the composition of only one ingredient, sludge. All the material found on the tank bottom is generally considered sludge. If a sample of sludge were taken and analyzed, it would contain all or any combination of the following items:

1. It may be a mixture of water and heavy fuel oil.
2. There may be insoluble solid compounds formed due to the chemical reaction of the fuel oil with the surrounding air.
3. There may be rust, dirt, and scale from many sources.
4. There may be organic precipitation caused by the blending of distillate oil with residual oil.
5. There may be some settling of suspended heavy chemical compounds due to the high cracking of residuals. This causes some coke and free carbon along with a small amount of heavy insoluble compounds.

When we consider the foregoing elements of sludge, only one source of trouble, we can readily conceive that the treatment of fuel oil is complicated. In view of this problem, the following suggestions are recommended:

1. Apply common sense in considering cure-alls.
2. Have a chemical analysis test made on the fuel oil in question.
3. Obtain the help of reputable fuel oil dealers.
4. Know exactly what the problem is. Be sure that fuel treatment will solve the problem.

PIPING SYSTEMS

Two types of piping systems are used in conveying the fuel oil to the burner. They are the *single-pipe* system and the *two-pipe system* (Figure 4–32). The single-pipe system is recommended whenever the bottom of the fuel tank is above the burner or

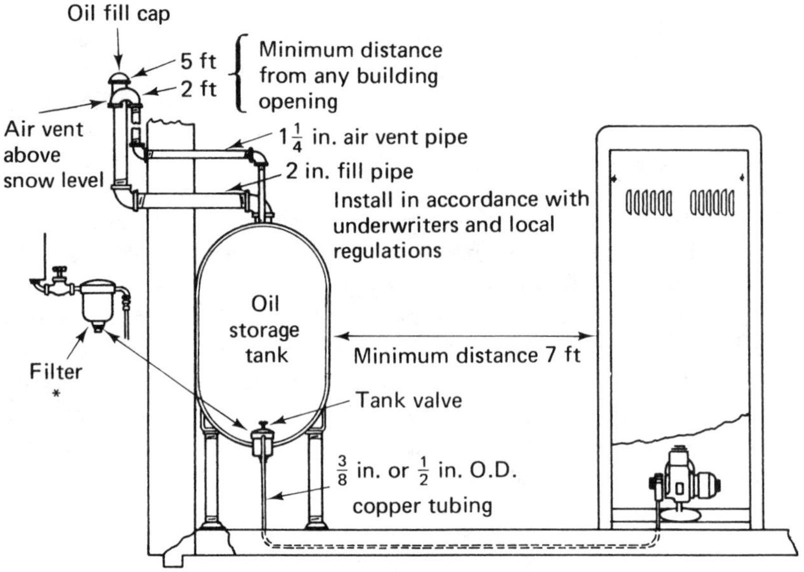
Oil fill cap
5 ft
2 ft
Minimum distance from any building opening
Air vent above snow level
1 1/4 in. air vent pipe
2 in. fill pipe
Install in accordance with underwriters and local regulations
Oil storage tank
Filter
*
Minimum distance 7 ft
Tank valve
3/8 in. or 1/2 in. O.D. copper tubing

(a)

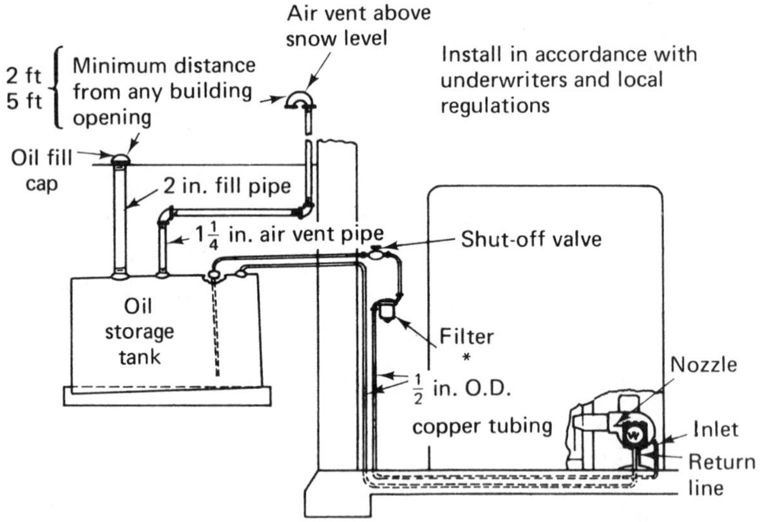
Air vent above snow level
2 ft
5 ft
Minimum distance from any building opening
Install in accordance with underwriters and local regulations
Oil fill cap
2 in. fill pipe
1 1/4 in. air vent pipe
Shut-off valve
Oil storage tank
Filter
*
Nozzle
1/2 in. O.D. copper tubing
Inlet
Return line

(b)

FIGURE 4–32 Oil piping system. (*Courtesy The Carlin Co.*)

at the same level as the burner. This includes outdoor fuel tanks that are at such levels. The two-pipe system is recommended when the fuel tank is below the level of the burner, and the fuel unit must pull (lift) the fuel up to the burner. For two-pipe installations, the bypass plug must be installed.

Table 4–3 shows, for the standard single-stage fuel unit, the allowable lift and lengths of 1/2- and 5/8-inch outside diameter (OD) tubing for both suction and return lines in the two-pipe systems.

When using the optional two-stage fuel unit, a greater amount of lift is attainable, as shown in Table 4–4.

TABLE 4–3 Single-stage units two-pipe systems (Sundstrand JA2BB-100)

LIFT	LENGTH OF TUBING (FEET)	
(FEET)	½ IN. OD	⅝ IN. OD
0	100	100
2	100	100
4	84	100
6	66	100
8	48	100
10	30	83

(Courtesy The Carlin Co.)

TABLE 4–4 Two-stage units two-pipe systems (Sundstrand H2PB-100)

LIFT	LENGTH OF TUBING (FEET)	
(FEET)	½ IN. OD	⅝ IN. OD
0	100	100
2	88	100
4	78	100
6	69	100
8	59	100
10	49	100
12	39	100
14	29	82
15	24	68

(Courtesy The Carlin Co.)

Be sure that all oil line connections are absolutely airtight. Check all connections and joints. Flared fittings are recommended. Do not use compression fittings.

For typical component and connection data, see Figure 4–33.

For typical combustion head and electrode adjustments, see Figure 4–34.

Check the equipment manufacturer's installation data for the specific requirements.

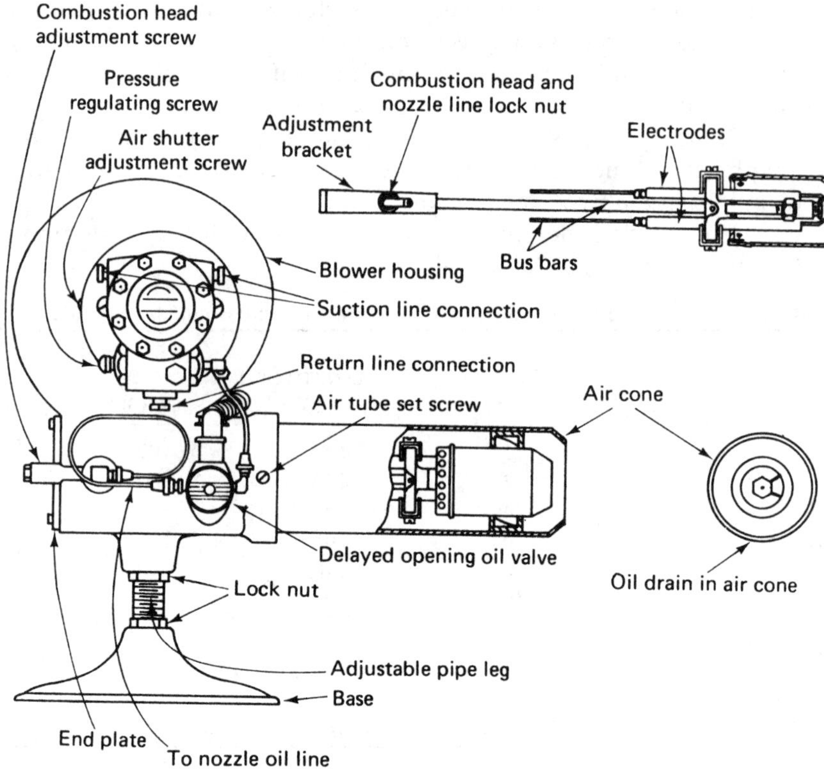

Combustion head adjustment screw
Pressure regulating screw
Air shutter adjustment screw
Adjustment bracket
Combustion head and nozzle line lock nut
Electrodes
Bus bars
Blower housing
Suction line connection
Return line connection
Air tube set screw
Air cone
Delayed opening oil valve
Oil drain in air cone
Lock nut
Adjustable pipe leg
Base
End plate
To nozzle oil line

FIGURE 4–33 Typical component location. (*Courtesy The Carlin Co.*)

GAS-OIL BURNERS

Power pressure burners are fully automatic for natural gas or light fuel oil. Fully automatic combination gas-oil pressure burners are manufactured to cover a wide range of heating requirements. They have particular value when continuous uninterrupted heat is desired (Figure 4–35). Within their capacity range, they are ideal for homes, apartments, churches, schools, hospitals, stores, shops, power boilers, and for many other heating applications. This type of burner is indispensable when gas utilities demand standby oil heating facilities.

Combination burners furnish 100% of the combustion air through the burner for firing rates up to their rated capacity. The oil is atomized by pressures generated by a gear-type high speed fuel pump assembled as a unit with an oil strainer and pressure regulating valve, which is an integral part of the burner. Injection of the oil by pressure through the nozzles produces a very fine oil spray. The air is delivered by a multivane fan through a steel nonreverberating blower tube equipped with a combustion head and is intimately mixed with the oil mist for efficient combustion. The gas is introduced through a double wall blower tube and injected into

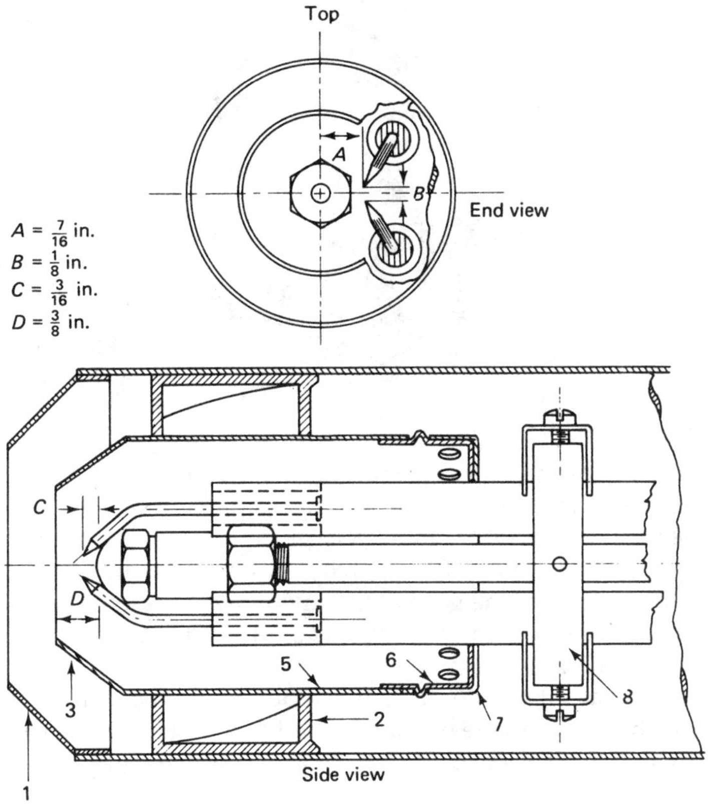
Top
End view
A = 7/16 in.
B = 1/8 in.
C = 3/16 in.
D = 3/8 in.
A
B
C
D
1
2
3
5
6
7
8
Side view

Part number	
1	Secondary air cone
2	Finned collar
3	Outer primary air cone
5	Primary air tube
6	Primary air tube holder
7	Primary air control ring
8	Electrode holder

FIGURE 4–34 Combustion head. (*Courtesy The Carlin Co.*)

the air just before it enters the combustion chamber. Because of the use of larger diameter fans and the air pressure differential through the combustion head, draft fluctuations have little effect on the operation and efficiency of this type of burner. Extremely high CO_2 settings can be realized with this equipment if desired.

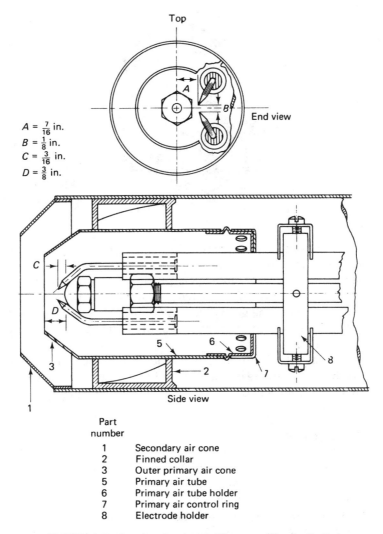

The igniter for oil firing is an electric spark from a 10,000-V transformer mounted on the burner. These burners use a separate gas pilot system for ignition when gas is being used as the main fuel. This pilot system consists of the pilot assembly, pilot solenoid valve, and a 6,000-V ignition transformer all mounted and wired on the burner. This unique pilot is supplied with air under pressure from the

FIGURE 4–35
Typical gas-oil burner. (*Courtesy Ray Burner Co.*)

burner blower, so its operation does not depend on firebox draft for its air supply. This also allows the pilot to operate even in forced draft conditions where positive firebox pressure would normally preclude use of an atmospheric-type pilot.

An air supply safety switch is mounted on these burners to ensure operation of the fan before the gas valve can operate. It assures the air supply for safe combustion. This switch immediately shuts off the main gas valve in the event of air supply failure.

A gas-oil selector switch is mounted on the burner, and the selection of fuel to be burned is made by merely moving the selector switch to either GAS or OIL. No other operation is necessary to change from one fuel to another.

REVIEW QUESTIONS

1. What must be done to fuel oil before it will burn?
2. Define a fuel oil burner.
3. What is the most popular method of fuel oil preparation for burning?
4. What controls the pressure of the fuel oil in the burner?
5. What is the approximate voltage output of a fuel oil ignition transformer?
6. What is the operating oil pressure in a high pressure oil burner?
7. What is the operating pressure range of a low pressure oil burner?
8. What type of oil burner uses a cup to atomize the fuel oil?
9. What is the first thing that must be done to fuel oil before it can be ignited?
10. What is the main purpose of a fuel oil burner nozzle?
11. What device is sized to deliver a fixed amount of fuel oil to be burned?
12. What device delivers the fuel oil in a pattern best suited for the combustion chamber?
13. How does an increase in fuel oil pressure affect the spray angle?
14. Name the two basic classifications of nozzles.
15. What is the purpose of the oil burner spray nozzle?

16. Which type of fuel oil filter provides the best filtering?
17. Name the moving parts of a rotary oil burner.
18. What are the purposes of the nozzle in a Ray horizontal rotary oil burner?
19. On the Ray rotary oil burner, how much primary air should be used through the nozzle?
20. What causes carbon to form on the refractories in the front part of the firebox using a rotary oil burner?
21. What is usually the cause of flame pulsations?
22. What can be done to the atomizing cup to promote good fuel atomization?
23. What can be done to a rotary cup to help stabilize the flame when using a light viscosity fuel oil?
24. What must be done to the fan and its parts to maintain proper firing capacity of an oil burner?
25. How is the amount of air regulated by the Ray rotary oil burner?
26. What type of adjustments are required on the worm gear shaft of a Ray rotary oil burner?
27. What should be done before replacing a damaged worm gear or driving pin in a Ray rotary oil burner?
28. What would be an indication of an out-of-adjustment fuel feed assembly on a Ray rotary oil burner?
29. What method is used to determine if the fuel tube is absolutely straight?
30. When the hollow shaft and bearings are fitted on the assembly of a Ray rotary oil burner, how much end play should there be?
31. Where should the drain groove be located when replacing the bearing cover on an oil burner?
32. What could be the cause of lubricating oil being drawn from the front shaft bearing?
33. What components are used in conjunction with the Ray viscosity valve to adjust automatically the fuel oil rate to match the load requirements?
34. What is the purpose of the single rotary gear pump used on manual or semiautomatic burners?
35. To where does the primary oil pump discharge?
36. What type of threads are used on the pump seal gland nut?
37. What device is used to maintain an initial supply of oil at a suitable temperature for starting purposes?
38. What is used to cool the nozzle on an oil burner?
39. What device operates when an overheating or external fire causes the safety string to burn apart?
40. What is the minimum gap between properly operating ignition system electrodes?

5

heat exchangers

The objectives of this chapter are:

- To show you the purpose of a heat exchanger.
- To acquaint you with the flow of the heating medium through a heat exchanger.
- To introduce the different types of heat exchangers to you.
- To familiarize you with the gas flow through the heat exchanger.

PRIMARY HEAT EXCHANGERS

A *heat exchanger* is defined as a device used to transfer heat from one medium to another.

The heat exchanger (Figure 5–1) is the heart of the heating plant. It is designed to transfer heat from the combustion gases to the heating medium flowing through the passages. It also serves as the combustion area. Openings in the lower section permit installation of the main burners and allow secondary air to reach the flame. As the fuel is burned, the flue gases rise through the vent passages. Restrictions are incorporated in the heat exchanger to control the flow of combustion air and reduce the amount of excess air to the limits of 35 to 50%. The restriction is produced in several ways: (1) the top section is reduced in cross-sectional width; (2) baffles in the exhaust openings at the top of the heat exchanger are used (Figure 5–2); and (3) a combination of both. These restrictions also permit maximum heat transfer by reducing the vent gas velocity that allows the maximum heat to be extracted from it. The gases then leave the heat exchanger and enter the venting system.

There are two general classifications of heat exchangers: (1) the barrel and (2) the sectionalized (Figure 5–3). They may be constructed of either cast iron or

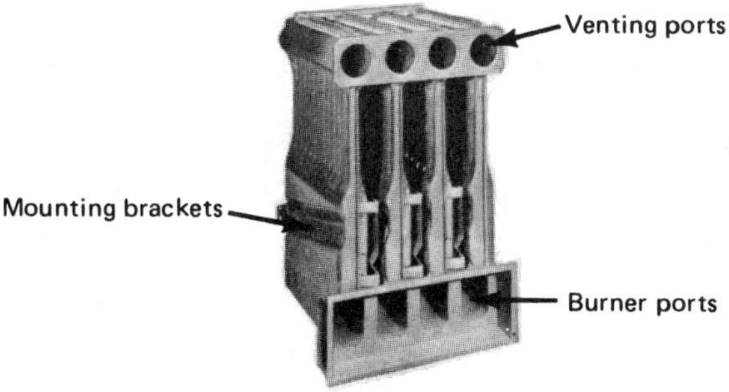
Venting ports
Mounting brackets
Burner ports

FIGURE 5–1
A forced-air gas heat exchanger.

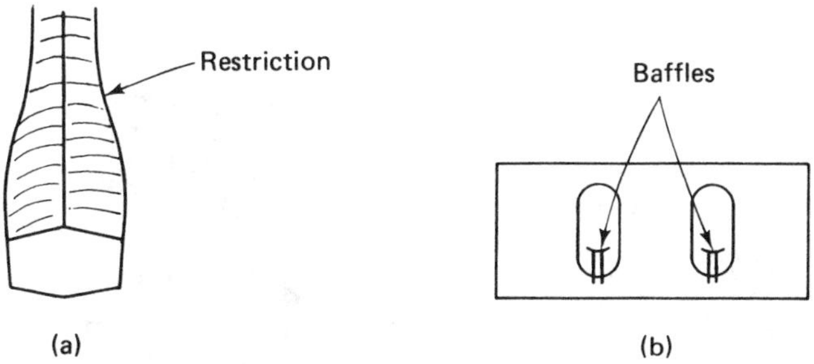
Restriction
Baffles
(a)
(b)

FIGURE 5–2 Gas vent restrictions.

steel. Due to the weight, expense, and slow response to temperature change of cast iron, steel has almost dominated the forced air furnace heating industry in recent years.

The major problem with steel is the noise that sometimes accompanies the expansion and contraction associated with the heating and cooling of a heat exchanger. This noise, however, can be reduced by placing dimples and ribs in the

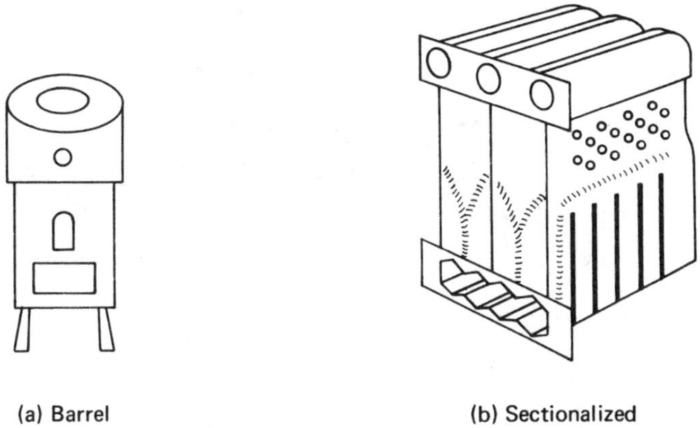
(a) Barrel
(b) Sectionalized

FIGURE 5–3 Types of heat exchangers.

metal during the stamping operation. Ceramic coating is also used to help eliminate noise while providing corrosion protection. This ceramic coating does not hinder the transfer of heat from the heat exchanger.

The application of the heat exchanger determines its physical shape. The heat exchanger in a horizontal forced-air furnace (Figure 5–4) would need to be shaped differently from one used in an upflow furnace (Figure 5–5). In most instances, there is very little difference from the upflow to the downflow heat exchanger.

There are two flow paths through a furnace heat exchanger. Both the heated air and combustion gases pass through the unit without any mixing of the two. The heated air flows around the outside of the heat exchanger (Figure 5–6), while the flue gases pass through the inside.

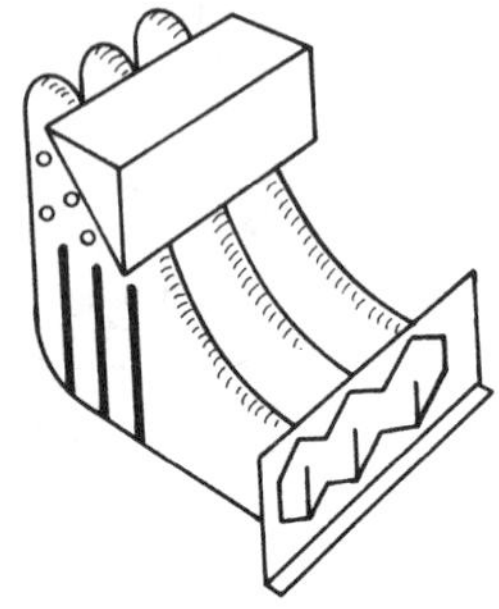

FIGURE 5–4
Horizontal heat exchanger.

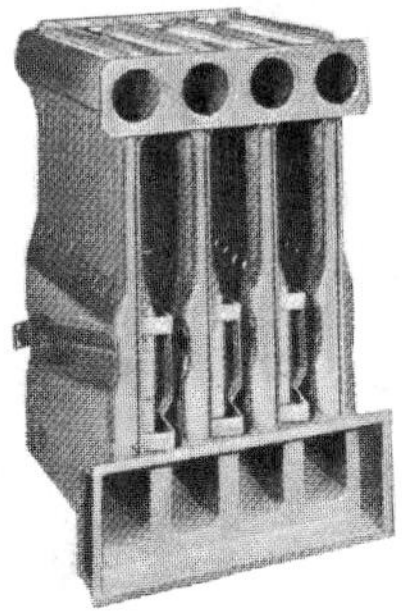

FIGURE 5–5
Upflow heat exchanger. (*Courtesy of Dearborn Stove Co.*)

The heat exchanger is normally an extremely trouble-free apparatus, especially when the burners are installed and adjusted properly. From time to time, however, due to the continuous expansion and contraction, a hole will develop in the metal. A hole in the heat exchanger presents a dangerous situation. When the

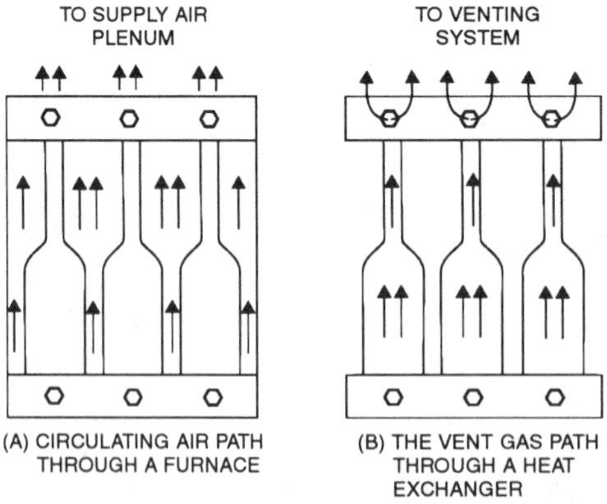
TO SUPPLY AIR PLENUM
TO VENTING SYSTEM
(A) CIRCULATING AIR PATH THROUGH A FURNACE
(B) THE VENT GAS PATH THROUGH A HEAT EXCHANGER

FIGURE 5–6
Flow paths through a heat exchanger.

flame is burning before the blower starts, carbon monoxide could be admitted to the air stream. After the blower starts, the flame will be agitated due to the greater pressure in the duct system. The amount of agitation will, naturally, depend on the size of the hole. A large hole could cause the flame to be blown from the front of the furnace and catch fire to anything in the immediate area, along with depositing the products of combustion in the home. The only difference between a large hole and a small hole is that the small hole is more difficult to detect. In any case, a leaking heat exchanger must be replaced.

When main burners that are dirty or out of adjustment have been used for a period of time, the flue passages in the heat exchanger may become clogged with soot. This soot is hazardous because it retards the combustion process and does not allow the removal of all the products of combustion. Soot is a deposit of unburned fuel; therefore, it is combustible. Because soot is combustible, caution should be exercised in the use of highly combustible sprays or fogs for its removal. Sometimes when these products are used the soot will float out of the combustion area and be a hazard to the surroundings. At times the smoldering soot can be seen floating from the vent above the roof. To desoot a heat exchanger properly, the main burners must be removed and cleaned. While the burners are removed, the flue passages must be cleaned of all soot deposits. This is a costly, time consuming, and dirty job that could be avoided by keeping the main burners properly adjusted. An example of the proper alignment of burner to heat exchanger is shown in Figure 5–7.

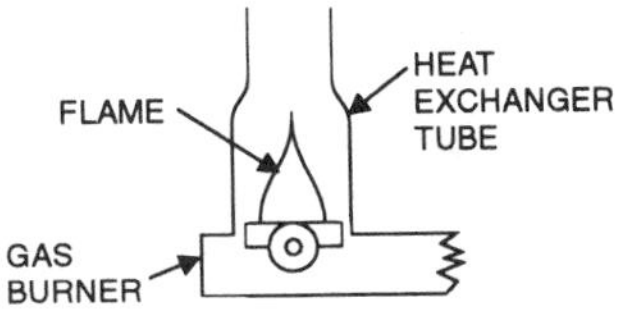
FLAME
HEAT EXCHANGER TUBE
GAS BURNER

FIGURE 5–7
Alignment of burner and heat exchanger. (*Courtesy of Modine Manufacturing Co.*)

HIGH EFFICIENCY HEAT EXCHANGERS

The higher Btu output of a furnace is gained in two ways: (1) Additional sections are added to the heat exchanger, and (2) a finned coil is added in the exhaust of the heat exchanger. These are generally referred to as *secondary heat exchangers.* They usually use either forced draft or induced draft combustion blowers so that the combustion air ratios which are set to the factory settings will maintain maximum combustion efficiency.

Additional Sections (Secondary Heat Exchangers)

When additional heat exchangers are used, the efficiency is increased from around 65% to around 98%, depending on how many and what type of sections are added (Figure 5–8). The products of combustion pass through the heat exchanger, as in

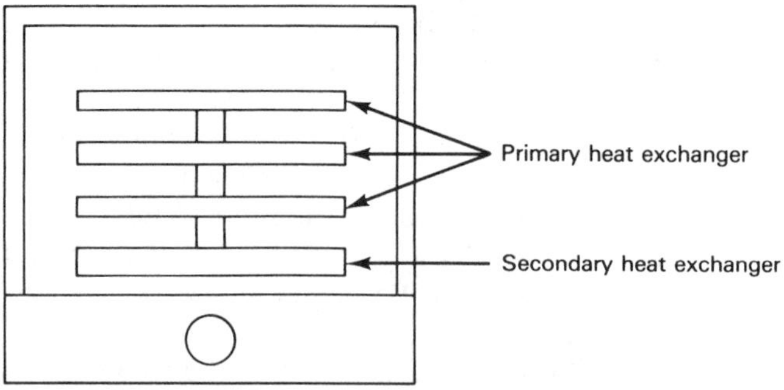
Primary heat exchanger
Secondary heat exchanger

FIGURE 5–8 Additional heat exchanger for increased efficiency. (*Courtesy of Billy C. Langley,* High Efficiency Gas Furnace Troubleshooting Handbook, ©1991, *p. 3; reprinted by permission of Prentice-Hall, Inc., Englewood Cliffs, NJ*).

the standard heat exchanger, and then into the secondary part of the heat exchanger. When this type of heat exchanger is used, some of the circulating air passes over the primary part of the heat exchanger and some passes over the secondary part of the heat exchanger.

The circulating air does not pass over all of the sections on every pass through the furnace. However, enough circulating air comes into contact with the additional section, or sections, to increase the efficiency to about 80 to 85% when only one additional section is used. The second section increases the efficiency to the 90 to 98% efficiency rating. When the efficiency is increased to this range, the furnace is generally termed a *condensing gas furnace.*

Finned Coils (Secondary Heat Exchangers)

The finned coil is located in the air stream ahead of the regular heat exchanger. See Figure 5–9. The products of combustion enter the primary heat exchanger from the combustion zone, and heat is transferred to the circulating air the same as in the standard gas heating furnace. The products of combustion then pass through a tail pipe, or other type of design, into the secondary heat exchanger. The cooler entering air passes over the finned heat exchanger first and causes the water vapor to condense, giving up the latent heat of condensation to the circulating air. The circulating air then flows over the warmer primary heat exchanger, where it absorbs more heat.

Condensate Drain

On condensing gas furnaces, the secondary heat exchangers are equipped with a condensate drain connection that allows the condensate to be drained into the city drain system. This corrosive condensate should not be allowed to drain onto the ground or other places where damage can be caused. Plastic pipe is normally used as the drain line because of the corrosive nature of the condensate. Be sure to follow

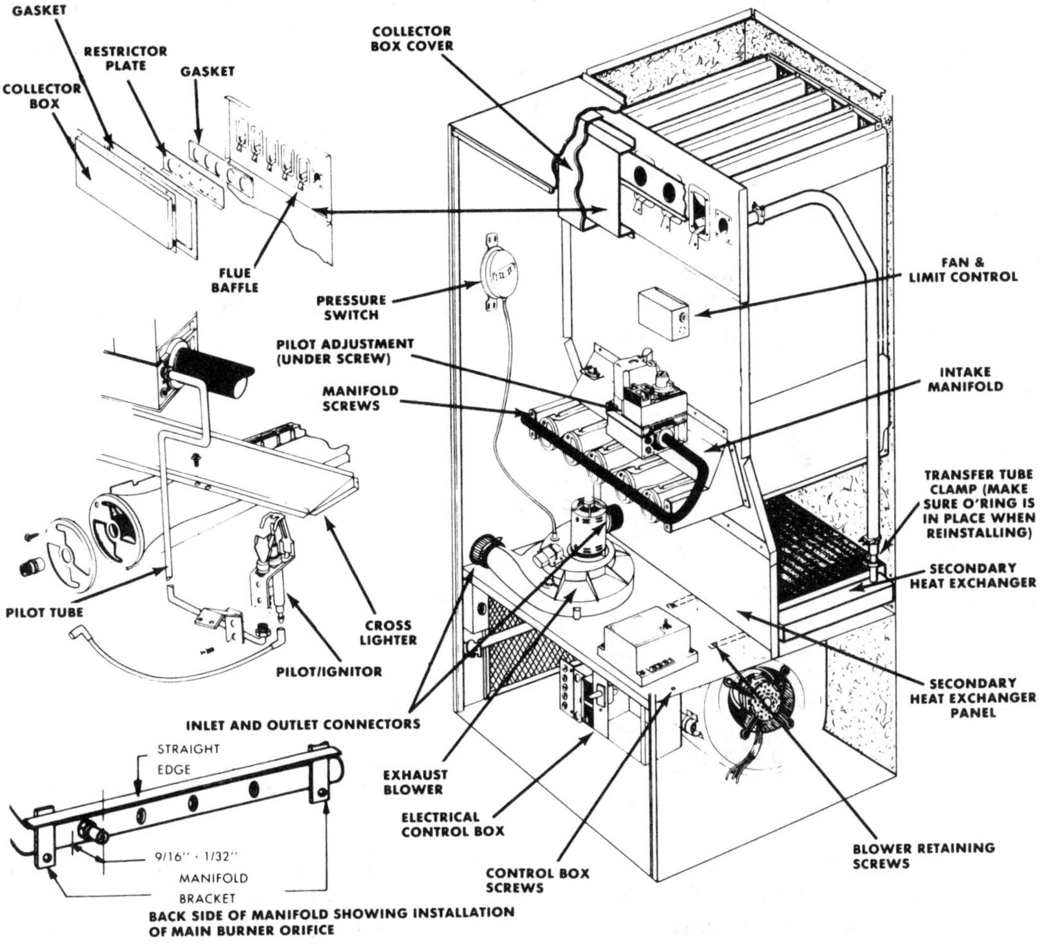
GASKET
RESTRICTOR PLATE
GASKET
COLLECTOR BOX
COLLECTOR BOX COVER
FLUE BAFFLE
PRESSURE SWITCH
PILOT ADJUSTMENT (UNDER SCREW)
MANIFOLD SCREWS
FAN & LIMIT CONTROL
INTAKE MANIFOLD
TRANSFER TUBE CLAMP (MAKE SURE O'RING IS IN PLACE WHEN REINSTALLING)
SECONDARY HEAT EXCHANGER
PILOT TUBE
CROSS LIGHTER
PILOT/IGNITOR
SECONDARY HEAT EXCHANGER PANEL
INLET AND OUTLET CONNECTORS
STRAIGHT EDGE
EXHAUST BLOWER
ELECTRICAL CONTROL BOX
9/16" - 1/32"
MANIFOLD
BRACKET
CONTROL BOX SCREWS
BLOWER RETAINING SCREWS
BACK SIDE OF MANIFOLD SHOWING INSTALLATION OF MAIN BURNER ORIFICE

FIGURE 5–9 Finned coil secondary heat exchanger location [*Courtesy of Inter-City Products, Inc. (USA)*].

the proper procedure for installing the drain. This is generally covered in detail in the manufacturer's installation literature. These instructions should be followed unless there is a conflict with the local ordinance, in which case the local ordinance should be used.

Heat Pipe Heat Exchangers

Heat pipe heat exchangers increase the efficiency rating of a furnace by increasing the amount of heat exchanger surface. However, they do not increase the efficiency to the point of being considered a condensing gas furnace. This type of heat exchanger simply uses a group of pipes as the heat exchanger (Figure 5–10). There is a combustion zone, as with any other heat exchanger. From the combustion zone, the products of combustion are directed into the pipes, which become the heat exchanger. The circulating air passes over the outside of the heat pipes. These types of heat exchangers are considered to be about 85% efficient.

HYDRONIC SECONDARY HEAT EXCHANGERS

Up to this point, only heat exchangers used on gas- or oil-fired units have been discussed. They are called *primary heat exchangers.* This type of heat exchanger by no means comprises all the elements used for heat transfer.

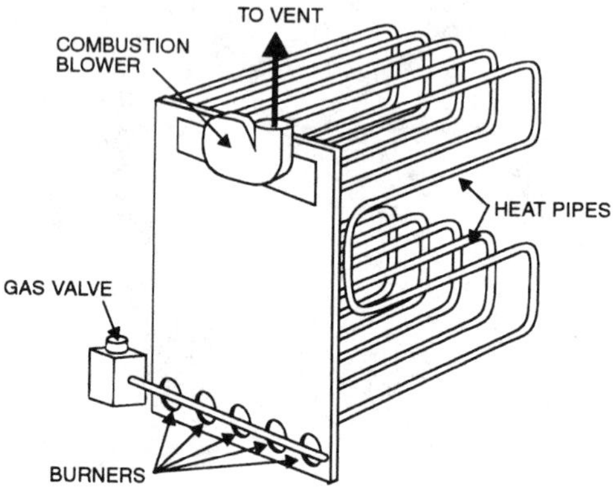
TO VENT
COMBUSTION BLOWER
HEAT PIPES
GAS VALVE
BURNERS

FIGURE 5–10
Typical heat pipe heat exchanger.

Hydronic heating also enjoys a fair share of the heating industry. The boilers used for hydronic heating also employ the same type of primary heat exchanger as forced-air units. However, water is forced through the boiler's passages in place of air (Figure 5–11).

When a boiler is used, a single heat exchanger will not transfer the desired heat to the structure. There must be another (secondary) heat transfer element placed in the room. When hydronic heating (hot water or steam) is used, the heating medium is passed from the boiler to the room heating element through a system of piping and returned to the boiler, where it is reheated and recirculated. In some installations the secondary heating element is attached to a duct system and used as a central heating unit.

Natural draft hydronic heating elements, sometimes referred to as radiators, are manufactured for mounting in a variety of locations. They are an assembly of increments, each composed of vertical heating fins and a dividing plate that is reinforced with a steel tube beading for strength and protection.

The heating element consists of a heavy-wall copper tube formed into a continuous serpentine coil. In turn, the coil passes through each heating fin and is

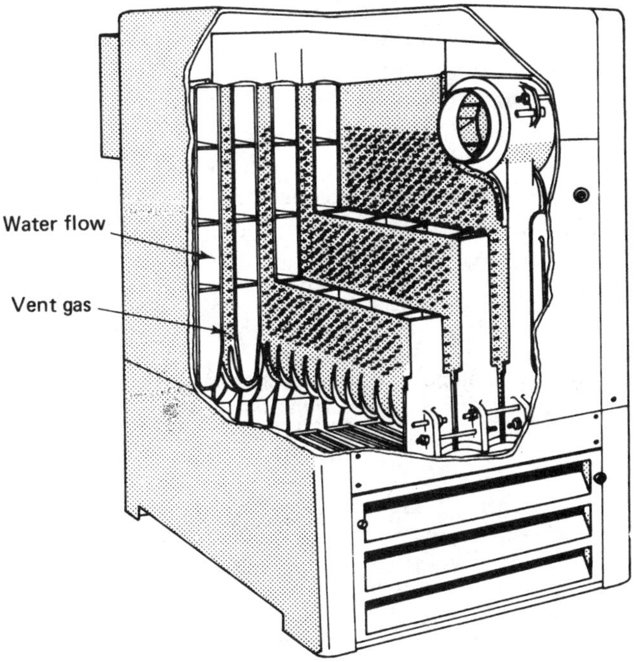
Water flow
Vent gas

FIGURE 5–11
Water and gas flow through a boiler. (*Courtesy of Weil-McLain Co., Hydronic Division.*)

expanded under 3,750 psig of hydraulic pressure to lock all elements into a one-piece unit. Such bonding is ideal for high pressure, high temperature systems since there is no solder or soft metal to break away.

Another type of secondary heat exchanger is the *embedded coil.* This type of element has a pipe embedded in the floor, wall, or ceiling (Figure 5–12). This system of piping is connected to the boiler in the conventional manner. The coil has no fins and employs the surrounding structure to radiate the heat. It is ideal in play rooms or dens where a warm floor is desired. These elements, naturally, should be installed during construction of the room.

The efficient operation of any hydronic heating element requires that all air be kept from collecting in the piping. Most installations have air vents installed on top of the coil as well as in the highest point of the entire system. Should these

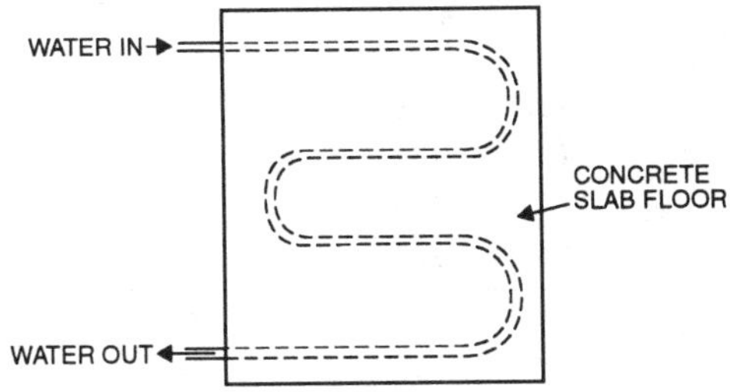
WATER IN
CONCRETE SLAB FLOOR
WATER OUT

FIGURE 5–12
Embedded hydronic element.

elements become air locked, the heating will be reduced and sometimes completely stopped. To maintain maximum output from these units, the air must be continuously removed from the piping, coils, and boilers.

Each manufacturer designs and fabricates its heat exchanger to fit a definite set of conditions. Manufacturers will publish the conditions for a specific output rating of their heat exchanger (Table 5–1). Sometimes the heat exchangers of the same manufacturer will have different conditions and performance ratings. It can be seen that when considering a heat exchanger, several factors must be considered. The difficulty in selecting a heat exchanger is being sure which will deliver the most overall satisfaction.

TABLE 5-1 Typical radiator selection chart, wall mounted position, radiator heat outputs in square feet equivalent direct radiation. *(Courtesy of Shaw-Perkins Manufacturing Co.)*

Rad. No.	Length L See Note	Ship. Weight in Pounds	200° Water	220° Water	1 Lb Steam 240° Water	15 Lb Steam 280° Water	35 Lb Steam	50 Lb Steam	100 Lb Steam	150 Lb Steam
Height 14½ inches										
10	25	54	10	12	14	18	23	25	31	35
15	37.5	78	16	20	24	31	38	42	52	59
20	50	102	23	28	33	43	53	59	73	83
25	62.5	127	30	36	43	56	68	76	94	107
30	75	151	37	45	52	68	84	94	116	132
35	87.5	175	43	53	62	81	98	111	137	156
40	100	199	50	61	71	93	114	128	158	180
45	112.5	224	56	69	81	106	128	145	179	204
50	125	248	63	77	90	118	144	162	200	228
Height 23½ inches										
10	25	76	16	20	23	30	37	41	51	58
15	37.5	109	26	32	37	49	60	67	84	95
20	50	142	37	45	52	68	84	94	116	132
25	62.5	176	47	57	67	87	107	120	149	169
30	75	209	57	70	82	107	131	147	181	206
35	87.5	242	67	82	96	126	154	173	214	243
40	100	275	78	94	111	145	178	199	246	280
45	112.5	309	88	107	126	164	201	226	279	317
50	125	342	98	119	141	183	225	252	311	354

REVIEW QUESTIONS

1. What is the purpose of restrictions in a heat exchanger?
2. What can be done to steel heat exchangers to reduce noise?
3. Name the paths of flow through a forced-air heat exchanger.
4. What is deposit of soot in a heat exchanger an indication of?
5. Why is a soot accumulation in a heat exchanger hazardous?
6. Why is a finned coil added to a furnace?
7. What type of draft is used on higher efficiency gas furnaces?
8. What process allows a condensing gas furnace to be more efficient than standard gas furnaces?
9. Other than gas furnaces, where are secondary heat exchangers used?
10. In a hydronic heating system, what must be done regarding air inside the pipes?
11. Why is a cracked heat exchanger dangerous?
12. When are embedded coils installed for a hydronic heating system?
13. How are steam and hot water radiators sometimes referred to?

6

venting

The objectives of this chapter are:

- To acquaint you with the operating principles of the venting system.
- To point out the safety precautions for a venting system.
- To introduce you to the problems encountered in designing a venting system.
- To acquaint you with barometric draft control design and installation procedures.
- To instruct you in gas burner combustion testing.
- To acquaint you with oil burner draft.
- To introduce you to the operation of draft regulators.

American Gas Association (AGA) design certified gas equipment must be capable of venting products of combustion through the draft diverter opening without connection to a vent or chimney. Therefore, the only purpose of a vent for an AGA design certified heating unit is to convey the flue gas products from the draft diverter of the unit to the outside of the building. The vent may be defined as a system of piping used to remove all the products of combustion to the outside air.

VENTING

After the fuel is burned in the combustion area, it is passed through the flue passages of the heat exchanger. On leaving the heat exchanger, these products of combustion enter the draft diverter, or the draft control, whichever is used. At this point, the flue gases are mixed with an amount of air equal to that required for combustion air. For example, when 1 ft^3 of natural gas is burned, 15 ft^3 of combustion air are required. At the draft diverter, another 15 ft^3 of air per ft^3 of natural gas are required for dilution air.

DRAFT DIVERTER

The purpose of the draft diverter (Figure 6–1) is to neutralize excessive drafts and downdrafts through the heating unit and to produce an emergency outlet for relieving the flue gases in the event that the chimney or breeching should become obstructed. Chimney draft is not necessary for the proper operation of the unit, which has, as a part of its design, a constant natural draft through the heat exchanger for the proper performance of the burner and pilots. Excessive updraft or downdraft could otherwise adversely affect combustion, resulting in the generation of carbon monoxide (CO), reducing combustion efficiency, causing hazardous ignition, or even extinguishing the safety pilot, thus causing a nuisance shut-down.

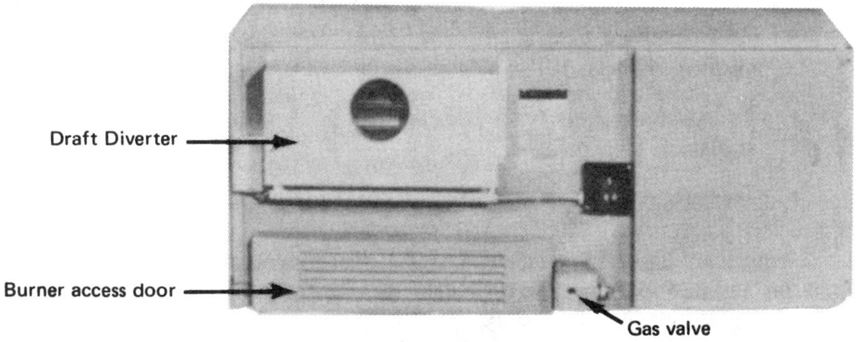
Draft Diverter
Burner access door
Gas valve

FIGURE 6–1 Draft diverter installed on a horizontal furnace. (*Courtesy of Dearborn Stove Co.*)

The correct installation of the heating unit and its draft diverter is vital for safe operation. All heating units are tested and listed by AGA with their draft diverter in position. Check the following list to be sure that the draft diverter is correctly installed.

1. Do install the draft diverter free of obstructions below and to the side.
2. Do install exactly as specified by the manufacturer's instructions.
3. Do obtain correct draft diverter from the manufacturer if it is missing. If assistance is needed, check with the fuel supplier.
4. Do not alter the pipe lengths between the draft diverter and the unit flue outlet.
5. Do not change the position of the draft diverter on the heating unit.

6. Do not change the design of the draft diverter by adjusting or cutting the baffles, skirt, etc.
7. Do not locate the draft diverter in a different room from the unit.
8. Do not substitute or exchange draft diverters.

VENT PIPING

The venting system not only conveys the gases, it also conveys the air taken in through the relief opening of the draft diverter. This dilution air imposes an added volume of gases that must be handled by the vent system. The vent, therefore, must be capable of conveying the flue gases and dilution air to the outside atmosphere with a minimum of draft resistance. If the vent system has insufficient flue area or height, some flue gases will spill from the relief opening of the draft diverter. Flue gas spillage is a hazard that must be avoided by the proper design and construction of the venting system. It is obvious that the vent system having the least resistance would be a vertical vent taken directly from the outlet of the draft diverter. This low cost hook-up is highly recommended, where applicable, and can be easily used on multiple-outlet units as well as on single-outlet units.

The installation of a double-wall UL-listed gas unit is not normally a complicated matter. Most vent installations are relatively simple, requiring the observation of only a few basic rules. Some complex multiunit or multistory vent systems, however, will require the engineered capacities and configurations which are clearly set forth in tables. These tables give recommended vent size for various configurations and unit draft diverter outlet diameters according to the rated Btu input of the unit and the height of the vent. These Gas Vent Institute (GVI) vent capacity tables are based on the use of UL-listed double-wall gas vent pipes and fittings from the unit to the vent cap (Figure 6–2). The concept of using UL-

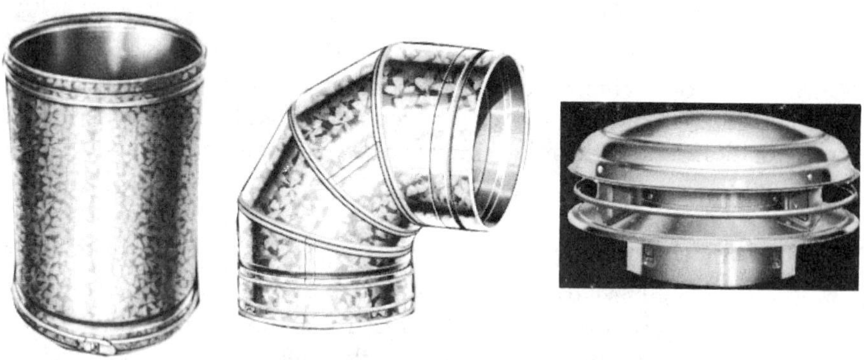
Round pipe
90° adjustable elbow
Metlcap (vent cap)

FIGURE 6–2 Common vent pipe and fittings. (*Courtesy of Hart and Cooley Manufacturing Co.*)

listed materials for the entire venting system is consistent with the "General Information" statement issued by the Underwriters' Laboratories on August 5, 1960 (Figure 6–3).

When installing or servicing heating equipment, make sure that all heating units and vents have an adequate supply of air for combustion and venting. Air, as it flows into and through a heating unit and is vented up the stack, serves three vital functions. First, on entering the heating unit as combustion air, it furnishes the oxygen required for the flame. Second, while circulating through the heating unit, air serves as the medium through which the heat is transferred from the flame to the heat exchanger. Third, air is introduced directly into the stack as dilution air.

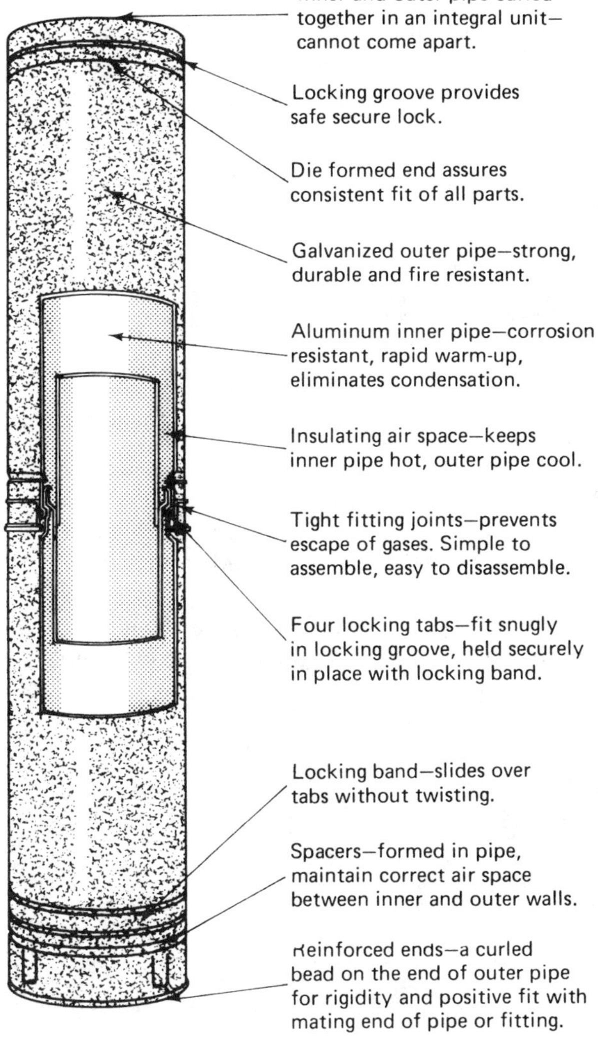

Inner and outer pipe curled together in an integral unit—cannot come apart.
Locking groove provides safe secure lock.
Die formed end assures consistent fit of all parts.
Galvanized outer pipe—strong, durable and fire resistant.
Aluminum inner pipe—corrosion resistant, rapid warm-up, eliminates condensation.
Insulating air space—keeps inner pipe hot, outer pipe cool.
Tight fitting joints—prevents escape of gases. Simple to assemble, easy to disassemble.
Four locking tabs—fit snugly in locking groove, held securely in place with locking band.
Locking band—slides over tabs without twisting.
Spacers—formed in pipe, maintain correct air space between inner and outer walls.
Reinforced ends—a curled bead on the end of outer pipe for rigidity and positive fit with mating end of pipe or fitting.

FIGURE 6–3
Vent pipe construction. (*Courtesy of Hart and Cooley Manufacturing Co.*)

Dilution air serves a double purpose. It carries the spent gases up the vent, and it serves as a variable through which overfire draft is governed.

Draft is the force that is applied to the movement of a column of air into and through a heating unit. The force involved is static pressure—the ton-per-square-foot pressure—exerted by the weight of the earth's atmosphere.

The movement of air through a heating unit requires a static pressure differential. To produce this draft, the pressure of the air surrounding the heating unit and vent system must be greater than the pressure exerted by the gases within the unit and vent system.

This pressure differential, and therefore, the degree of draft force exerted at a given barometric reading, depends on two factors.

1. The amount of temperature differential both inside and outside the heating unit and vent system.
2. The height to which this temperature differential is stacked, or the vent height (Figure 6–4).

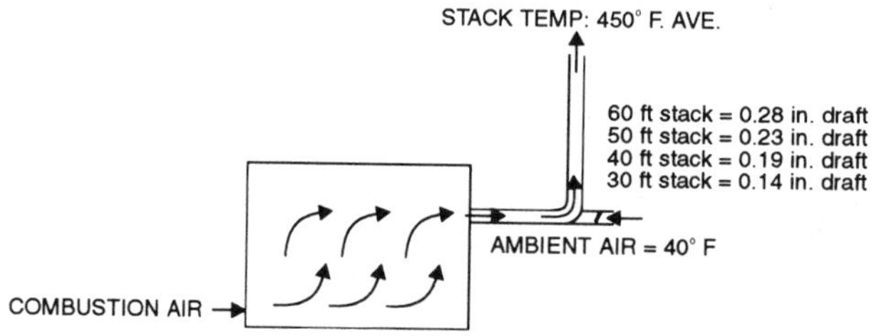
STACK TEMP: 450° F. AVE.
60 ft stack = 0.28 in. draft
50 ft stack = 0.23 in. draft
40 ft stack = 0.19 in. draft
30 ft stack = 0.14 in. draft
AMBIENT AIR = 40° F
COMBUSTION AIR

FIGURE 6–4
Comparison of vent height and draft.

A natural draft cannot be produced without a vent system. Draft cannot be used as a source of oxygen unless the vent is connected to the unit. A vent, no matter how high, can provide no overfire draft if it is separated from rather than connected to the heating unit.

The draft force produced by a vent system of any given height varies constantly. This is because the conditions which create the draft are subject to the following continuous changes (Figure 6–5).

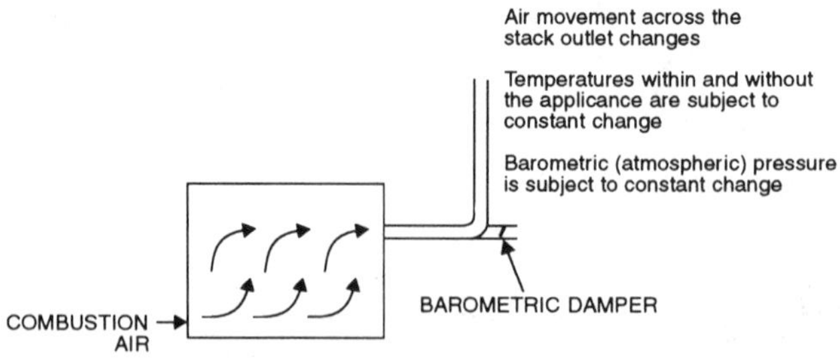
Air movement across the stack outlet changes
Temperatures within and without the applicance are subject to constant change
Barometric (atmospheric) pressure is subject to constant change
BAROMETRIC DAMPER
COMBUSTION AIR

FIGURE 6–5
Representative changing of vent conditions.

1. *Temperature change within a heating unit.* The weight of air, and therefore the pressure it exerts, varies directly with the temperature. Air at 30° has two times the weight of air at 400° F. Thus, the pressure of air within a heating unit changes as the temperature of the gases change. The vent system should be designed to maintain a temperature of at least 300° F. Under no condition should the inside vent temperature drop below 225° F above ambient temperature.
2. *Temperature change outside a heating unit.* The pressure exerted by the air around a heating unit changes as the ambient air temperatures change. Vent piping and fittings are designed for a 60° F outside ambient temperature.
3. *Barometric pressure changes.* As the barometric pressure rises and falls, the draft produced by a vent system of any given height, at any given temperature differential, also rise and falls proportionately.
4. *Air turbulence.* The capturing and retarding action of the wind gusts that alternate with calm periods at the stack outlet causes constant, abrupt, and severe changes in the draft action of the vent system.

It should be noted that these four forces governing draft production are subject to constant changes. Only the stack height remains fixed, while air turbulence changes, ambient air temperatures change, stack temperatures change, and the barometric pressure changes.

As stated before, the capacities given in the tables apply specifically to UL-listed double-wall B gas vents constructed entirely of listed materials from the draft diverter outlet to the vent cap. Recommendations for venting system design and capacity in gas vent tables are based on the insulating qualities and fluid flow characteristics of gas vent pipe and fittings. These tables do not apply to materials having lower values such as masonry, asbestos-cement, or single-wall metal pipe. Nor do they consider vents less than 6 ft in height.

VENT DESIGN

The following is a listing of the factors that will affect the operation of a gas vent system. The capacities given in the tables for type B gas venting are consistent with those found in publications by the Gas Appliance Manufacturers Association (GAMA) and National Fire Protection Association (NFPA) 54 National Fuel Gas Code.

Systems

The tables generated by most vent manufacturers apply to system design using type B gas vents and to model TD chimneys when used for venting AGA-listed category I gas-fired, draft-hood-equipped or fan-assisted combustion appliances. At no time should a venting system for a listed category II, III, or IV appliance be sized with these tables.

Clearance

Vent installations must provide the proper clearances to combustible materials as specified in the appropriate Underwriters Laboratories Inc. conditions for listing provided in the product catalogs and embossed on the vent pipe. The following is a listing of the recommended clearances:

Type B-1: Type B pipe sizes through 12 inches in diameter require 1-inch clearance throughout the entire length of the vent.

Type B-2 × 4: Type B size 4-inch oval is listed for use within a 2 × 4 stud wall with the specified firestop spacers at the plate locations.

Type B-W: Type B oval pipe 4-inch size for use within 2 × 4 stud walls for venting AGA-listed wall furnaces when using the appropriate spacers and firestops.

Air Supply

For satisfactory performance of the appliance and the venting system, an adequate supply of fresh air must be provided. When proper air supply has been provided for other appliance such as clothes driers, range hoods, fireplaces, etc., then the following method as provided by NFPA Standard 54, American National Standards Institute (ANSI) Z223.1 will provide the additional air needed for the furnace and water heater. The two grills specified in Figure 6–6 (a), (b), (c), and (d) must be installed so that one is at or below the combustion air inlet of the appliance and the other is above the relief opening of the draft hood. In no case should the vertical distance between them be less than 3 2/3 ft. Any ducting used should have at least the same free area as the grills to be used.

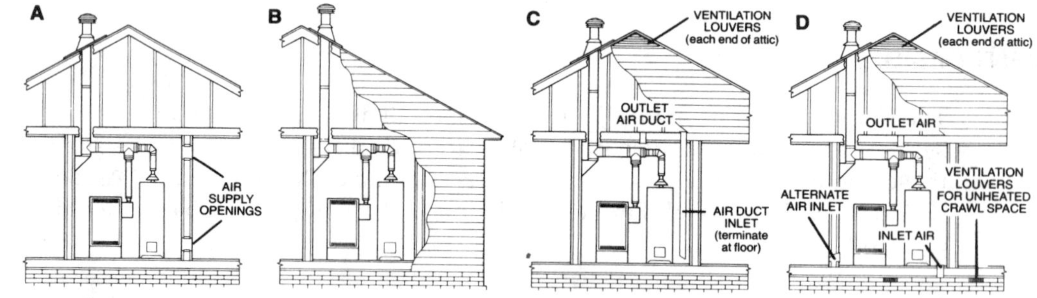
A
AIR SUPPLY OPENINGS
B
C
VENTILATION LOUVERS (each end of attic)
OUTLET AIR DUCT
AIR DUCT INLET (terminate at floor)
D
VENTILATION LOUVERS (each end of attic)
OUTLET AIR
VENTILATION LOUVERS FOR UNHEATED CRAWL SPACE
ALTERNATE AIR INLET
INLET AIR

A. ALL AIR FROM INSIDE BUILDING

Free Area of Each Grill = Total Input* / 1000

(Use 2 Grills Facing Into Large Interior Room)

B. ALL AIR FROM OUTDOORS

Free Area of Each Duct = Total Input* / 2000

Free Area of Each Grill = Total Input* / 4000

C. ALL AIR FROM VENTILATED ATTIC

Free Area of Each Duct or Grill = Total Input* / 4000

D. AIR IN FROM CRAWL SPACE; OUT INTO ATTIC

Free Area of Each Grill = Total Input* / 4000

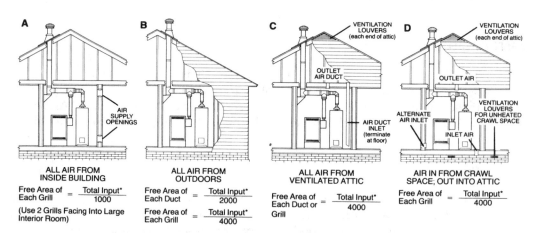

*Total Input = Total of combined appliance input ratings in BTU; (Free Area in square inches)

FIGURE 6–6 Recommended vent grill locations. (*Courtesy of Hart & Cooley, Inc.*)

Local Building Code

Should the local building code differ from the recommendations given in this text, consult with the local building inspector or other local administrative authority. The information given is based on the latest scientific data, which have been verified by a long and satisfactory use history. The data and practices given herein will invariably provide better results than practices required by an obsolete code.

Correction for Altitude

The vent system should always be designated for the sea level nameplate rating (greatest input when the unit has a modulated input) of the furnace, regardless of the actual derated operating input required by the local altitude.

Outside Vents

The sizing tables are not applicable to outside (exposed) chimneys or vents. A type B vent lining within an exposed masonry chimney is considered to be an enclosed vent system, and as such these tables may be used.

Connectors

Single-wall pipe (stove pipe) is not recommended for use in type B venting systems. Because of the higher heat loss from the flue products, the draft is reduced and condensation can occur. The resulting moisture may corrode the pipe and will likely leak out on the building and contents, causing damage.

When single-wall connector pipe usage is accepted local practice, the following considerations must be followed:

1. The minimum clearance to combustibles is 6 inches instead of the 1 inch required for type B vents.
2. The heat loss is roughly double that for type B vent pipe; therefore, do not use it in any cold or concealed space as condensation will result, leading to venting failure and possible other damage.
3. The common venting tables were generated using a maximum vent connector length of 1 ½ ft (18 inches) for each inch of connector diameter. See Table 6–1.
4. When using Table 6–1 for type B vent connectors, remember that doubling the length shown results in reducing the capacity in the various tables by 10%.

TABLE 6–1 Connector length allowance

SIZE	3"	4"	5"	6"	7"	8"	10"	12"	14"	24"
LENGTH (FEET)	4.5	6	7.5	9	10.5	12	15	18	21	36

5. If the vent connectors are combined prior to entering the common vent, the maximum common vent capacity listed in the common venting tables must be reduced by 10%.

Connector Rise

The immediate vertical height from the flue collar to the first turn (connector rise) will have an important effect on the proper functioning of a vent system. For a venting system to prime (for flow up the vent to begin), the vent liner must be heated by the flue gases. If it is easier (less resistance) for the flue products to spill out the draft hood relief opening than to flow into the vent, priming can be delayed or prevented. By using all of the vertical height (head room) available (never less than 1 ft), a venting system will usually prime within 8 to 10 seconds.

Vent Cap Termination

Use only companion AGA-listed caps or roof assemblies. The capacity and wind resistance depend on the correct termination of the vent system. Terminations on any false chimney or other support or surface must comply with the instructions for a roof surface.

Appliance type limitations. Appliances which are not to be connected to type B gas vents are:

1. Gas incinerators: Use the model TD chimney for these appliances. When so connected with other appliances, no additional allowance for their input is required.
2. Gas clothes dryers: These devices produce a pressure discharge into the vent system and will cause backflow to the other connected appliances. They will also discharge lint, which may eventually block the discharge.
3. Power burners: These are not classified as category I with a positive draft and, therefore, are not allowed on type B venting systems. Use special vent systems such as Ultravent ® for categories III and IV, when applicable.
4. All condensing type: Category IV (high efficiency) appliances are not allowed on type B venting systems. Condensation will leak out and damage the structure and contents. Use a vent system such as Ultravent.

Table limitations. The vent sizing tables presented include the following considerations:

1. Lateral lengths.
2. Low resistance vent caps.
3. AGA-listed category I appliances.
4. Two 90° turns except for 0 lateral.

The vent connector must be routed to the vent using the shortest possible route. Longer connectors than those listed in Table 6–1 are possible but require an increase in the size, rise, or total vent height to compensate for the additional length.

Condensation

The condensing of water vapor from the products of combustion of gas fuels can be minimized by the use of sizing tables. When the vent system is designed properly, dilution air entering a draft hood (if available) reduces the temperature at which water vapor will condense (dew point). Exceptions that will cause condensation are as follows:

1. Temporarily (for a few seconds) after burner ignition, condensation will form on the cold inner liner of the vent. Before it increases to drop size, the liner will have been heated above the dew point temperature causing this small amount of condensate to reevaporate. If the vent is located outdoors and the temperature is very low, condensation may continue. This is a good reason to avoid outdoor installations. It is also important not to extend the vent above the roof more than the rules require.
2. Extremely long vents or long laterals in unheated spaces can allow the flue products to cool to the dew point temperature. Do not wrap insulation around type B vents in an effort to prevent condensation. This method is not reliable and may contribute to other problems.
3. The air supply, discussed earlier, is of great importance for the proper operation of a vent system. Again, if sufficient make-up air is not available to replace that required by the burner and the draft hood, the system is starved. The first result is that less air enters the draft hood, allowing the dew point temperature to rise inside the vent pipe. In other words, condensation can occur at a higher temperature. At some point in the dilution process, condensation will start appearing in the vent. Further starving for air can result in water running out of the vent and damage to the structure and its contents.

When in doubt for any reason, such as questionable dimensions, a borderline chart selection, or undetermined head clearance for maximum connector rise, always use the next larger size vent system. This does not apply to table minimums.

Vent Caps

Listed vent caps for use on double-wall type B vent systems are designed to serve two purposes: (1) prevent rain and debris from entering the vent, and (2) help prevent a downdraft condition in the vent due to adverse wind conditions. The vent sizing tables apply to vents, vent caps, or roof housing of the same make and style as the vent pipe. For safe, efficient operation, do not use combination roof jacks or caps or termination designs fabricated by anyone other than the vent manufacturer.

Always install an approved vent cap immediately after installation of the vent to prevent debris and damage.

Wall Furnace Vents

Wall furnaces (vented recess heaters) require a 12-ft minimum vent height measured from the floor to the top of the vent or, in the case of combined vents, to the top of the vent connector. Many vented wall furnaces require connection to type BW vents.

Flashing and top assembly using an RM cap. Certified cap sizes 3 to 12 inches round and 4 inches oval are listed by UL for installation on gas vents terminating a sufficient distance from the roof, so that no discharge opening is less than 2 ft horizontally from the roof surface. The lowest discharge opening shall be no closer than the minimum height shown in Figure 6–7. These minimum heights may be used provided that the vent is not less than 8 ft from any vertical wall. This alone means that no installation shall terminate by piercing a wall with a short piece of pipe and a cap.

Metal cap terminations. Cap sizes 14 to 24 inches in diameter are for gas vents which extend at least 2 ft above the highest point where they pass through a building and at least 2 ft higher than any portion of the building within 10 ft of the

Termination Dimensions for Type RM Caps

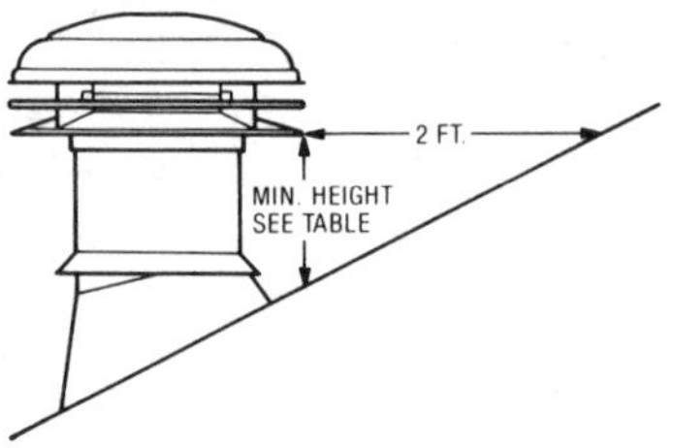
2 FT.
MIN. HEIGHT
SEE TABLE

Roof Pitch	Minimum Height from Roof to Lowest Discharge Opening Foot
Flat to 7/12	1.0
Over 7/12 to 8/12	1.5
Over 8/12 to 9/12	2.0
Over 9/12 to 10/12	2.5
Over 10/12 to 11/12	3.25
Over 11/12 to 12/12	4.0
Over 12/12 to 14/12	5.0
Over 14/12 to 16/12	6.0
Over 16/12 to 18/12	7.0
Over 18/12 to 20/12	7.5
Over 20/12 to 21/12	8.0

Model RM caps are listed under the "Draft Loss and Wind Effect" requirements of U/L Standard 441.

FIGURE 6–7
Termination dimensions for type RM caps. (*Courtesy of Hart & Cooley, Inc.*)

vent pipe. If any adjacent structures are within 10 ft of the vent and are higher, then the vent must terminate at least 2 ft above these structures. This recommendation should be followed unless local code requirements state otherwise.

Offsets in the attic space should be used to minimize the amount of vent pipe that must be exposed above the roof. No gas vent should be terminated less than 5 ft in vertical height above the highest connected appliance draft hood outlet.

General termination considerations. A cap or chimney housing offers protection against the entrance of rain, snow, and debris as well as birds. It also will minimize the effect of wind on the vent. It will protect the vent from downdrafts due to wind impinging directly on the vent. However, no vent cap, cowl, or top can overcome the adverse effect of a region of high static pressure around the vent terminal or the effect of an interior region of low pressure. Regions of high static pressure around the vent terminal can be avoided by following the aforementioned general rule for the vent termination. Low or negative interior pressures in the building may be caused by (1) failure to provide for combustion air, (2) excessive use of exhaust fans, and (3) tight building construction resulting in the lack of infiltration air. Vented clothes dryers and fireplaces will also remove large amounts of air from the interior of the building, tending to produce a low interior pressure.

This also means that no type B vent installation should terminate by piercing a wall with a short piece of vertical or horizontal pipe and a cap.

TYPES OF SYSTEMS AND CAPACITY TABLES

The following are definitions of the various types of venting systems:

1. Single appliance vent: A single appliance vent is an independent vent for one appliance. See Figure 6–8.

2. Total height (*H*): The total height (*H*) is the vertical distance measured between the appliance collar connection and the vent termination. See Figure 6–8.

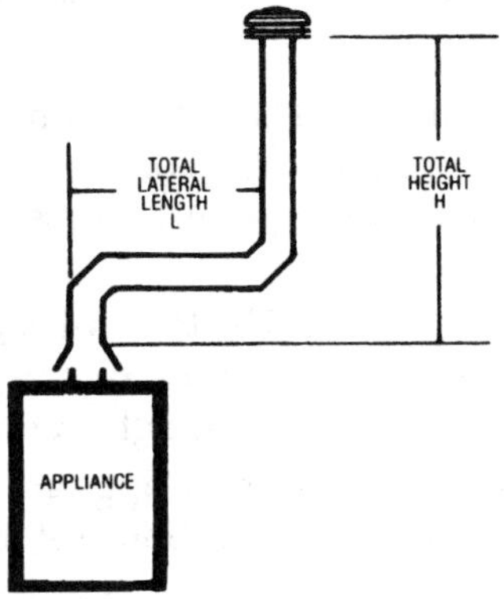
TOTAL
LATERAL
LENGTH
L
TOTAL
HEIGHT
H
APPLIANCE

FIGURE 6–8
Single appliance unit. (*Courtesy of Hart & Cooley, Inc.*)

3. Total lateral length (L): The total lateral length (L) is the horizontal distance or length of offset between the appliance collar and the main vertical portion of the vent. See Figure 6–8.
4. Multiple (combined) appliance vent: The multiple appliance vent is a venting system combining the connectors of two or more furnaces at one floor level to a common vertical vent. The connector in a combined vent system connects an individual furnace flue collar to the common vent or manifold.
5. Minimum total height: The minimum total height is the vertical distance measured from the highest appliance flue collar outlet in the system to the termination of the vent. See Figure 6–9. This minimum height is a fixed dimension for any one vent system regardless of the number or placement of appliances in the system.
6. Connector rise: The connector rise for any appliance in a vent system is the vertical distance from the flue collar outlet to the point where the next connector joins the system. See Figure 6–9.

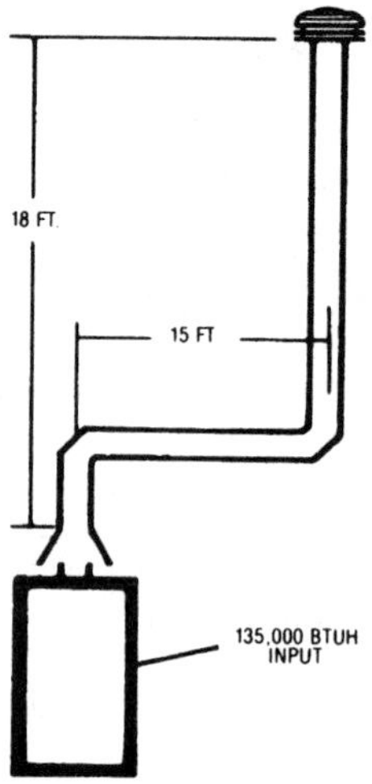
18 FT
15 FT
135,000 BTUH
INPUT

FIGURE 6–9
Total vent height. (*Courtesy of Hart & Cooley, Inc.*)

7. Common vent: The common vent is that portion of the venting system above the lowest interconnection. When the common vent is entirely vertical, the system is called a vertical or V type. All others are called lateral or L type. See Figure 6–10.
8. Fan-assisted combustion system: The fan-assisted combustion system is an appliance equipped with a fan to draw or force products of combustion through the combustion chamber and/or heat exchanger.
9. Fan min: Fan min refers to the minimum appliance input rating of a category I appliance with a fan-assisted combustion system that could be attached to the vent.

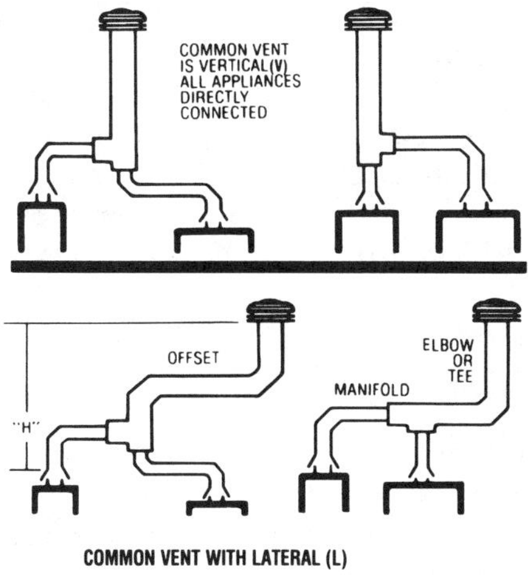
COMMON VENT
IS VERTICAL(V)
ALL APPLIANCES
DIRECTLY
CONNECTED
OFFSET
ELBOW
OR
TEE
MANIFOLD
"H"
COMMON VENT WITH LATERAL (L)

FIGURE 6–10
Common vent. (*Courtesy of Hart & Cooley, Inc.*)

10. Fan max: Fan max refers to the maximum appliance input rating of a category I appliance with a fan-assisted combustion system that could be attached to the vent.
11. Nat max: Nat max refers to the maximum appliance input rating of a category I appliance equipped with a draft hood that could be attached to the vent. There are no minimum appliance input ratings for draft-hood-equipped appliances.
12. Fan + fan: Fan + fan refers to the maximum combined input rating of two or more fan-assisted appliances attached to a common vent.
13. Fan + nat: Fan + nat refers to the maximum combined input rating of two or more fan-assisted appliances and one or more draft-hood-equipped appliances attached to the common vent.
14. Nat + nat: Nat + nat refers to the maximum combined input rating of one or more draft-hood-equipped appliances attached to the common vent.
15. NR: NR means not recommended due to potential for condensate formation and/or pressurization of the venting system.
16. NA: NA means not applicable due to physical or geometric constraints.
17. Draft hood (draft diverter): The draft hood is a device built into an appliance, or made a part of the vent connector from an appliance, which is designed to (1) provide for the ready escape of the flue gases from the appliance in the event of no draft, backdraft, or vent stoppage beyond the draft hood; (2) prevent a backdraft from entering the appliance; and (3) neutralize the effect of stack action of the chimney or gas vent upon the operation of the appliance.

    The correct installation of the heating unit and its draft diverter is vital for safe operation. All heating units are tested and listed by AGA with their draft diverter in position. Check the following list to be sure that the draft diverter is correctly installed.

a. Install the draft diverter free of obstructions below and to the side.
b. Install it exactly as specified by the manufacturer's instructions.
c. Obtain the correct draft diverter from the equipment manufacturer if it is missing. If assistance is needed, check with the fuel supplier.
d. Do not alter the pipe lengths between the draft diverter and the unit flue outlet.
e. Do not change the position of the draft diverter on the heating unit.
f. Do not change the design of the draft diverter by adjusting or cutting the baffles, skirt, etc.
g. Do not locate the draft diverter in a different room from the heating unit.
h. Do not substitute or exchange draft diverters.

18. Vent: The vent is a passageway used to convey the flue gases from the gas using equipment, or their vent connector, to the outside atmosphere.

19. Vent connector: The vent connector is the pipe or duct which connects a fuel-gas-burning appliance to a vent or chimney.

20. Flue collar: The flue collar is that portion of an appliance that is designed for the attachment of a draft hood, vent connectors, or the venting system.

Single-Appliance Vent Systems

The following are general rules for venting single appliances:

Normally, a vent equal to the size of the draft hood outlet can be considered satisfactory for venting a single appliance. It is important to note that this rule may not apply to cases where an extra high vent is required. It may be desirable to calculate the system to determine whether it is possible to reduce the size of the vent.

How to use the single-appliance vent table. Use the following procedure to determine the proper vent size for a single appliance vent. Refer to Tables 6–2 and 6–3.

1. Determine the total height (*H*) and the total lateral length (*L*) based on the location of the appliance, the vent, and the height to the vent termination.
2. Read down the total height (*H*) column at the left side of the table to a height equal to the total height.
3. Select the horizontal row for the appropriate length of lateral (*L*) of the vent pipe (zero for straight vertical vents).
4. Read across to the column showing a capacity equal to or greater than the appliance nameplate input rating.
5. If the vent size shown at the top of the column containing the correct capacity is equal to or greater than the appliance draft hood, use the vent size shown in the table.

6. If the vent shown is smaller than the draft hood size, see the section "Draft Hood to Vent Reduction" (p. 119).

Heights and laterals other than shown in the table. To use the vent capacity tables to estimate in-between operating conditions, take the fractional difference between the next smaller and the next larger figures in both the total height (*H*) and the total length (*L*) in Btu.

TABLE 6–2 Type B double wall gas vent capacities with Type B double wall connector. (*Courtesy of Hart & Cooley, Inc.*)

		Vent and Connector Diameter - D																	
		3"			4"			5"			6"			7"			8"		
		Appliance Input Rating in Thousands of BTU Per Hour																	
Height H (ft)	Lateral L (ft)	FAN Min	FAN Max	NAT Max	FAN Min	FAN Max	NAT Max	FAN Min	FAN Max	NAT Max	FAN Min	FAN Max	NAT Max	FAN Min	FAN Max	NAT Max	FAN Min	FAN Max	NAT Max
6	0	0	78	46	0	152	86	0	251	141	0	375	205	0	524	285	0	698	370
	2	13	51	36	18	97	67	27	157	105	32	232	157	44	321	217	53	425	285
	4	21	49	34	30	94	64	39	153	103	50	227	153	66	316	211	79	419	279
	6	25	46	32	36	91	61	47	149	100	59	223	149	78	310	205	93	413	273
8	0	0	84	50	0	165	94	0	276	155	0	415	235	0	583	320	0	780	415
	2	12	57	40	16	109	75	25	178	120	28	263	180	42	365	247	50	483	322
	5	23	53	38	32	103	71	42	171	115	53	255	173	70	356	237	83	473	313
	8	28	49	35	39	98	66	51	164	109	64	247	165	84	347	227	99	463	303
10	0	0	88	53	0	175	100	0	295	166	0	447	255	0	631	345	0	847	450
	2	12	61	42	17	118	81	23	194	129	26	289	195	40	402	273	48	533	355
	5	23	57	40	32	113	77	41	187	124	52	280	188	68	392	263	81	522	346
	10	30	51	36	41	104	70	54	176	115	67	267	175	88	376	245	104	504	330
15	0	0	94	58	0	191	112	0	327	187	0	502	285	0	716	390	0	970	525
	2	11	69	48	15	136	93	20	226	150	22	339	225	38	475	316	45	633	414
	5	22	65	45	30	130	87	39	219	142	49	330	217	64	463	300	76	620	403
	10	29	59	41	40	121	82	51	206	135	64	315	208	84	445	288	99	600	386
	15	35	53	37	48	112	76	61	195	128	76	301	198	98	429	275	115	580	373
20	0	0	97	61	0	202	119	0	349	202	0	540	307	0	776	430	0	1057	575
	2	10	75	51	14	149	100	18	250	166	20	377	249	33	531	346	41	711	470
	5	21	71	48	29	143	96	38	242	160	47	367	241	62	519	337	73	697	460
	10	28	64	44	38	133	89	50	229	150	62	351	228	81	499	321	95	675	443
	15	34	58	40	46	124	84	59	217	142	73	337	217	94	481	308	111	654	427
	20	48	52	35	55	116	78	69	206	134	84	322	206	107	464	295	125	634	410
30	0	0	100	64	0	213	128	0	374	220	0	587	336	0	853	475	0	1173	650
	2	9	81	56	13	166	112	14	283	185	18	432	280	27	613	394	33	826	535
	5	21	77	54	28	160	108	36	275	176	45	421	273	58	600	385	69	811	524
	10	27	70	50	37	150	102	48	262	171	59	405	261	77	580	371	91	788	507
	15	33	64	NR	44	141	96	57	249	163	70	389	249	90	560	357	105	765	490
	20	56	58	NR	53	132	90	66	237	154	80	374	237	102	542	343	119	743	473
	30	NR	NR	NR	73	113	NR	88	214	NR	104	346	219	131	507	321	149	702	444
50	0	0	101	67	0	216	134	0	397	232	0	633	363	0	932	518	0	1297	708
	2	8	86	61	11	183	122	14	320	206	15	497	314	22	715	445	26	975	615
	5	20	82	NR	27	177	119	35	312	200	43	487	308	55	702	438	65	960	605
	10	26	76	NR	35	168	114	45	299	190	56	471	298	73	681	426	86	935	589
	15	59	70	NR	42	158	NR	54	287	180	66	455	288	85	662	413	100	911	572
	20	NR	NR	NR	50	149	NR	63	275	169	76	440	278	97	642	401	113	888	556
	30	NR	NR	NR	69	131	NR	84	250	NR	99	410	259	123	605	376	141	844	522
100	0	NR	NR	NR	0	218	NR	0	407	NR	0	665	400	0	997	560	0	1411	770
	2	NR	NR	NR	10	194	NR	12	354	NR	13	566	375	18	831	510	21	1155	700
	5	NR	NR	NR	26	189	NR	33	347	NR	40	557	369	52	820	504	60	1141	692
	10	NR	NR	NR	33	182	NR	43	335	NR	53	542	361	68	801	493	80	1118	679
	15	NR	NR	NR	40	174	NR	50	321	NR	62	528	353	80	782	482	93	1095	666
	20	NR	NR	NR	47	166	NR	59	311	NR	71	513	344	90	763	471	105	1073	653
	30	NR	NR	NR	NR	NR	NR	78	290	NR	92	483	NR	115	726	449	131	1029	627
	50	NR	NR	NR	NR	NR	NR	NR	NR	NR	147	428	NR	180	651	405	197	944	575

TABLE 6–2 Continued

Vent and Connector Diameter - D

Appliance Input Rating in Thousands of BTU Per Hour

Height H (ft)	Lateral L (ft)	10" FAN Min	10" FAN Max	10" NAT Max	12" FAN Min	12" FAN Max	12" NAT Max	14" FAN Min	14" FAN Max	14" NAT Max	16" FAN Min	16" FAN Max	16" NAT Max
6	0	0	1121	570	0	1645	850	0	2267	1170	0	2983	1530
	2	75	675	455	103	982	650	138	1346	890	178	1769	1170
	4	110	668	445	147	975	640	191	1338	880	242	1761	1160
	6	128	661	435	171	967	630	219	1330	870	276	1753	1150
8	0	0	1261	660	0	1858	970	0	2571	1320	0	3399	1740
	2	71	770	515	98	1124	745	130	1543	1020	168	2030	1340
	5	115	758	503	154	1110	733	199	1528	1010	251	2013	1330
	8	137	746	490	180	1097	720	231	1514	1000	289	2000	1320
10	0	0	1377	720	0	2036	1060	0	2825	1450	0	3742	1925
	2	68	852	560	93	1244	850	124	1713	1130	161	2256	1480
	5	112	839	547	149	1229	829	192	1696	1105	243	2238	1461
	10	142	817	525	187	1204	795	238	1669	1080	298	2209	1430
15	0	0	1596	840	0	2380	1240	0	3323	1720	0	4423	2270
	2	63	1019	675	86	1495	985	114	2062	1350	147	2719	1770
	5	105	1003	660	140	1476	967	182	2041	1327	229	2696	1748
	10	135	977	635	177	1446	936	227	2009	1289	283	2659	1712
	15	155	953	610	202	1418	905	257	1976	1250	318	2623	1675
20	0	0	1756	930	0	2637	1350	0	3701	1900	0	4948	2520
	2	59	1150	755	81	1694	1100	107	2343	1520	139	3097	2000
	5	101	1133	738	135	1674	1079	174	2320	1498	219	3071	1978
	10	130	1105	710	172	1641	1045	220	2282	1460	273	3029	1940
	15	150	1078	688	195	1609	1018	248	2245	1425	306	2988	1910
	20	167	1052	665	217	1578	990	273	2210	1390	335	2948	1880
30	0	0	1977	1060	0	3004	1550	0	4252	2170	0	5725	2920
	2	54	1351	865	74	2004	1310	98	2786	1800	127	3696	2380
	5	96	1332	851	127	1981	1289	164	2759	1775	206	3666	2350
	10	125	1301	829	164	1944	1254	209	2716	1733	259	3617	2300
	15	143	1272	807	187	1908	1220	237	2674	1692	292	3570	2250
	20	160	1243	784	207	1873	1185	260	2633	1650	319	3523	2200
	30	195	1189	745	246	1807	1130	305	2555	1585	369	3433	2130
50	0	0	2231	1195	0	3441	1825	0	4934	2550	0	6711	3440
	2	41	1620	1010	66	2431	1513	86	3409	2125	113	4554	2840
	5	90	1600	996	118	2406	1495	151	3380	2102	191	4520	2813
	10	118	1567	972	154	2366	1466	196	3332	2064	243	4464	2767
	15	136	1536	948	177	2327	1437	222	3285	2026	274	4409	2721
	20	151	1505	924	195	2288	1408	244	3239	1987	300	4356	2675
	30	183	1446	876	232	2214	1349	287	3150	1910	347	4253	2631
100	0	0	2491	1310	0	3925	2050	0	5729	2950	0	7914	4050
	2	30	1975	1170	44	3027	1820	72	4313	2550	95	5834	3500
	5	82	1955	1159	107	3002	1803	136	4282	2531	172	5797	3475
	10	108	1923	1142	142	2961	1775	180	4231	2500	223	5737	3434
	15	126	1892	1124	163	2920	1747	206	4182	2469	252	5678	3392
	20	141	1861	1107	181	2880	1719	226	4133	2438	277	5619	3351
	30	170	1802	1071	215	2803	1663	265	4037	2375	319	5505	3267
	50	241	1688	1000	292	2657	1550	350	3856	2250	415	5289	3100

Height H (ft)	Lateral L (ft)	18" FAN Min	18" FAN Max	18" NAT Max	20" FAN Min	20" FAN Max	20" NAT Max	22" FAN Min	22" FAN Max	22" NAT Max	24" FAN Min	24" FAN Max	24" NAT Max
6	0	0	3802	1960	0	4721	2430	0	5737	2950	0	6853	3520
	2	225	2250	1480	296	2782	1850	360	3377	2220	426	4030	2670
	4	300	2242	1475	390	2774	1835	469	3370	2215	555	4023	2660
	6	341	2235	1470	437	2767	1820	523	3363	2210	618	4017	2650
8	0	0	4333	2220	0	5387	2750	0	6555	3360	0	7838	4010
	2	212	2584	1700	278	3196	2110	336	3882	2560	401	4634	3050
	5	311	2563	1685	398	3180	2090	476	3863	2545	562	4612	3040
	8	354	2552	1670	450	3163	2070	537	3850	2530	630	4602	3030
10	0	0	4782	2450	0	5955	3050	0	7254	3710	0	8682	4450
	2	202	2868	1890	264	3556	2340	319	4322	2840	378	5153	3390
	5	300	2849	1871	382	3536	2318	458	4301	2818	540	5132	3371
	10	364	2818	1840	459	3504	2280	546	4268	2780	641	5099	3340
15	0	0	5678	2900	0	7099	3620	0	8665	4410	0	10393	5300
	2	186	3467	2260	239	4304	2800	290	5232	3410	346	6251	4080
	5	283	3442	2235	355	4278	2777	426	5204	3385	501	6222	4057
	10	346	3402	2193	432	4234	2739	510	5159	3343	599	6175	4019
	15	385	3363	2150	479	4192	2700	564	5115	3300	665	6129	3980
20	0	0	6376	3250	0	7988	4060	0	9785	4980	0	11753	6000
	2	175	3955	2570	220	4916	3200	269	5983	3910	321	7154	4700
	5	270	3926	2544	337	4885	3174	403	5950	3880	475	7119	4662
	10	334	3880	2500	413	4835	3130	489	5896	3830	573	7063	4600
	15	372	3835	2465	459	4786	3090	541	5844	3795	631	7007	4575
	20	404	3791	2430	495	4737	3050	585	5792	3760	689	6953	4550
30	0	0	7420	3770	0	9341	4750	0	11483	5850	0	13848	7060
	2	159	4734	3050	199	5900	3810	241	7194	4650	285	8617	5600
	5	252	4701	3020	312	5863	3783	373	7155	4622	439	8574	5552
	10	316	4647	2970	386	5803	3739	456	7090	4574	535	8505	5471
	15	354	4594	2920	431	5744	3695	507	7026	4527	590	8437	5391
	20	384	4542	2870	467	5686	3650	548	6964	4480	639	8370	5310
	30	440	4442	2785	540	5574	3565	635	6842	4375	739	8239	5225
50	0	0	8774	4460	0	11129	5635	0	13767	6940	0	16694	8430
	2	141	5864	3670	171	7339	4630	209	8980	5695	251	10788	6860
	5	234	5826	3639	283	7295	4597	336	8933	5654	394	10737	6818
	10	295	5763	3585	355	7224	4542	419	8855	5585	491	10652	6749
	15	330	5701	3534	396	7155	4511	465	8779	5546	542	10570	6710
	20	361	5641	3481	433	7086	4479	506	8704	5506	586	10488	6670
	30	412	5523	3431	494	6953	4421	577	8557	5444	672	10328	6603
100	0	0	10485	5300	0	13454	6700	0	16817	8600	0	20578	10300
	2	120	7591	4600	138	9577	5800	169	11803	7200	204	14264	8800
	5	208	7548	4566	245	9528	5769	293	11748	7162	341	14204	8756
	10	268	7478	4509	318	9447	5717	374	11658	7100	436	14105	8683
	15	304	7409	4451	358	9367	5665	418	11569	7037	487	14007	8610
	20	330	7341	4394	387	9289	5613	452	11482	6975	523	13910	8537
	30	378	7209	4279	446	9136	5509	514	11310	6850	592	13720	8391
	50	486	6956	4050	572	8841	5300	659	10979	6600	752	13354	8100

Example:

A typical use of the tables for single-appliance venting is shown in Figure 6–10. The furnace has an input rating of 135,000 Btu per hour and is equipped with a 5-inch draft hood. The total height (*H*) of the vent is 18 ft with a 15-ft total lateral length (*L*).

Procedure: Go down the total height (*H*) column of Table 6–2 to 20-ft height with a 15-ft lateral under the Nat column, giving 142,000 Btu for a 5-inch vent. For a 15-ft total height with a 15-ft lateral, the maximum capacity for a 5-inch vent is 128,000 Btu.

TABLE 6–3 Type B double wall gas vent capacities with single wall connector. (*Courtesy of Hart and Cooley, Inc.*)

Vent and Connector Diameter - D; Appliance Input Rating in Thousands of BTU Per Hour

Height H (ft)	Lateral L (ft)	3" FAN Min	3" FAN Max	3" NAT Max	4" FAN Min	4" FAN Max	4" NAT Max	5" FAN Min	5" FAN Max	5" NAT Max	6" FAN Min	6" FAN Max	6" NAT Max	7" FAN Min	7" FAN Max	7" NAT Max	8" FAN Min	8" FAN Max	8" NAT Max	10" FAN Min	10" FAN Max	10" NAT Max	12" FAN Min	12" FAN Max	12" NAT Max
6	0	38	77	45	59	151	85	85	249	140	126	373	204	165	522	284	211	695	369	371	1118	569	537	1639	849
	2	39	51	36	60	96	66	85	156	104	123	231	156	159	320	213	201	423	284	347	673	453	498	979	648
	4	NR	NR	33	74	92	63	102	152	102	146	225	152	187	313	208	237	416	277	409	664	443	584	971	638
	6	NR	NR	31	83	89	60	114	147	99	163	220	148	207	307	203	263	409	271	449	656	433	638	962	627
8	0	37	83	50	58	164	93	83	273	154	123	412	234	161	580	319	206	777	414	360	1257	658	521	1852	967
	2	39	56	39	59	108	75	83	176	119	121	261	179	155	363	246	197	482	321	339	768	513	486	1220	743
	5	NR	NR	37	77	102	69	107	168	114	151	252	171	193	352	235	245	470	311	418	754	500	598	1104	730
	8	NR	NR	33	90	95	64	122	161	107	175	243	163	223	342	225	280	458	300	470	740	486	665	1089	715
10	0	37	87	53	57	174	99	82	293	165	120	444	254	158	628	344	202	844	449	351	1373	718	507	2031	1057
	2	39	61	41	59	117	80	82	193	128	119	287	194	153	400	272	193	531	354	332	849	559	475	1242	848
	5	52	56	39	76	111	76	105	185	122	148	277	186	190	388	261	241	518	344	409	834	544	584	1224	825
	10	NR	NR	34	97	100	68	132	171	112	188	261	171	237	369	241	296	497	325	492	808	520	688	1194	788
15	0	36	93	57	56	190	111	80	325	186	116	499	283	153	713	388	195	966	523	336	1591	838	488	2374	1237
	2	38	69	47	57	136	93	80	225	149	115	337	224	148	473	314	187	631	413	319	1015	673	457	1491	983
	5	51	63	44	75	128	86	102	216	140	144	326	217	182	459	298	231	616	400	392	997	657	562	1469	963
	10	NR	NR	39	95	116	79	128	201	131	182	308	203	228	438	284	284	582	381	470	966	628	664	1433	928
	15	NR	NR	NR	NR	NR	72	158	186	124	220	290	192	272	418	269	334	568	367	540	937	601	750	1399	894
20	0	35	96	60	54	200	118	78	346	201	114	537	306	149	772	428	190	1053	573	326	1751	927	473	2631	1346
	2	37	74	50	56	148	99	78	248	165	113	375	248	144	528	344	182	708	468	309	1146	754	443	1689	1098
	5	50	68	47	73	140	94	100	239	158	141	363	239	178	514	334	224	692	457	381	1126	734	547	1665	1074
	10	NR	NR	41	93	129	86	125	223	146	177	344	224	222	491	316	277	666	437	457	1092	702	646	1626	1037
	15	NR	NR	NR	NR	NR	80	155	208	136	216	325	210	264	469	301	325	640	419	526	1060	677	730	1587	1005
	20	NR	NR	NR	NR	NR	NR	186	192	126	254	306	196	309	448	285	374	616	400	592	1028	651	808	1550	973
30	0	34	99	63	53	211	127	76	372	219	110	584	334	144	849	472	184	1168	647	312	1971	1056	454	2996	1545
	2	37	80	56	55	164	111	76	281	183	109	429	279	139	610	392	175	823	533	296	1346	863	424	1999	1308
	5	49	74	52	72	157	106	98	271	173	136	417	271	171	595	382	215	806	521	366	1324	846	524	1971	1283
	10	NR	NR	NR	91	144	98	122	255	168	171	397	257	213	570	367	265	777	501	440	1287	821	620	1927	1243
	15	NR	NR	NR	115	131	NR	151	239	157	208	377	242	255	547	349	312	750	481	507	1251	794	702	1884	1205
	20	NR	NR	NR	NR	NR	NR	181	223	NR	246	357	228	298	524	333	360	723	461	570	1216	768	780	1841	1166
	30	NR	NR	NR	NR	NR	NR	NR	NR	NR	NR	NR	NR	389	477	305	461	670	426	704	1147	720	937	1759	1101
50	0	33	99	66	51	213	133	73	394	230	105	629	361	138	928	515	176	1292	704	295	2223	1189	428	3432	1818
	2	36	84	61	53	181	121	73	318	205	104	495	312	133	712	443	168	971	613	280	1615	1007	401	2426	1509
	5	48	80	NR	70	174	117	94	308	198	131	482	305	164	696	435	204	953	602	347	1591	991	496	2396	1490
	10	NR	NR	NR	89	160	NR	118	292	186	162	461	292	203	671	420	253	923	583	418	1551	963	589	2347	1455
	15	NR	NR	NR	112	148	NR	145	275	174	199	441	280	244	646	405	299	894	562	481	1512	934	668	2299	1421
	20	NR	NR	NR	NR	NR	NR	176	257	NR	236	420	267	285	622	389	345	866	543	544	1473	906	741	2251	1387
	30	NR	NR	NR	NR	NR	NR	NR	NR	NR	315	376	NR	373	573	NR	442	809	502	674	1399	848	892	2159	1318
100	0	NR	NR	NR	49	214	NR	69	403	NR	100	659	395	131	991	555	166	1404	765	273	2479	1300	395	3912	2042
	2	NR	NR	NR	51	192	NR	70	351	NR	98	563	373	125	828	508	158	1152	698	259	1970	1168	371	3021	1817
	5	NR	NR	NR	67	186	NR	90	342	NR	125	551	366	156	813	501	194	1134	688	322	1945	1153	460	2990	1796
	10	NR	NR	NR	85	175	NR	113	324	NR	153	532	354	191	789	486	238	1104	672	389	1905	1133	547	2938	1763
	15	NR	NR	NR	132	162	NR	138	310	NR	188	511	343	230	764	473	281	1075	656	447	1865	1110	618	2888	1730
	20	NR	NR	NR	NR	NR	NR	168	295	NR	224	487	NR	270	739	458	325	1046	639	507	1825	1087	690	2838	1696
	30	NR	NR	NR	NR	NR	NR	231	264	NR	301	448	NR	355	685	NR	418	988	NR	631	1747	1041	834	2739	1627
	50	NR	NR	NR	NR	NR	NR	NR	NR	NR	NR	NR	NR	540	584	NR	617	866	NR	895	1591	NR	1138	2547	1489

To obtain the correct capacity for an 18-ft total height (*H*), take three-fifths of the difference between 128,000 and 142,000 ($3/5 \times 14{,}000 = 8{,}400$) and add it to the 15-ft figure of 128,000 ($8{,}400 + 128{,}000 = 136{,}000$), indicating that even with a 15-ft lateral the 5-inch vent would have sufficient capacity for a 135,000-Btu furnace if the vent has an 18-ft total height (*H*).

Draft Hood to Vent Reduction: If the vent size determined from the appropriate tables is less than the size of the draft hood outlet or flue collar, the smaller vent may be used provided that the following conditions are met:

1. The vent is at least 10 ft in height. When the vent is less than 10 ft in height, it should be at least as large as the flue collar outlet.
2. Vents for draft hoods measuring 12 inches in diameter or less should not be reduced more than one pipe size. A 6- to 5-inch or a 12- to 10-inch reduction is one pipe size reduction. For larger gas-burning equipment, such a boilers, with draft hood sizes from 14 to 24 inches in diameter, reductions of more than two pipe sizes are not recommended (24 to 20 inches is a two-pipe-size reduction).
3. The maximum capacity listed in the tables for a fan-assisted appliance is reduced by 10%.
4. Regardless of the size of vent shown by the tables for such appliances, do not connect any 3-inch vents to 4-inch draft hoods. This provision does not apply to fan-assisted combustion appliances.

Multiple-Appliance Vent Systems

The following is a method used to determine each vent connector size for these types of installations. Refer to Tables 6–4 and 6–5.

1. Determine the minimum total vent height for the system from a sketch of the proposed system.
2. Determine the connector rise for each appliance.
3. Enter the reference from the vent connector Table 6–4 or 6–5 at the line showing the minimum total height equal to or less than that determined in step 2 above. Continue horizontally on that line for the first appliance connector rise using the appliance name plate Btu rating at sea level. Always use the next higher figure in the table. Read the connector vent size for that appliance at the top of the column.
4. Using the same minimum total height, repeat the procedure for each appliance using its connector rise and Btu/h rating. Never use a connector size smaller than the draft hood outlet size. (This does not apply to appliances approved for use with special venting systems.)

How to Determine Common Vent Size

Use the following procedure to determine common vent sizes. Refer to Tables 6–6 and 6–7.

1. Total all appliance Btu/h output ratings which are to be connected to this common vent.
2. Enter the reference from the common vent Table 6–6 or 6–7 at the same minimum total height used to determine the vent connector sizes in step 3 above.
3. Move horizontally across from this minimum total height figure using either the *L* or the *V* line as determined from the preceding definitions.
4. Select the first value which is equal to or greater than the total Btu/h ratings.
5. The size of the required common vent is found at the top of this column.

TABLE 6–4 Type B vent connector capacities. (*Courtesy of Hart & Cooley, Inc.*)

Vent Connector Diameter - D

Appliance Input Rating in Thousands of BTU Per Hour

Vent Height H (ft)	Connector Rise R (ft)	3"			4"			5"			6"			7"			8"			10"		
		FAN Min	FAN Max	NAT Max	FAN Min	FAN Max	NAT Max	FAN Min	FAN Max	NAT Max	FAN Min	FAN Max	NAT Max	FAN Min	FAN Max	NAT Max	FAN Min	FAN Max	NAT Max	FAN Min	FAN Max	NAT Max
6	1	22	37	26	35	66	46	46	106	72	58	164	104	77	225	142	82	296	185	128	466	289
	2	23	41	31	37	75	55	48	121	86	60	183	124	79	253	168	95	333	220	131	526	345
	3	24	44	35	38	81	62	49	132	96	62	199	139	82	275	189	97	363	248	134	575	386
8	1	22	40	27	35	72	48	49	114	76	64	176	109	84	243	148	100	320	194	138	507	303
	2	23	44	32	36	80	57	51	128	90	66	195	129	86	269	175	103	356	230	141	564	358
	3	24	47	36	37	87	64	53	139	101	67	210	145	88	290	198	105	384	258	143	612	402
10	1	22	43	28	34	78	50	49	123	78	65	189	113	89	257	154	106	341	200	146	542	314
	2	23	47	33	36	86	59	51	136	93	67	206	134	91	282	182	109	374	238	149	596	372
	3	24	50	37	37	92	67	52	146	104	69	220	150	94	303	205	111	402	268	152	642	417
15	1	21	50	30	33	89	53	47	142	83	64	220	120	88	298	163	110	389	214	162	609	333
	2	22	53	35	35	96	63	49	153	99	66	235	142	91	320	193	112	419	253	165	658	394
	3	24	55	40	36	102	71	51	163	111	68	248	160	93	339	218	115	445	286	167	700	444
20	1	21	54	31	33	99	56	46	157	87	62	246	125	86	334	171	107	436	224	158	681	347
	2	22	57	37	34	105	66	48	167	104	64	259	149	89	354	202	110	463	265	161	725	414
	3	23	60	42	35	110	74	50	176	116	66	271	168	91	371	228	113	486	300	164	764	466
30	1	20	62	33	31	113	59	45	181	93	60	288	134	83	391	182	103	512	238	151	802	372
	2	21	64	39	33	118	70	47	190	110	62	299	158	85	408	215	105	535	282	155	840	439
	3	22	66	44	34	123	79	48	198	124	64	309	178	88	423	242	108	555	317	158	874	494
50	1	19	71	36	30	133	64	43	216	101	57	349	145	78	477	197	97	627	257	144	984	403
	2	21	73	43	32	137	76	45	223	119	59	358	172	81	490	234	100	645	306	148	1014	478
	3	22	75	48	33	141	86	46	229	134	61	366	194	83	502	263	103	661	343	151	1043	538
60-100	1	18	82	37	28	158	66	40	262	104	53	442	150	73	611	204	91	810	266	135	1285	417
	2	19	83	44	30	161	79	42	267	123	55	447	178	75	619	242	94	822	316	139	1306	494
	3	20	84	50	31	163	89	44	272	138	57	452	200	78	627	272	97	834	355	142	1327	555

Vent Connector Diameter - D

Appliance Input Rating in Thousands of BTU Per Hour

Vent Height H (ft)	Connector Rise R (ft)	12"			14"			16"			18"			20"			22"			24"		
		FAN Min	FAN Max	NAT Max	FAN Min	FAN Max	NAT Max	FAN Min	FAN Max	NAT Max	FAN Min	FAN Max	NAT Max	FAN Min	FAN Max	NAT Max	FAN Min	FAN Max	NAT Max	FAN Min	FAN Max	NAT Max
6	2	174	764	496	223	1046	653	281	1371	853	346	1772	1080	NA	NA	NA	NA	NA	NA	NA	NA	NA
	4	180	897	616	230	1231	827	287	1617	1081	352	2069	1370	NA	NA	NA	NA	NA	NA	NA	NA	NA
	6	NA	NA	NA	NA	NA	NA	NA	NA	NA	NA	NA	NA	NA	NA	NA	NA	NA	NA	NA	NA	NA
8	2	186	822	516	238	1126	696	298	1478	910	365	1920	1150	NA	NA	NA	NA	NA	NA	NA	NA	NA
	4	192	952	644	244	1307	884	305	1719	1150	372	2211	1460	471	2737	1800	560	3319	2180	662	3957	2590
	6	198	1050	772	252	1445	1072	313	1902	1390	380	2434	1770	478	3018	2180	568	3665	2640	669	4373	3130
10	2	196	870	536	249	1195	730	311	1570	955	379	2049	1205	NA	NA	NA	NA	NA	NA	NA	NA	NA
	4	201	997	664	256	1371	924	318	1804	1205	387	2332	1535	486	2887	1890	581	3502	2280	686	4175	2710
	6	207	1095	792	263	1509	1118	325	1989	1455	395	2556	1865	494	3169	2290	589	3849	2760	694	4593	3270
15	2	214	967	568	272	1334	790	336	1760	1030	408	2317	1305	NA	NA	NA	NA	NA	NA	NA	NA	NA
	4	221	1085	712	279	1499	1006	344	1978	1320	416	2579	1665	523	3197	2060	624	3881	2490	734	4631	2960
	6	228	1181	856	286	1632	1222	351	2157	1610	424	2796	2025	533	3470	2510	634	4216	3030	743	5035	3600
20	2	223	1051	596	291	1443	840	357	1911	1095	430	2533	1385	NA	NA	NA	NA	NA	NA	NA	NA	NA
	4	230	1162	748	298	1597	1064	365	2116	1395	438	2778	1765	554	3447	2180	661	4190	2630	772	5005	3130
	6	237	1253	900	307	1726	1288	373	2287	1695	450	2984	2145	567	3708	2650	671	4511	3190	785	5392	3790
30	2	216	1217	632	286	1664	910	367	2183	1190	461	2891	1540	NA	NA	NA	NA	NA	NA	NA	NA	NA
	4	223	1316	792	294	1802	1160	376	2366	1510	474	3110	1920	619	3840	2365	728	4681	2860	847	5606	3410
	6	231	1400	952	303	1920	1410	384	2524	1830	485	3299	2340	632	4080	2875	741	4976	3480	860	5961	4150
50	2	206	1479	689	273	2023	1007	350	2659	1315	435	3548	1665	NA	NA	NA	NA	NA	NA	NA	NA	NA
	4	213	1561	860	281	2139	1291	359	2814	1685	447	3730	2135	580	4601	2633	709	5569	3185	851	6633	3790
	6	221	1631	1031	290	2242	1575	369	2951	2055	461	3893	2605	594	4808	3208	724	5826	3885	867	6943	4620
60-100	2	192	1923	712	254	2644	1050	326	3490	1370	402	4707	1740	NA	NA	NA	NA	NA	NA	NA	NA	NA
	4	200	1984	888	263	2731	1346	336	3606	1760	414	4842	2220	523	5982	2750	639	7254	3330	769	8650	3950
	6	208	2035	1064	272	2811	1642	346	3714	2150	426	4968	2700	539	6143	3350	654	7453	4070	786	8892	4810

TABLE 6–5 Single wall vent connector capacities for multiple category I appliances connected to a common vent. (*Courtesy of Hart & Cooley, Inc.*)

Vent Height H (ft)	Connector Rise R (ft)	3" FAN Min	3" FAN Max	3" NAT Max	4" FAN Min	4" FAN Max	4" NAT Max	5" FAN Min	5" FAN Max	5" NAT Max	6" FAN Min	6" FAN Max	6" NAT Max	7" FAN Min	7" FAN Max	7" NAT Max	8" FAN Min	8" FAN Max	8" NAT Max	10" FAN Min	10" FAN Max	10" NAT Max
		Vent Connector Diameter - D — Appliance Input Rating Limits in Thousands of BTU Per Hour																				
6	1	NR	NR	26	NR	NR	46	NR	NR	71	NR	NR	102	207	223	140	262	293	183	447	463	286
	2	NR	NR	31	NR	NR	55	NR	NR	85	168	182	123	215	251	167	271	331	219	458	524	344
	3	NR	NR	34	NR	NR	62	121	131	95	174	198	138	222	273	188	279	361	247	468	574	385
15	1	NR	NR	29	79	87	52	116	138	81	177	214	116	238	291	158	312	380	208	556	596	324
	2	NR	NR	34	83	94	62	121	150	97	185	230	138	246	314	189	321	411	248	568	646	387
	3	NR	NR	39	87	100	70	127	160	109	193	243	157	255	333	215	331	438	281	579	690	437
30	1	47	60	31	77	110	57	113	175	89	169	278	129	226	380	175	296	497	230	528	779	358
	2	50	62	37	81	115	67	117	185	106	177	290	152	236	397	208	307	521	274	541	819	425
	3	54	64	42	85	119	76	122	193	120	185	300	172	244	412	235	316	542	309	555	855	482
50	1	46	69	33	75	128	60	109	207	96	162	336	137	217	406	188	284	604	245	507	951	384
	2	49	71	40	79	132	72	114	215	113	170	345	164	226	473	223	294	623	293	520	983	458
	3	53	72	45	83	136	82	119	221	128	178	353	186	235	486	252	304	640	331	535	1013	518

Regardless of the common vent size determined by the foregoing procedure, the vent must be at least as large as the largest connector. If more than one connector is this same size, then use a common vent one size larger.

Example:

Connect a 45,000-Btu/h water heater with a draft hood and 1-ft connector rise with a 100,000-Btu/h fan-equipped furnace with a 2-ft connector rise to a common vent with a minimum total vent height of 18 ft. Refer to Figure 6–11.

TABLE 6–6 Type B vent capacities when using type B connectors. (*Courtesy of Hart & Cooley, Inc.*)

Vent Height H (ft)	Vent Type	4" FAN +FAN	4" FAN +NAT	4" NAT +NAT	5" FAN +FAN	5" FAN +NAT	5" NAT +NAT	6" FAN +FAN	6" FAN +NAT	6" NAT +NAT	7" FAN +FAN	7" FAN +NAT	7" NAT +NAT	8" FAN +FAN	8" FAN +NAT	8" NAT +NAT	10" FAN +FAN	10" FAN +NAT	10" NAT +NAT
		Common Vent Diameter - D — Combined Appliance Input Rating in Thousands of BTU Per Hour																	
6	L	74	65	52	112	93	82	163	129	117	247	198	160	323	251	210	538	416	328
	V	92	81	65	140	116	103	204	161	147	309	248	200	404	314	260	672	520	410
8	L	81	72	58	124	103	91	179	142	130	271	220	178	355	278	230	592	462	372
	V	101	90	73	155	129	114	224	178	163	339	275	223	444	348	290	740	577	465
10	L	88	78	63	135	113	98	194	155	142	294	239	193	382	302	250	640	502	396
	V	110	97	79	169	141	124	243	194	178	367	299	242	477	377	315	800	627	495
15	L	100	90	73	156	131	114	226	182	164	342	282	224	445	355	290	739	586	452
	V	125	112	91	195	164	144	283	228	206	427	352	280	556	444	365	924	733	565
20	L	109	98	81	172	146	127	251	204	182	380	315	250	497	399	325	828	661	512
	V	136	123	102	215	183	160	314	255	229	475	394	310	621	499	405	1035	826	640
30	L	122	110	94	195	168	147	289	238	211	438	367	290	576	468	375	967	780	592
	V	152	138	118	244	210	185	361	297	266	547	459	360	720	585	470	1209	975	740
50	L	134	122	107	223	195	171	337	282	248	513	438	338	683	565	440	1161	950	688
	V	167	153	134	279	244	214	421	353	310	641	547	423	854	706	550	1451	1188	860
100	L	140	130	NR	249	222	NR	391	337	NR	601	526	383	820	698	500	1427	1202	780
	V	175	163	NR	311	277	NR	489	421	NR	751	658	479	1025	873	625	1784	1502	975

TABLE 6–6 Continued

Common Vent Diameter - D

Combined Appliance Input Rating in Thousands of BTU Per Hour

Vent Height H (ft)	Vent Type	12" FAN +FAN	12" FAN +NAT	12" NAT +NAT	14" FAN +FAN	14" FAN +NAT	14" NAT +NAT	16" FAN +FAN	16" FAN +NAT	16" NAT +NAT	18" FAN +FAN	18" FAN +NAT	18" NAT +NAT	20" FAN +FAN	20" FAN +NAT	20" NAT +NAT	22" FAN +FAN	22" FAN +NAT	22" NAT +NAT	24" FAN +FAN	24" FAN +NAT	24" NAT +NAT
6	L	720	557	470	1027	792	652	1388	1069	825	1802	1386	1076	2270	1744	1328	2790	2142	1576	3365	2581	1912
	V	900	696	588	1284	990	815	1735	1336	1065	2253	1732	1345	2838	2180	1660	3488	2677	1970	4206	3226	2390
8	L	795	618	522	1138	882	730	1542	1193	952	2006	1549	1208	2530	1951	1488	3112	2398	1760	3756	2893	2144
	V	994	773	652	1423	1103	912	1927	1491	1190	2507	1936	1510	3162	2439	1860	3890	2998	2200	4695	3616	2680
10	L	861	673	570	1234	960	796	1674	1300	1040	2182	1690	1316	2755	2132	1624	3393	2622	1920	4098	3166	2336
	V	1076	841	712	1542	1200	995	2093	1625	1300	2727	2113	1645	3444	2665	2030	4241	3278	2400	5123	3957	2920
15	L	998	789	660	1435	1128	926	1952	1528	1208	2547	1987	1525	3221	2506	1888	3977	3090	2232	4813	3736	2720
	V	1247	986	825	1794	1410	1158	2440	1910	1510	3184	2484	1910	4026	3133	2360	4971	3862	2790	6016	4670	3400
20	L	1124	893	733	1605	1270	1032	2178	1718	1352	2849	2238	1712	3638	2842	2112	4458	3482	2496	5399	4209	3040
	V	1405	1116	916	2006	1588	1290	2722	2147	1690	3561	2798	2140	4548	3552	2640	5573	4352	3120	6749	5261	3800
30	L	1326	1062	820	1898	1514	1220	2576	2046	1592	3358	2661	2016	4242	3354	2488	5231	4126	2944	6352	4998	3584
	V	1658	1327	1025	2373	1892	1525	3220	2558	1990	4197	3326	2520	5303	4193	3110	6539	5157	3680	7940	6247	4480
50	L	1619	1312	1024	2329	1878	1490	2371	2546	1944	4147	3319	2460	5254	4192	3040	6493	5166	3600	7870	6250	4380
	V	2024	1640	1280	2911	2347	1863	3964	3183	2430	5184	4149	3075	6567	5240	3800	8116	6458	4500	9837	7813	5475
100	L	2055	1705	1336	2986	2461	1960	4100	3362	2560	5399	4407	3240	6878	5589	4000	8545	6918	4736	10403	8399	5760
	V	2569	2131	1670	3732	3076	2450	5125	4202	3200	6749	5509	4050	8597	6986	5000	10681	8648	5920	13004	10499	7200

Water Heater Vent Connector Size: Use the vent connector Table 6–4 under Nat. Read down the minimum total vent height column to 15 ft and read across the 1-ft connector rise line to a Btu/h rating equal to or larger than the water heater input rating. This figure shows 53,000 Btu and is in the column for a 4-inch connector. Since this size is in excess of the water heater input, it is not necessary to find the maximum input for an 18-ft minimum total vent height. Use a 4-inch connector. See Figure 6–11(a).

TABLE 6–7 Type B common vent capacities when using single wall connectors. (*Courtesy of Hart & Cooley, Inc.*)

Common Vent Diameter - D

Combined Appliance Input Rating in Thousands of BTU Per Hour

Vent Height H (ft)	Vent Type	4" FAN +FAN	4" FAN +NAT	4" NAT +NAT	5" FAN +FAN	5" FAN +NAT	5" NAT +NAT	6" FAN +FAN	6" FAN +NAT	6" NAT +NAT	7" FAN +FAN	7" FAN +NAT	7" NAT +NAT	8" FAN +FAN	8" FAN +NAT	8" NAT +NAT	10" FAN +FAN	10" FAN +NAT	10" NAT +NAT
6	L	71	62	51	109	90	80	160	126	115	243	195	157	318	248	206	532	412	326
	V	89	78	64	136	113	100	200	158	144	304	244	196	398	310	257	665	515	407
8	L	78	70	57	121	101	90	174	138	127	265	215	174	349	274	228	584	455	368
	V	98	87	71	151	126	112	218	173	159	331	269	218	436	342	285	730	569	460
10	L	85	75	61	130	110	96	190	151	139	286	234	189	374	295	247	630	494	390
	V	106	94	76	163	137	120	237	189	174	357	292	236	467	369	309	787	617	487
15	L	97	86	70	151	127	112	220	177	160	333	274	219	435	347	286	724	574	442
	V	121	108	88	189	159	140	275	221	200	416	343	274	544	434	357	905	718	553
20	L	105	94	78	166	142	124	244	198	178	370	306	242	485	390	316	810	646	501
	V	131	118	98	208	177	155	305	247	223	436	383	302	606	487	395	1013	808	626
30	L	116	106	90	189	162	143	280	229	206	426	357	279	562	456	367	946	762	578
	V	145	132	113	236	202	179	350	286	257	533	446	349	703	570	459	1183	952	723
50	L	127	116	102	214	186	163	325	270	237	498	423	328	666	549	428	1134	926	670
	V	159	145	128	268	233	204	406	337	296	622	529	410	833	686	535	1418	1157	838

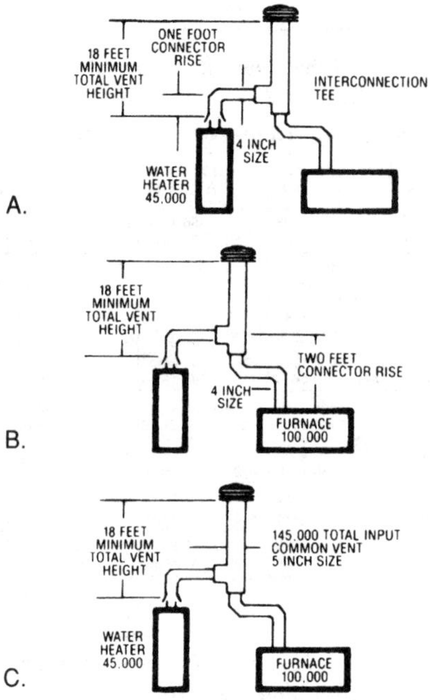
A.
ONE FOOT
CONNECTOR
RISE
18 FEET
MINIMUM
TOTAL VENT
HEIGHT
INTERCONNECTION
TEE
4 INCH
SIZE
WATER
HEATER
45,000
B.
18 FEET
MINIMUM
TOTAL VENT
HEIGHT
TWO FEET
CONNECTOR RISE
4 INCH
SIZE
FURNACE
100,000
C.
18 FEET
MINIMUM
TOTAL VENT
HEIGHT
145,000 TOTAL INPUT
COMMON VENT
5 INCH SIZE
WATER
HEATER
45,000
FURNACE
100,000

FIGURE 6–11
Determining common vent size. (*Courtesy of Hart & Cooley, Inc.*)

Furnace Vent Connector: Use the vent connector Table 6–4. Read down the total vent height column to 15 ft and read across the 2-ft connector rise line to the fan column. Note that a 4-inch vent size shows 96,000 Btu per hour or less than the furnace input rating. However, with a 20-ft total height, read across the 2-ft connector rise line. Note that the 4-inch vent size shows 105,000 Btu/h. An 18-ft height is three-fifths of 105,000 – 96,000 = 5,400. 96,000 + 5,400 = 101,400, which is the maximum input for an 18-ft minimum total vent height. Therefore, a 4-inch connector would be the correct size for the furnace, providing that the furnace had a 4-inch or smaller draft hood outlet. See Figure 6–11(b).

Common Vent Size: The total input to the common vent is 145,000 Btu/h. The vent goes straight through the roof, so use the *V* line of Table 6–4 under Fan + Nat column. Note that for a 15-ft minimum total vent height, the maximum Btu/h for a 5-inch vent is 164,000 Btu/h, which is greater than the total input to the common vent. Therefore, the common vent can be 5 inches in diameter. See Figure 6–11(c).

Additional Guides for Multiple-Appliance Venting

Always use the available headroom for the maximum connector rise after allowing for the listed clearance to combustibles. Obtain maximum connector rise by such methods as extending the connectors between the floor joists. The gained increase in venting power and efficiency of the system permits reduction of vent connector sizes.

Alternate ways for increasing vent height and connector size. If a combined vent cannot be used because of limitations in connector rise or total vent height, alternatives such as those illustrated in Figure 6–12 may be used to secure greater connector rise or greater total vent height.

The configurations of the vent connector are not important as long as the connector rise and length requirements are met. All of the illustrated methods in Figure 6–13 permit correct vent operation.

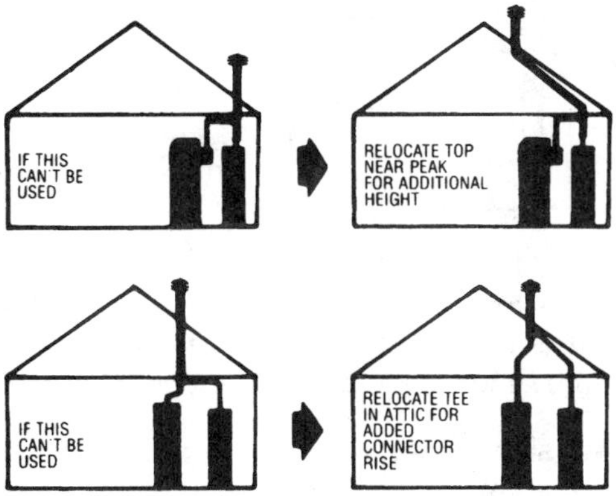
IF THIS CAN'T BE USED
RELOCATE TOP NEAR PEAK FOR ADDITIONAL HEIGHT
IF THIS CAN'T BE USED
RELOCATE TEE IN ATTIC FOR ADDED CONNECTOR RISE

FIGURE 6–12
Alternative venting methods. (*Courtesy of Hart & Cooley, Inc.*)

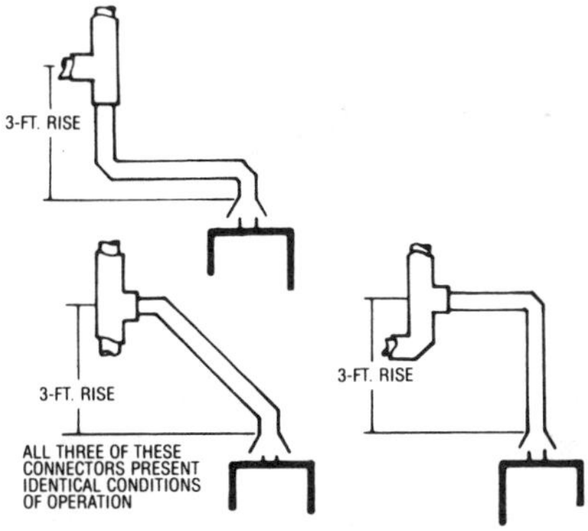
3-FT. RISE
3-FT. RISE
3-FT. RISE
ALL THREE OF THESE CONNECTORS PRESENT IDENTICAL CONDITIONS OF OPERATION

FIGURE 6–13
Total vent height. (*Courtesy of Hart & Cooley, Inc.*)

For economy, consider all alternatives. It is important in a combined vent system that the cost of the individual vent versus combined vents be considered, especially if the system is short or there are many fittings needed.

Frequently, individual vents will prove more economical than a combined system. See Figure 6–14.

Self-venting connectors sized from single-appliance vent tables. When a vent connector as a part of a combined vent system has a rise of 5 ft or more,

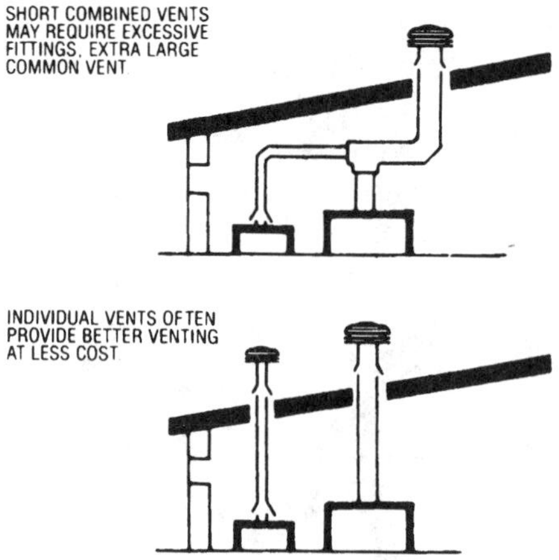
SHORT COMBINED VENTS MAY REQUIRE EXCESSIVE FITTINGS, EXTRA LARGE COMMON VENT
INDIVIDUAL VENTS OFTEN PROVIDE BETTER VENTING AT LESS COST

FIGURE 6–14
Individual vs. multiple gas venting. (*Courtesy of Hart & Cooley, Inc.*)

it can be installed as though it were an individual vent by using the appropriate single-appliance vent tables. It is important when sizing self-venting connectors that allowances be made for lateral length and the number of turns involved.

When in doubt, use one size larger vent. It is neither possible nor practical in some cases to anticipate all installation or operational contingencies in designing a vent system. A safe rule is, when in doubt use one size larger connectors and common vents than required by the vent tables.

Size of interconnecting tees. Interconnecting tees must be the same size as the common vent, as shown in Figure 6–15.

Use of Manifolds

To size manifolds, use the following procedures:

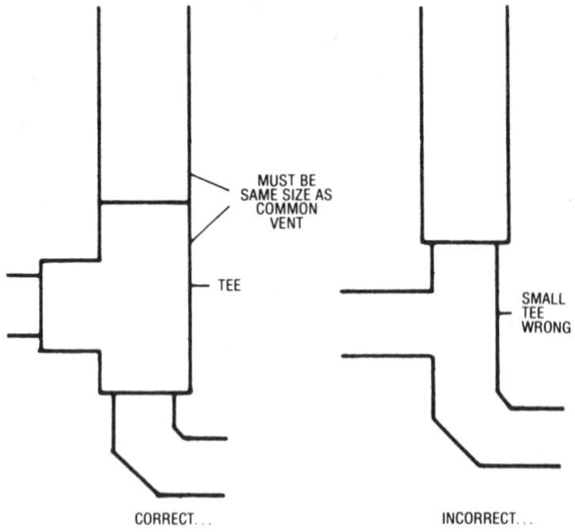
MUST BE SAME SIZE AS COMMON VENT
TEE
CORRECT...
SMALL TEE WRONG
INCORRECT...

FIGURE 6–15
Correct methods of increasing the vent size. (*Courtesy of Hart & Cooley, Inc.*)

Use of line *L* capacities for manifold sizing. A manifold is merely a vent system which is a horizontal extension of the lower end of a common vent. The connection of a manifold to a common vent may be made by either a 90° elbow or a tee. A manifold should be sized as a common vent using the combined total capacity and applicable total height of the vent system. The *L* lines in the common vent table must be used to determine the capacity of the manifold and common vent.

Horizontal versus sloped manifolds. Some codes require pitched or sloped manifolds. The requirements for sloped manifolds or connectors are a holdover from vent systems having low insulating values where condensation may occur. Adequate connector rise is necessary for proper venting of all appliances; therefore, lateral manifolds should not be excessively sloped. Too much manifold slope may cause insufficient connector rise at the appliance farthest from the common vent, increasing the chance of draft hood spillage. See Figure 6–16.

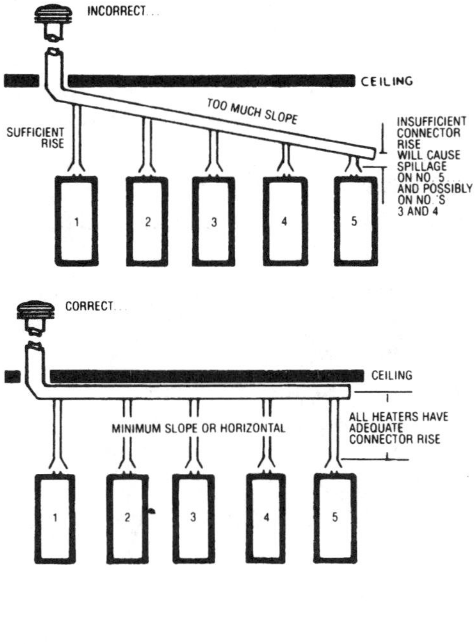
INCORRECT...
CEILING
TOO MUCH SLOPE
SUFFICIENT RISE
INSUFFICIENT CONNECTOR RISE WILL CAUSE SPILLAGE ON NO. 5 AND POSSIBLY ON NO.'S 3 AND 4
1
2
3
4
5
CORRECT...
CEILING
MINIMUM SLOPE OR HORIZONTAL
ALL HEATERS HAVE ADEQUATE CONNECTOR RISE
1
2
3
4
5

FIGURE 6–16
Proper manifold sloping. (*Courtesy of Hart & Cooley, Inc.*)

Manifold connectors. The vent connectors from a group of appliances on one level may enter from below or from the side of the manifold. In either case, the connector rise should be measured as the vertical distance from the draft hood outlet to the lowest level at which the connector enters the manifold. Care must be exercised in designing these systems, especially with connector turns and lengths because the heat loss is apt to be greater for such systems and cause accompanying vent capacity reduction.

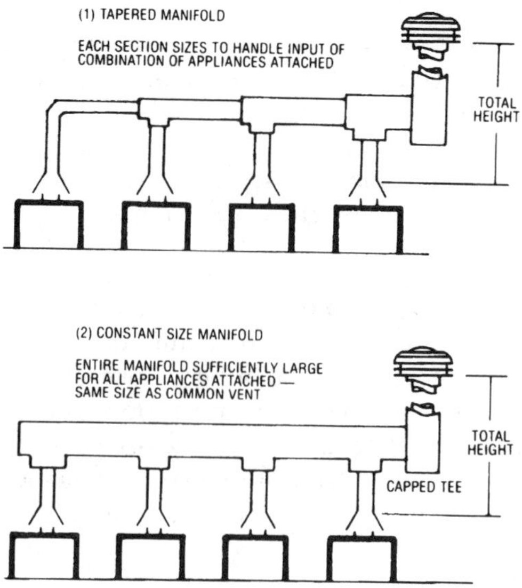
(1) TAPERED MANIFOLD
EACH SECTION SIZES TO HANDLE INPUT OF COMBINATION OF APPLIANCES ATTACHED
TOTAL HEIGHT
(2) CONSTANT SIZE MANIFOLD
ENTIRE MANIFOLD SUFFICIENTLY LARGE FOR ALL APPLIANCES ATTACHED — SAME SIZE AS COMMON VENT
TOTAL HEIGHT
CAPPED TEE

FIGURE 6–17
Methods of sizing manifolds. (*Courtesy of Hart & Cooley, Inc.*)

Sizing manifolds. As shown in Figure 6–17, manifolds may be designed either as (1) tapered or (2) constant size. The choice is dictated on the basis of convenience and cost.

Tapered: Use the total heat input to each portion of the manifold under construction, using the *L* capacities from the common vent table for the total vent height.

Constant size: Determine the required size of the common vent, based on the total input and the total vent height using the *L* capacities from the common vent table. Then use this size for the entire manifold.

Table limitations: When three or more appliances are connected to the same manifold, the largest appliance cannot exceed seven times the input of the smallest one. No more than eight appliances of identical input may be connected to the same manifold unless the size indicated by the table is reduced 10%, or by using a connector rise of at least 3 ft. Manifold lengths should not exceed 10 ft or 50% of the vertical height.

Multiple-Story Venting

A multiple-story vent system serves gas appliances at two or more different levels of a building. In designing multiple-story vent systems, use the vent connector and multiple appliance vent tables. When properly designed, such multiple-story vent systems will function satisfactorily when combinations of from one appliance to all appliances in this system are operating.

Figures 6–18 and 6–19 illustrate the major principles of multiple-story installation, which are as follows.

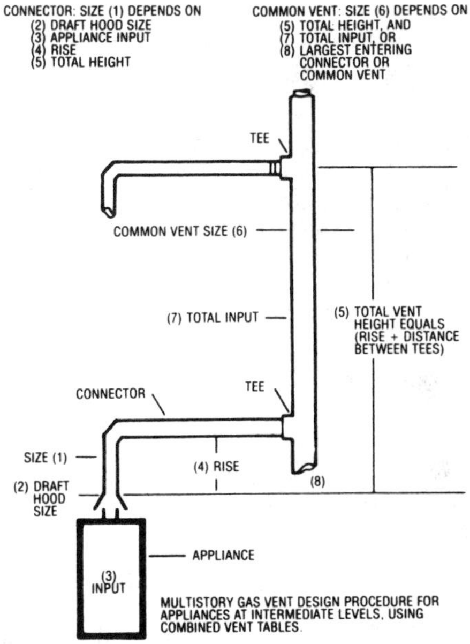
CONNECTOR: SIZE (1) DEPENDS ON
(2) DRAFT HOOD SIZE
(3) APPLIANCE INPUT
(4) RISE
(5) TOTAL HEIGHT
COMMON VENT: SIZE (6) DEPENDS ON
(5) TOTAL HEIGHT, AND
(7) TOTAL INPUT, OR
(8) LARGEST ENTERING CONNECTOR OR COMMON VENT
TEE
COMMON VENT SIZE (6)
(7) TOTAL INPUT
(5) TOTAL VENT HEIGHT EQUALS (RISE + DISTANCE BETWEEN TEES)
CONNECTOR
TEE
SIZE (1)
(4) RISE
(8)
(2) DRAFT HOOD SIZE
APPLIANCE
(3) INPUT
MULTISTORY GAS VENT DESIGN PROCEDURE FOR APPLIANCES AT INTERMEDIATE LEVELS, USING COMBINED VENT TABLES.

FIGURE 6–18
Multistory gas vent design procedure for appliances at intermediate levels, using combined vent tables. (*Courtesy of Hart & Cooley, Inc.*)

1. The overall system should be divided into smaller simple combined vent systems for each level, each using minimum total vent height for each level as illustrated.
2. Each vent connector from the appliance to the vertical common vent should be designed from the vent connector tables as in multiple-appliance vent systems.
3. For sizing of the vertical stack section, the common vent table is used. The vertical common vent for each system must be large enough to accommodate the accumulated total input of all the appliances discharging into it, but should never be smaller in diameter than the largest section below it.
4. The vent connector from the first floor or the lowest appliance to the common vent is designed as if terminating at the first tee or interconnection. The next lowest appliance is considered to have a combined vent which terminates at the second interconnection. The same principle continues on to the highest connecting appliance, with the top-floor appliance having a total vent height measured to the outlet of the common vent. The multiple-story system has no limit as to height as long as the common vent is sized to accommodate the total input.

It is important to keep the following points in mind.

1. The common vent height must always be computed as the distance from the outlet from the draft hood to the lowest part of the opening from the next interconnection above.

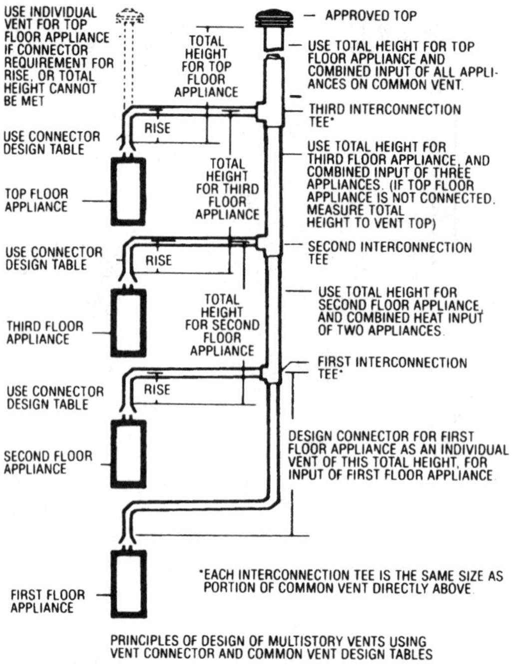
USE INDIVIDUAL VENT FOR TOP FLOOR APPLIANCE IF CONNECTOR REQUIREMENT FOR RISE, OR TOTAL HEIGHT CANNOT BE MET
APPROVED TOP
TOTAL HEIGHT FOR TOP FLOOR APPLIANCE
USE TOTAL HEIGHT FOR TOP FLOOR APPLIANCE AND COMBINED INPUT OF ALL APPLIANCES ON COMMON VENT.
THIRD INTERCONNECTION TEE*
USE CONNECTOR DESIGN TABLE
RISE
TOTAL HEIGHT FOR THIRD FLOOR APPLIANCE
USE TOTAL HEIGHT FOR THIRD FLOOR APPLIANCE, AND COMBINED INPUT OF THREE APPLIANCES. (IF TOP FLOOR APPLIANCE IS NOT CONNECTED, MEASURE TOTAL HEIGHT TO VENT TOP)
TOP FLOOR APPLIANCE
SECOND INTERCONNECTION TEE
USE CONNECTOR DESIGN TABLE
RISE
TOTAL HEIGHT FOR SECOND FLOOR APPLIANCE
USE TOTAL HEIGHT FOR SECOND FLOOR APPLIANCE, AND COMBINED HEAT INPUT OF TWO APPLIANCES.
THIRD FLOOR APPLIANCE
FIRST INTERCONNECTION TEE*
USE CONNECTOR DESIGN TABLE
RISE
DESIGN CONNECTOR FOR FIRST FLOOR APPLIANCE AS AN INDIVIDUAL VENT OF THIS TOTAL HEIGHT, FOR INPUT OF FIRST FLOOR APPLIANCE
SECOND FLOOR APPLIANCE
*EACH INTERCONNECTION TEE IS THE SAME SIZE AS PORTION OF COMMON VENT DIRECTLY ABOVE
FIRST FLOOR APPLIANCE
PRINCIPLES OF DESIGN OF MULTISTORY VENTS USING VENT CONNECTOR AND COMMON VENT DESIGN TABLES

FIGURE 6–19
Principles of multistory vents using vent connector and common vent design tables. (*Courtesy of Hart & Cooley, Inc.*)

2. If the connector rise is inadequate, increase the connector size, always making sure of the maximum available connector rise.
3. Be sure that the air supply to each appliance is adequate for proper operation. A separation of appliance rooms from an occupied area and some provision for outside air supply are necessary.
4. If an air shaft is used for installation of the common vent, be sure that sufficient space is provided for the fittings, clearance to combustibles and access for proper assembly.
5. These calculations apply only when the entire system is constructed of listed double-wall type B vent materials.

Ratio of connector size to common vent size. Whenever the area of the common vent becomes more than seven times the area of the vent connector entering it, the connector rise must be increased 1 ft above the allowable vent connector rise shown in the tables. For example, when an appliance input is 90,000 Btu/h using a 5-inch vent connector (area = 20 square inches) in a system having a minimum vent height of 10 ft, the vent connector rise must be 2 ft on the lower units where the common vent size is required, such as 14 inches (area = 154 square inches), the vent connector rise must be increased to 3 ft to avoid draft hood spillage.

This requirement does not apply when the connector rise is originally over 5 ft and consequently self-venting.

Economy of parallel systems. It may frequently prove more economical to group appliances to upper and lower common vent systems so that smaller vent sizes can be used. Even though many appliances may be connected to a single multiple-story common vent, the increase in size caused by this may prove uneconomical because of the required space for access and the need for numerous fittings. An alternate procedure is to use parallel common vents with staggered connections at alternate floors, thereby greatly increasing the minimum total vent height available to each connected appliance.

Example of multiple-story systems. Consider Figure 6–18 as a four-story apartment with each heater having a 90,000-Btu/h input and a 5-inch draft hood. The minimum total vent height is 10 ft for the three lower floors and 6 ft for the top floor. The common vent is vertical, so use the *V* lines of Table 6–6 under the Nat + Nat column for figuring the common vent size. Table 6–8 shows the calculations for venting all four floors into the common vent.

However, if the heater on the top floor is vented separately, Table 6–9 shows the result of increasing the minimum total vent height of the third-floor appliance to 16 ft and decreasing the total input to the common vent to 270,000 Btu/h.

Table 6–9 indicates the economics of venting the top floor separately, which eliminates the larger sizes of vent pipe and the use of costly increasing fittings.

Special considerations and additional precautions. The following items require additional precautions:

TABLE 6–8 Common height (multistory). (*Courtesy of Hart & Cooley, Inc.*)

Appliance	Input Total BTUH To Common Vent	Available Connector Rise	Min. Total Vent Height	Connector Size	Common Vent Size
1	90,000	10'	10'	5"	self venting connector
2	180,000	1'	10'	6"	7"
3	270,000	1'	10'	6"	8"
4	360,000	1'	6'	6"	10"

TABLE 6–9 Minimum total vent height (multistory). (*Courtesy of Hart & Cooley, Inc.*)

Appliance	Input Total BTUH To Common Vent	Available Connector Rise	Min. Total Vent Height	Connector Size	Common Vent Size
1	90,000	10'	10'	5"	self venting connector
2	180,000	1'	10'	6"	7"
3	270,000	1'	10'	6"	7"
4	90,000	8'	6'	5"	self venting connector

1. Combustion air: The combustion air requirements must be supplied from outside the living areas from sources such as hallways, service areas, or outdoor balconies in accordance with the information in Section 5.3 of the NFPA Standard 54 ANSI Z223.1. It is preferred that this air be taken into the appliance room directly from outdoors. This is important because any restriction in the common vent or termination will cause the flue products of all the appliances below this obstruction to spill out the draft hoods of other appliances just below this obstruction.
2. Other cautions:
   a. Provide proper clearance to combustibles around the common vent in its chase or shaft.
   b. Use the highest connector possible, and if the capacity is borderline use the next larger size connector.
   c. The only draft effect to be considered available is the result of the vertical height from the draft hood relief opening of the highest appliance on that level to the point where the connector for the appliance above enters the common vent. Never use the height to the termination except for the top floor.
   d. The appliances on the first floor are considered to be self-venting (vertical height 5 ft or more), and therefore sizing is calculated using Table 6–2 for single-appliance venting.

SPECIAL TYPES OF POWER VENTING

The following is a description of the venting system used for low temperature and low pressure forced- or induced-draft appliances.

Definitions

The following are definitions for the power vent system and the components that are used in power vent systems:

Forced draft: Forced draft indicates that the combustion air fan or blower is located ahead of the burner compartment.

Induced draft: Induced draft indicates that the combustion air fan or blower is located at or after the exit of the flue products from the heat exchanger.

High efficiency appliance: A high efficiency appliance is one which has efficiency of 78% AFUE (Annual Fuel Use Efficiency) or greater. A category I appliance may be suitable for use with either a type B vent or a special gas vent (Ultravent ®) system. A category III appliance should be vented with a special gas vent only. These appliances may be either condensing or noncondensing; however, they are typically used without draft hoods and may have a positive pressure at the flue collar.

Condensing appliance: A condensing appliance is one which has a large amount of heat removed from its products of combustion, causing the water vapors to condense in its heat exchanger and continue to condense in the venting system.

Ultravent: Ultravent is an Underwriters Laboratories, Inc. listed system of straight pipe and fittings which, when assembled in the prescribed manner, form a pressure- and liquid-tight conveyance for flue gases while resisting their corrosive properties and temperature. The Ultravent system, since it is gas and liquid tight as well as temperature and corrosion resistant, may be used to vent categories III and IV appliances and category I appliances as a special gas vent, if the appliance is certified for such use.

Definition of ANSI Categories of High Efficiency Appliances

The following are definitions of high efficiency appliance categories:

Category I: The vent pressure is nonpositive and the flue gas dew point temperature is greater than 140° F (noncondensing).

Category II: The vent pressure is nonpositive and the flue gas temperature is less than 140° F (condensing).

Category III: The vent pressure is positive and the flue gas temperature is greater than 140° F (noncondensing).

Category IV: The vent pressure is positive and the flue gas dew point temperature is less than 140° F (condensing).

Materials Used

Ultravent pipe and fittings are made from Ultem® resin, a new engineered thermoplastic from General Electric that resists high temperature and corrosion.

When assembled using the prescribed sealant, the system is gas and liquid tight and is resistant to corrosion from gaseous fuel condensate at temperatures up to 480° F.

Clearances

The Ultravent system is an uninsulated, single-wall vent. The flue gas temperatures of appliances which may be connected to the Ultravent system very greatly. As a general rule, maintain a minimum of 5 inches, clearance to all combustible materials unless the appliance manufacturer's instructions specify less clearance. In most instances a lesser clearance will be specified.

Firestopping

Firestopping must be provided whenever the vent passes through a floor or ceiling. Use the Ultravent combination support/firestop. See the installation instructions for construction details. For roofs, use an Ultravent storm collar, roof flashing. See the installation instructions for construction details.

Additional Cautions

1. Locate the terminators to prevent the flue products from being picked up by any fan or gravity air inlet to the building.
2. Locate the terminators so that any condensate dripping will not cause staining of buildings or damage to plantings, or so winter conditions will not freeze the water vapor on painted building surfaces.
3. Should it be necessary to terminate on an outside wall, keep the termination point at least 7 feet above any public walkways, above any snow drifting or piling level. Screen the opening if animals are likely to try to enter the vent.
4. Do not discharge this vent into an existing masonry or metal chimney.
5. Do not use Type B vent materials anywhere in this system.
6. Do not connect any other appliances into this system.
7. Use the installation instructions supplied with the appliance.
8. Remember that these definitions apply to the appliance and do not necessarily reflect the performance of the connected vent system.

Air Supply

For satisfactory performance of the appliance and venting system, an adequate supply of fresh air must be provided. This subject was covered earlier on page 108. A review of that material is suggested. One additional comment is necessary. Due to the absence of a draft hood on appliances in these categories, the required make-up air is reduced; therefore, all make-up air supply grills determined in that section can be reduced by 50% of their free area.

Correction for Altitude

As with all vent systems, forced draft vent systems are to be designed using the sea level nameplate rating regardless of any actual derated input required for local altitudes.

Condensate Removal

All appliances which condensate water vapor from the flue gas either full time, part time, or infrequently must be installed with the following precautions:

1. Horizontal runs must be pitched upward or downward from the furnace for condensate removal. Use a slope of 1/4 inch per foot minimum.
2. Horizontal runs must be supported at 10-ft intervals to prevent sagging and pocketing of the condensate.
3. Seal all joints as instructed in the installation manual. *Do not use screws.*

4. Dispose of the condensate as instructed by the appliance manufacturer. Some allow the condensate to return into the vent where it is collected with the condensate from the heat exchanger; others provide a drain fitting at the flue collar or a special adaptor.

Terminations

Terminations can be either vertically upward without offsets which terminate above the roof, or horizontal runs with a termination just outside a wall. See Figure 6–20.

The positive pressure caused by the wind hitting on the wall can be sufficient to overcome the ability of the vent blower to discharge the flue products. Under this condition the flow safety switch (required on AGA-listed appliances) will shut down the burner, stopping the heat flow into the building. Use only the proper termination recommended by the appliance manufacturer.

Fittings Equivalent

The total amount of pipe and fittings that can be used depends on such things as flue gas temperature, the number of fittings used, and the available vent pressure. See the appliance manufacturer's installation instructions for the recommended vent length limitations. See Table 6–10 for equivalent lengths of fittings.

MODEL TD CHIMNEY SYSTEM

Description

The Metalvent Model TD Chimney Systems ® consist of straight sections and other necessary fittings, which are constructed of a stainless steel outer jacket, and a stainless steel inner liner spaced 2 inches smaller (1 inch on each side) in diameter to provide an enclosure for solid pack insulation. This chimney system may be fully enclosed by the structure when the minimum clearance of 2 inches is maintained to all materials of the structure contents. This means no insulation is wrapped or packed around this chimney system closer than 2 inches. Follow the installation instructions.

Appliances Which May Be Connected

This system is to be used with all neutral or negative draft gas, liquid, or solid fuel-fired residential appliances and other building heating appliances; steam boilers operating at not over 50 psi; and pressing machine boilers which produce flue products up to 1,000° F during normal operation, up to 1,400° F during short periods of unusual firing, and to 2,100° F for short duration.

This chimney system is listed by Underwriters Laboratories as complying with Standard UL 103-HT. For proper installation, read and follow the installation

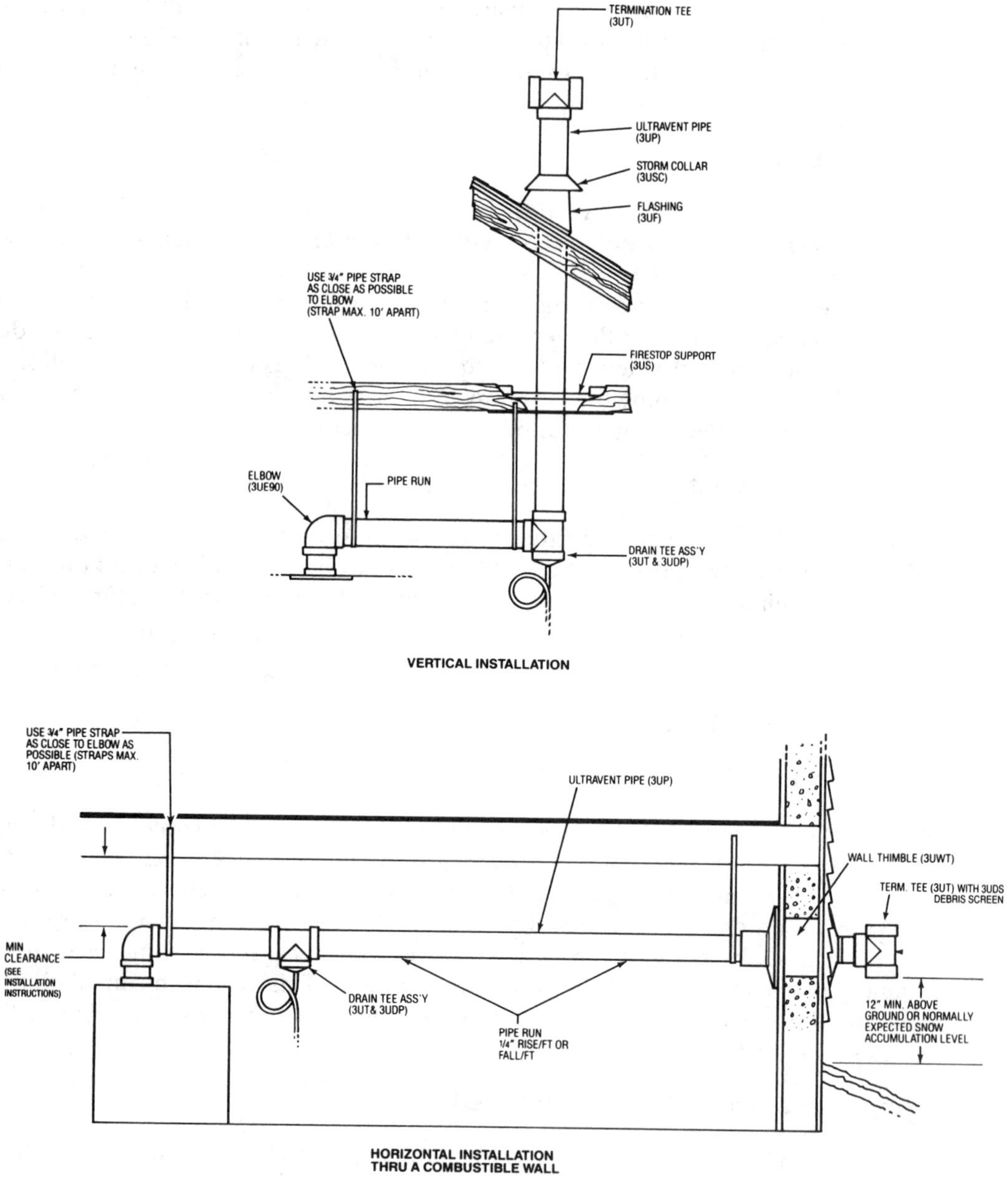
TERMINATION TEE
(3UT)
ULTRAVENT PIPE
(3UP)
STORM COLLAR
(3USC)
FLASHING
(3UF)
USE 3/4" PIPE STRAP
AS CLOSE AS POSSIBLE
TO ELBOW
(STRAP MAX. 10' APART)
FIRESTOP SUPPORT
(3US)
ELBOW
(3UE90)
PIPE RUN
DRAIN TEE ASS'Y
(3UT & 3UDP)
VERTICAL INSTALLATION
USE 3/4" PIPE STRAP
AS CLOSE TO ELBOW AS
POSSIBLE (STRAPS MAX.
10' APART)
ULTRAVENT PIPE (3UP)
WALL THIMBLE (3UWT)
TERM. TEE (3UT) WITH 3UDS
DEBRIS SCREEN
MIN
CLEARANCE
(SEE
INSTALLATION
INSTRUCTIONS)
DRAIN TEE ASS'Y
(3UT& 3UDP)
PIPE RUN
1/4" RISE/FT OR
FALL/FT
12" MIN. ABOVE
GROUND OR NORMALLY
EXPECTED SNOW
ACCUMULATION LEVEL
HORIZONTAL INSTALLATION
THRU A COMBUSTIBLE WALL

FIGURE 6–20 Vertical and horizontal installation through a combustible wall. (*Courtesy of Hart & Cooley, Inc.*)

TABLE 6–10 Equivalent length of fittings. (*Courtesy of Hart & Cooley, Inc.*)

Fitting	Equivalent Feet of Pipe
3 UE90 Elbow	10
3UES90 Elbow	2.25
3UE45 Elbow	4
3 UT TEE (as Elbow)	10
3 UT TEE (through flow)	7
3.5 UR3 Reducer	2.25
4 UR3 Reducer	15

instructions packed with the product. This information is also printed in the product catalog.

Preliminary Planning

Check the local building code for additional installation requirements for your area. The National Fire Protection Association Standards 31 and 211 require that the chimney extend at least 3 ft above the highest side of the roof opening through which the chimney passes and at least 2 ft higher than any portion of the building within a 10-ft horizontal distance. See Figure 6–21.

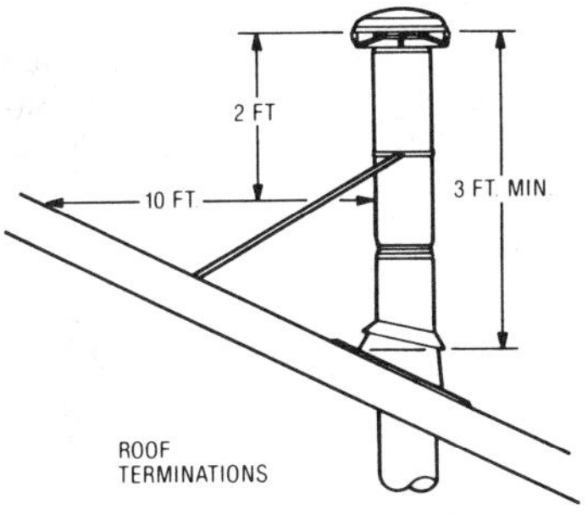
2 FT
10 FT
3 FT MIN
ROOF
TERMINATIONS

FIGURE 6–21
Roof termination point. (*Courtesy of Hart & Cooley, Inc.*)

Make a sketch of the proposed chimney system. Locate the chimney near the appliance, taking care that all structural and other obstructions are considered. Measure and note the horizontal and vertical sections needed plus all elbows and other fittings. Show the type of appliance and the input ratings in Btu/h.

From this sketch, enter either the appropriate reference from Chart 6–1 or Table 6–9 to determine the required chimney size. These are two separate methods of chimney sizing yielding the same results. Some installers prefer tables to charts.

Chimney Sizing General Information

The following data are based on these conditions:

1. A flue gas temperature rise of 600° F at 9% CO_2.
2. The connector length allowance is given in Table 6–8. The cap or other termination is also included.
3. The number of turns shown in Chart 6–1 is for the connector pipe. Two elbows or a tee and an elbow connected to each other must be considered three turns.

The draft shown at the top right of Chart 6–1 is the draft at the flue collar (see Figure 6–22). The relation between the outlet draft and the overfire draft is affected by the heat exchanger design. When in doubt, consult the appliance manufacturer.

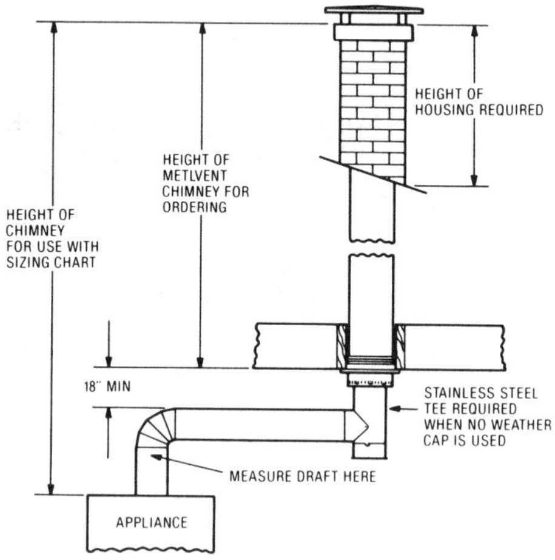
HEIGHT OF
HOUSING REQUIRED
HEIGHT OF
METLVENT
CHIMNEY FOR
ORDERING
HEIGHT OF
CHIMNEY
FOR USE WITH
SIZING CHART
18" MIN
STAINLESS STEEL
TEE REQUIRED
WHEN NO WEATHER
CAP IS USED
MEASURE DRAFT HERE
APPLIANCE

FIGURE 6–22
Principle of design of multistory vents using vent connector and common vent design tables. (*Courtesy of Hart & Cooley, Inc.*)

Chimney sizing for appliances using Chart 6–1.

Example:

Note the dashed line:
Given:

1. The input is 5 gph or 700,000 Btu/h
2. The draft required at the flue collar given by the furnace manufacturer is 0.04 inches water column.
3. The flue collar size is 10 inches, and a tee is required on the elbow. Consider for this example that the installation is like that shown in Figure 6–22.

Solution Proceed vertically upward from 700 MBtu/h to intersect the 10-inch flue size, then horizontally to the right to intersect 0.04 inches water draft, then vertically downward to read approximately 7 ft of height required. This must then be adjusted for the number of turns existing in the

Minimum Chimney Heights

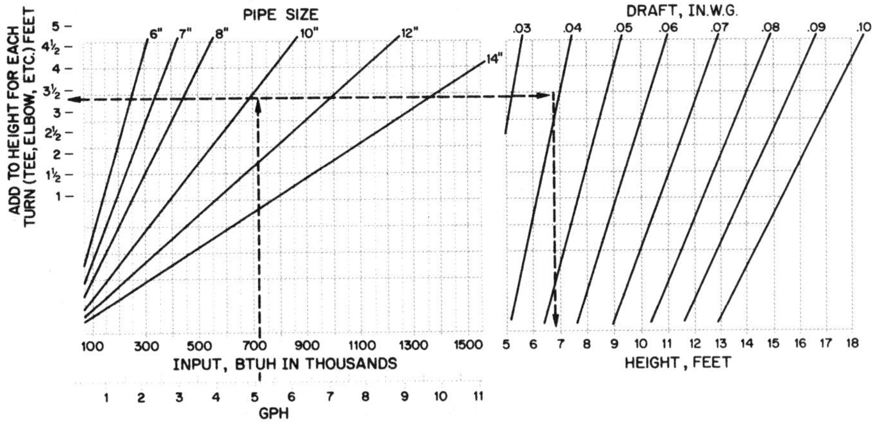
PIPE SIZE
6"
7"
8"
10"
12"
14"
DRAFT, IN.W.G.
.03
.04
.05
.06
.07
.08
.09
.10
ADD TO HEIGHT FOR EACH TURN (TEE, ELBOW, ETC.) FEET
5
4½
4
3½
3
2½
2
1½
1
100 300 500 700 900 1100 1300 1500
INPUT, BTUH IN THOUSANDS
1 2 3 4 5 6 7 8 9 10 11
GPH
5 6 7 8 9 10 11 12 13 14 15 16 17 18
HEIGHT, FEET

CHART 6–1 Minimum chimney heights. (*Courtesy of Hart & Cooley, Inc.*)

chimney system. Repeat from the starting point, moving vertically upward to intersect the 10-inch flue size, then horizontally to the left to read approximately 3 1/2 ft to be added for each tee or elbow. With one tee and one elbow, the total required height is 7 + 7 = 14 ft. By increasing the flue by one size, the required height can be reduced slightly. Checking the 12-inch flue size will show 6 1/2 + 3 = 9 1/2 ft.

Chimney Sizing for Fireplaces

Refer to Figure 6–23.

Example:

Note the blue line.

Given:

1. The fireplace opening is 30 inches by 24 inches.
2. The desired height from the top of the opening to the top of the chimney is 15 ft.

Solution At the right-hand section of Chart 6–1, find the intersection of the 30-inch and the 24-inch lines. Proceed horizontally to the left until the 15-ft height line is intersected. This intersection occurs within the limits of a 10-inch chimney size.

Chimney Sizing for Appliances Using Table 6–11

Example:

Given:

1. The input is 700,000 Btu/h.
2. The draft required is 0.04 inches water column.
3. The system contains one 90° elbow and one tee.
4. The collar size of the appliance is 10 inches.

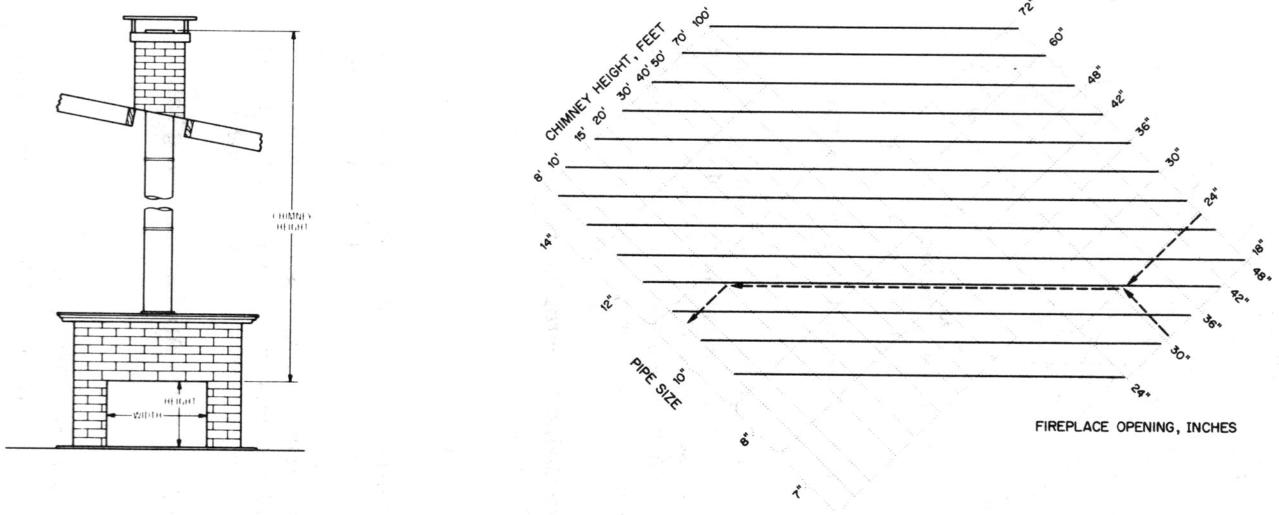
CHIMNEY HEIGHT
HEIGHT
WIDTH
CHIMNEY HEIGHT, FEET
8' 10' 15' 20' 30' 40' 50' 70' 100'
72" 60" 48" 42" 36" 30" 24" 18"
48" 42" 36" 30" 24"
FIREPLACE OPENING, INCHES
PIPE SIZE
14" 12" 10" 8" 7"

FIGURE 6–23 Chimney sizing for fireplaces. (*Courtesy of Hart & Cooley, Inc.*)

Solution Find the input 700,000 Btu/h in the left column and 0.04 inches water column draft at the top of the table. Next to the input column is a column giving multiple choices of chimney sizes. For a trial, use the flue collar size of 10 inches. Next to the 10-inch size is a figure, 3 1/2, which is explained as the feet that must be added for each 90° turn. Proceed right from the 700 figure, then follow the 10-inch size horizontally to the right until it intersects the 0.04 vertical column. The height shown is 7 ft. There are two 90° turns required (the 90° elbow and the tee given); therefore 7 ft must be added (2 times 3 1/2) to the 7 ft given in the table. The required chimney height is then 14 ft (7 + 7). A larger size chimney will result in a lower total height if such is required.

TABLE 6–11 Chimney size and height. (*Courtesy of Hart & Cooley, Inc.*)

INPUT BTUH x 1000	CHIMNEY SIZE	90° TURNS EQUIV FT*	CHIMNEY DRAFT SPECIFIED, inches w.c.						
			0.04	0.05	0.06	0.07	0.08	0.09	0.10
	6	1	6	7 1/2	9	10 1/2	12	13 1/2	15
100	7	1	5 1/2	7	8 1/2	10	11 1/2	12 1/2	14
	8	1	5 1/2	7	8	9 1/2	11	12	13 1/2
	6	2	6 1/2	8 1/2	10	11 1/2	13 1/2	15	16 1/2
200	7	1 1/2	6 1/2	8	9 1/2	11	12 1/2	14	16
	8	1	6	7 1/2	9	10 1/2	12	13 1/2	15
	10	1	6	7	8 1/2	10	11 1/2	13	14 1/2
	6	4 1/2	7 1/2	9 1/2	11	12 1/2	15	16 1/2	18 1/2
300	7	3	7	8 1/2	10	12	13 1/2	15	17
	8	1 1/2	6 1/2	8	9 1/2	11	13	14 1/2	16
	10	1	6	7 1/2	9	10 1/2	12	13 1/2	15
	7	4 1/2	7 1/2	9 1/2	11	12 1/2	15	16 1/2	18 1/2
400	8	3	7	8 1/2	10	12	13 1/2	15	17
	10	1 1/2	6 1/2	8	9 1/2	11	13	14 1/2	16
	12	1	6	7 1/2	9	10 1/2	12	13 1/2	15
	8	4	7 1/2	9	11	12 1/2	14 1/2	16 1/2	18
500	10	2	6 1/2	8	9 1/2	11 1/2	13	14 1/2	16
	12	1	6	7 1/2	9	10 1/2	12	13 1/2	15
	14	1	6	7 1/2	8 1/2	10	11 1/2	13	14 1/2
	10	2 1/2	7	8 1/2	10	12	13 1/2	16	17
600	12	1 1/2	6 1/2	8	9 1/2	11	12 1/2	14	15 1/2
	14	1	6	7 1/2	9	10 1/2	12 1/2	14	15 1/2
	10	3 1/2	7	9	10 1/2	12 1/2	14	16	17 1/2
700	12	1 1/2	6 1/2	8	9 1/2	11	13	14 1/2	16
	14	1	6	7 1/2	9	10 1/2	12 1/2	14	15 1/2
	10	4 1/2	7 1/2	9 1/2	11	13	14 1/2	16 1/2	18
800	12	2 1/2	7	8 1/2	10	11 1/2	13 1/2	15	16 1/2
	14	1	6 1/2	8	9 1/2	11	12 1/2	14	15 1/2
900	12	3	7	8 1/2	10	12	14	15 1/2	17
	14	2	6 1/2	8	9 1/2	11	13	14 1/2	16
1000	12	3 1/2	7	9	10 1/2	12 1/2	14	16	17 1/2
	14	2	6 1/2	8	9 1/2	11 1/2	13	14 1/2	16
1100	12	4	7 1/2	9	10 1/2	12 1/2	14 1/2	16	18
	14	2	6 1/2	8 1/2	10	11 1/2	13 1/2	15	16 1/2
1200	12	4 1/2	7 1/2	9 1/2	11	13	15	16 1/2	18 1/2
	14	2 1/2	7	8 1/2	10	12	13 1/2	15 1/2	17
1300	14	3	7	9	10 1/2	12	14	15 1/2	17 1/2
1400	14	3 1/2	7	9	10 1/2	12 1/2	14	16	17 1/2
1500	14	4	7 1/2	9	11	12 1/2	14 1/2	16 1/2	18

*For EACH 90° turn (tee, elbow) ADD this value to Tabled Height figure to get ACTUAL vertical height required.

BASIC METHOD FOR SIZING ALL TYPES AND CONDITIONS OF VENTS AND CHIMNEYS

The following are simplified procedures for the solving of systems in general. All the previous information has been solutions of general methods for specific types and conditions; that is, it has been simplified for common conditions and restraints resulting in shorter solutions for the user.

For special cases, when one of the previous solutions does not apply, the designer can use the following method.

Sizing Method

This method supplies a worksheet that permits an orderly procedure in the sizing of vents. The step-by-step explanation and the example will assist the user in becoming familiar with this method.

Procedure for worksheet. Use Figure 6–24. By following this procedure, any vent, chimney, or air-handling vent system can be determined.

Prepare a sketch of the required system, including all pertinent capacities and dimensions.

Line 1. Always enter the theoretical draft d_t corrected from Table 6–12 using the temperature rise (flue gas temperature minus 60° F) and the vent or chimney height as measured from the appliance collar to the termination.

$$d_t = d_t \times \text{height}/100$$

Enter d_t as a positive value on Line 1. When the unit has a draft hood, leave lines 2 and 3 blank and proceed to Line 4.

Line 2. When the appliance is forced draft (having an internal forced- or induced-draft fan or blower), enter the available pressure measured at the flue collar on Line 2. Skip Line 3 and proceed to Line 4.

Line 3. When the appliance specifications indicate that a negative pressure is required at the flue collar, the appliance requires a draft greater (more negative) than d_t on Line 1. Enter this value as a negative amount on Line 3.

Line 4. When a draft inducer is added in the venting system after the appliance flue collar, enter the available draft on Line 4.

Line 5. Add the lines above that have numbers entered. Only the number on Line 3 should be negative (when used). Enter the sum on Line 5.

Line 6. Refer to Table 6–16 for values to use on lines 6 through 10. When using the appropriate low resistance vent cap, enter the equivalent length of straight pipe given. Notice that this equivalent length is not a constant, but it can be related to the draft in Line 5 for a satisfactory approximation. For example, if Line 5 is between 0.10 and 0.15, then the termination cap resistance is approximately equal to 15 ft of straight pipe. Enter the estimated equivalent feet for the termination cap on Line 6.

D_p		INPUT FACTOR	
Use either 2 or 3 Add 4 IF applicable. 1. For natural draft ADD d_t.	0.15	14. Nameplate INPUT for unit, in 1000 BTUH.	500
2. Forced draft available at flue collar, ADD d_f.		15. Flue gas temperature rise, °F.	300
3. Negative at collar and requires natural draft, SUBTRACT d_n.		16. Percent CO_2. Use Table 3 if NOT known.	5.3
4. Draft inducer added in the flue, ADD d_b.		17. Factor from Table 10, 11, 12 for fuel specified (use data Lines 16, 17).	.57
5. ADD #1 and all other numbers, d_p.	0.15	18. Altitude correction MULTIPLY by Line 17.	—
USE FIGURE 2 FOR BELOW: 6. Equivalent length for Cap or Terminal	15	19. MULTIPLY appropriate Line 14 by Line 17.	285
7. Equivalent length for 90° elbow (times number used).		20. CHIMNEY OR VENT SIZE, INCHES.	10"
8. Equivalent length for 45° elbow (times number used).			
9. Equivalent length for tee(s).	50		
10. Equivalent length for draft hoods or barometric damper.	30		
11. TOTAL length vent vertical plus horizontal	45		
12. TOTAL EQUIV. LENGTH.* 140 x 1.5 = 210		*For combined gas vents, multiply equivalent length by factor of 1.5 to determine common vent.	
13. D_P = Line 5X $\frac{100}{\text{Line 12}}$	0.07		

FIGURE 6–24 Worksheet (example). (*Courtesy of Hart & Cooley, Inc.*)

Line 7. Following the same evaluation as described for Line 6, enter the estimated equivalent length for the 90° elbows on Line 7. Remember to multiply the value from Table 6–13 by the number of elbows.

Line 8. Enter the equivalent length for all 45° elbows. When two elbows are adjacent, multiply the equivalent length by 1.5 for the correct value.

Line 9. Enter the equivalent length for all tees.

Line 10. Enter the equivalent length for a draft hood or barometric damper when used.

Line 11. Enter the total length (vertical plus horizontal) of vent or chimney, in feet.

Line 12. Add lines 6 through 11 and enter here. When the system contains two or more appliances on the same vent or chimney, multiply the total equivalent length by 1.5.

D_p		INPUT FACTOR	
Use either 2 or 3 Add 4 IF applicable. 1. For natural draft ADD d_T.		14. Nameplate INPUT for unit, in 1000 BTUH.	
2. Forced draft available at flue collar, ADD d_f.		15. Flue gas temperature, °F.	
3. Negative at collar and requires natural draft. SUBTRACT d_n.		16. Percent CO_2. Use Table 17 if NOT known.	
4. Draft inducer added in the flue, ADD d_b		17. Factor from Table 10, 11, or 12 for fuel specified (use data Lines 15, 16).	
5. ADD #1 and all other numbers, d_p.		18. Altitude correction MULTIPLY by Line 17.	
USE TABLE 16 FOR BELOW: 6. Equivalent length for Cap or Terminal		19. MULTIPLY appropriate Line 14 by Line 17.	
7. Equivalent length for 90° elbow (times number used).		20. CHIMNEY OR VENT SIZE, INCHES.	
8. Equivalent length for 45° elbow(s) (times number used).			
9. Equivalent length for tee(s).			
10. Equivalent length for draft hoods or barometric damper.			
11. TOTAL length vent vertical plus horizontal			
12. TOTAL EQUIV. LENGTH.*			
13. $Dp = \text{Line 5} \times \frac{100}{\text{Line 12}}$			

* For combined gas vents, multiply equivalent length by factor of 1.5 to determine common vent.

FIGURE 6–24a Chimney or vent sizing worksheet. (*Courtesy of Hart & Cooley, Inc.*)

Line 13. To determine the draft factor D_p, multiply Line 5 by 100 divided by Line 12 or,

$$D_p = \text{Line } 5 \times 100/\text{Line } 12.$$

A more precise determination will result when the values on lines 6 through 10 are figured using D_p, Line 13. When Line 12 is 85 to 120, the change will not be justified.

Line 14. Enter the sea level nameplate input rating for the appliance in thousands of Btu/h. That is Btu/h ÷ 1,000.

Line 15. Enter, from the appliance specifications, the flue gas temperature at the flue collar. When the information for lines 15 and 16 is not available, use the following:

1. Natural gas with draft hood: 5.3% CO_2 @ 300° F temperature rise.
2. LP gas with draft hood: 6% CO_2 @ 300° F temperature rise.
3. Either gas, no draft hood: 8% CO_2 @ 400° F temperature rise.
4. Oil: 9% CO_2 @ 500° F temperature rise.

Line 16. Enter, from the appliance specifications sheet, the percent of CO_2 in the flue gas.

Line 17. Using the values from lines 15 and 16 and the appropriate Table 6–10, 6–11, or 6–12, determine the flow factor. The tables are straightforward in use, and values can be interpolated by proportions.

Line 18. When the installation is to be located at an altitude of 2,000 ft or above, multiply Line 17 by the appropriate altitude correction from Table 6–13 and enter here.

Line 19. Multiply Line 17 or Line 18 by Line 14 to determine the input factor (cubic feet per minute (cfm) of flue gases at the actual temperature).

Line 20. Vent or chimney size. Using the input factor from Line 19 and draft D_p from Line 13, enter Chart 6–2 or Chart 6–3 depending on the values. (Chart 6–3 is a continuation of Chart 6–2 for larger sizes.) Determine the vent size and enter that on Line 20.

Example

Given: A sea level system has five appliances (natural gas) connected to a manifold. Each appliance has an input of 100,000 Btu/h. The manifold is to be a constant size and the vertical vent is to have a total height of 30 ft and a manifold length of 15 ft. See Figure 6–25. The appliances have draft hoods. The flue gas temperature rise is 300° F, and the CO_2 is 5.3%.

Following the procedure for worksheet, Figure 6–23:

Line 1: Because these appliances have draft hoods, they operate with a natural draft. Use Table 6–14. For a flue gas temperature rise of 300° F, the theoretical draft D_t = 0.5 inches water column per 100 ft of height.

Then

$$D_t = 0.5 \times 30/100 = 0.15 \text{ inches water column.}$$

TABLE 6–12 Natural gas flow factors, CFM for 1000 BTUH Input. (*Courtesy of Hart & Cooley, Inc.*)

CO_2	Flue Gas Temperatures °F										
%	60	100	200	300	400	500	600	800	1000	1500	2000
3	.67	.72	.85	.98	1.11	1.24	1.37	1.62	1.88	2.52	3.17
4	.50	.54	.63	.73	.83	.92	1.02	1.21	1.40	1.88	2.37
5	.41	.44	.56	.60	.68	.76	.84	1 .03	1.15	1.55	1.94
6	.34	.37	.43	.49	.56	.63	.69	.86	.95	1.28	1.61
7	.30	.32	.38	.44	.50	.55	.61	.76	.84	1.13	1.42
8	.27	.29	.34	.39	.45	.50	.55	.88	.75	1.02	1.28
9	.24	.26	.30	.35	.40	.44	.49	.61	.67	.90	1.14
10	.23	.25	.29	.34	.38	.42	.47	.58	.65	.87	1.09
11	.21	.23	.27	.31	.35	.39	.43	.52	.39	.79	.99
12	.19	.20	.24	.28	.31	.35	.39	.48	.53	.72	.90

Enter 0.15 on Line 1.

Line 2. Skip.

Line 3. Skip.

Line 4. Not applicable.

Line 5. Add (there is only one figure). Enter 0.15.

Lines 6–12. Checking Figure 6–25 again, it can be seen that there are two tees in each appliance vent system and there is a cap at the termination point. Checking Table 6–15, it is found that with Line 5 being 0.15 inches water column, the values in the center column apply. Therefore, use a cap equal to 15 equivalent feet of pipe, a tee equal to 25 equivalent feet of pipe, and a draft hood equal to 30 equivalent feet of pipe. There are 15 + 30, or 45, running feet of vent pipe. Enter these on the appropriate lines and total them on Line 12. Because there are more than two appliances on the same common vent, multiply by 1.5. Thus, 1.5 × 140 = 210 equivalent feet of vent pipe. Enter this on Line 12.

Line 13. Determine D_p by the formula given:

$$D_p = 0.15 \times 100/210 = 0.07.$$

Line 14. The input was given as 100,000 Btu/h for each of five appliances. The total input to use then is 500,000 Btu/h. Enter 500,000/1,000 = 500 on Line 14.

Line 15. Flue gas temperature rise is given as 300° F.

Line 16. Percent CO_2 is given as 5.3.

Line 17. Using Table 6–10 at 300° F and 5.3% CO_2, determine that the interpolated value is 0.57 CFM.

Line 18. Since this was given as a sea level installation, no correction is needed; however, if it was located at a higher elevation, the factor would be found in Table 6–13 and Line 17 would be multiplied by this factor.

Line 19. To obtain the total flue products, multiply Line 14 by Line 17. 500 × 0.57 = 285. Enter this amount on Line 19.

Line 20. Finally, using Chart 6–2,

Line 19 = 285

Line 13 = 0.07.

The intersection of these two values will be above the 9-inch vent size. Therefore, a 10-inch vent will be required. Enter 10-inch diameter for the manifold and common vertical vent.

TABLE 6–13 L.P. gas flow factors, cfm for 1000 Btu/h input. (*Courtesy of Hart & Cooley, Inc.*)

CO_2	Flue Gas Temperatures °F										
%	60	100	200	300	400	500	600	800	1000	1500	2000
3	.70	.75	.89	1.01	1.16	1.29	1.43	1.70	1.97	2.64	3.31
4	.53	.57	.67	.77	.88	.98	1.08	1.28	1.49	2.00	2.51
5	.43	.46	.55	.63	.71	.79	.88	1.04	1.21	1.62	2.03
6	.37	.40	.47	.54	.61	.68	.75	.90	1.04	1.39	1.75
7	.32	.34	.41	.47	.53	.59	.65	.78	.90	1.21	1.51
8	.29	.31	.37	.42	.48	.54	.59	.70	.81	1.09	1.37
9	.27	.29	.34	.39	.45	.50	.55	.65	.76	1.02	1.28
10	.24	.26	.30	.35	.40	.44	.49	.58	.67	.90	1.14
11	.23	.25	.29	.34	.38	.42	.47	.56	.65	.87	1.09
12	.21	.23	.27	.31	.35	.39	.43	.51	.59	.79	.99

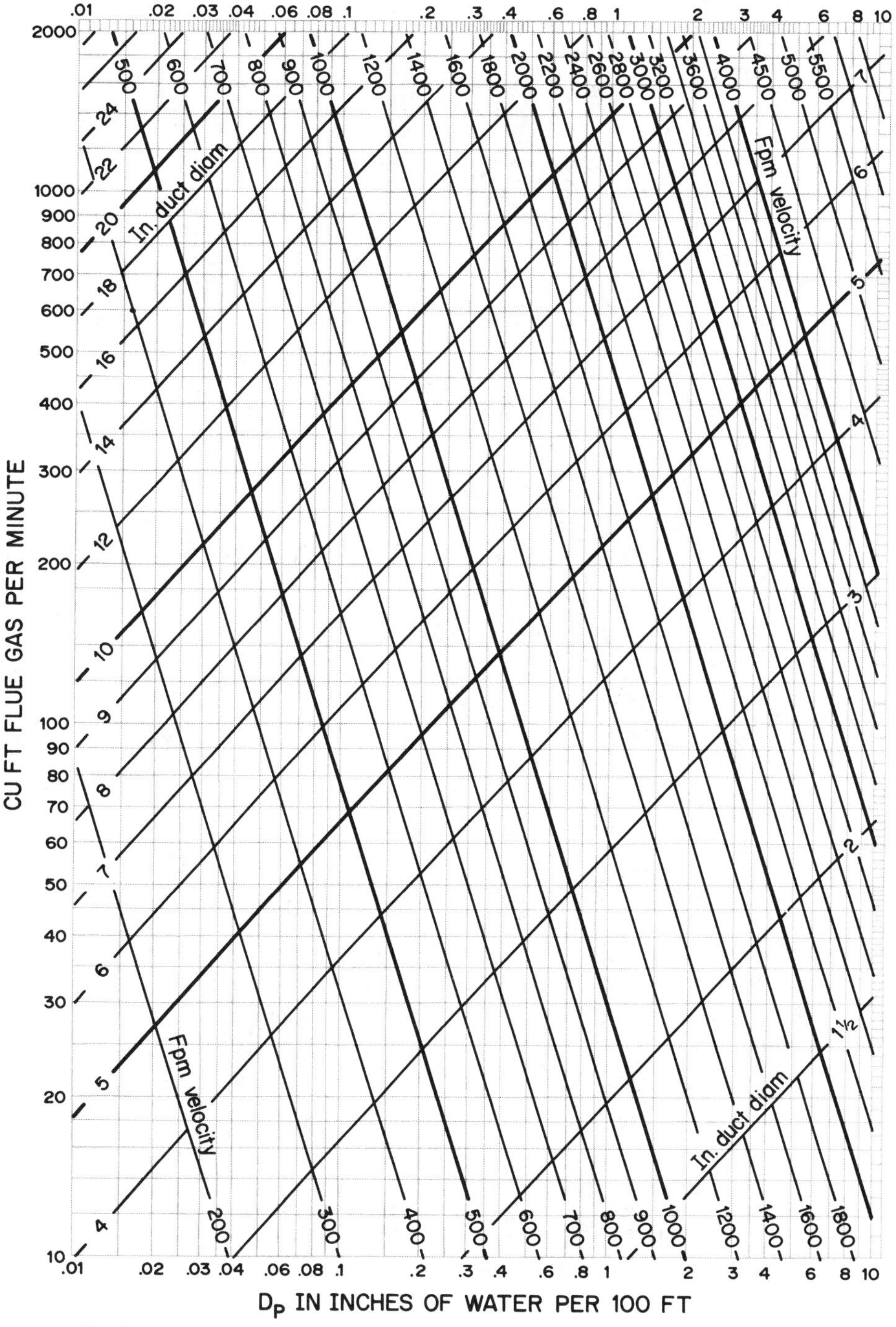
CU FT FLUE GAS PER MINUTE
Dp IN INCHES OF WATER PER 100 FT
In. duct diam
Fpm velocity
In. duct diam
Fpm velocity

CHART 6–2 Basic methods, sizing all types and conditions of vents and chimneys. (*Courtesy of Hart & Cooley, Inc.*)

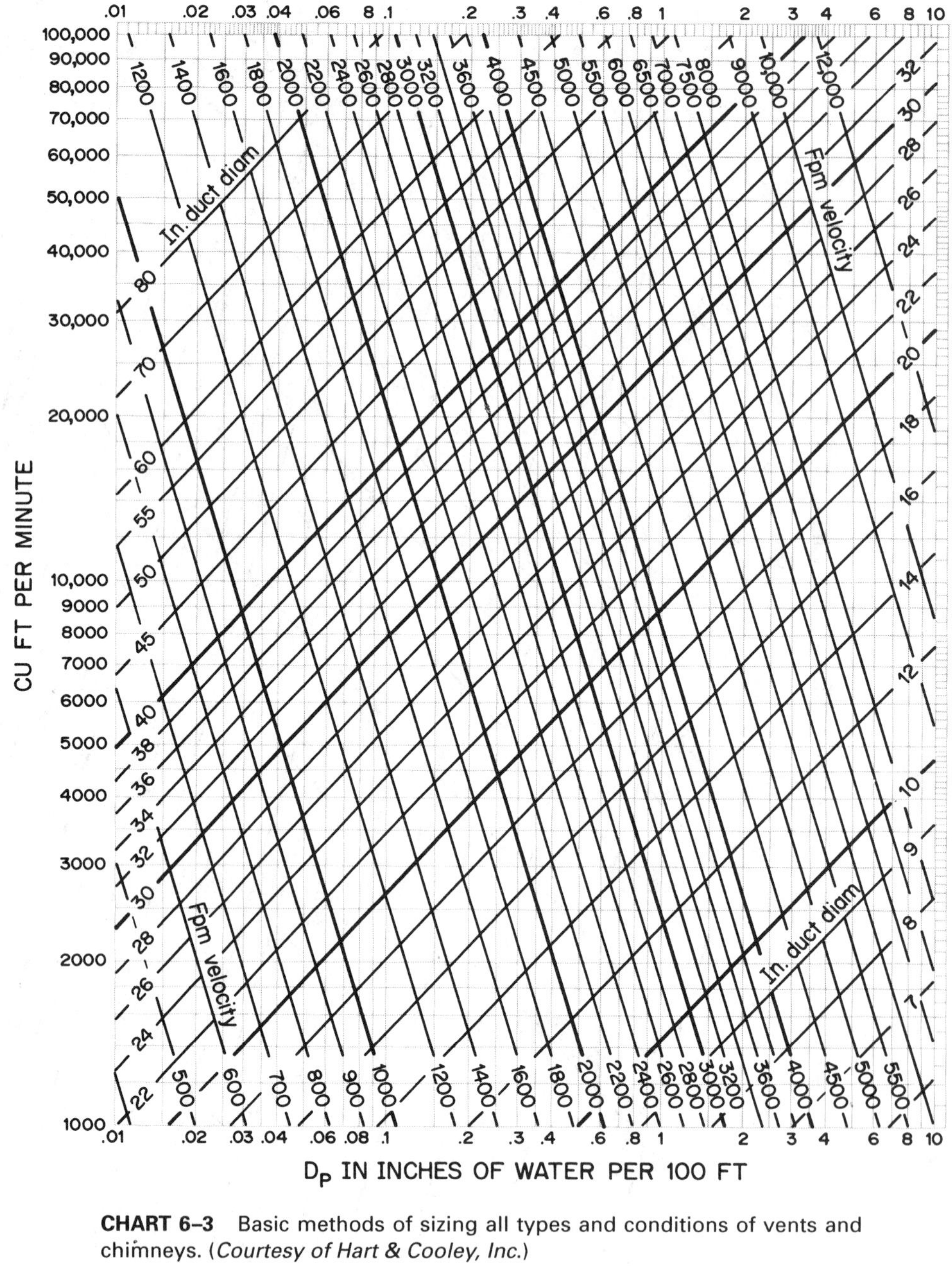
CU FT PER MINUTE
Dp IN INCHES OF WATER PER 100 FT
In. duct diam
Fpm velocity
In. duct diam
Fpm velocity

CHART 6–3 Basic methods of sizing all types and conditions of vents and chimneys. (*Courtesy of Hart & Cooley, Inc.*)

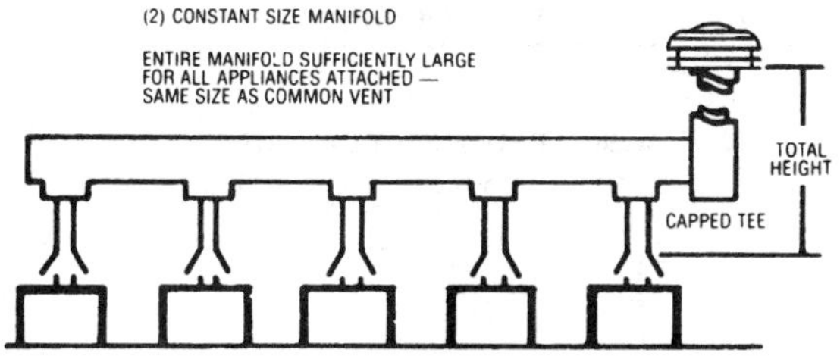
(2) CONSTANT SIZE MANIFOLD
ENTIRE MANIFOLD SUFFICIENTLY LARGE FOR ALL APPLIANCES ATTACHED — SAME SIZE AS COMMON VENT
TOTAL HEIGHT
CAPPED TEE

FIGURE 6–25
Constant size manifold. (*Courtesy of Hart & Cooley, Inc.*)

It is interesting, as an example, to check Table 6–4 for the common vent size in the special cases (type B) simplification of this procedure.

Enter Table 6–4 at the given least total height of 30 ft and move horizontally in the lateral *L* row until intersecting a number 500 (500,000 Btu/h) or larger. That number is found to be 590 (590,000 Btu/h).

Checking the Least Total Height column heading shows a 10-inch vent size. The type B gas vent Table 6–4 is the simplified special case of basic manifold.

TABLE 6–14 Oil (no. 2) gas flow factors, cfm for 1000 Btu/h input. (*Courtesy of Hart & Cooley, Inc.*)

CO_2	Flue Gas Temperatures °F										
%	60	100	200	300	400	500	600	800	1000	1500	2000
3	.75	.81	.95	1.10	1.24	1.38	1.53	1.82	2.11	2.83	3.55
4	.58	.62	.74	.85	.96	1.03	1.18	1.41	1.63	2.19	2.74
5	.47	.51	.60	.69	.78	.87	.96	1.14	1.32	1.77	2.22
6	.40	.43	.51	.58	.66	.74	.82	.97	1.12	1.51	1.89
7	.35	.38	.44	.51	.58	.65	.71	.85	.98	1.32	1.66
8	.32	.34	.41	.47	.53	.59	.65	.78	.90	1.21	1.51
9	.29	.31	.37	.42	.48	.54	.59	.70	.81	1.09	1.37
10	.26	.28	.33	.38	.43	.48	.53	.63	.73	.98	1.23
11	.24	.26	.30	.34	.40	.44	.49	.58	.67	.90	1.14
12	.23	.25	.29	.34	.38	.42	.47	.56	.65	.87	1.09

VENT DRAFT

Vent draft is extremely susceptible to the influence of the atmosphere. While it is general knowledge that outside atmospheric conditions, especially winds, bring about enormous changes in drafts, it is often disregarded that drafts change whenever the flue gas temperatures change. What this means is that drafts are changing almost constantly during the on and off cycles of the burner. Also, when two or more units are connected to a single vent, they affect each other, increasing or decreasing the draft as they start and stop individually.

Without some means for accurately controlling drafts, the fuel losses in a heating plant may be between 4% and 8%, and in some cases may reach 15%. This loss is occurring today in hundreds of thousands of existing heating plants.

If the heating plant is to operate without this waste, a definite and exact amount of air must be supplied to the fuel. This is the prime function of the barometric draft control.

TABLE 6–15 System resistance. (*Courtesy of Hart & Cooley, Inc.*)

FITTING	EQUIVALENT LENGTH FEET OF VENT		
	$d_p > .10$	$d_p < .10$ $> .15$	$d_p < .15$
Termination Cap (UL)	9	15	20
Round 4 pc. Elbow 90°	9	15	20
Tee, side inlet to run	15	25	33
Round 2 pc. Elbow 45°	4	7	9
Barometric Damper Regulator	9	15	20
Draft Hood	18	30	40
Straight Vent Pipe	ACTUAL running feet		

Example:

Cap	9
Tee	15
Inlet	18
90° Elbow	9
Pipe	37
	88 equivalent feet

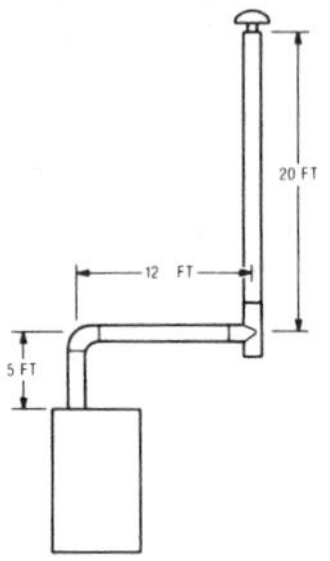
20 FT
12 FT
5 FT

When the correct draft is maintained automatically:

1. The weight or volume of air supplied to the fire is held to a minimum, reducing the weight and volume of the flue gas produced.
2. Combustion chamber temperatures are higher because the vent gases are not diluted with excess air.
3. A smaller volume of flue gases results in lower velocities through the heating unit, which increases the amount of time that the gases remain in contact with the heating surfaces.
4. More heat is absorbed and the flue gases are cooler when they enter the venting system than when the drafts are higher than needed.
5. Automatic draft control will also guard against incomplete combustion that will occur when manual dampers are closed too much. This is important with oil and coal fuels, but is tremendously important with gas as a fuel.

Because the draft needs of a heating unit are rigidly fixed, and because the draft forces being exerted are constantly varying, it is essential that a precisely accurate control be installed between the vent producing the draft and the unit using the draft. Note, however, that such a control must join the vent to its heating unit

(Figure 6–26). Any device that simply isolates a vent system from its heating unit does not control draft. It simply eliminates it as an oxygen source.

When the barometric draft control is installed on gas-fired equipment, the preferred location is part of the bull head tee [Figure 6–27 (a), (b), and (c)]. During normal operation the flue gases make a right angle turn behind the control but do not impinge on it. Should a downdraft occur, the air flowing in the opposite direction strikes the control directly, causing it to open outwardly, and venting the air into the room with a minimum of resistance. Entrained products of combustion are thus provided a greater relief.

field

FIGURE 6–26
A barometric draft control. (*Courtesy Conco, Field Control Division.*)

With oil or solid fuels, locate the draft control as shown in Figure 6–27 (d)–(j). The locations are recommended for normal updraft operation. The bull head tee is not recommended, except for gas-fired furnaces and boilers.

Parts (d)–(j) are acceptable for gas units where equipment room configuration does not permit the use of a bull head tee.

Parts (k)–(p) show the wrong way or poor locations. Even though (l), (m), and (n) appear to be bull head tees, they are incorrect. On correct installations, gases are directed toward the gate when the normal direction of flow is reversed. Part (k) shows the draft controls on top and bottom of a vent pipe, a placement that would cause the control to be inoperative. A draft control may be placed in a vent system at some point higher than the unit outlet. Part (p) would result in poor draft control operation.

The draft control should be placed as close as possible to the heating equipment except on forced-draft systems. The draft control must be in the same room as the heating unit and at least 12 inches beyond the stack switch on oil-fired units and a minimum of 18 inches from any combustible materials.

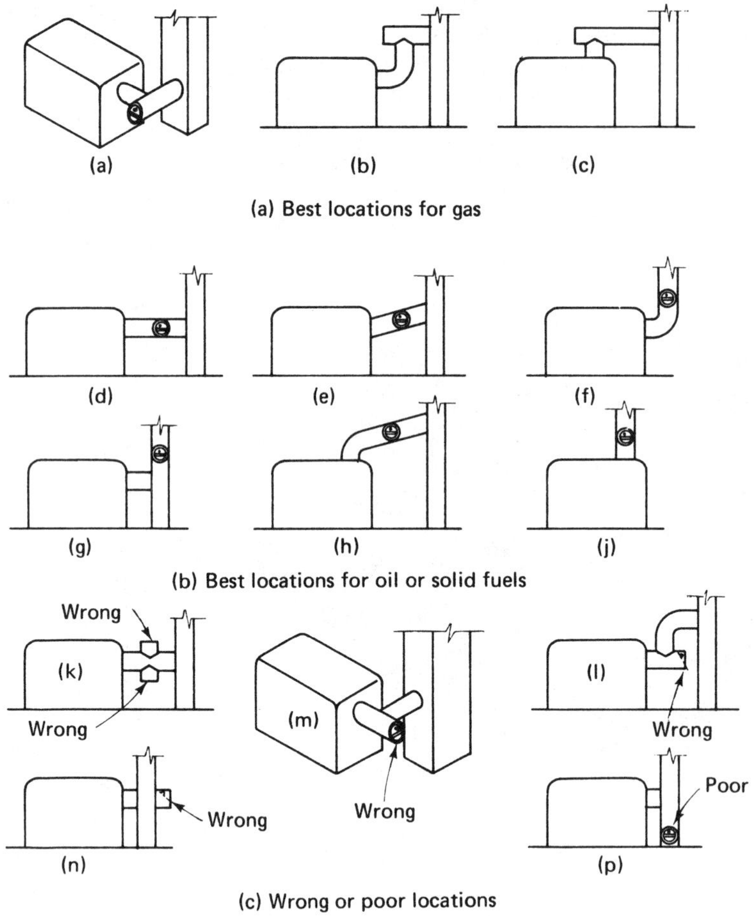
(a)
(b)
(c)
(a) Best locations for gas
(d)
(e)
(f)
(g)
(h)
(j)
(b) Best locations for oil or solid fuels
Wrong
(k)
Wrong
(m)
(l)
Wrong
Wrong
Wrong
Poor
(n)
(p)
(c) Wrong or poor locations

FIGURE 6–27 Draft control location. (Courtesy Conco, Field Control Division.)

OIL BURNER DRAFT

In the oil heating industry, the word *draft* is used to describe the slight vacuum, or suction, which exists inside most heating units. The amount of vacuum is *draft intensity. Draft volume,* on the other hand, specifies the volume (cubic feet) of gas that a chimney can handle in a given time. Draft intensity is measured in inches of water column. Just as a mercury barometer is used to measure atmospheric pressure in inches of mercury, a draft gauge is used to measure draft intensity (which is really pressure) in inches of water. *Natural draft* is actually thermal draft and occurs when gases that are heated expand so that a given volume of hot gas will weigh less than an equal volume of the same gas at a cooler temperature. Since hot combustion gases weigh less per volume than room air or outdoor air, they tend to rise.

The rising of these gases is contained and increased by enclosing the gases in a tall chimney. The vacuum or suction called draft is then created throughout this column of hot gases.

Currential draft occurs when high winds or air currents across the top of a chimney create a suction in the stack and draw the gases up. *Induced-draft* blowers can be used in the stack to supplement natural draft where necessary.

There are three factors which control how much draft a chimney can make:

- The height of the chimney: The higher the chimney, the greater the draft.
- The weight per unit volume of the hot combustion products: The hotter the gases, the greater the draft.
- The weight per unit volume of the air outside the home: The colder the outside air, the greater the draft.

Since the outside temperature and the flue gas temperature can change, the draft will not be constant. When the heating unit starts up, the chimney will be filled with cool gases. After the heating unit has operated for a while, the gases and the chimney surface will be warmer, and the draft will increase. Also, when the outside air temperature drops, the draft will increase. To indicate the effect of these changes, the information in Table 6–16 was determined for a 20-ft-high chimney. It can be seen that the draft produced by this chimney could be expected to vary from 0.11 to 0.136 inches of water column. The high draft is over 12 times more than the low draft. This large variation cannot be tolerated for the following reasons:

- Too little draft can reduce the combustion air delivery of the burner and can result in an increase in the production of smoke.
- Excessively high draft increases the air delivery of the burner fan and can increase air leakage into the heating plant, reducing CO_2 and raising the stack temperature, resulting in reduced operating efficiency.
- High draft during the burner off periods increases the standby heat losses up the chimney.

To understand why a varying draft causes these problems, it should be kept in mind that the air pressure (positive draft) caused by a flame retention burner fan

TABLE 6–16 Examples of draft changes in a chimney. (*Courtesy of R.W. Beckett Corp.*)

Condition	Outside Temperature, °F	Chimney Temperature, °F	Draft, "H_2O"
Winter start up	20	110	.050
Winter operation	20	400	.136
Fall start up	60	80	.011
Fall operation	60	400	.112

averages about 0.40 inches of water column in the air tube. If the combustion chamber has a draft of 0.10 inches of water and the burner fan provides a pressure of 0.40 inches of water column, the total force causing the air flow will be 0.50 inches of water. If the combustion chamber drops to 0.10 inches of water, the total pressure becomes 0.40 + 0.01 = 0.41 inches of water. This is a reduction in draft of about 18%, which will cause a reduction in the amount of air flowing into the combustion chamber. Remember what happens when the excess air is not properly adjusted? The burner will likely smoke as a result of this change. It is for this reason that the proper draft should be obtained before the air adjustment is set.

Because draft will not exist to any great amount during a cold start-up, the burner should not depend on the additional combustion air caused by the draft. The best way to be sure that the burner does not depend on this air is to set the burner for smoke-free combustion with a low overfire draft (0.01 to 0.02 inches of water). If a burner cannot produce smoke-free combustion under low draft conditions, there is something wrong with the burner or combustion chamber, and it should be corrected. Using a high draft setting to obtain enough combustion air for clean burning is like depending on a crutch, which is not always there. A burner which gives clean combustion only with high draft will cause smoke and soot anytime the chimney is not producing high draft.

In Chapter 2, we described the effect of air leaks, and perhaps you realize now that air leaks occur because of draft inside the heating plant. It is easy to see that less draft will cause less air leakage and produce a higher efficiency. Therefore, sealing air leaks can aid in improving heating plant efficiency.

Draft Regulators

From the previous information, it is clear that a constant draft is needed, and this draft should be no more than that which will just prevent the escape of combustion products into the home. Since natural draft as obtained from a chimney will vary, it is necessary to have some sort of regulation. The normal draft regulator for home heating plants is the so-called bypass, or air bleed, type as shown in Figure 6–28. This type of regulator is simply a swinging door which is counterweighted so that anytime the draft in the flue is higher than the regulator setting, the door is pulled in. When the door is pulled in, it allows room air to flow into the flue and mix with the combustion gases. Because the room air is much cooler than the flue gases, it cools them. When the cooler gases fill the chimney, there is less temperature difference between the chimney gases and the outside air and less draft is produced. If the draft is less than the regulator setting, the counterweight keeps the swinging door closed and only flue gases flow into the chimney. This gives the highest draft possible under those conditions.

It is important to understand that the function of a draft regulator is to maintain a stable or fixed draft through the heating equipment within the limits of available draft of the chimney, by means of an adjustable barometric damper.

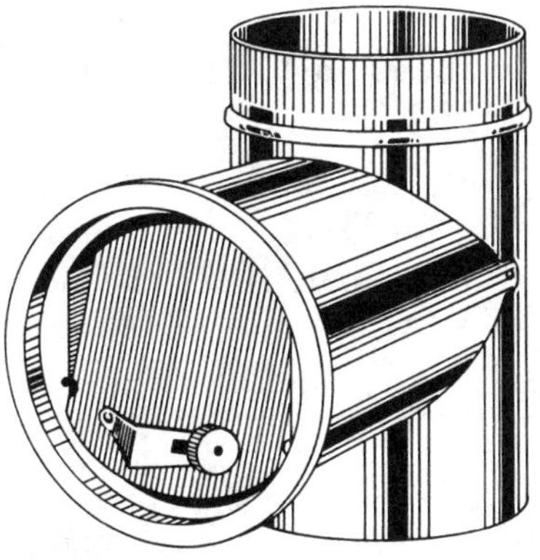

FIGURE 6–28
Draft control. (*Courtesy of R.W. Beckett Corp.*)

Draft can be measured by using a draft gauge. It cannot be estimated or "eyeballed." The draft should be checked at two different locations in the heating plant: (1) over the fire, which indicates the firebox draft condition; and (2) in the breech connection.

Draft over the fire. The draft over the fire is the most important and should be measured first. The overfire draft must be constant so that the burner air delivery will not change. The overfire draft must be at the lowest level which will just prevent the escape of combustion products into the home under all operating conditions. Normally an overfire draft of 0.01 to 0.02 inches of water column (WC) will be high enough to prevent leakage of combustion products and still not cause large air leaks or standby losses.

If the overfire draft is higher than 0.02 inches, the draft regulator weight should be adjusted to allow the regulator door to open more. If the regulator door is already wide open, a second regulator should be installed in the stack pipe and adjusted. If the draft is below 0.01 inches the draft regulator weight should be adjusted to just close the regulator door. Do not move the weight more than necessary to close the door. Never wire or weight a regulator so that it can never open. There may be times when the outside air is colder, or the chimney hotter, or high wind is affecting the draft, and the draft needs regulation.

The overfire draft is also affected by soot buildup on the heat exchanger surfaces. As the soot builds up, the heat exchanger passages are reduced and a greater resistance to the flow of gases is created. This causes the overfire draft to drop. As the overfire draft drops, the burner air delivery is reduced and the flame becomes even more smoky. It is a vicious cycle which gets increasingly worse.

Draft at the breech connection. After the overfire draft is set, the draft at the breech connection should be measured. The breech draft will normally be slightly more than the overfire draft because the flow of gases is restricted (slowed down) in the heat exchanger. This restriction, or lack of it, is a clue to the design and condition of the heat exchanger. A clean heat exchanger of good design will cause the breech draft to be in the range of 0.03 to 0.06 inches when the overfire draft is 0.01 to 0.02 inches.

Flue and Chimney Exhaust

The following are some recommendations for the flue and chimney exhaust piping:

1. Flue pipe: The flue pipe should be the same size as the breech connection on the heating plant. For modern oil heating plants, this should cause no problem in the sizing of the flue pipe. The sizes generally are 6 inches for firing rates under 1 gph, 7 inches for firing rates to 1.5 gph, and 8 inches for 1.5 to 2.00 gph firing rates. The flue pipe should be as short as possible and installed so that it has a continuous rise from the heating plant to the chimney. Elbows should be minimized and the pipe should be joined with sheetmetal screws and straps.

   The draft regulator should be installed in the flue pipe before it contacts the chimney and after the stack primary control, if one is used. Make sure that the draft regulator diameter is at least as large as the flue pipe diameter.
2. Chimney: Table 6–17 shows the recommended size and height for chimneys based on Btu input. Figure 6–27 provides information about some of the common chimney problems and their possible solution.

TABLE 6–17 Provided to inform you of some of the common chimney problems and their possible solutions. (*Courtesy of R.W. Beckett Corp.*)

Gross Btu Input	Rectangular Tile	Round Tile	Minimum Height
144,000	8½″ x 8½″	8″	20 feet
235,000	8½″ x 13″	10″	30 feet
374,000	13″ x 13″	12″	35 feet
516,000	13″ x 18″	14″	40 feet
612,000	—	15″	45 feet
768,000	18″ x 18″	—	50 feet
960,000	20″ x 20″	18″	55 feet

POWER VENTING

A power-supplied pressure increase can add to gravity draft or deliver all the energy required for venting. Many different types of heating equipment have forced- or induced-draft burner components capable of producing enough pressure for satisfactory power venting. For any common gas equipment depending on gravity for combustion air flow, power venting can be obtained with a fan or booster at the vent outlet, inlet, or anywhere in between.

Power venting provides safe, reliable equipment operation under a number of adverse conditions. It can also solve a number of common problems. For example:

1. Long horizontal vents for which access to the outside was prohibited by the building construction.
2. Too small a vent size due to space limitations or because of increasing the requirements of existing venting systems.
3. An excessive number of turns that create high pressure losses and reduce gravity flow.
4. A need for more draft than a short vent can supply, such as on roof mount units.
5. Erratic or inadequate venting caused by winds, adverse internal pressures, restricted air supply, or indoor-outdoor temperature changes.
6. A need for more dilution air to lower vent gas temperature, thus reducing vent heat losses to surroundings.
7. An improper design of a combined vent system, such as with insufficient connector rise.
8. Roof location problems brought about by penthouses, nearby mechanical installations, smoke dissipation, etc.

The location recommended for any power vent device or draft booster is near the vent outlet, preferably outdoor above the roof. Thus, the entire venting system is kept under a negative pressure, and whether the system is an individual or combined vent, draft is ensured to all interconnecting draft diverters.

GAS BURNER COMBUSTION TESTING

Automatically controlled gas heating represents the ultimate in trouble-free, economical heating comfort. The types of gas burners used in most domestic heating units are termed *atmospheric injection burners*. This term is derived from the fact that gas pressure is used at only 2 to 5 inches of water column pressure and the air is furnished by atmospheric pressure. The following discussion will cover only the atmospheric injection burner and will not apply to power gas burners.

The atmospheric injection burner is tremendously flexible under adverse conditions. However, full advantage of this diversity can be obtained only if proper use is made of the adjustment means on the burner. Easy to follow, well proven procedures of adjusting and testing gas heating equipment have been organized by the American Gas Association in cooperation with a large group of authorities, including the American Standards Association. Furthermore, many cities and towns have already legislated, or are considering legislating, ordinances governing the installation and adjustment of gas-fired heating equipment. All these codes include requirements for a combustion test. The principal objective of the combustion test on gas burning equipment is to analyze the flue gases for (1) safe combustion, and (2) maximum operation of the heating equipment.

Air for Combustion

In an atmospheric injection-type burner, there are two sources of combustion air–primary and secondary (Figure 6–29). Primary air is the air that is mixed with the gas before it burns. The air is drawn into the burner by the flow of gas through the venturi. The air starts moving into the burner the instant the gas is turned on because of the action of the gas flow through the venturi. As combustion of this gas-air mixture starts burning, additional air is sucked into the furnace by the draft through the furnace. This additional air is called secondary air. It is supplied around the burner head and mixes with the gas after it has been ignited.

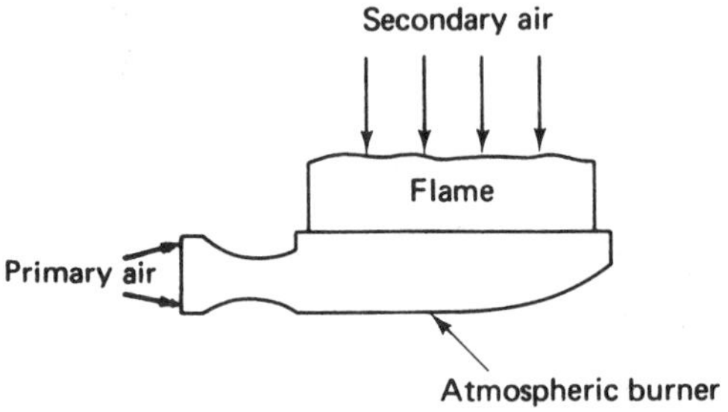
Secondary air
Flame
Primary air
Atmospheric burner

FIGURE 6–29
Atmospheric gas burner.

None of the atmospheric injection-type burners used at present induce enough primary air for complete combustion. Therefore secondary air must be provided in sufficient quantity to complete combustion and thus eliminate the possibility of CO gas in the products of combustion. Since CO is an extremely dangerous gas, the basic requirement in installing and servicing any gas burning equipment is that no unit will be left with any CO in the flue products.

Air Adjustments and Excess Air

The primary air is adjusted by means of the primary air shutter for the proper flame characteristic. If, even with the air shutter completely open, the flame characteristics indicate insufficient primary air, this lack may be corrected by reducing the size of the orifice and increasing the manifold gas pressure, thereby maintaining the correct gas input. For atmospheric burners with normal limits of adjustment, this primary air setting has little effect on carbon dioxide (CO_2). It is the secondary air setting that greatly establishes the percentage of CO_2 in the flue products.

The secondary air should be adjusted to prevent excessive agitation of the flame and yet provide sufficient air for a CO-free combustion. On gas-designed equipment, the amount of secondary air is fixed, and no secondary air adjustment can be made. However, many service technicians find it desirable to check the CO_2 content and flue gas temperature as an added check on the fuel input adjustment since these tests will indicate any error in metering or reading during this important adjustment.

The CO_2 in the combustion products is really an indication of the amount of excess air supplied for the combustion of the gas being burned. Excess air is the quantity of air admitted to the furnace in excess of that required for perfect combustion, which would produce the ultimate percentage of CO_2 in the flue products. Table 6–18 indicates the ultimate CO_2 percentage for common types of gaseous fuel. The addition of some excess air is required for all fuel gases to ensure safe combustion; thus a gas burner is never adjusted to provide the ultimate percentage of CO_2; levels from 70% of the ultimate to 80% of the ultimate CO_2 are common.

TABLE 6–18 Ultimate CO_2 percentage, characteristics of fuel gases used straight or mixed for domestic service. (*Courtesy of Bacharach Instrument Co.*)

	Specific Gravity	Btu per Ft^3—Gross	Ultimate Percent CO_2
Natural Gas	0.57–0.70	900–1200	11.7–12.2
Carbureted Water Gas	0.65–0.71	500–600	15.8–17.3
Coke Oven Gas	0.35–0.49	500–600	10.3–12.1
Coal Gas (Retort Process)	0.41–0.49	500–600	11.5–12.1
Refinery Oil Gas	0.88–1.00	1470–1660	13.6–14.4
Butane Gas	1.95–2.04	3180–3260	14.0
Propane Gas	1.52–1.57	2500–2580	13.7

If the ultimate CO_2 is known, the excess air may be determined by measuring either percentage of CO_2 or percentage of oxygen. The percentage of oxygen at a given excess air value varies less for different fuel gases than does the percentage of CO_2.

The amount of excess air required for safe heating unit operation cannot be stated specifically because most units have considerable tolerances. Table 6–19 lists generally accepted limits of excess air for flue-connected heating units.

TABLE 6–19 Limits of percent excess air for flue-connected gas appliances. (*Courtesy of Bacharach Instrument Co.*)

Central Heating Furnace or Boiler	25–50%
Gas Water Heater	25–50%
Space Heater	50–100%

Too much excess air is the principal reason for inefficient combustion. If the CO_2 test shows an abnormally high excess air, the secondary air adjustment means alone may not provide the desired results; it may be necessary also to restrict the flow of combustion products. This may be accomplished by reducing the flue pipe size connecting the equipment to the draft diverter. The purpose is to restrict the discharge of combustion products from the equipment to such a point that there is a slight positive pressure in the bottom portion of the furnace.

CO_2 Check of Secondary Air Adjustment

When the neutral pressure point has been adjusted to the proper position, either by using a vent pipe of smaller diameter or by inserting into the flue pipe a neutral pressure point slide, the combustion air shutter should be adjusted so that the CO_2 percentage in the flue products is in accordance with the applicable codes or standards. The CO_2 test must be made at the inlet side of the draft hood (Figure 6–30). If the CO_2 reading is not within the recommended limits, the combustion air shutter should be readjusted and then the neutral pressure point should be rechecked and reset if necessary.

FIGURE 6–30
Checking CO_2 of furnace. (*Courtesy of Bacharach Instrument Co.*)

The primary air is adjusted by means of the primary air shutter for proper flame characteristics in accordance with the burner manufacturer's instructions. If even with the air shutter wide open the flame appearance indicates insufficient primary air, this deficiency may be corrected by decreasing the orifice size and increasing the manifold pressure, thereby maintaining the proper gas input. For atmospheric-type burners with normal limits of adjustment, the primary air setting has little affect on CO_2. It is the secondary air setting which largely establishes the percentage of CO_2. It is the secondary air setting which largely establishes the percentage of CO_2 in the flue gas products.

The secondary air should be adjusted to prevent excessive aeration of the flue products and yet provide sufficient air for CO-free combustion. On gas-designed equipment the amount of secondary air is fixed, and no secondary air adjustment is required. On conversion burners, however, the secondary air setting is a necessary step in the adjustment procedure, and the CO_2 test provides a convenient means of making this adjustment quickly and accurately. Many service technicians find that it is desirable to check the CO_2 and flue gas temperature on gas-designed equipment, as an added check on the fuel input adjustment, since these tests will show up any error in metering or reading during this important adjustment.

The CO_2 in the combustion products (in comparison with the CO_2 which is theoretically produced by perfect combustion) is really a measure of the amount of excess air supplied for the combustion of the gas being burned. This amount of air is introduced into the furnace in excess of that required for perfect combustion (no excess air), which would produce the ultimate CO_2 in the flue gas products. Table 6–18 shows the ultimate percent CO_2 for common types of fuel gases. The addition of some excess air is required for all fuel gases to ensure safe combustion; thus a gas burner is never adjusted to produce the ultimate percent CO_2. Generally accepted practice is to adjust the air supply for CO_2 levels from 70 to 80% of the ultimate CO_2. For natural gas (11.8% ultimate CO_2), this corresponds to a CO_2 percent in the flue gas products between approximately 8 1/4 and 9 1/2 %.

If the ultimate CO_2 is known, the excess air may be determined by measuring either percent CO_2 or the percent oxygen. The two graphs in Figure 6–31 show the excess air percentage in terms of CO_2 and oxygen, for natural gas and carburetted water gas. Percent oxygen at a given excess air value varies less for different fuel gases than does the percent CO_2. If the ultimate percent of CO_2 is constant, it is thoroughly feasible to use CO_2 measurements of excess air. This is almost universal because approximately 95% of all domestic gas installations are supplied with natural gas or mixed gas, both of which have a practically constant ultimate percent of CO_2. However, in areas where manufactured gas is supplied the ultimate CO_2 in the flue products may vary appreciably because of differences in manufacturing processes and mixtures. In such areas the municipal codes usually specify that O_2 shall be measured in the flue products instead of, or in addition to, the CO_2.

The amount of excess air required for safe gas appliance operation cannot be stated specifically because most appliances have considerable tolerances. Table 6–19 lists generally accepted limits of excess air for flue connected domestic appliances.

Too much excess air is the principal reason for inefficient combustion. If the CO_2 test shows an unnecessary high excess air, the secondary air adjustment means alone may not provide the desired results; it may be necessary also to restrict the flow of combustion products. On conversion burner installations this may be accomplished by either reducing the flue pipe size connecting the furnace (boiler) to the draft diverter or by installing a neutral pressure point adjuster between the flue collar and the draft diverter. See Figure 6–32. In either case, the purpose is to restrict the discharge of combustion products from the furnace (boiler) to such a degree that there is slightly positive pressure in the top portion and a slightly negative pressure in the bottom portion of the furnace (boiler).

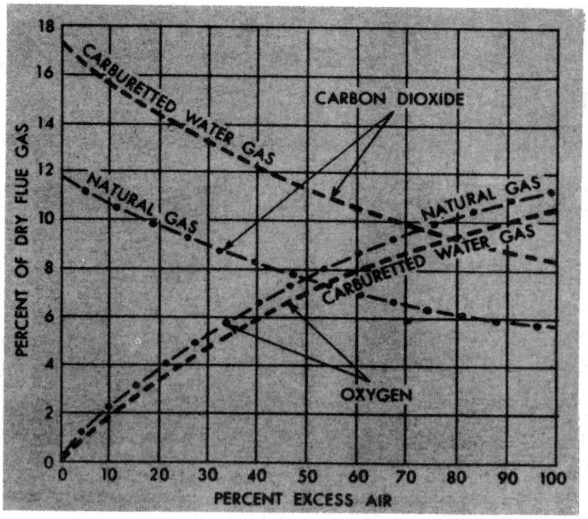
CARBURETTED WATER GAS
CARBON DIOXIDE
NATURAL GAS
NATURAL GAS
CARBURETTED WATER GAS
OXYGEN
PERCENT OF DRY FLUE GAS
PERCENT EXCESS AIR
0
2
4
6
8
10
12
14
16
18
0
10
20
30
40
50
60
70
80
90
100

FIGURE 6–31
Relationship of excess air to CO_2 and oxygen in flue gases for natural gas and carburetted water gas. *(Courtesy of Bacharach Instrument Co.)*

Combustion Efficiency

The percentage of CO_2 and the temperature of the flue gases are the two measurements that establish the percentage of combustion efficiency. These two measurements should be taken on the inlet side of the draft diverter. The flue gas thermometer is inserted into the same hole for the CO_2 test.

Any heat contained in the flue gases at this point will be mostly wasted to the vent. Obviously, the higher the temperature of these flue gases, the greater will be the loss of heat that has not been absorbed by the heat exchanger. Furthermore, these flue gases cannot be cooled below the room air temperature, so that the difference between the actual flue gas temperature and room air temperature (known as the *net flue gas temperature*) represents a measure of the useable heat loss in the vent. The relationship between net flue gas temperature, percent CO_2, and combustion efficiency can be shown in a curve or table. However, the use of a slide rule calculator for obtaining the combustion efficiency is usually preferred. By means of such a calculator (Figure 6–33), setting one slide to the net stack temperature and the other to the percent CO_2 will permit direct readings of percent combustion efficiency. The percent combustion efficiency is an indication of the useful heat obtained from the gas being burned. It is expressed in a percentage of the total heat produced by the burning of the gas, assuming complete combustion with no excess air. For example, a combustion efficiency of 60% means that 60% of the total heat input has been absorbed by the furnace. The percentage of stack loss is an indication of the heat that is wasted in the hot vent gases instead of being absorbed by the furnace. The percentage of stack loss is obtained by subtracting the percentage of combustion efficiency from 100%.

A word of caution: The stack loss percentage is not a measure of what can be saved by readjustment of the equipment since the flue gases must still be allowed to escape. Furthermore, combustion efficiency must not be confused with overall

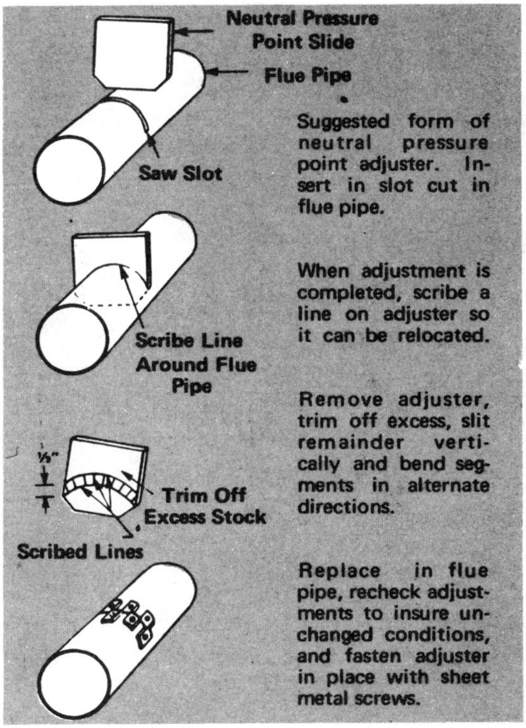
Neutral Pressure
Point Slide
Flue Pipe
Suggested form of neutral pressure point adjuster. Insert in slot cut in flue pipe.
Saw Slot
When adjustment is completed, scribe a line on adjuster so it can be relocated.
Scribe Line
Around Flue
Pipe
Remove adjuster, trim off excess, slit remainder vertically and bend segments in alternate directions.
½"
Trim Off
Excess Stock
Scribed Lines
Replace in flue pipe, recheck adjustments to insure unchanged conditions, and fasten adjuster in place with sheet metal screws.

FIGURE 6–32
Typical installation of neutral pressure point adjuster. (*Courtesy of Bacharach Instrument Co.*)

efficiency of the furnace, which must include allowance for radiation and miscellaneous heat losses.

As the percentage of CO_2 decreases and stack temperature increases, the heat wasted to the vent increases and combustion efficiency decreases. With ultimate CO_2 (no excess air) and flue gas temperature at room temperature, all available heat has been extracted and combustion efficiency is 100%. Gas-burning equipment should always be capable of 75% combustion efficiency. Anything higher than 80% may adversely affect vent draft. In some cases it may cause a corrosive moisture to condense in the flue and at an extreme may result in incomplete combustion.

CO-Free Combustion

As has already been stated, it is important to make sure that the products of combustion do not contain any CO. It is possible that overfired equipment could have CO in the flue gases, especially with flame impingement on cold surfaces even with excess air, CO_2, and O_2 within acceptable limits. The flue gas sample for the carbon monoxide check must be taken at the inlet side of the draft diverter to make sure that the sample is not diluted by air being drawn in at the draft diverter.

The monoxor (Figure 6–34) is a practical, reliable instrument for testing CO concentrations in flue gas samples. According to gas industry standards, CO-free combustion is defined as that which produces less than 0.04% CO in an air-free sample of the flue gas.

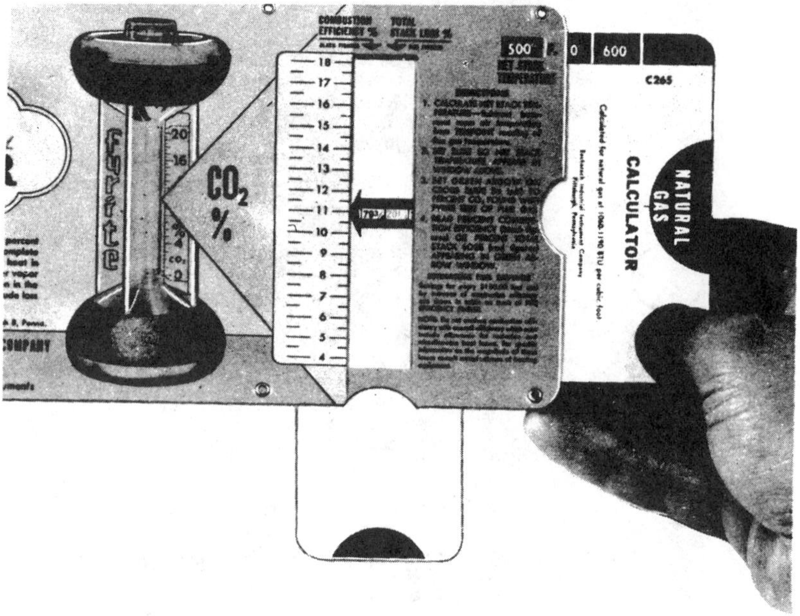
COMBUSTION EFFICIENCY %
TOTAL STACK LOSS %
500
600
C265
CO2 %
CALCULATOR
NATURAL GAS
COMPANY

FIGURE 6–33 Combustion efficiency and stack loss calculator. (*Courtesy of Bacharach Instrument Co.*)

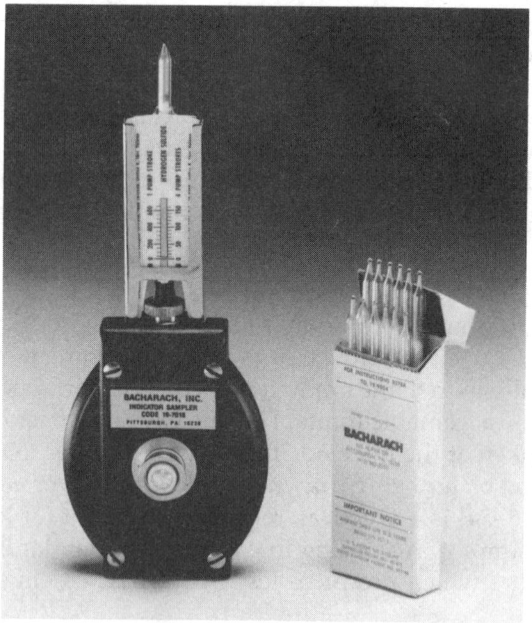
BACHARACH, INC.
INDICATOR SAMPLER
BACHARACH
IMPORTANT NOTICE

FIGURE 6–34
Monoxor indicator. (*Courtesy of Bacharach Instrument Co.*)

The CO test is similar to the CO_2 test but, of course, is made with the monoxor instead of the "firite". A flue gas sample is taken through the same hole used for the CO_2 test. The flue gas sample is first drawn into the collecting bladder of the sampling assembly and subsequently drawn through the glass indicator tube of the monoxor. This indicator tube is inserted into the rubber connector of the monoxor sampler that contains a suction pump operated by a push button on the outside of the sampler housing. These two steps are the flue gas sampling procedure shown in Figure 6–35.

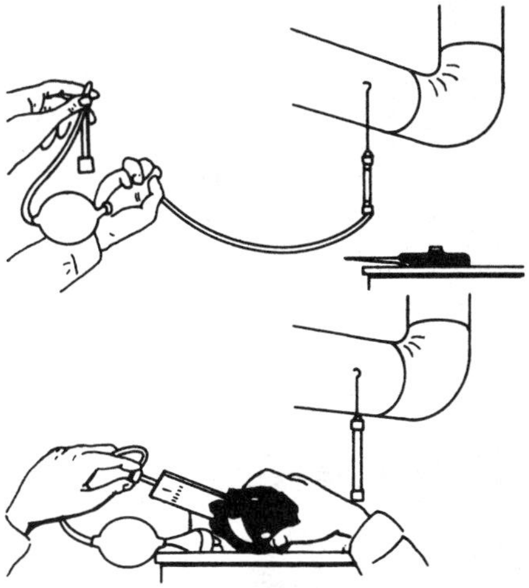

FIGURE 6–35
Sampling flue gas for monoxor CO test. (*Courtesy of Bacharach Instrument Co.*)

The monoxor indicator tube is filled with a yellow-colored CO-sensitive chemical. The CO content of the sample is absorbed by the yellow-colored chemical causing the production of a dark-brownish stain. The length of this stain is directly proportional to the CO concentration in the sample, and the stain length is measured on the instrument's etched metal scale. The scale is calibrated directly in CO percentage. The monoxor indicator tube can be used for two tests if stain from the first test penetrates less than one half the length of the yellow chemical. The unstained end of the tube may be used for the second test by reversing the tube's position in the monoxor sampler.

Draft Diverter and Flue

After the air adjustments and the flue gas tests have been completed, the draft diverter should be checked with the neutral pressure point gauge. (See Figure 6–36.) This can be done by holding the fish tail tip of the instrument just inside the edge of the diverter opening (Figure 6–37). A negative pressure (draft) should be indicated. It is also sound practice to check the flue draft on the vent side of the diverter to be certain that the vent is adequate for moving the products of combustion from the draft diverter to the outside atmosphere under all weather conditions. Otherwise,

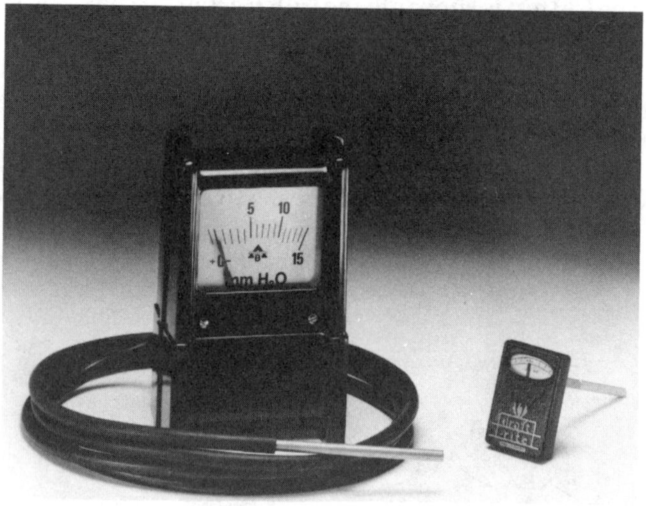
5
10
15
cm H₂O

FIGURE 6–36 Neutral pressure point gauge. (*Courtesy of Bacharach Instrument Co.*)

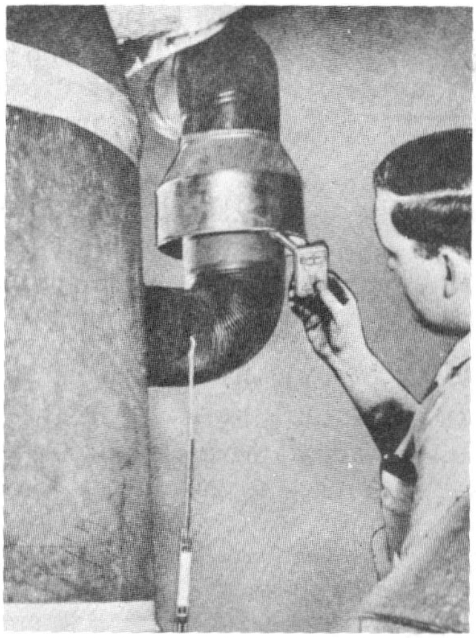

FIGURE 6–37
Using neutral pressure point indicator. (*Courtesy of Bacharach Instrument Co.*)

there will be a smothering of the flame that will produce CO. The location of the check point is above the draft diverter (Figure 6–38). The gauge illustrated has a range of 0.14 inch water column updraft, which is the preferred range for this test. The draft check must be made with the burner operating. Inadequate venting capacity of the vent will be indicated by (1) very low updraft, (2) temporary downdraft, or (3) at an extreme, a steady downdraft reading.

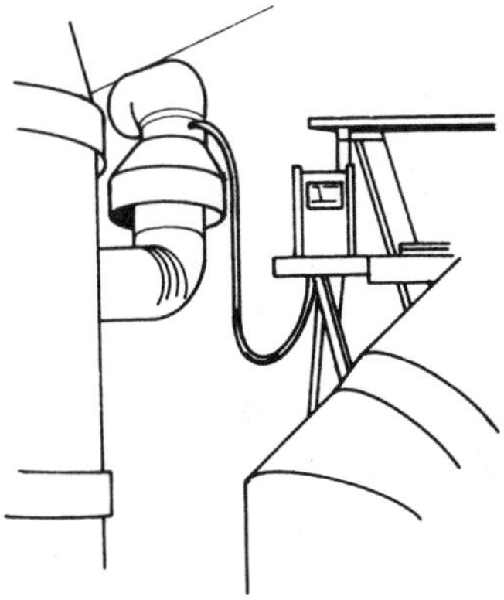

FIGURE 6–38
Using draft gauge. (*Courtesy of Bacharach Instrument Co.*)

All of these will indicate either inadequate vent size or height, a blocked vent, or vent downdraft caused by roof obstructions or nearby structures. Appropriate steps must be taken to remedy these defective conditions.

AUTOMATIC FLUE DAMPER

When the furnace blower is operating, air must be drawn from inside the home and be allowed to pass up through the venting system to the atmosphere with the products of combustion. This flow of air is not required when the burner is not functioning. The automatic flue damper prevents this wasteful flow of heated household air by opening and closing on demand from the thermostat. The automatic flue damper can reduce fuel consumption as much as 16%.

The automatic flue damper is installed on top of the draft diverter before the venting system is connected (Figure 6–39). When the thermostat calls for heat, the automatic flue damper blade moves to the full open position before the burner can ignite (Figure 6–40). While the burner is on, the damper blade stays in the full open position. A spring-loaded mechanism will open the damper fully in case of a power failure.

When the burner shuts off, an electric motor automatically closes the automatic flue damper to stop the flow of heated air out the venting system (Figure 6–41), thus providing a savings during the heating season.

REVIEW QUESTIONS

1. What is the purpose of a properly designed venting system?
2. Outline the flow of air through a heating unit, from the combustion air to the vent termination point.
3. Where is dilution air taken into the vent system?
4. How much air is required for dilution air?

FIGURE 6–39
Automatic flue damper location.
(*Courtesy Carrier Corporation.*)

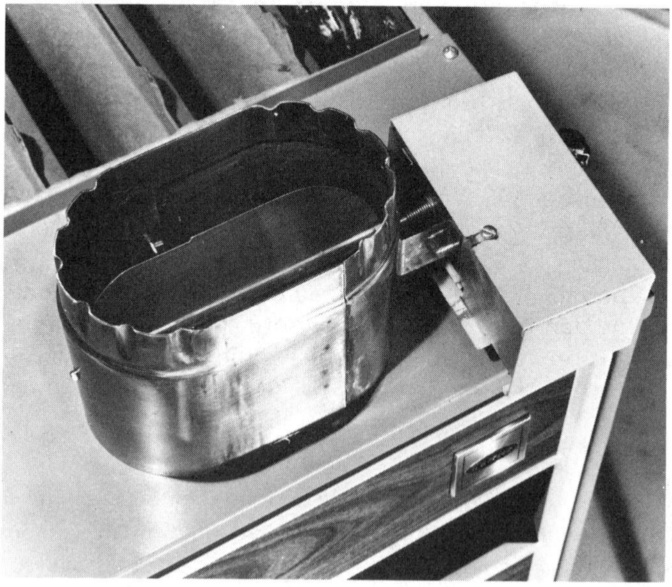

FIGURE 6–40 Automatic flue damper open. (*Courtesy Carrier Corporation.*)

FIGURE 6–41 Automatic flue damper closed. (*Courtesy Carrier Corporation*).

5. What is the most efficient venting system?
6. What is the most desirable route for a venting system?
7. Name the three functions of air to a gas heating unit.
8. What causes air to flow into and through a heating unit?
9. What are the purposes of dilution air?
10. Name the factors that govern the degree of draft force exerted at a given barometric reading.
11. Name the four variables that affect the draft of a venting system.
12. What is the required clearance between the type B-1 and type B vent pipe sizes through 12 inches in diameter and a combustible material?
13. What is the minimum required distance between the two combustion air grills to a furnace closet?
14. To what rating should a vent system be designed?
15. Why is a single-wall vent pipe not recommended for use in type B venting systems?
16. What is the minimum distance from the flue collar to the first turn in a vent system?
17. Are condensing gas furnaces allowed to be connected to a type B vent system?
18. How is the condensing of water vapor from the products of combustion minimized by the use of a sizing table?
19. When should the next larger size vent system be used to prevent any problems?
20. What part of the vent system prevents rain and debris from entering the vent and helps prevent downdrafts?

21. What is the minimum vent height required on wall furnaces?
22. What is the minimum horizontal distance from the roof to the vent termination point?
23. What is the minimum distance from a vertical wall and a vent termination point?
24. What refers to the maximum appliance input rating of a category I appliance with a fan-assisted combustion system that could be attached to the vent?
25. What part of the vent system provides for ready escape of flue gases from the appliance in the event of no draft, backdraft, or vent stoppage beyond the draft hood?
26. What part of the venting system connects a fuel gas-burning appliance to a vent or chimney?
27. Generally, at what size is a single-appliance vent system designed?
28. In a combined vent system, what size must the interconnecting tees be?
29. From where must combustion air always be taken?
30. Where is the combustion air fan located on forced-draft systems?
31. At what AFUE (annual fuel use efficiency) is a furnace considered to be high efficiency?
32. Define a condensing gas furnace.
33. Define an ANSI category IV furnace.
34. How are the products of combustion disposed of?
35. What is the prime function of a barometric draft control?
36. What will an automatic draft control guard against when manual dampers are closed too much?
37. Where must a draft control be located?
38. What is the recommended location for a power vent device or draft control?
39. What are the two sources of combustion air on an atmospheric injection-type burner?
40. Why must secondary air be provided to a burner?
41. What does the presence of CO_2 in combustion products really indicate?
42. At what percentage of the ultimate CO_2 is the generally accepted practice to adjust the air supply?
43. What is the principal reason for insufficient combustion?
44. Define net flue gas temperature.
45. When checking for carbon monoxide, where must the flue gas sample be taken?
46. Define draft as it is used in combustion.
47. What could be the problem if a burner cannot produce smoke-free combustion under low fire conditions?
48. At what point should the draft first be measured?
49. What should the breach draft be?

7

electric heating

The objectives of this chapter are:

- To provide you with the operating principles of electric heating.
- To describe heating element construction to you.
- To cause you to become aware of the different methods of using electricity for heating purposes.
- To introduce you to the method of repairing broken heating elements.

Electric heating is beneficial from the standpoint of ecology because the fuel burned in producing electricity is consumed in one remote area. At the electricity consumer level, there is no waste to be removed. There are no vent pipes to be installed or ashes to be removed. The homeowner realizes clean, efficient heat.

However, certain adjustments must be made for electric heating. The user must be willing to pay higher electric bills. There is also the increased initial cost of construction. This cost is consumed in the added electrical service and wiring and the extra insulation required when electric heating is used. The contractor must be more knowledgeable about construction codes and service required.

Electric heating is probably the most efficient form of heating known today. Theoretically, when 1 Btu of electrical energy is used, 1 Btu of heat energy is released. In other words, the Btu input equals the Btu output. When one kW of electricity is converted to heat, 3,410 Btu are released.

TYPES OF ELECTRIC HEATING EQUIPMENT

There are several operations that must be considered in a discussion of electric heating: resistance heating, heat pump, and a combination of the two.

Resistance Heating

Resistance heating involves a large number of operations, all of which are merely variations but are considered different in use and application. Under resistance heating, we may list resistance elements, baseboard elements, heating cable, portable and unit-type heaters, and duct heaters. However, regardless of the use or application, resistance heating wire is selected for high resistance per unit of length, and for the stability of this resistance at high temperatures. The type of wire used for this purpose is a high nickel chrome alloy resistance wire. Resistance heating does not use a centrally heated medium. The energy is expended directly into the air stream.

Durability also must be considered when determining the wire used for resistance heating. The high nickel content in this wire retards oxidation and separation of the wire. If, however, the resistance wire does become broken, it may be repaired by cleaning the two ends, overlapping them, and applying borax liberally to the two pieces. Apply heat to the cleaned area until the two metals fuse together. After the wire has cooled to room temperature, remove the excess borax with a wet cloth. If this borax is allowed to remain on the wire, an immediate burnout will occur because the borax causes excessive spot heating.

Resistance Elements

Resistance elements are more commonly used in forced-air central heating systems. They are termed such to differentiate them from duct heater strips, and are

FIGURE 7–1 Resistance element installed in a furnace frame. (*Courtesy NORDYNE*)

mounted directly in the air stream inside the furnace (Figure 7–1). The elements are staggered for more uniform heat transfer, to eliminate hotspots, and to ensure maximum heat transfer. Each element is protected by a temperature limit switch and a fuse link. These units are located inside the furnace proper (Figure 7–2) in an element compartment. These units emit approximately 3,410 Btu/kW input.

Baseboard Elements

Baseboard elements may be used for the whole house or for supplementary heating. These elements are usually enclosed in a housing that provides safety as well as efficient use of the available heat.

These units are available in a wide variety of sizes and may be joined together, usually end to end, to provide the Btu required for any given installation. The ideal location of installation is at the point of greatest heat loss. The electrical element in these units is normally of the enclosed type. That is, the electric resistance has fins

FIGURE 7–2 Electric element compartment.

added for increasing the surface area of the heating element and increasing efficiency. Baseboard units have a direct Btu output per W input. An example of this may be seen in Table 7–1, where we have the applied voltage, the element length, the wattage rating, and the corresponding Btu output rating. A similar table should be consulted when baseboard heating is planned. All manufacturers have such tables representing these values for their equipment.

TABLE 7–1 Wattage and Btu/h rating (typical)

Volts	Length	Watts	Btu/hr.	Cat. No.
120	28 in.	500	1707	BB-SC-281
	40 in.	750	2360	BB-SC-401
	48 in.	1000	3413	BB-SC-481
	60 in.	1250	4266	BB-SC-601
208/240	28 in.	375/500	1280/1707	BB-SC-284
	40 in.	560/750	1927/2360	BB-SC-404
	48 in.	750/1000	2360/3413	BB-SC-484
	60 in.	935/1250	3108/4266	BB-SC-604
	72 in.	1125/1500	3840/5118	BB-SC-724
	96 in.	1500/2000	5118/6824	BB-SC-964
	120 in.	1875/2500	6399/8547	BB-SC-1204
240/277	28 in.	375/500	1280/1707	BB-SC-287
	40 in.	560/750	1927/2360	BB-SC-407
	48 in.	750/1000	2360/3413	BB-SC-487
	60 in.	935/1250	3108/4266	BB-SC-607
	72 in.	1125/1500	3840/5118	BB-SC-727
	96 in.	1500/2000	5118/6824	BB-SC-967
	120 in.	1875/2500	6399/8547	BB-SC-1207

(Courtesy of Chromalox Comfort Conditioning Division, Emerson Electric Co.)

Duct Heaters

Electric duct heaters are factory-assembled units consisting of a steel frame, open coil heating elements, and an integral control compartment. These matched combinations are fabricated to order in a wide range of standard and custom sizes, heating capacities, and control modes. They are prewired at the factory, inspected, and ready for installation at the job site.

The duct heater frame is made in two basic configurations—slip-in (Figure 7–3) or flanged (Figure 7–4). The slip-in model is normally used in ducts up to 72 inches wide by 36 inches high, while the flanged model is used in larger ducts or where duct layout would make it impossible or impractical to use the slip-in type.

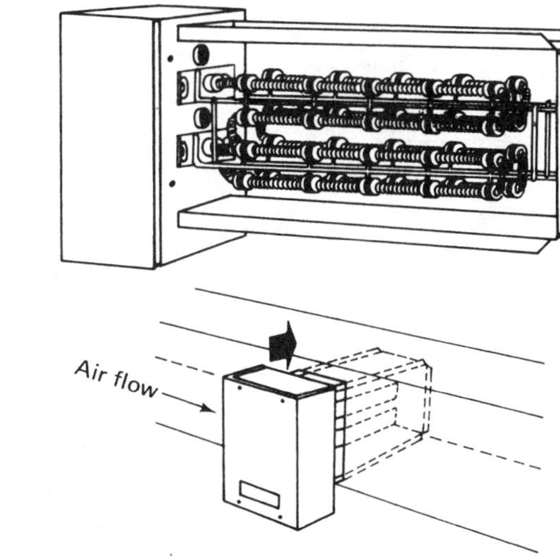
Air flow

FIGURE 7–3
Slip-in duct heater. (*Courtesy Gould Inc., Heating Element Division*)

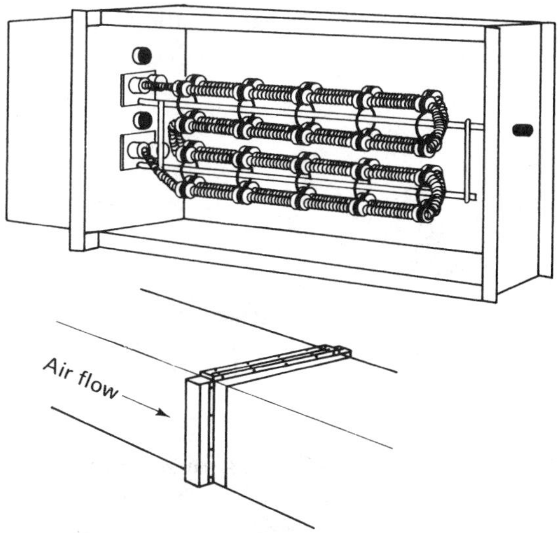
Air flow

FIGURE 7–4
Flange type duct heater. (*Courtesy Gould Inc., Heating Element Division*)

The slip-in heater, which is standard, is designed so that the heater frame is slightly smaller than the duct dimensions. The frame is inserted through a rectangular opening cut in the side of the duct, with the face of the heater at a right angle to the air stream. It is secured in place by sheetmetal screws running through the control compartment.

The flanged-type heater is designed so that the frame matches the duct dimensions. The frame is then attached directly to the external flanges of the duct.

Open coil heating elements (Figure 7–5) are a high nickel chrome alloy wire (usually 80% nickel and 20% chromium) physically designed to meet the application requirements of duct heaters. Essentially 100% of the electrical energy input to

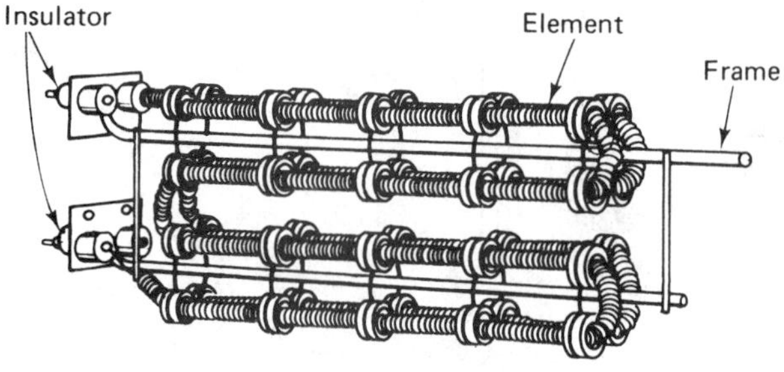
Insulator
Element
Frame

FIGURE 7–5
Heater coils and rack. (*Courtesy Gould Inc., Heating Element Division*)

the coil is converted to heat energy, regardless of the temperature of the surrounding air or the velocity with which it passes over the heating element.

An iron-chrome-aluminum alloy is also used for coils in special low wattage applications. Regardless of the type of wire used, ceramic bushings (Figure 7–6) are used that insulate the coil from surrounding sheetmetal, float freely in embossed openings, and prevent binding and cracking as the heater cycles. The bushings are held in place by curved sheetmetal tabs. Their extra-heavy body enables the bushings to withstand high humidity conditions and a 2,000-V dielectric test.

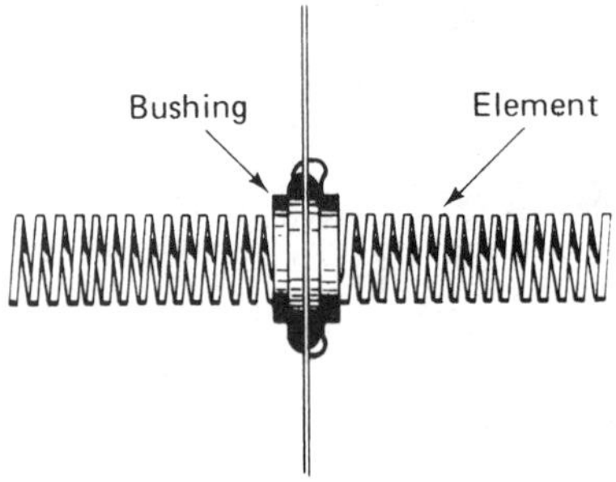
Bushing
Element

FIGURE 7–6
Coil and bushing. (*Courtesy Industrial Engineering and Equipment Co.*)

Stainless steel terminals (Figure 7–7) have threads for securing electrical connections. The coil is mechanically crimped into its terminal with closely adjusted tools to ensure cool, minimum resistance connections. High temperature molded phenolic terminal insulators are used that will not crack or chip during normal use.

Aluminized steel brackets support the coils (Figure 7–8). They are spot welded in place and spaced to prevent coil sag. The brackets are reinforced two ways: (1) by ribbing the bends at each end, and (2) by grooving the edges facing the air stream.

While UL-listed heaters are available for virtually any kW capacity specified, they are restricted to a maximum watt density of 156 W/inch of duct face area for both slip-in and flange-type heaters. To determine the maximum kW for given duct size, the following formula (courtesy of Gould Inc., Heating Element Division) may be used.

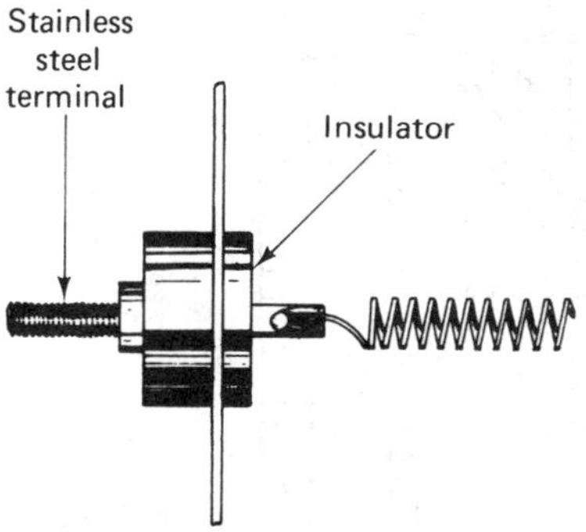
Stainless
steel
terminal
Insulator

FIGURE 7–7
Heater element terminal. (*Courtesy Industrial Engineering and Equip-*

Brackets
Insulating bushing
Element

FIGURE 7–8
Element brackets. (*Courtesy Industrial Engineering and Equipment Co.*)

$$\text{maximum kW} = \frac{156 \text{ duct width (in.) height (in.)}}{1{,}000}$$

Electric heaters differ from steam or hot water coils in that the Btu/h output is constant as long as the heater is energized. It is, therefore, necessary that sufficient and uniform air flow be provided to carry away this heat and to prevent overheating and nuisance tripping of the thermal cut-outs. The minimum velocity required is determined from Figure 7–9 on the basis of entering air temperature and W/ft^2 of cross-sectional duct area.

Example:

To find the minimum air velocity required for a 10,000-W heater installed in a duct 12 inches high and 24 inches wide and operating in a minimum inlet air temperature of 65° F.

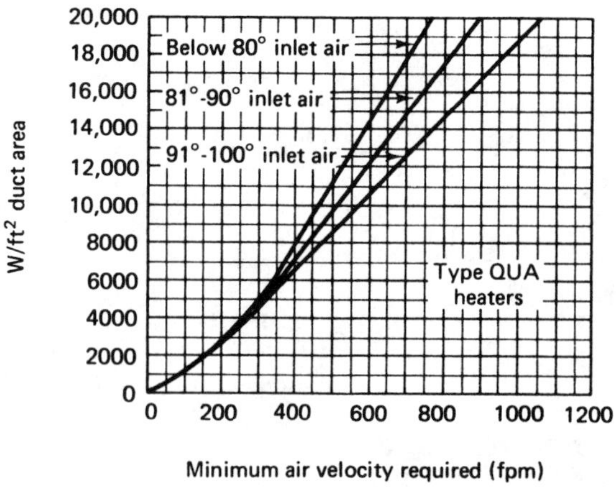
20,000
18,000
16,000
14,000
12,000
10,000
8000
6000
4000
2000
0
W/ft² duct area
Below 80° inlet air
81°-90° inlet air
91°-100° inlet air
Type QUA heaters
0 200 400 600 800 1000 1200
Minimum air velocity required (fpm)

FIGURE 7–9 Minimum air velocity requirements (typical). (*Courtesy Industrial Engineering and Equipment Co.*)

1. Use the top curve (below 80° inlet air).
2. W/ft^2 of duct area = $\dfrac{10{,}000}{1 \text{ ft } 2 \text{ ft}} = 5{,}000 \text{ w}/\text{ft}^2$.
3. This point on the curve corresponds to 310 fpm; thus, the minimum velocity required is 310 fpm. Since the duct area is 2 ft^2, the minimum cfm required is 620 cfm.

The minimum velocity should be uniformly distributed across the duct at the point of insertion of the heater. See Figure 7–10 for typical heater misapplications that result in nonuniform airflow. It is suggested that the heater be installed at least 48 inches from any change in duct direction, any abrupt changes in duct size, and any air-moving equipment.

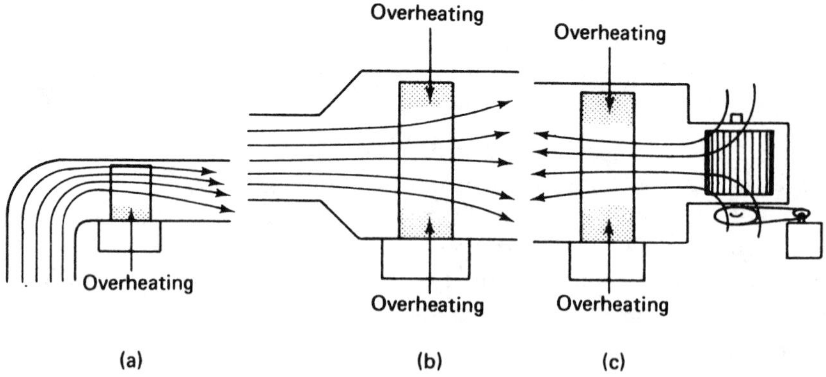
Overheating
Overheating
Overheating
Overheating
Overheating
(a)
(b)
(c)

FIGURE 7–10 Typical heater misapplications. (*Courtesy Industrial Engineering and Equipment Co.*)

KILOWATT CALCULATION

For space heating, the required kilowatt input may be calculated as outlined in the *ASHRAE Guide* or *NEMA Publication Number HE 1-1966* to arrive at the actual heat loss.

(typical equation)

$$\text{kW} = \frac{\text{Btu/h heat loss}^{*}}{3.412}$$

ELEMENT COMPARISON

Although it is less expensive, the *open coil* is far superior to the enclosed design for most space heating applications. Only for special applications, such as exposed heaters and hazardous areas, is the finned tubular construction recommended.

The reasons for recommending the open coil are as follows.

Longer heater life. The open coil releases its heat directly to the airstream. As a result, the open coil runs cooler than the coil in a finned tubular element, where it is isolated from the air by insulation and a metal sheath. Low coil temperatures mean long life.

Low pressure drop. Large open spaces between coils result in free air flow and negligible pressure drop (Figure 7–11). This important factor can result in a reduced fan size for a given cfm.

Accurate control. The lightweight coil reaches operating temperature very quickly when it is energized. This low thermal inertia results in more precise temperature control, as the air temperature responds rapidly to a call for heat. Comparing the open coil to the finned tubular construction (Figure 7–12), note that the air heated with an open coil reaches steady-state temperature in less than half the time required for an enclosed heater.

Easy handling. Their sturdy, lightweight construction makes open coils easier to mount and handle. For example, a 30,000-W heater to fit a 30-inch × 16-inch duct weighs only 30 pounds. Such a heater can be installed quickly without the use of elaborate hoisting equipment.

Large electrical clearances. Clearances between the coil and frame enable open coils to withstand severe applications such as vibration and high voltages.

Duct heaters are advantageous because they are adaptable to cooling, air cleaning, ventilation, and humidification. They are also easily applied to zone control.

*(Courtesy Gould Inc., Heating Element Division)

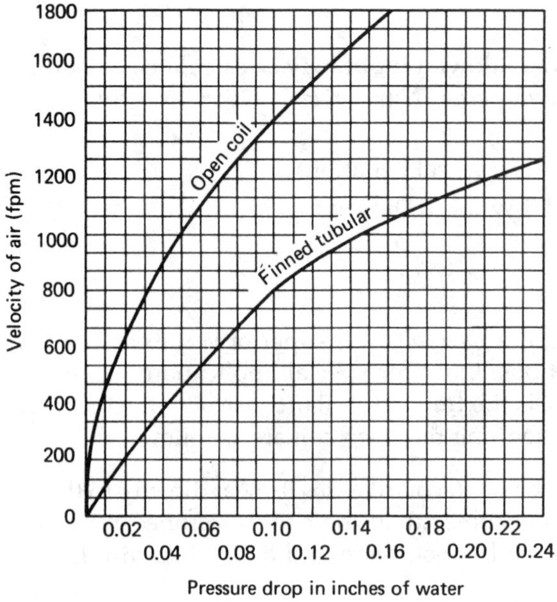

1800
1600
1400
1200
1000
800
600
400
200
0
Velocity of air (fpm)
Open coil
Finned tubular
0.02 0.06 0.10 0.14 0.18 0.22
0.04 0.08 0.12 0.16 0.20 0.24
Pressure drop in inches of water

FIGURE 7–11
Pressure drop comparison of an open coil and a finned tubular duct heater (typical). (*Courtesy Industrial Engineering and Equipment Co.*)

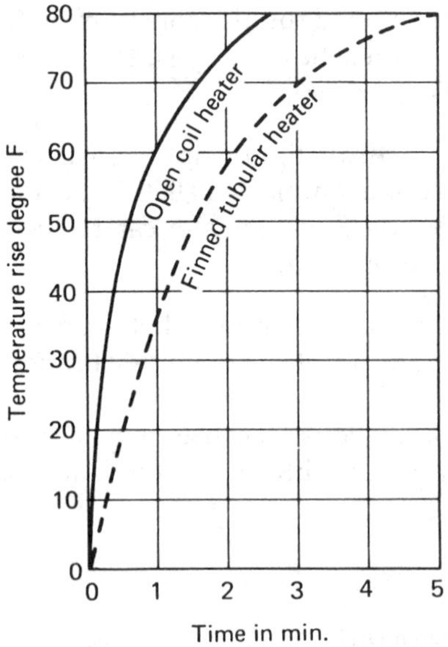

80
70
60
50
40
30
20
10
0
Temperature rise degree F
Open coil heater
Finned tubular heater
0 1 2 3 4 5
Time in min.

FIGURE 7–12
Response times of open coil and finned tubular duct heaters (typical). (*Courtesy Industrial Engineering and Equipment Co.*)

On the other hand, a tempering heater may be required at the fan location to prevent delivery of cold air to rooms when the primary heater is off, thus increasing the overall cost of the equipment and its operation.

REVIEW QUESTIONS

1. When considering electric heat, what must the user do?
2. How many Btu does each kilowatt of electricity produce?
3. How is the wire for electric heating elements selected?
4. Why is high nickel content of resistance wire necessary?
5. Why are the resistance wires in electric elements staggered?
6. Name the two types of duct heaters available.
7. What is the efficiency rating of electric heating elements?
8. What is the purpose, other than insulating, of the ceramic bushings used on electric heating elements?
9. What must be energized along with the electric resistance heating elements?
10. What design of electric heating element is the most efficient?

8

heat pump systems

The objectives of this chapter are:

- To provide you with the principles of heat pump operation.
- To acquaint you with the different operating cycles of heat pump systems.
- To introduce you to the operating controls of a heat pump system.
- To acquaint you with the different types of heat pump systems that are available.

Heat pump systems may be considered as part of the electric heating segment of the heating industry, because they operate through the use of electricity and the expended electrical energy contributes to the heating effect of the system. Generally, the heat pump may be defined as an air conditioning system that is used to remove heat to and from a conditioned space. The operation of the heat pump is based on the normal refrigeration cycle. In the winter it uses a reversed refrigerant flow through the use of flow control devices and check valves. Heat is removed from the outdoor air and released into the conditioned space.

This is possible because air always contains some heat, even down to absolute zero (–460° F). Therefore, at the temperatures encountered in normal heating conditions there is plenty of heat in the air. The problem is for the equipment to remove this heat and deliver it into the space.

Heat is energy; it is contained in all substances. Although heat cannot be picked up by hand like a rock or poured into a container like water, it can be moved from one place to another. For heat to be moved, it must be absorbed by a substance, except in the case of radiant energy. The substance is then relocated to where the heat is needed and the heat is released. Thus, the heat is moved by the intermediate substance. The intermediate substance used in a heat pump is known as a *refrigerant*.

The fundamentals of the basic refrigeration cycle must be understood before one can understand the operation of a heat pump. The same basic components are used in both systems (Figure 8–1). The difference is that the refrigerant flow is reversed.

The purpose of a standard refrigeration unit is to pick up heat from a place where it is not wanted and pump it to someplace where it is unobjectionable. Heat is absorbed into the system by the refrigerant as it changes from a liquid to a vapor in the indoor coil (Figure 8–2). The gaseous refrigerant is then pumped from the indoor coil to the outdoor coil by the compressor. To accomplish this, the refrigerant is compressed, resulting in heat being added to it. As the refrigerant passes through the outdoor coil, the total heat contained in the gaseous refrigerant is given up and the gaseous refrigerant turns to a liquid refrigerant. In a normal refrigeration cycle, the rejection of heat from the outdoor coil is not so important as the heat absorption process in the indoor coil.

In a heat pump system, the heat rejected by the indoor coil is just as important, if not more important, than the refrigeration effect provided by the indoor coil in a typical refrigeration system. A heat pump system operating in the cooling cycle absorbs heat into the indoor coil (Figure 8–3). If the indoor coil temperature is maintained at 45° F, as shown, air or water can flow over the outside of the indoor

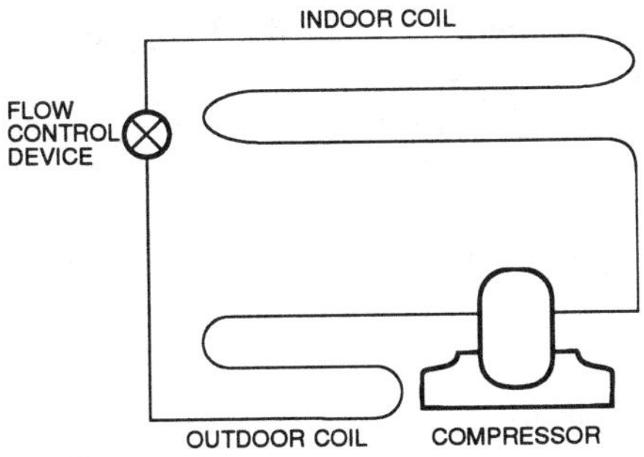
INDOOR COIL
FLOW
CONTROL
DEVICE
OUTDOOR COIL
COMPRESSOR

FIGURE 8–1
Refrigeration cycle.

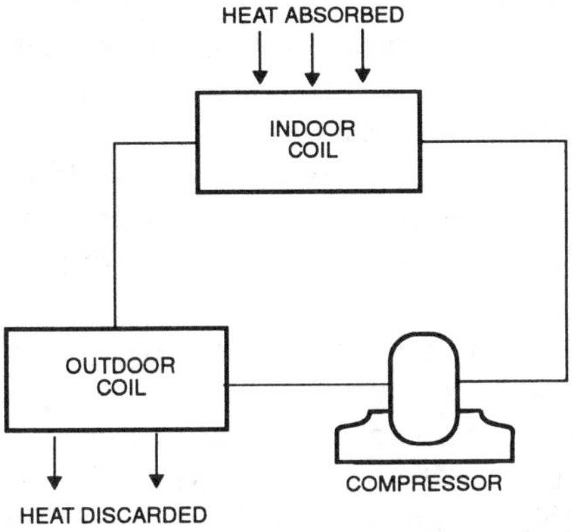
HEAT ABSORBED
INDOOR
COIL
OUTDOOR
COIL
COMPRESSOR
HEAT DISCARDED

FIGURE 8–2
Heat transferred in a refrigeration system.

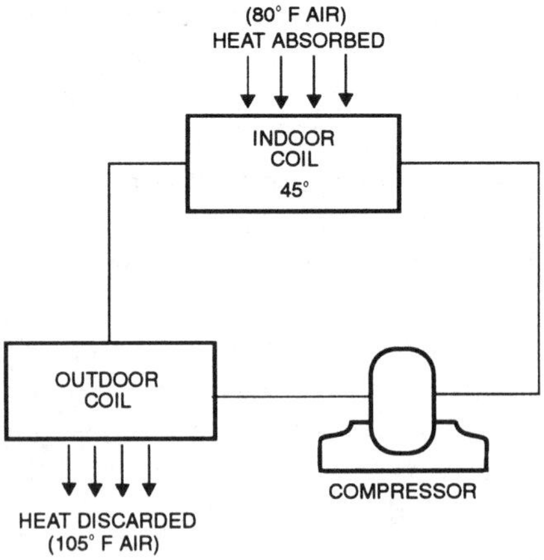
(80° F AIR)
HEAT ABSORBED
INDOOR
COIL
45°
OUTDOOR
COIL
COMPRESSOR
HEAT DISCARDED
(105° F AIR)

FIGURE 8–3
Temperature change at the evaporator.

coil and be cooled as the air or water gives up heat to the refrigerant inside the indoor coil. The refrigerant is then pumped into the outdoor coil, where it gives up its heat of condensation to some cooling medium, such as air or water, which carries away the heat. When the heat pump is in the cooling cycle, its operation is identical to a refrigeration system. Therefore, a refrigeration system is a heat pump. However, it is customary to use the term *heat pump* only for those refrigeration systems that are provided with special devices to use the heat which is rejected from the condensing (indoor) coil.

Heat pump units are classified by their heat source, which is the medium from which the outdoor coil obtains its heat. These heat sources are air to air, water to air, water to water, and ground to air.

OPERATION

The name *heat pump* is descriptive of the unit's operation. The first word, *heat,* is a physical property which has two factors—level and amount. Both factors are important in providing comfort. The level or intensity of heat is measured by the temperature of a substance. The amount or quantity is measured by the British thermal unit (Btu).

The second word, *pump,* indicates a device for making something go in a direction that it ordinarily would not go (for example, making water run uphill or making a boat go upstream against the current). The pump, or compressor, in a heat pump serves a similar function because it forces heat to move in a way it would not ordinarily move.

Some outside force is required to lift heat or energy from a lower temperature to a higher temperature. This force is provided by the compressor. The amount of force required is determined by how far the heat must be raised. For example, it is easier to remove heat from air at 40° F and release it at 75° F than it is to remove heat from air at 40° F and release it at 90° F. The compressor provides the force required to lift the heat—not only in the heat pump but in a refrigeration or air conditioning unit.

When the thermostat calls for cooling, the flow of refrigerant is in the same direction as with any other refrigeration system. Liquid refrigerant is evaporated inside the indoor coil, and the vapor flows to the compressor (Figure 8–4). The high pressure, high temperature discharge gas then goes to the outdoor coil, where it is again condensed into a liquid. The liquid is caused to bypass the first flow control device by the check valves and is controlled by the second flow control device.

When the thermostat calls for heating, the reversing valve is energized. The hot discharge gas from the compressor is directed to the indoor coil (Figure 8–5). The condensed liquid leaves the indoor coil and flows to the outdoor coil. It again is caused to bypass the first flow control device because of the check valve operation, and the flow is controlled by the second flow control device. The liquid is evaporated in the outdoor coil, where it absorbs heat from the outdoor air. The

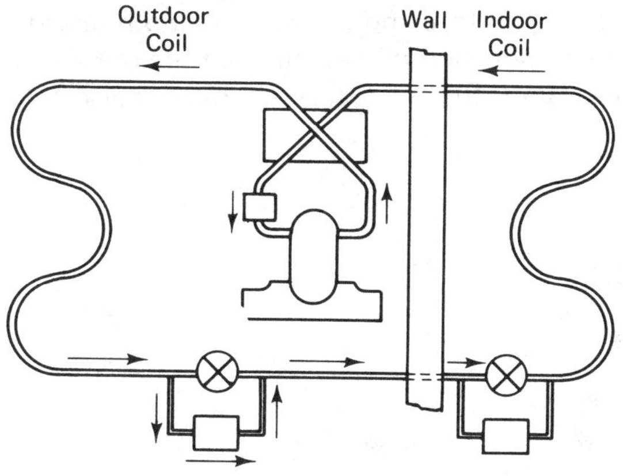
Outdoor Coil
Wall
Indoor Coil

FIGURE 8–4
Heat pump in cooling cycle.

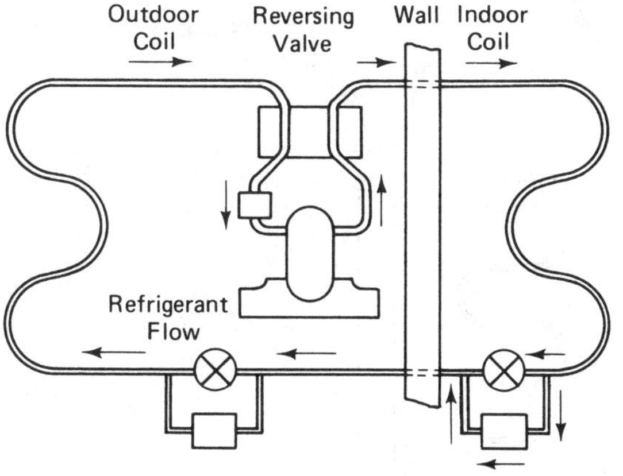
Outdoor Coil
Reversing Valve
Wall
Indoor Coil
Refrigerant Flow

FIGURE 8–5
Heat pump in heating cycle.

heat is then transferred into the conditioned space by the refrigerant when it gives up its latent heat of condensation in the indoor coil.

OPERATING CYCLES

There are three operating cycles for a heat pump system: the cooling cycle, which provides cooling in the summer; the heating cycle, which provides heat during the winter; and the defrost cycle, which removes frost and ice from the outdoor door coil during wintertime operation so that the system can maintain its efficiency.

Cooling Cycle

A typical air conditioning system has two coils, or heat exchangers—one inside the conditioned space and one outside. The indoor coil serves as the evaporating coil, where the liquid refrigerant evaporates and absorbs heat from the air passing over it. It is termed the *indoor coil.* (Figure 8–6).

The evaporated refrigerant leaves the evaporating (indoor) coil and returns to the compressor, where its temperature and pressure are both raised to a higher level. This hot, high pressure vapor then passes through the outdoor coil, where the excess heat is removed and the vapor is changed back into a liquid (Figure 8–7).

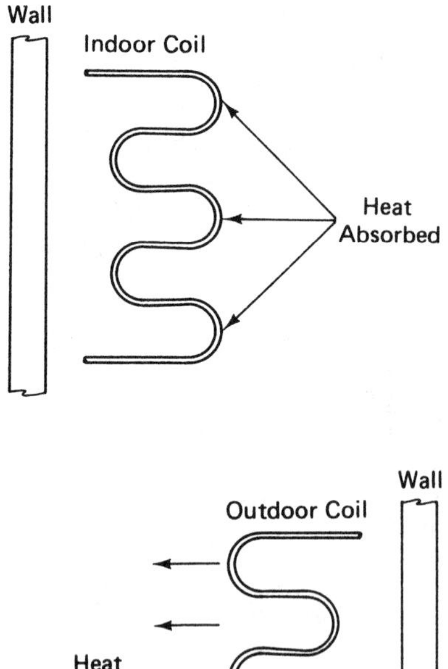
Wall
Indoor Coil
Heat
Absorbed

FIGURE 8–6
Indoor coil.

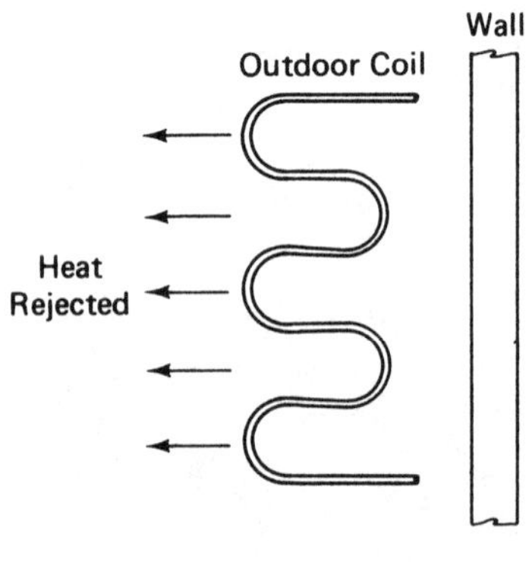
Wall
Outdoor Coil
Heat
Rejected

FIGURE 8–7
Outdoor coil.

The liquid refrigerant then flows back inside the conditioned space and is metered into the indoor coil by the flow control device. These devices are usually an expansion valve or a capillary tube (Figure 8–8). When the unit is in the cooling cycle, the refrigerant is evaporated at a fairly constant temperature, usually between 30° F and 40° F. The condensing temperature will vary depending on the temperature of the condensing medium, air or water.

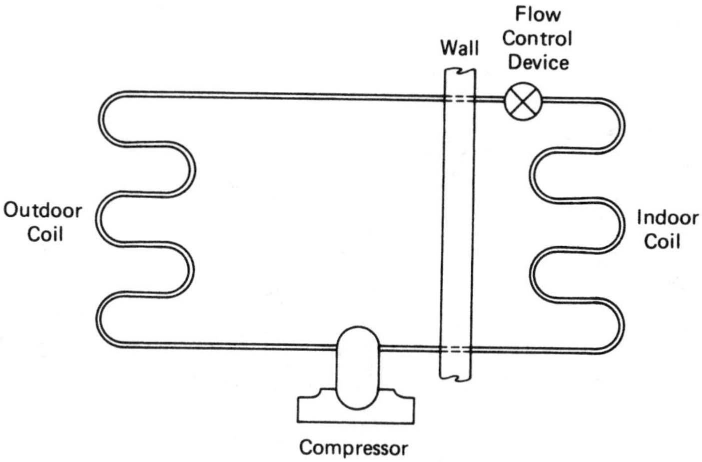
Flow
Control
Device
Wall
Outdoor
Coil
Indoor
Coil
Compressor

FIGURE 8–8 Flow control device.

Heating Cycle

The basic components used during the heating cycle are the same as those used during the cooling cycle, with the addition of some controls. Both the indoor and outdoor coils on a heat pump system will serve as either the evaporating coil or the condensing coil, depending on the direction of refrigerant flow.

Defrost Cycle

When the heat pump is operating in the heating cycle, the refrigerant is being evaporated in the outdoor coil. When the temperature of the outdoor coil falls below 32° F, frost will start appearing on the coil surface. If this frosting is allowed to continue, the deposit of ice will gradually build up until the flow of air through the coil will be restricted. This restriction of air will decrease the heat transfer and seriously affect the efficiency of the unit. Modern heat pump units, however, are designed to handle this problem automatically through the use of refrigerant flow controls and system controls.

HEAT PUMP REFRIGERANT SYSTEM COMPONENTS

For the heat pump system to operate automatically, certain other refrigeration system components are required. These extra components are required to reverse the flow of refrigerant and to protect the compressor.

Reversing Valve

A reversing valve, or four-way valve, is used to change the direction of the refrigerant flow. The direction of flow will depend on whether heating or cooling is needed indoors (Figure 8–9). The reversing valve is an important part of the heat pump cycle and it must operate smoothly, reliably, and efficiently. Valve manufacturers have designed reversing valves which meet these requirements (Figure 8–10).

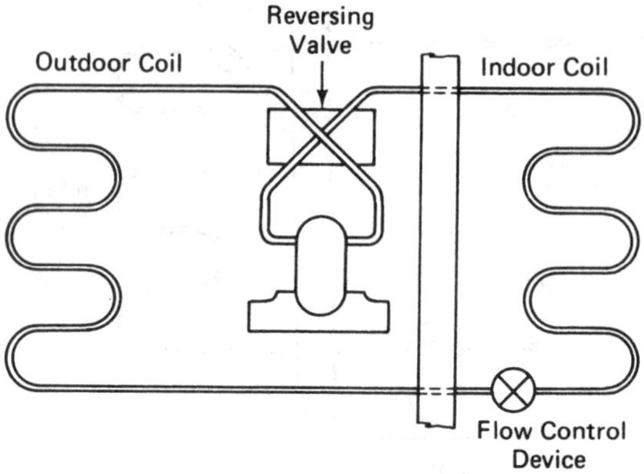
Reversing Valve
Outdoor Coil
Indoor Coil
Flow Control Device

FIGURE 8–9
Reversing valve directing refrigerant.

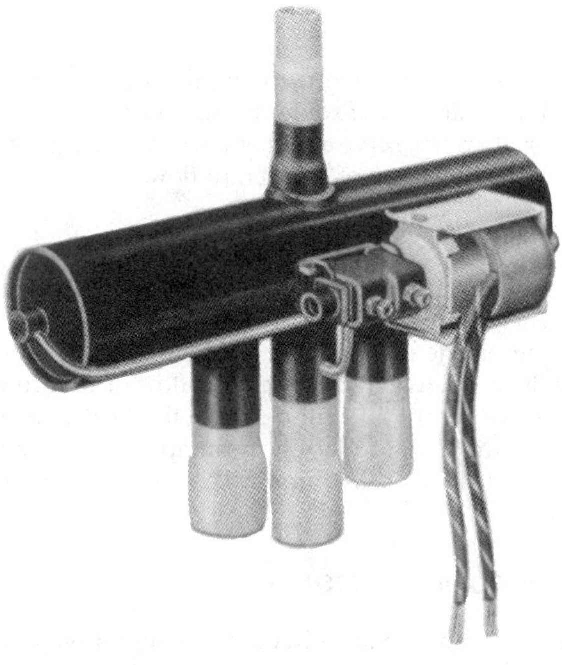

FIGURE 8–10
Reversing valve. (*Courtesy of Ranco Controls Division.*)

There are two type of reversing valves used on modern heat pump systems: the direct-acting and the pilot-operated type. In the direct-acting valve, the plunger is connected directly to the solenoid plunger. In the pilot–operated type, a solenoid valve controls the flow of discharge gas which operates the valve plunger. When the solenoid is energized, the discharge gas is directed to one end of the plunger and forces it to the other end of the reversing valve. When the solenoid is deenergized, the discharge gas is directed to the other end of the plunger and forces it to the opposite end of the reversing valve (Figure 8–11). The reversing valve may be in either the heating or cooling position when deenergized; different manufacturers use different positions. The trend, however, is toward having the valve in the heating position when deenergized.

Suction Line Accumulator

A suction line accumulator is normally used to catch any liquid refrigerant and prevent it from entering the compressor (Figure 8–12). These devices are located in the suction line between the reversing valve and the compressor. If liquid refrigerant is allowed to enter the compressor, damage to the bearings, valves, and all wearing surfaces may occur.

Suction line accumulators are usually equipped with some type of oil return that will allow any trapped oil to be returned to the compressor so that proper lubrication will be maintained.

Flow Control Devices

There are three types of flow control devices used on heat pump systems: expansion valves, capillary tubes, and refrigerant restrictors. These devices are used to control the flow of refrigerant into the coil. One is needed for each coil (Figure 8–13).

Thermostatic expansion valves. The most commonly used device for controlling the flow of refrigerant into the evaporating coil is the thermostatic expansion valve (Figure 8–14). An orifice in the valve meters the flow into the evaporating coil. The rate of refrigerant flow is regulated as required by the needle-type plunger and seat that vary the orifice opening.

Bidirectional expansion valves. These valves are designed for use on heat pump systems. They will allow the refrigerant to flow in either direction; thus one valve eliminates the need for two flow control devices and two check valves in the system, thereby simplifying the piping (Figure 8–15).

Capillary tubes. The capillary tube is the simplest type of refrigerant flow control device used on modern heat pump systems. However, its use is limited to certain applications. A capillary tube is a small-diameter tube through which the refrigerant flows to the evaporating coil (Figure 8–16). The capillary tube is not a true valve since it is not adjustable and cannot be regulated readily.

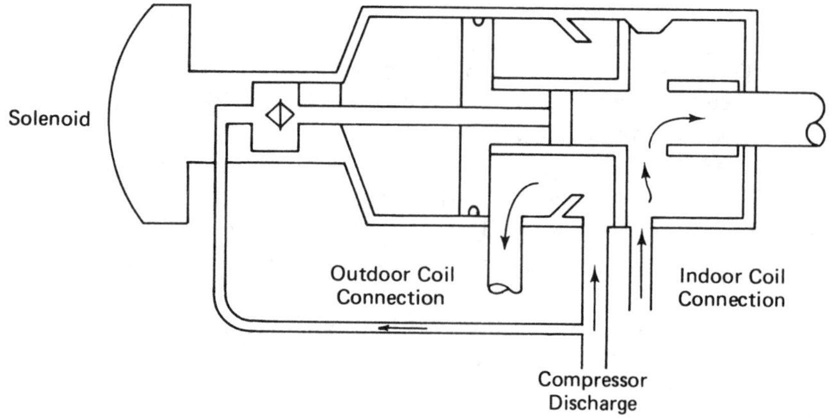
Solenoid
Outdoor Coil
Connection
Indoor Coil
Connection
Compressor
Discharge

(a) Direct-Acting Reversing Valve

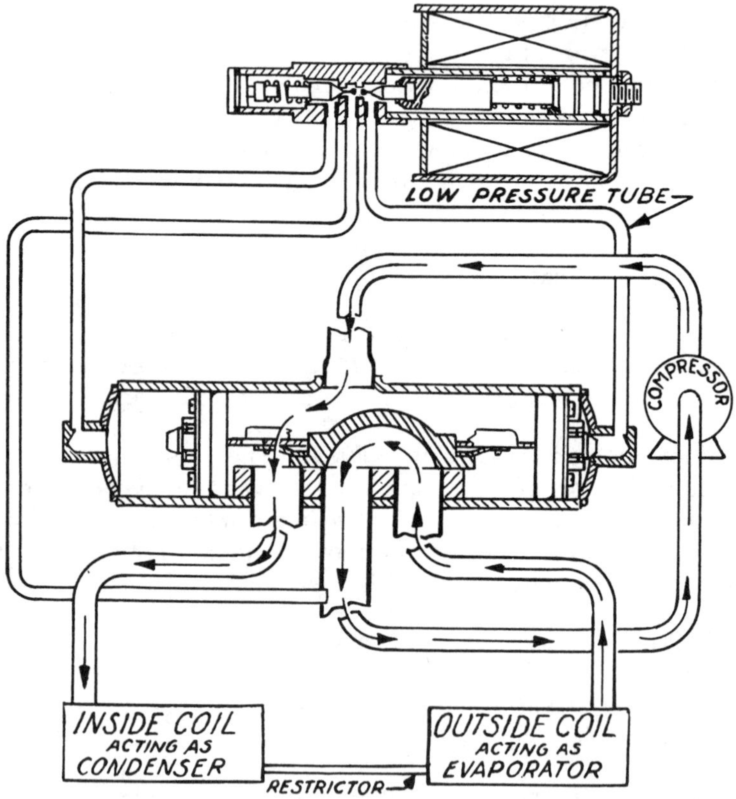
LOW PRESSURE TUBE
COMPRESSOR
INSIDE COIL
ACTING AS
CONDENSER
OUTSIDE COIL
ACTING AS
EVAPORATOR
RESTRICTOR

(b) Pilot-operated reversing valve

FIGURE 8–11
Reversing valve operating methods. (*Courtesy of Ranco Controls Division.*)

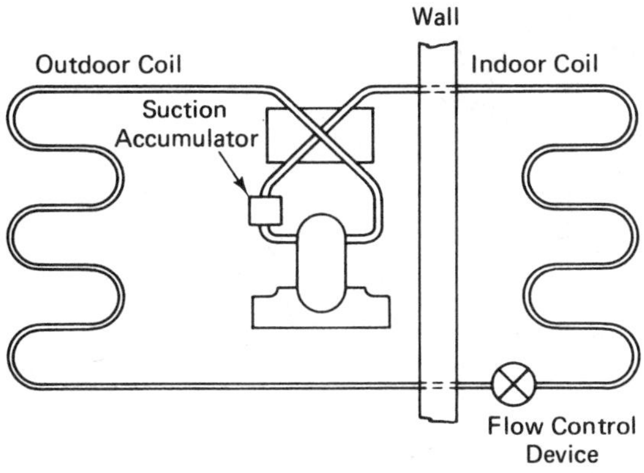
Wall
Outdoor Coil
Indoor Coil
Suction
Accumulator
Flow Control
Device

FIGURE 8–12
Suction line accumulator location.

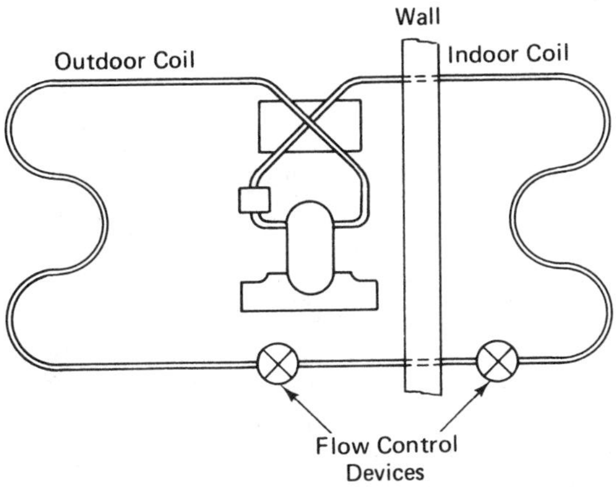
Wall
Outdoor Coil
Indoor Coil
Flow Control
Devices

FIGURE 8–13
Flow control device location.

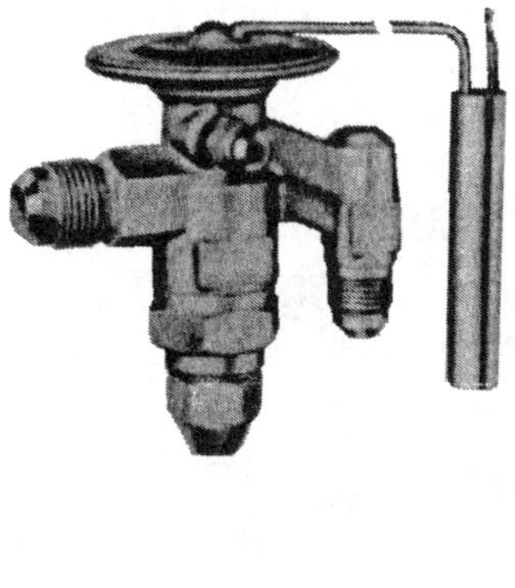

FIGURE 8–14
Thermostatic expansion valve. (*Courtesy of Alco Controls Division, Emerson Electric Co.*)

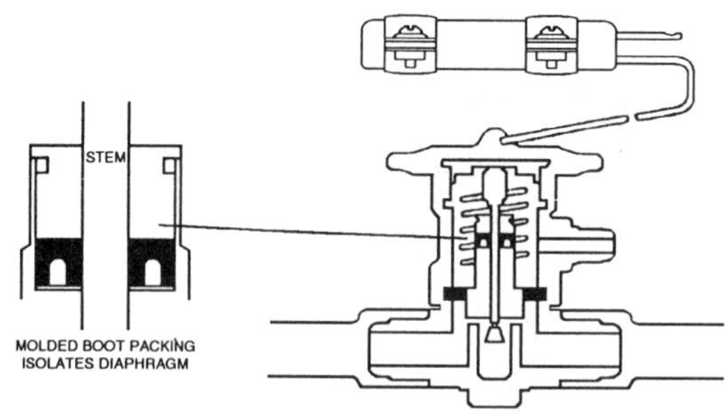
STEM
MOLDED BOOT PACKING
ISOLATES DIAPHRAGM

FIGURE 8–15
Bidirectional expansion valve. *(Courtesy of Alco Controls Division, Emerson Electric Co.)*

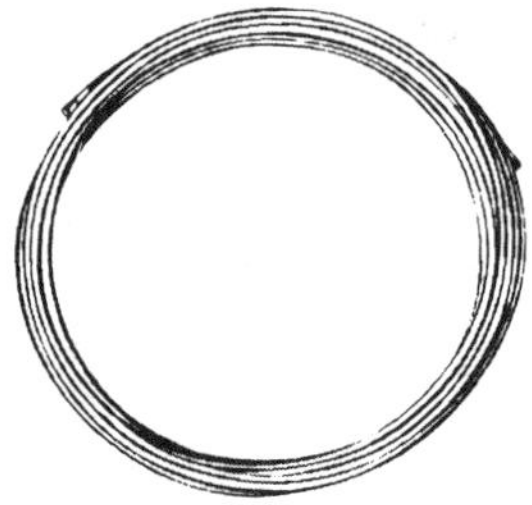

FIGURE 8–16
Capillary tubing.

Aeroquip flow control device. This is a primary refrigeration expansion device. When properly sized, it can be used to replace other types of flow control devices. The aeroquip flow control device is made up of three different components: the field connector, the restrictor, and the distributor (Figure 8–17). The field connector attaches to the liquid line. The restrictor fits inside and meters the refrigerant flow to the evaporating coil. The restrictor can be replaced easily in the field, providing the possibility to fine tune the system. The distributor provides refrigerant flow to the coil and is available in a wide variety of outlet configurations to match the system requirements (Figure 8–18).

FIGURE 8–17
Aeroquip flow control device components. (*Courtesy of Aeroquip.*)

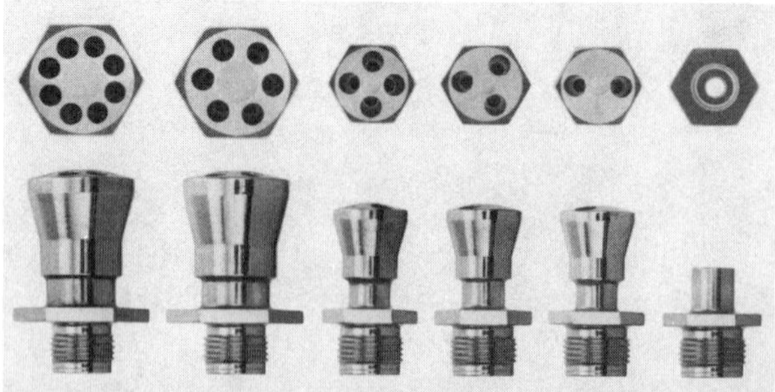

FIGURE 8–18
Aeroquip distributor configuration. (*Courtesy of Aeroquip.*)

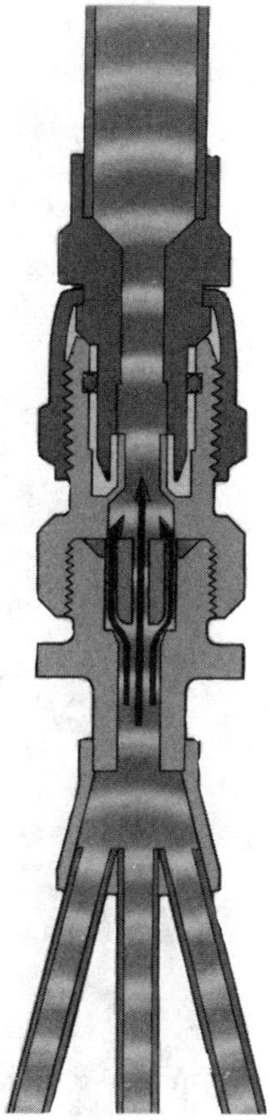

FIGURE 8–19
Refrigerant flow during cooling cycle. (*Courtesy of Aeroquip.*)

As liquid refrigerant flows through the device, the restrictor meters the flow into the distributor housing (Figure 8–19). This direction of flow results in a proper pressure drop to vaporize the refrigerant as it travels through the coils. The rate at which the refrigerant is vaporized is determined by the restrictor orifice size. The restrictor can be replaced quickly with another one of a different flow rate, allowing a system to be fine tuned in the field.

The design of the valve also permits reverse flow of the refrigerant during the heating cycle. As the liquid refrigerant flows in the reversed direction through the device, the restrictor moves back to a free-flow position, offering no, or very little, restriction to the flow of refrigerant. This reverse flow capability eliminates the need for check valves and their related plumbing in heat pump applications (Figure 8–20).

Check valves. There are two check valves used to direct the flow of refrigerant through the proper metering device (Figure 8–21). They are installed so that they will open and allow the refrigerant to pass or close and prevent passage of the refrigerant during a particular function of the unit. Check valves are not normally used on systems equipped with refrigerant restrictors.

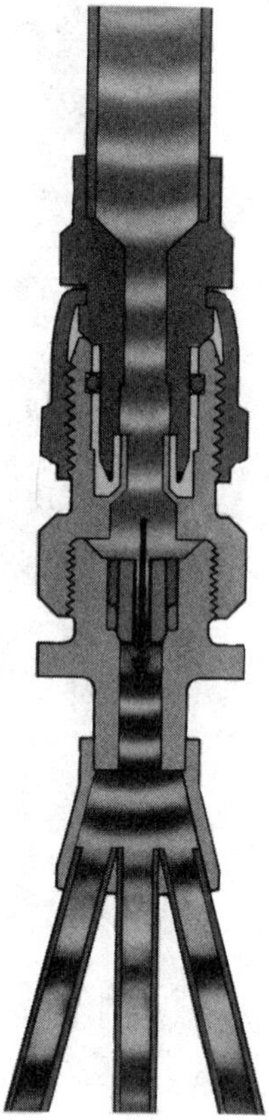

FIGURE 8–20
Refrigerant flow during heat pump heating cycle. (*Courtesy of Aeroquip.*)

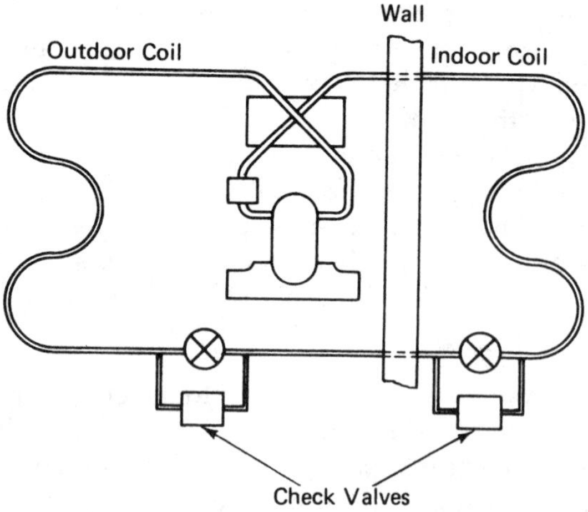
Wall
Outdoor Coil
Indoor Coil
Check Valves

FIGURE 8–21
Check valve locations.

DEFROST METHODS

The most popular and efficient method of removing frost from the outdoor coil is to direct the hot discharge gas from the compressor through the outdoor coil for a period of time long enough to melt the accumulated frost. This change in the direction of the refrigerant flow is caused by the change in position of the reversing valve.

At the same time that the reversing valve is energized, the outdoor blower is stopped to reduce the flow of cold, outside air over the outdoor coil, thus allowing the frost to melt much faster. Since no heat is provided for the conditioned space during the defrost cycle, supplementary electric resistance heaters are used for this purpose. Thus, when the defrost relay is energized, an electric signal is also sent to the supplementary heaters inside the conditioned space. These heaters temper the indoor air during the defrost period to prevent blowing cold air into the conditioned space.

The methods used for frost detection are many and varied. The following are some of the more common methods.

Air pressure. This method uses the air pressure differential across the outdoor coil (Figure 8–22). This control has a time delay to prevent false defrost initiation. When the pressure approaches a given drop, the defrost cycle is initiated. When the coil is de-iced and the air pressure differential is reduced to a given pressure, the system is placed back into the heating mode.

Time termination. This termination method is usually used in conjunction with any of the previously mentioned initiation methods. After the system has been in the defrost cycle for a given period of time, it will automatically return to the heating cycle.

Pressure termination. This type of termination uses the fact that the refrigerant pressure will increase when the frost is gone.

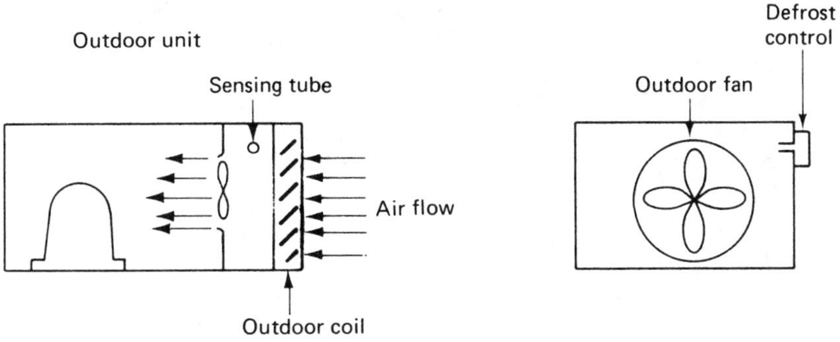
Outdoor unit
Sensing tube
Air flow
Outdoor coil
Defrost control
Outdoor fan

FIGURE 8–22 Air pressure differential defrost initiation.

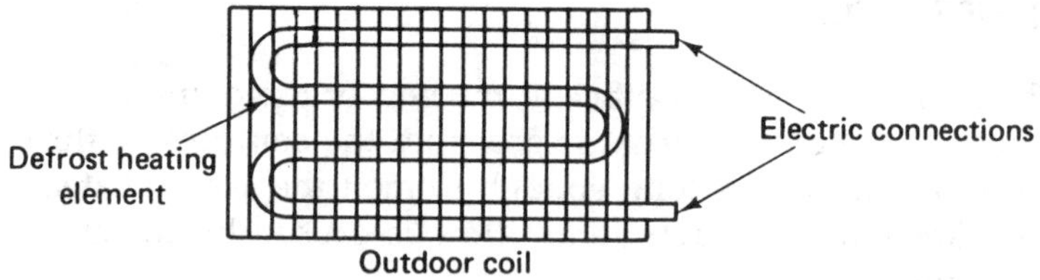
Defrost heating element
Electric connections
Outdoor coil

FIGURE 8–23 Electric defrost heater.

Some manufacturers incorporate electric strip heaters attached to the outdoor coil to speed up the defrosting of the coil. This heater is energized by the defrost initiation control (Figure 8–23).

Time and temperature. The time and temperature method of automatically initiating the defrost cycle is very popular with heat pump manufacturers. A clock timer is used in conjunction with a defrost thermostat. The two together determine when a defrost cycle is needed and then initiate the cycle. The defrost thermostat contacts are normally open, and they close on a fall in temperature of the outdoor coil. The thermostat sensing element is located on a refrigerant outlet of the outdoor coil. When the coil temperature drops to about 32° F, the contacts close. The contacts are in electrical series with the normally open contacts on the clock timer. The clock motor is wired in electrical parallel with the outdoor fan motor and runs when the compressor runs.

In operation, after a predetermined period of time has passed (usually 90 minutes), the timer motor causes the normally open set of contacts to close. These contacts remain closed for only a few seconds during each cycle of the clock mechanism. If the defrost thermostat contacts are closed at the same time as the clock timer contacts, the defrost relay coil is energized and a defrost cycle is initiated. If the defrost thermostat contacts are not closed, it indicates that the outdoor coil does not need to be defrosted, and the clock timer will start another cycle. The amount of time required for the clock timer to cycle can be reduced to 30 minutes if necessary to help keep the unit operating with a frost-free outdoor coil. Depending on the area weather conditions, the timing cycle may be varied to allow proper coil defrosting. The clock timer motor is deenergized along with the outdoor fan motor during the defrost cycle.

The unit will stay in the defrost cycle until the defrost thermostat warms up to approximately 65° F. The thermostat contacts will then open, which deenergizes the defrost relay and terminates the defrost cycle.

Solid state. Some units are equipped with a solid-state board that initiates the defrost cycle. Thermistors within the solid-state system sense the difference between the ambient air temperature and the refrigerant temperature. When the temperature differential exceeds the differential band, the defrost cycle is initiated and continues until the defrost pressure, or temperature, switch terminates it.

In operation, when the ice on the coil has reached a predetermined point, the defrost cycle is initiated. The unit is returned to the cooling cycle so that the hot discharge gas passes through the outdoor coil. The outdoor fan is stopped so that the cold outdoor air will not flow through the coil and adversely affect the defrost efficiency. The indoor fan is kept running to evaporate the refrigerant before it returns to the compressor. Also, the auxiliary heat is turned on to prevent cool air from blowing into the conditioned area. When all the ice has melted, the defrost termination control will return the unit back to the normal heating cycle.

SUPPLEMENTARY HEAT

Heat pump systems are generally sized to fit the cooling load. Therefore, supplementary heat is added to bring the unit to the desired heating capacity. Each manufacturer will provide charts indicating the cooling and heating capacity in Btu for any given piece of equipment. From these charts and the calculated heat load of the building, the amount of supplementary heat needed may be determined. This supplementary heat is usually in the form of resistance heaters installed in the discharge air of the equipment.

Supplementary heat is used only when the heat pump cannot supply the desired amount of heat to the conditioned space. It is controlled from two points: (1) the second stage on the indoor thermostat, and (2) the outdoor thermostats, which are generally mounted in the outdoor unit. When both of these thermostats demand heat, the supplementary heat will be energized. The design engineer will determine the temperature setting of the outdoor thermostats. If these thermostats are set too high, the electric bill may be excessively high. On the other hand, if they are set too low, an insufficient amount of heat may be supplied to the conditioned space. There must be a point where comfort and economy are maintained.

REHEATING

A heat pump should not be used for reheating purposes, especially in the summer. The compressors used in heat pump systems operate with a higher compression ratio than normal air conditioning compressors. It is not difficult for these units to operate with a 300° F discharge gas temperature. This is the maximum allowable discharge gas temperature for a compressor. Many manufacturers will include controls on their units to protect the compressor from operating in this temperature range. This high discharge gas temperature is caused by too low an air flow over the indoor coil. An indoor air temperature of 90° F, or higher, will require more air to be moved across the indoor coil. Caution must be exercised to make certain that any supplementary heating is located in the air stream after the indoor coil.

WATER SOURCE HEAT PUMP SYSTEMS

Because of recent research, the development and manufacture of materials that transfer heat have brought about the evolution of water source heat pump systems. Water source heat pump systems are used in closed-loop, earth-coupled applications. When the closed-loop, earth-coupled system is used, the water is circulated through special plastic tubing that is buried in the ground. The circulating solution may be a mixture of water and antifreeze or just plain water. In these systems there is no waste water that must be properly disposed of. Once the water circuit is properly charged, there is no need for additional water or antifreeze solution unless there is a leak.

Water source heat pump systems are also used on open-loop applications. However, there is the problem of disposing of the waste water that has passed through the system. The open-loop system takes water from some source (such as a well, a pond or lake, or other similar source), and then it is discharged in some appropriate area. Sometimes the disposal of the water is a bigger and more expensive problem than obtaining it.

The basic operating principle of this system is that the heat is absorbed by the refrigerant in the indoor air. It is then rejected to the circulating water from the refrigerant during the cooling cycle. During the heating cycle, heat is taken from the water by the refrigerant and given to the air passing over the indoor coil.

Dehumidification of the indoor air is accomplished in the same manner as with any cooling unit. It is removed in the form of condensate. The medium used to transfer the heat is the refrigerant inside the system. The refrigeration system components, with the exception of the water to refrigerant heat exchanger, are the same as those used on any other type of heat pump system. This type of system is most popular for heating and cooling residential buildings, where an adequate source of water or ground area is available.

CLOSED-LOOP, EARTH-COUPLED SYSTEMS

In the closed-loop, earth-coupled heat pump system a solution is circulated through special pipes buried in the ground. This solution may be water or a water-antifreeze mixture. These are three methods of installation for these types of heat pump systems: (1) horizontal, (2) vertical, and (3) lake loops.

Typical Closed-Loop, Earth-Coupled Methods

These types of systems may be either vertical or horizontal tubing buried in the ground. The vertical system is the most popular when the area is too small to accommodate the amount of piping needed for the horizontal type of system. In these types of systems, the required number of boreholes are drilled into the ground of sufficient size and depth to allow the proper amount of piping to be buried for the size of system being installed (Figure 8–24). Note that the letters shown on Figures 8–24 and 8–25 match the corresponding "Feet per Nominal Ton" category in Table 8–1. The specifications and installation instructions for all methods shown are available from the manufacturer of the equipment being used. The earth-coupling length information given here applies only to and has been specifically designed for Command–Aire Earth Energy Heat Pump Systems® to be installed in residential and light commercial applications in the Waco, Texas area.

The horizontal system is used in applications where the building site is large enough to allow the proper amount of pipe to be buried horizontally (Figure 8–25).

The choice of vertical, horizontal, or lake loop earth coupling should be based on the characteristics of the particular application. Horizontal and vertical systems are designed to produce the same fluid temperature under a given set of conditions. The lake loop system will provide slightly lower fluid temperatures, but the reduced installation cost should compensate for any minor reduction in performance. The three earth-coupling methods should be considered at each application, with the most cost-effective method chosen after all methods have been evaluated.

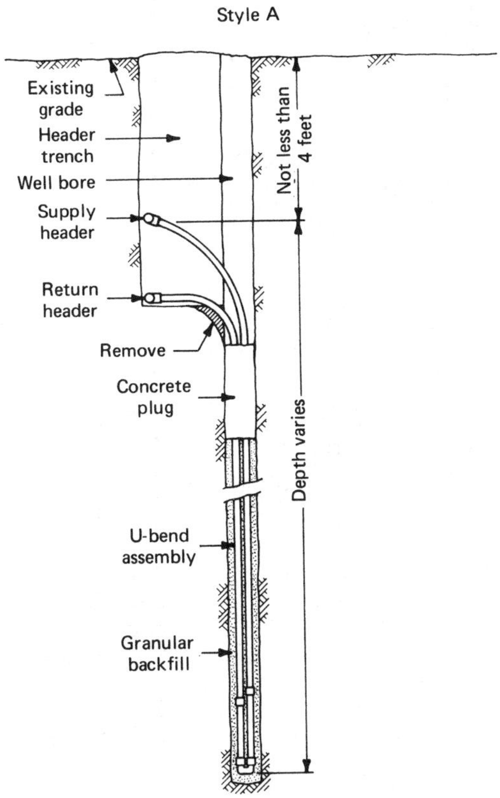
Style A
Existing grade
Header trench
Well bore
Supply header
Return header
Remove
Concrete plug
U-bend assembly
Granular backfill
Not less than 4 feet
Depth varies

FIGURE 8–24
Vertical piping application. (*Courtesy of Command-Aire Corporation.*)

TABLE 8–1 Command-Aire earth coupling design information. *(Courtesy of Command-Aire Corporation.)*

Zone	Feet Per Nominal Ton[a]				Anti-freeze[b]	Pumping Unit		
	A	B	C	D		LPK1	LPK2	LPK3
1	200	550	650	875	0	060–130	190–350	410–610
2	210	575	675	900	0	060–130	190–350	410–610
3	220	600	725	975	0	060–130	190–350	410–610
4	240	675	800	1075	0	060–130	190–350	410–610
5	260	750	900	1200	0	060–130	190–350	410–610
6	280	850	1025	1350	0	060–130	190–350	410–610
7								
8								
9								

Source: Command-Aire Corporation.

[a]Horizontal length is lineal feet of pipe per ton; vertical length is lineal feet of borehole per ton.

[b]Percentage of volume.

Earth-Coupled System Design

The design of an earth-coupled system is divided into the following steps: (1) Determine the structure design heating load in Btu/hr loss and design cooling load in Btu/hr gain, (2) select a water source heat pump, and (3) select an earth–type coil and materials to be used.

Determine the structure design heating and cooling loads. An accurate heat load calculation is a necessity if the system is to perform satisfactorily. Therefore, it is recommended that some nationally accepted method be used for these calculations.

Select a water source heat pump. The water source heat pump selected for use in an earth-coupled system may be required to operate at entering water temperatures that may vary between 30 and 100° F. Because of this, it is necessary that the minimum and maximum temperatures of the water source selected remain within that range through all seasons. There are several models on the market which have a much smaller operating range of entering water temperature. Some of these units are not satisfactory for use on earth-coupled heat pump installations.

The heating and cooling capacities of water source heat pumps should be determined from the manufacturer's specifications for the local groundwater temperatures and design conditions.

Select the earth–type coil and materials. In earth-coupled systems, the earth coupling is a particular method which uses water or another type of solution that is circulated through special pipes buried in the ground. The heat is transferred to and from the water through the walls of the pipes. These types of systems are used in areas where there is an insufficient amount of available groundwater, or where the drilling of a well would be impractical. The required piping may be buried by either the vertical or horizontal method.

Of particular importance when designing an earth-coupled heat pump system is the balance between the heat pump unit and the earth-coupled loop. When this balance is achieved, the earth-coupled loop will remove all the heat that is transferred to the water by the heat pump during the cooling cycle, and will provide all the heat that is required to heat the water during the heating cycle. The final result of a perfectly balanced system is that the change in the water temperature through the heat pump system is offset by an equal and opposite change in temperature through the earth-coupled loop. For example, if when operating in the cooling cycle the heat pump unit causes a temperature rise in the circulating water of 15° F, the loop must respond with a corresponding but opposite temperature drop of 15° F.

Even though the earth-coupled loop is designed for balanced rise and fall of the water temperature (which suggests that the net average loop water temperature remains constant), because the ground temperature may vary ±15° F from season to season, the loop water temperature may vary ±20° F from the balance point

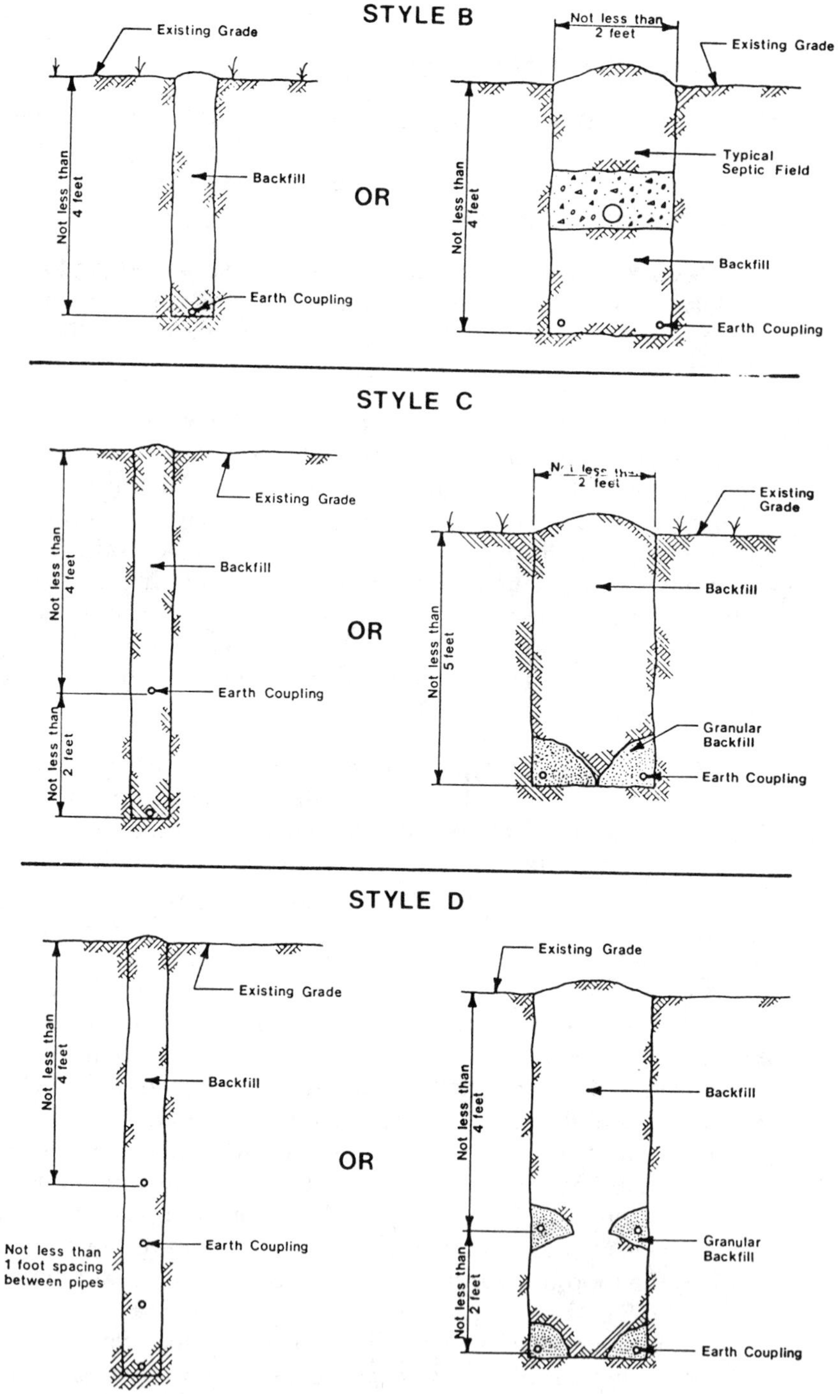
STYLE B
Existing Grade
Not less than 4 feet
Backfill
Earth Coupling
OR
Not less than 2 feet
Existing Grade
Typical Septic Field
Not less than 4 feet
Backfill
Earth Coupling
STYLE C
Existing Grade
Not less than 4 feet
Backfill
Earth Coupling
Not less than 2 feet
OR
2 feet
Existing Grade
Backfill
Not less than 5 feet
Granular Backfill
Earth Coupling
STYLE D
Existing Grade
Not less than 4 feet
Backfill
Earth Coupling
Not less than 1 foot spacing between pipes
OR
Existing Grade
Backfill
Not less than 4 feet
Not less than 2 feet
Granular Backfill
Earth Coupling

FIGURE 8–25 Horizontal piping applications. (*Courtesy of Command-Aire Corporation.*)

temperature. This is because the ground is able to overheat the loop water during the heating season and may overcool it during the cooling season. For this reason, the water temperature entering the water source heat pump system may drop below 30° F during the heating season or rise above 100° F during the cooling season. This range in the temperature of the entering water is extremely important because water source heat pump systems are designed to operate within specific operating temperature ranges. The manufacturer's specifications must be checked to make certain that the unit is designed to operate within the entering water temperature range at the point of application. The temperature ranges are established to protect both the heat pump unit and water loop piping. Also, these temperature ranges are generally based on water only passing through the system. The low temperature limit of 40° F in a water source heat pump is established to protect the loop from freezing. Again, this low limit is assumed for water only circulating through the system. However, if the circulating water is mixed with a nontoxic antifreeze solution, the entering water temperature can be allowed to fall to 30° F.

Reasons for Using an Earth-Coupled System

Some of the major reasons for using earth-coupled systems are as follows:

1. Unlike a standard solar system, the loop operates day or night, rain or shine, all year long, delivering heat to and from the heat pump unit.
2. It is cost effective in either northern or southern climates.
3. Because the water circulates through a sealed, closed loop of high-strength plastic pipe, it eliminates scaling, corrosion, water shortage, pollution, waste, and disposal problems which are possible in some open-well water systems.

Vertical Earth-Coupled System

A vertical earth-coupled system consists of one or more vertical boreholes through which water flows in a plastic pipe. A distinct advantage of a vertical system over a horizontal system is that the vertical system requires less surface area (acreage). In areas where the ambient groundwater (average well water) temperature is less than 60° F, the use of an antifreeze solution such as propylene to avoid freezing is recommended.

The boreholes are drilled 5 to 6 inches in diameter for 1½-inch-diameter pipe. For ¾-inch-diameter pipe loop systems, the boreholes are 3 to 4 inches in diameter, thus lowering the drilling costs (Table 8–2). The vertical loops are connected in parallel to a 1½-inch-diameter pipe header. The ¾-inch diameter pipe also costs less per ton of heat pump capacity. The smaller pipe is easier to handle, yet there is no sacrifice in the pressure rating. Also, having two loops in one hole reduces the borehole length. The depth for these systems is usually between 80 and 180 ft.

FIGURE 8–25 Horizontal piping applications. (*Courtesy of Command-Aire Corporation.*)

The basic components of a vertical earth-coupled system are shown in Figure 8-26. Each borehole contains a double length of pipe with a U-bend fitting at the bottom. Several boreholes may be joined in series or parallel (Figure 8 –27). Sand or gravel packing is required around the piping to assure heat transfer. In addition, the bore around the pipes immediately below the service (connecting) lines must be cemented closed to prevent surface water contamination of an aquifer, in accordance with local health department regulations (Figure 8–28).

Series U-bend. A series U-bend earth coupling is one in which all the water flows through all the pipe, progressively traveling down and then up each well bore. Series wells need not be of equal length.

Pipe. The 1 ½- inch CTS (copper tube size) or IPS (iron pipe size) polybutylene (PB) pipe is commonly used in 5- and 6-inch boreholes. IPS PB pipe is used with insert fittings and clamps. Turn the clamps so that they face inward and will not be chafed by the well bore. Tape the clamped section of the U-bend with duct tape to provide added protection to the clamps while the pipe is being installed into the well. CTS pipe is heat fused with fittings. Polyethylene pipe is heat fused with butt joints.

Stiffener. Tape the last 10 to 15 ft of pipe above the U-bend together and to a ridged piece of pipe or conduit. This will make it easier to install the pipe in the well.

Fill and pressure test. Fill the system with water and pressure test before lowering the U-bend into a well bore. When drilling with air, a bore can be completed that contains no water. If unfilled plastic pipe is lowered into the bore, it will be crushed as the hole slowly fills.

Multiple wells. Multiple 100-ft wells connected in series are the easiest to drill and install in most areas. It will be difficult to sink water-filled plastic U-bends into mud-filled holes over 150 ft deep without weights. Wells are generally spaced 10 ft apart in residential systems.

Service lines. Follow the guidelines for the horizontal earth coil when installing service lines to and from a U-bend.

Parallel U-bend. A parallel U-bend earth coupling is one in which water flows out through one header, is divided equally, and flows simultaneously down two or more U-bends. It then returns to the other header. Headers are reverse-return plumbed so that equal-length U-bends have equal flow rates. Lengths of individual parallel U-bends must be within 10% of each other to ensure equal flow in each well. Either 1½-inch polybutylene or polyethylene pipe is used for the headers with 1-inch or ¾-inch pipe used for U-bends. Four-inch boreholes are sufficient for placement of the 1-inch U-bends. Follow the instructions given previously under "Series U-Bend" for the stiffener, fill, and pressure test,

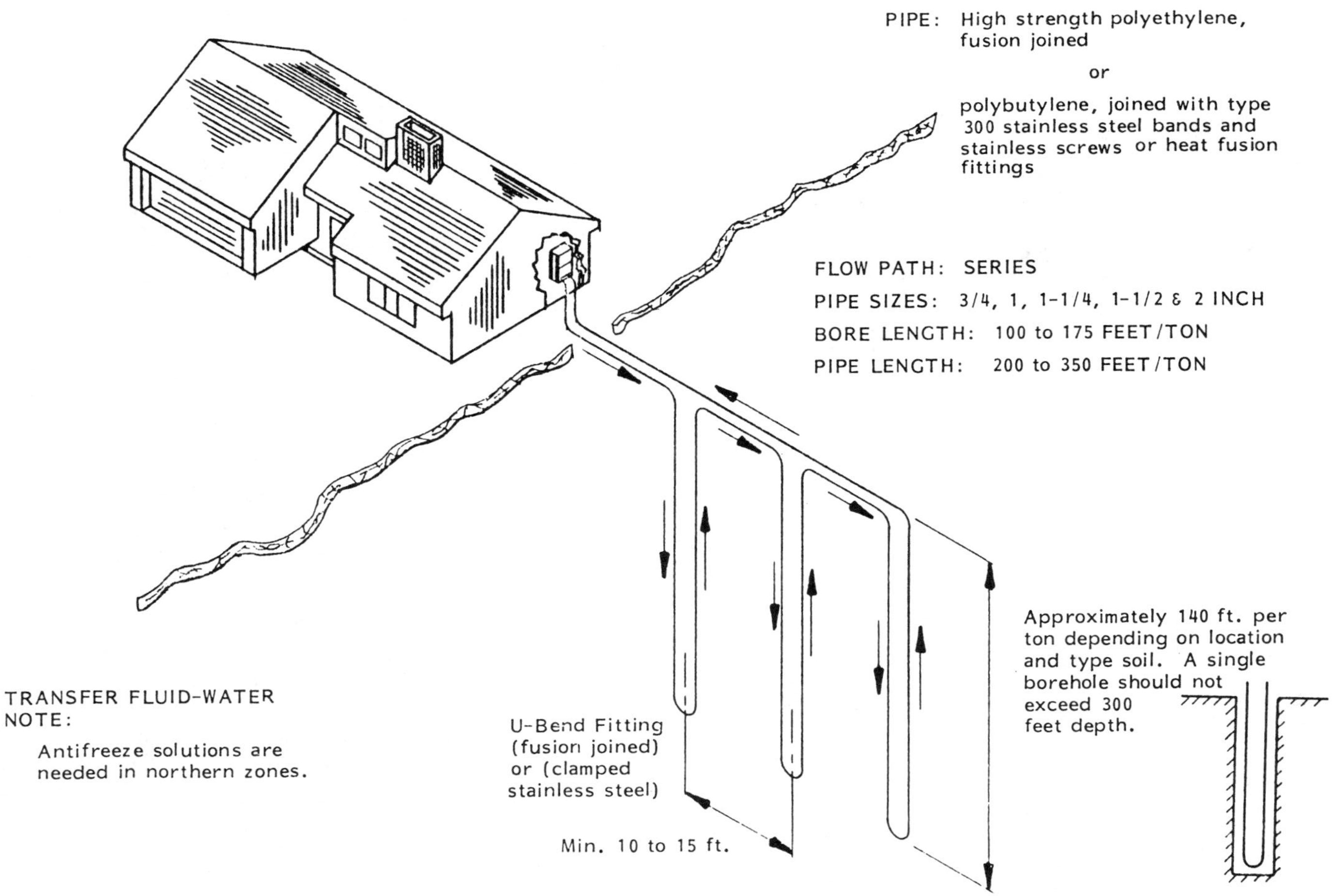
PIPE: High strength polyethylene, fusion joined
or
polybutylene, joined with type 300 stainless steel bands and stainless screws or heat fusion fittings
FLOW PATH: SERIES
PIPE SIZES: 3/4, 1, 1-1/4, 1-1/2 & 2 INCH
BORE LENGTH: 100 to 175 FEET/TON
PIPE LENGTH: 200 to 350 FEET/TON
Approximately 140 ft. per ton depending on location and type soil. A single borehole should not exceed 300 feet depth.
TRANSFER FLUID-WATER
NOTE:
Antifreeze solutions are needed in northern zones.
U-Bend Fitting (fusion joined) or (clamped stainless steel)
Min. 10 to 15 ft.

FIGURE 8–26 Vertical (series) system. (*Courtesy of Bard Manufacturing Company.*)

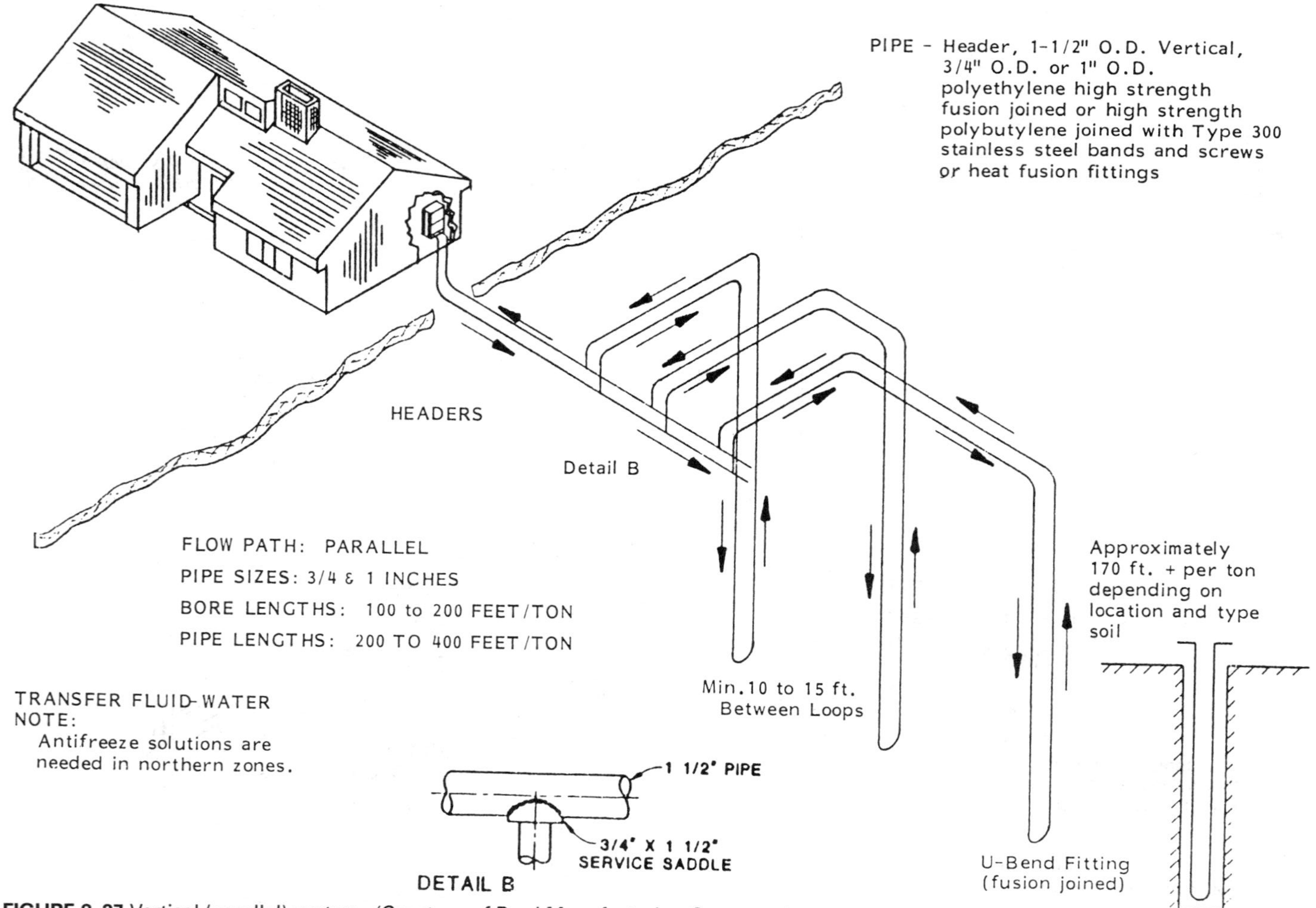
PIPE - Header, 1-1/2" O.D. Vertical, 3/4" O.D. or 1" O.D. polyethylene high strength fusion joined or high strength polybutylene joined with Type 300 stainless steel bands and screws or heat fusion fittings
HEADERS
Detail B
FLOW PATH: PARALLEL
PIPE SIZES: 3/4 & 1 INCHES
BORE LENGTHS: 100 to 200 FEET/TON
PIPE LENGTHS: 200 TO 400 FEET/TON
TRANSFER FLUID-WATER
NOTE:
Antifreeze solutions are needed in northern zones.
Min. 10 to 15 ft. Between Loops
Approximately 170 ft. + per ton depending on location and type soil
U-Bend Fitting (fusion joined)
1 1/2" PIPE
3/4" X 1 1/2" SERVICE SADDLE
DETAIL B

FIGURE 8–27 Vertical (parallel) system. *(Courtesy of Bard Manufacturing Company.)*

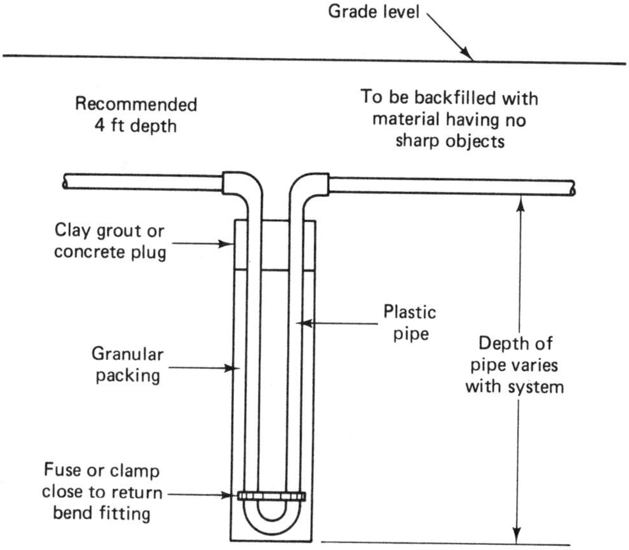
Grade level
Recommended
4 ft depth
To be backfilled with
material having no
sharp objects
Clay grout or
concrete plug
Plastic
pipe
Granular
packing
Depth of
pipe varies
with system
Fuse or clamp
close to return
bend fitting

FIGURE 8–28
Vertical borehole and piping detail for earth-coupled system. (*Courtesy of Billy Langley,* Heating, Ventilating, Air Conditioning and Refrigeration, *©1990, p. 575; reprinted by permission of Prentice-Hall, Inc., Englewood Cliffs, N.J.)*

TABLE 8–2 Minimum diameters (in.) for boreholes. (*Source: Bard Manufacturing Company.*)

Nom Pipe Size	Single U-Bend	Double U-Bend
3/4	3 1/4	4 1/2
1	3 1/2	5 1/2
1 1/4	4	5 3/4
1 1/2	4 3/4	6
2	6	7

multiple wells, and service lines. For the minimum diameters for boreholes, see Table 8–2.

Horizontal Earth-Coupled System

A horizontal earth-coupled system is similar to a vertical system in that the water circulates through underground piping. However, the piping in this type of system is buried in a trench (Figures 8–29 to 8–31). The pipe depths in the northern zone of the United States should be 3 to 5 ft (Figure 8–32). Burying the pipes at excessive depths will reduce the ability of the sun to recharge the heat used during the heating cycle. Pipe depths in the southern zone of the United States should be 4 to 6 ft so that the high temperature of the soil in late summer will not seriously affect the performance of the system.

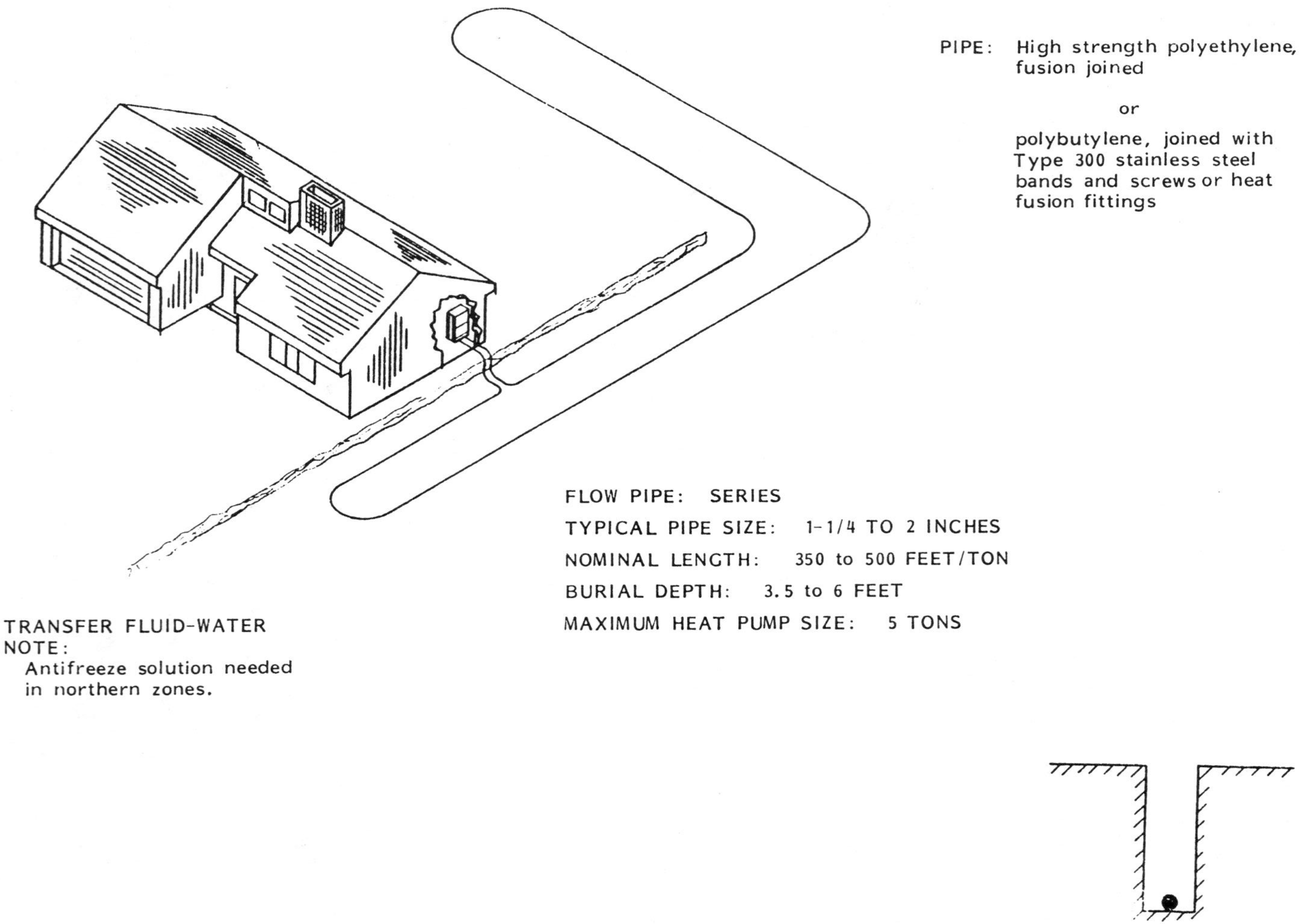
PIPE: High strength polyethylene, fusion joined
or
polybutylene, joined with Type 300 stainless steel bands and screws or heat fusion fittings
FLOW PIPE: SERIES
TYPICAL PIPE SIZE: 1-1/4 TO 2 INCHES
NOMINAL LENGTH: 350 to 500 FEET/TON
BURIAL DEPTH: 3.5 to 6 FEET
MAXIMUM HEAT PUMP SIZE: 5 TONS
TRANSFER FLUID-WATER
NOTE:
Antifreeze solution needed in northern zones.

FIGURE 8–29 Horizontal (series) system—one pipe in trench. (*Courtesy of Bard Manufacturing Company.*)

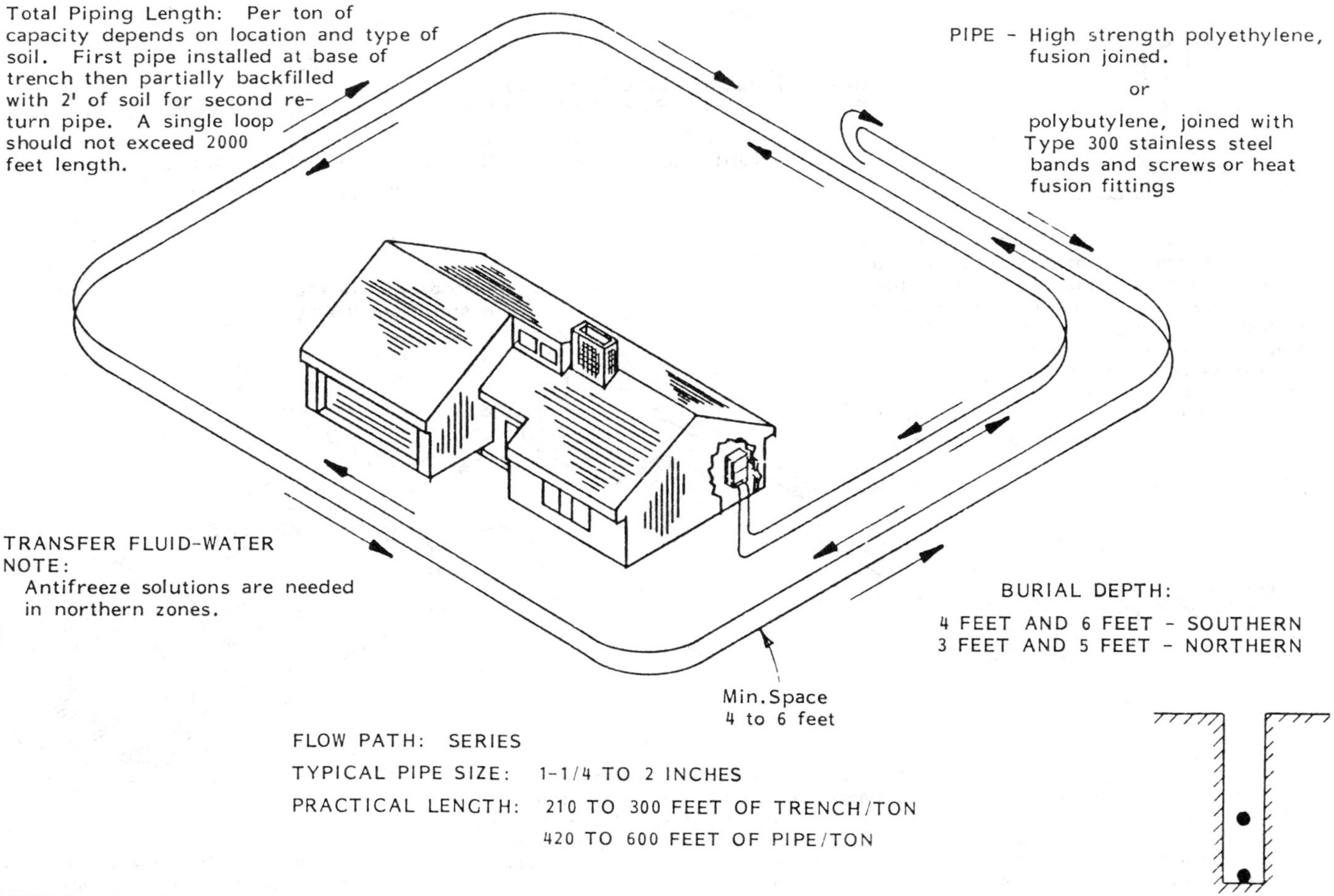
Total Piping Length: Per ton of capacity depends on location and type of soil. First pipe installed at base of trench then partially backfilled with 2' of soil for second return pipe. A single loop should not exceed 2000 feet length.
PIPE - High strength polyethylene, fusion joined.
or
polybutylene, joined with Type 300 stainless steel bands and screws or heat fusion fittings
TRANSFER FLUID-WATER
NOTE:
Antifreeze solutions are needed in northern zones.
BURIAL DEPTH:
4 FEET AND 6 FEET - SOUTHERN
3 FEET AND 5 FEET - NORTHERN
Min. Space
4 to 6 feet
FLOW PATH: SERIES
TYPICAL PIPE SIZE: 1-1/4 TO 2 INCHES
PRACTICAL LENGTH: 210 TO 300 FEET OF TRENCH/TON
420 TO 600 FEET OF PIPE/TON

FIGURE 8–30 Horizontal (series) system—two pipes in same trench. (*Courtesy of Bard Manufacturing Company.*)

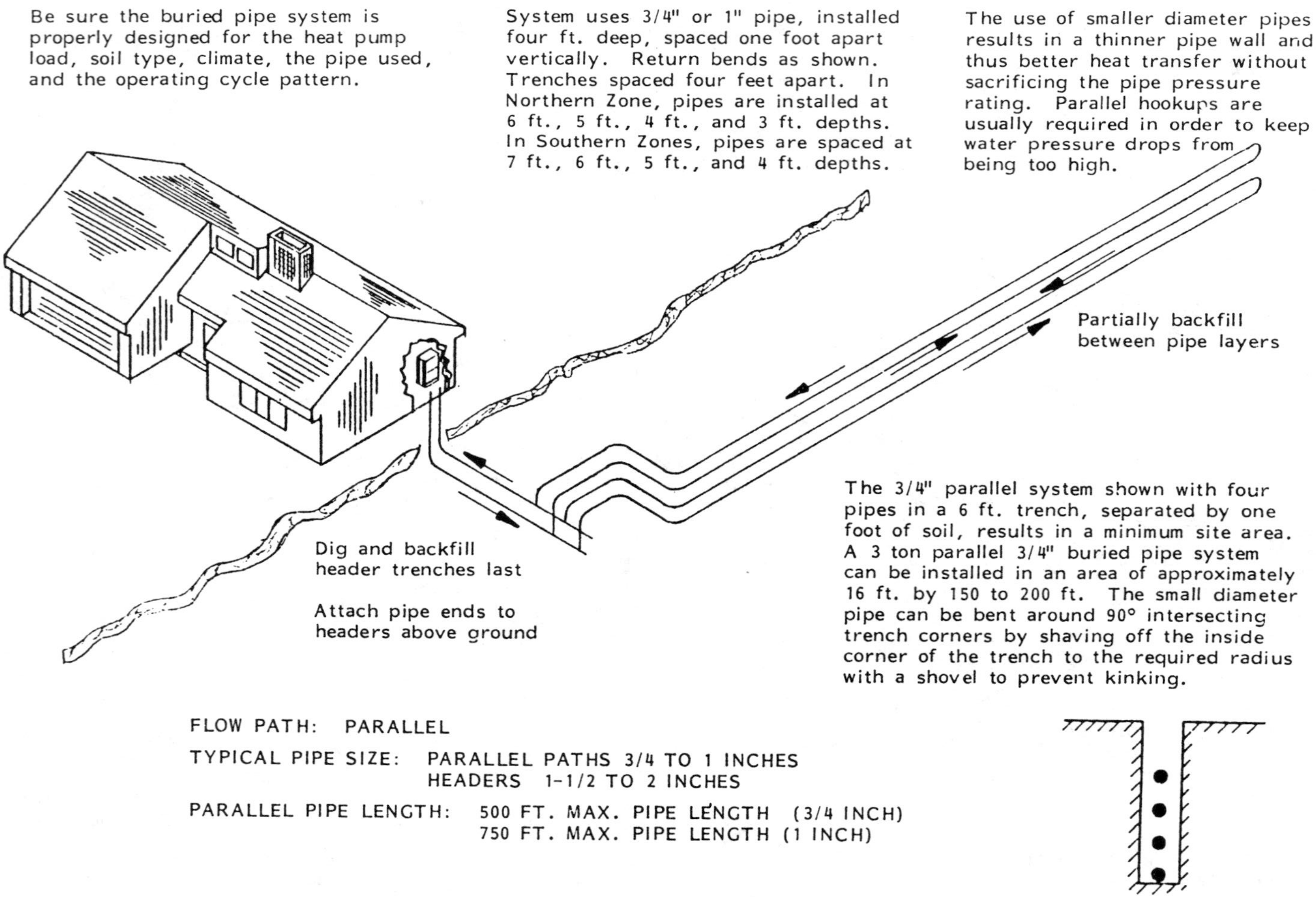
Be sure the buried pipe system is properly designed for the heat pump load, soil type, climate, the pipe used, and the operating cycle pattern.
System uses 3/4" or 1" pipe, installed four ft. deep, spaced one foot apart vertically. Return bends as shown. Trenches spaced four feet apart. In Northern Zone, pipes are installed at 6 ft., 5 ft., 4 ft., and 3 ft. depths. In Southern Zones, pipes are spaced at 7 ft., 6 ft., 5 ft., and 4 ft. depths.
The use of smaller diameter pipes results in a thinner pipe wall and thus better heat transfer without sacrificing the pipe pressure rating. Parallel hookups are usually required in order to keep water pressure drops from being too high.
Partially backfill between pipe layers
Dig and backfill header trenches last
Attach pipe ends to headers above ground
The 3/4" parallel system shown with four pipes in a 6 ft. trench, separated by one foot of soil, results in a minimum site area. A 3 ton parallel 3/4" buried pipe system can be installed in an area of approximately 16 ft. by 150 to 200 ft. The small diameter pipe can be bent around 90° intersecting trench corners by shaving off the inside corner of the trench to the required radius with a shovel to prevent kinking.
FLOW PATH: PARALLEL
TYPICAL PIPE SIZE: PARALLEL PATHS 3/4 TO 1 INCHES
HEADERS 1-1/2 TO 2 INCHES
PARALLEL PIPE LENGTH: 500 FT. MAX. PIPE LENGTH (3/4 INCH)
750 FT. MAX. PIPE LENGTH (1 INCH)

FIGURE 8–31 Horizontal multilevel (parallel) system. (*Courtesy of Bard Manufacturing Company.*)

Northern zone = 3 to 5 ft
Southern zone = 4 to 6 ft

FIGURE 8–32 Pipe depths for different regions. (*Courtesy of Billy Langley,* Heating, Ventilating, Air Conditioning, and Refrigeration, *1990, p. 578; reprinted by permission of Prentice-Hall, Inc., Englewood Cliffs, N.J.*)

Antifreeze will be necessary in the northern zone to prevent freezing of the circulated water and to allow the system to gain capacity and efficiency by using the large amount of heat released when the water contained in the soil is frozen. The antifreeze solution used in these systems is nontoxic propylene glycol or calcium chloride.

The use of multiple pipes in a single trench substantially reduces the total trench length. If a double layer of pipe is laid in the trench, the two layers should be set 2 ft apart to minimize thermal interference (Figure 8–33). For example, consider a 1 ½-inch series horizontal system with pipes at 5- and 3-ft depths. After installing the first pipe at 5 feet, partially backfill to the 3-ft depth using a depth gauge stick before installing the second pipe in the trench. Install the pipes with the return line running closest to the surface and the supply line running below it. This arrangement will maximize the overall system efficiency by providing warmer water in the heating mode and cooler water for the cooling mode. Connect the pipe ends to the heat pump after the pipe temperature has stabilized, so that shrinkage will not pull the pipe loose.

Two pipes in the same trench, one above the other, separated by 2 ft of earth require a trench 60% as long as that of a single-pipe system. The total length of pipe would be 120% as long as a single-pipe system, due to the heat transfer effect between the pipes. In addition, when laying a double layer of pipe, be careful to avoid kinks when making the return bend (Figure 8–34).

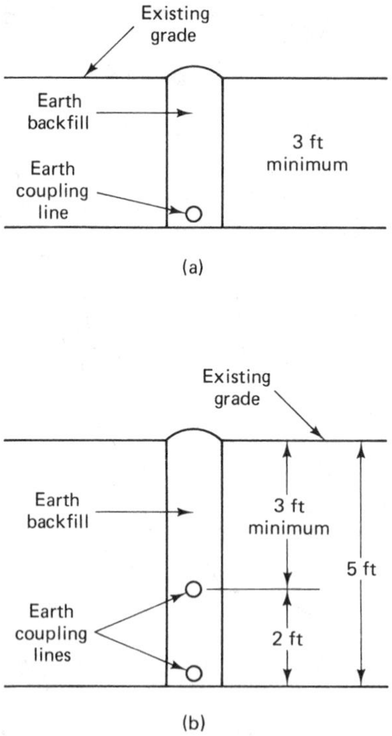
Existing grade
Earth backfill
3 ft minimum
Earth coupling line
(a)
Existing grade
Earth backfill
3 ft minimum
5 ft
2 ft
Earth coupling lines
(b)

FIGURE 8–33
Separation of the horizontal earth-coupled pipes in the trench for (a) single- and (b) double-layer designs. (*Courtesy of Billy Langley,* Heating, Ventilating, Air Conditioning, and Refrigeration, *©1990, p. 578; reprinted by permission of Prentice-Hall, Inc., Englewood Cliffs, N.J.*)

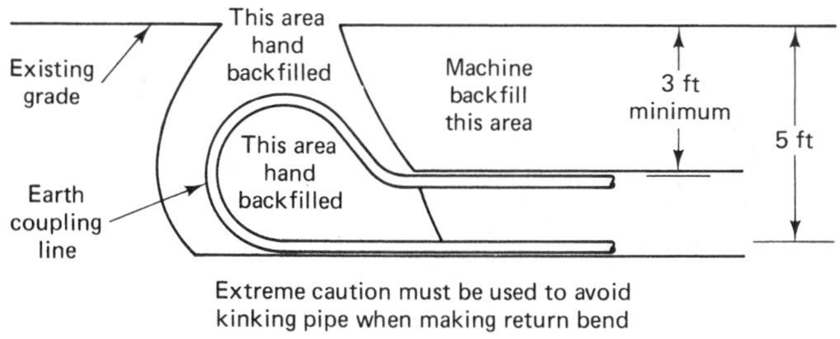
This area hand backfilled
Existing grade
Machine backfill this area
3 ft minimum
5 ft
This area hand backfilled
Earth coupling line
Extreme caution must be used to avoid kinking pipe when making return bend

FIGURE 8–34
Detail of a narrow trench return bend used on double layer horizontal earth-coupling system. (*Courtesy of Billy Langley,* Heating, Ventilating, Air Conditioning, and Refrigeration, *©1990, p. 579; reprinted by permission of Prentice-Hall, Inc., Englewood Cliffs, N.J.*)

Backfill the trench by hand when changing direction. If it is necessary to join two pipes together in the trench, use the fusion technique for IPS 304 stainless steel or brass fittings for greater strength and durability, and then mark the fitting locations for future reference by inserting a steel rod just below the grade (Figure 8–35). The steel rod will aid later in locating the fittings with a metal detector.

Trenches can be located closer together if the pipe in the previous trench can be tested and covered before the next trench is started. This also makes backfilling easier. Spacing of 4 to 5 ft between the pipes is good.

In areas with dry climates and heavy clay soil, any heat that is dissipated into the soil may significantly reduce the thermal conductivity of the soil. In such cases the designer may specify additional feet of pipe per ton of heat pump capacity. A few inches of sand may also be put in with the pipe, or a drip irrigation pipe may be buried with the top pipe to add occasional amounts of water to the soil.

In series horizontal earth couplings, all the water flows through all the pipe. These may be made of 1-, 1 ½- and 2-inch pipe either insert coupled or fused.

Narrow trenches. Narrow trenches are dug with trenching machines. The trenches are usually 6 inches wide. Generally speaking, the trencher will require about 5 ft between trenches. This is sufficient spacing for horizontal earth coils.

The pipe can then be coiled into an adjoining trench. Since the trencher spaces the trenches about 5 ft apart, looping the coil from one trench to another will give a return of large enough diameter. The end trench should be backhoed to give enough room for the large-diameter bend.

If the pipe is brought back in the same trench, bend the pipe over carefully to avoid kinking it, and handfill the area around the return bend (Figure 8–34). To reduce the bend radius, elbows may be used. However, keeping the number of fittings underground to a minimum may be preferable since the potential for leaks is reduced.

If a double layer of pipe is used, the incoming water to the heat pump unit should be from the deepest pipe (Figure 8–36). This provides the heat pump with the coolest water during the cooling season and the warmest water during the heating season.

Backhoe trenches. If a backhoe is used, the trench will probably be about 2 ft wide. In a wide backhoe trench, two pipes may be placed side by side, one on each side of the trench. The pipes in the trench must be at least 2 ft apart.

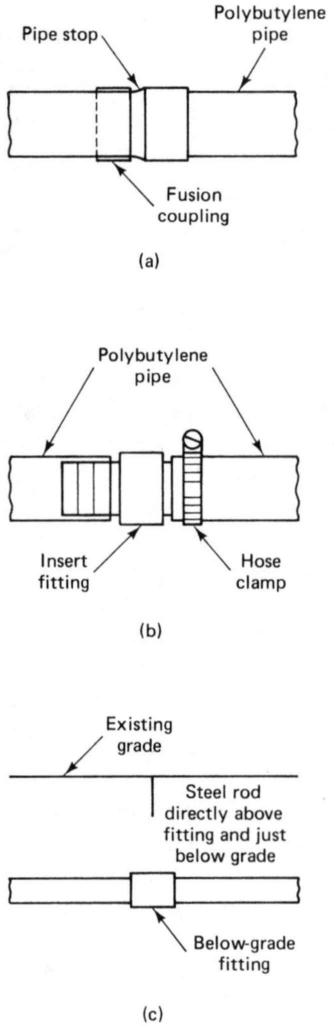
Pipe stop
Polybutylene pipe
Fusion coupling
(a)
Polybutylene pipe
Insert fitting
Hose clamp
(b)
Existing grade
Steel rod directly above fitting and just below grade
Below-grade fitting
(c)

FIGURE 8–35
Hose fitting and location detail: (a) socket-fused coupling; (b) clamp installation; (c) fitting location indicator. *(Courtesy of Billy Langley,* Heating, Ventilating, Air Conditioning, and Refrigeration, *©1990, p. 579; reprinted by permission of Prentice-Hall, Inc., Englewood Cliffs, N.J.)*

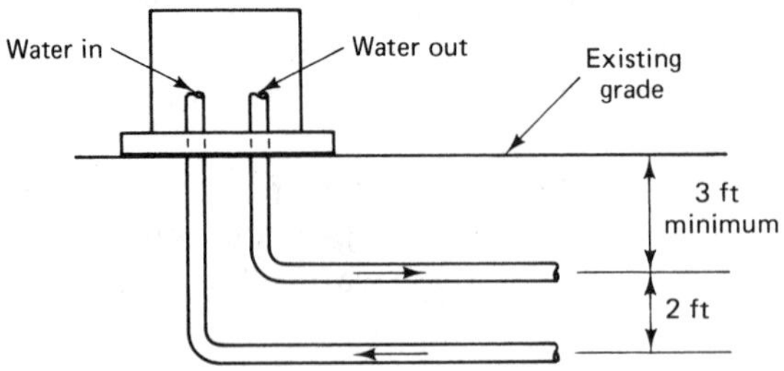
Water in
Water out
Existing grade
3 ft minimum
2 ft

FIGURE 8–36
Double-layer horizontal earth coil water flow connection schematic. (*Courtesy of Billy Langley,* Heating, Ventilating, Air Conditioning, and Refrigeration, *©1990, p. 580; reprinted by permission of Prentice-Hall, Inc., Englewood Cliffs, N.J.*)

Backhoe carefully around the pipe with fine soil or sand. Do not drop clumps of clay or rock onto the pipe (Figure 8–37). A pit may be dug at the end of the trench to accommodate a 4-inch-diameter return bend (Figure 8–38).

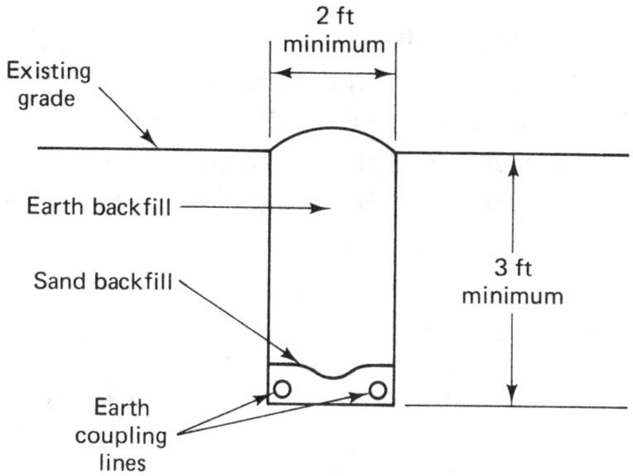
2 ft
minimum
Existing
grade
Earth backfill
Sand backfill
3 ft
minimum
Earth
coupling
lines

FIGURE 8–37
Backhoed trench. *(Courtesy of Billy Langley,* Heating, Ventilating, Air Conditioning, and Refrigeration, *©1990, p. 580; reprinted by permission of Prentice-Hall, Inc., Englewood Cliffs, N.J.)*

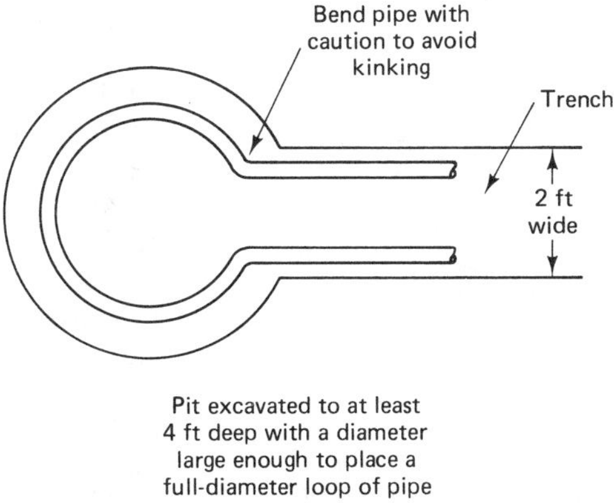
Bend pipe with
caution to avoid
kinking
Trench
2 ft
wide
Pit excavated to at least
4 ft deep with a diameter
large enough to place a
full-diameter loop of pipe

FIGURE 8–38
Detail of wide trench return for either a single- or double-layer horizontal earth coil. *(Courtesy of Billy Langley,* Heating, Ventilating, Air Conditioning, and Refrigeration, *©1990, p. 580; reprinted by permission of Prentice-Hall, Inc., Englewood Cliffs, N.J.)*

Service lines. The recommendations for the horizontal earth coils also apply for the installation of the service lines to and from the U-bend wells and pond or lake heat exchanger. Bury the service lines a minimum of 3 ft for single-layer pipe and 3 to 5 ft for double-layer installations. If two pipes are buried in the same trench, keep them 2 ft apart.

A parallel horizontal earth coupling is one in which the water flows out through a supply header, is divided equally, and then flows simultaneously into two or more earth coils. The water then returns to the other header. Headers are reverse-return plumed so that equal-length earth coils have equal flow rates. The length of individual parallel earth coils must be within 10% of each other to ensure equal flow in each coil.

Freeze protection. The antifreeze solutions that are used in earth-loop systems must be nontoxic and noncorrosive: nontoxic because should a leak develop in the loop system, the groundwater will not be contaminated and noncorrosive for the protection of the metal components of the system that come into contact with the circulating solution.

When the local well water temperature is below 60° F, the water in the earth loop should be protected from freezing temperatures down to 18° F. The recommended antifreeze solution is propylene glycol. The amount of antifreeze needed for an earth loop is determined as follows: Calculate the approximate volume of water in the loop system by using Table 8–3. This table gives the gallons of water per 100 ft of pipe. Add 2 gallons for the equipment room devices and the heat pump.

TABLE 8–3 Water capacity of pipe. (*Source: Bard Manufacturing Company.*)

Pipe Material	Nominal Pipe Size (in.)	Gallons per 100 ft of Pipe
Polyethylene		
SDR-11	¾	3.02
SDR-11	1	4.73
SDR-11	1¼	7.52
SDR-11	1½	9.85
SDR-11	2	15.40
SCH 40	¾	2.77
SCH 40	1	4.49
SCH 40	1¼	7.77
SCH 40	1½	10.58
SCH 40	2	17.43
Polybutylene		
SDR-17 IPS	1½	11.46
SDR-17 IPS	2	17.91
SDR-13.5 CTS	1	3.74
SDR-13.5 CTS	1¼	5.59
SDR-13.5 CTS	1½	7.83
SDR-13.5 CTS	2	13.38
Copper	1	4.3

Propylene glycol. When the groundwater at a depth of 100 ft is 66° F or less, a 20% solution by volume of propylene glycol is required to provide the necessary protection. The percentage of antifreeze solution depends on the geographical location. A solution of 20% by volume of propylene glycol will provide freeze protection down to 18° F. For example, for a system that holds 100 gallons of water in the loop, 20 gallons of propylene glycol are needed to provide the desired protection.

To add the antifreeze, two small pieces of hose, a bucket, and a small submersible pump are needed. Block the system by closing a ball valve. Blocking the flow prevents the antifreeze from being pumped into one boiler drain and out the other.

Attach the hoses to the boiler drains. Run the uppermost hose to a drain. Connect the other hose to the submersible pump in the bucket. Put full-strength propylene glycol in the bucket and pump the desired amount into the system. When the required amount has been pumped into the system, turn off the pump, close the boiler drains, disconnect the hoses, and open the isolation flange or gate valve.

Calcium chloride. A 20% by weight solution of calcium chloride and water may also be used as an antifreeze in the earth-coupled heat pump system. It is also nontoxic, a better heat conductor, and less expensive than propylene glycol. However, it is mildly corrosive. To determine the amount of calcium chloride needed for the system, multiply the gallons of water in the loop system by 1.4841 to find the pounds of 94 to 98% pure calcium chloride required to provide the desired 18° F freeze protection.

Closed Loop Lake Heat Exchanger

The closed-loop, earth-coupled heat pump systems that use a pond or lake for their heat source are by far the most economical of all such systems. The use of pond loops for residential application is restricted to bodies of water having at least 1 acre of surface area and a depth of 10 ft or more at the location of the loop during the worst-case low water conditions. The acceptability of any body of water varies depending on the size, average depth, and water source (springs, river or creek, runoff, etc.). Most manufacturers recommend that they be consulted to discuss any application larger than 5 tons, or any commercial installation.

Also, there are special considerations that must be given to pond loop applications. Never, under any circumstances, place a loop in a river or stream, because the drifting objects at flood conditions will cause damage to the loop. The pond loop gets colder than either a horizontal or vertical earth coupling, so regardless of the location, the system must use a 20% by volume antifreeze solution. Evaluate the location of the pond loop in relation to the building. There is no need to install a pond loop when the trench length to the pond and back to the building is equal to or greater than that required for a horizontal earth coil.

Heat exchanger. The length of the loop is determined to a great extent by the temperature of the water in the pond. For example, if the well water temperature is between 50 and 68° F, use 60 ft of ¾- inch copper pipe per nominal ton of refrigeration. The water temperature information is available from the local water well association or the National Water Well Association. In areas where the well water temperature is indicated to be either above 68° F or below 50° F, use 80 ft of ¾-inch copper pipe per nominal ton of refrigeration (Figure 8-39).

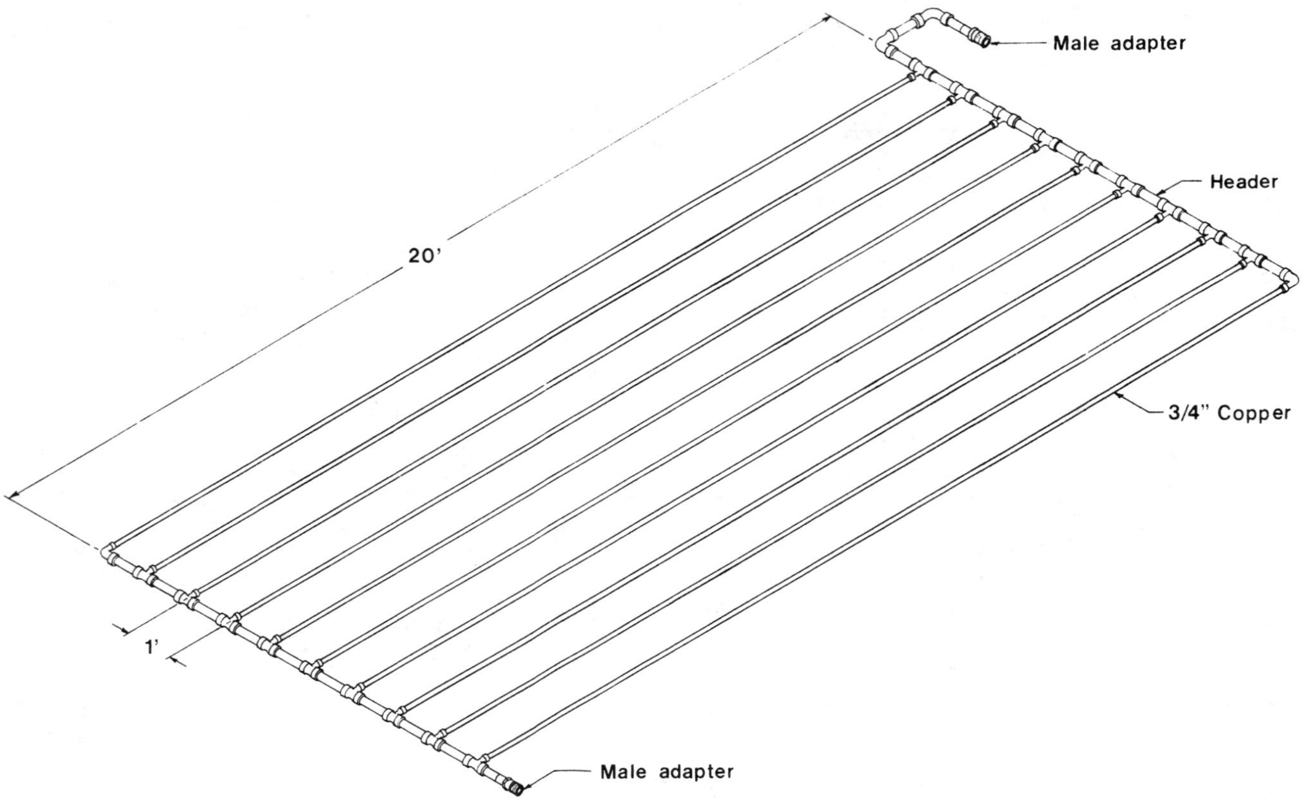
Male adapter
Header
20'
3/4" Copper
1'
Male adapter

FIGURE 8–39 Typical lake heat exchanger. (*Courtesy of Command-Aire Corporation.*)

Location. The pond loop will not perform correctly if it is allowed to settle down into the mud. It should be supported at least 1 ft off the bottom, and at least 9 ft below the surface of the water. Suspending the pond loop under a pier is an easy application method. Another is to secure an old tire to each of the corners and allow the pond loop to settle to the top of the tires. The tires act as spacers to keep the loop about 1 ft or so off the bottom of the pond.

Installation. The service lines for the pond loop should be buried 4 ft deep or below the frostline, whichever is deeper. The lines should be about 2 ft apart in a wide trench. If a narrow trench is used, the pipes should be at depths of 4 and 6 ft. Do not attempt to use the pond loop if the service lines cannot be buried across the shore of the lake and out into the water. Never bring service lines up from the loop and run them underneath the pier—always leave the lines in the water and as deep as possible.

OPEN-LOOP (GROUNDWATER) APPLICATION

In an open-loop system the water is taken into the system, passed through the heat pump, and then discharged. It is not recirculated as in the closed-loop ground water systems.

Since the water is a crucial element in the operation of these types of systems, many manufacturers recommend that the services of an experienced local well driller be obtained. His or her expertise is valuable in assessing an existing well and equipment or in making recommendations for a new installation. If no local driller is available, contact your state well drilling association or the National Water Well Association at 6375 Riverside Drive, Dublin, Ohio 43017 for referral. Their telephone number is (614) 761–1711.

The water used in these systems does not need to be suitable for human consumption; however, it should meet certain standards (Table 8–4). The local or state health department can run water tests to see if it will meet the manufacturer's recommendations.

Most heat pumps require a water flow rate of approximately 2 gpm per nominal ton of refrigeration. The required flow rate may need to be somewhat higher in northern climates during winter operation. Under extreme weather conditions, the heat pump(s) may run continuously for extended periods of time. Therefore, the water supply must be able to deliver the appropriate continuous water flow for 24 hours or more. A thorough "drawdown" testing of the well, according to the methods described by the National Water Well Association, will indicate if it has the capacity to handle the extreme conditions. If the well is intended to supply both the heat pump and the household water supply, be sure to conduct the drawdown test using the combined water requirements, not just the heat pump water flow. It is the responsibility of the driller to be knowledgeable as to all applicable water well codes and regulations and to perform the installation accordingly.

TABLE 8–4 Water standards for open-loop (groundwater) heat pump application. (*Courtesy of Command-Aire Corporation.*)

Scaling	
Calcium and magnesium salts (hardness)	Less than 350 ppm
Iron oxide	Low
Corrosion	
pH	5–10
Hydrogen	Less than 50 ppm
Carbon dioxide	Less than 75 ppm
Chloride	Less than 600 ppm
Total dissolved solids	Less than 1500 ppm
Biological growth	
Iron bacteria	Low
Suspended solids	Low

Source: Command-Aire Corporation.

A final consideration for the water well is for its pump sizing and depth. Careful evaluation of the pumping power requirements may prevent an installation where all the energy saved by the heat pump is consumed by the well pump.

After an adequate water supply is assured, the water must be delivered to the heat pump. When tapping into an existing pressure line, the tap should be made immediately after the pressure tank prior to down-line taps. The tank should be sufficiently large to prevent rapid cycling of the well pump. If the tank is not of sufficient size, a new tank must be installed, or it might be desirable to bypass the tank and cycle the pump with the heat pump unit. Remember, the electrical power consumed by the pump is an important consideration. These are decisions that need to be made for each installation, based on the existing equipment and water availability.

EFFICIENCY

Properly designed heat pump systems are more efficient than almost any other form of heating. Two related methods are used to calculate heat pump efficiency: (1) coefficient of performance (COP), and (2) seasonal energy efficiency ratio (SEER). The COP of a heat pump is the efficiency of the unit during the heating mode. SEER is the efficiency of the unit in the cooling mode. You should be familiar with the method used to calculate the efficiency of the unit in both operating modes.

Coefficient of Performance

This method has been used for many years when discussing the efficiency of heat pump systems in the heating mode. To calculate the COP of a heat pump, use the following formula:

$$\begin{aligned} \text{COP} &= \text{Btu out} \div \text{Btu paid for} \\ &= \text{Btu/h capacity} \div \text{unit wattage} \times 3.413\ \text{Btu/W}. \end{aligned}$$

Example:

Given a unit with 40,000 Btu/h heating capacity and consuming 4,380 W/h at 40° F, the COP would be

$$\begin{aligned} \text{COP} &= 40{,}000 \div 4{,}380 \times 3.413\ \text{Btu/W} \\ &= 40{,}000 \div 14{,}948.94 = 2.67. \end{aligned}$$

This is an indication of how efficient the units are when compared to electric resistance heating. Electric resistance heating produces 3.413 Btu of heat for each watt consumed. Thus the COP of electric resistance heating is 1 at maximum output. When we compare the COP of a heat pump to an electric resistance heater, we see that the heat pump can provide more heat per watt of power consumed. The heat pump is, therefore, more efficient. With the operating conditions in our previous example, the heat pump will deliver 2.67 times the amount of heat delivered by a resistance heater using the same wattage.

Seasonal Energy Efficiency Ratio

The SEER is a way of calculating the efficiency of heat pumps in the cooling mode and all other types of air conditioning units. The SEER rating of a complete system is made available to all persons interested in unit efficiency. It is an addition to the older energy efficiency ratio (EER) that takes into consideration the average number of seasonal operating hours.

The EER calculates how many Btu are produced for every watt of electric power consumed. It is calculated by dividing the total unit capacity, in Btu, of the system by the total electric power, in watts per hour, consumed by the equipment. To calculate the SEER of the unit, use the following formula:

Using the same unit as in the previous example, given a unit with 40,000 Btu/h cooling capacity and consuming 4,380 W/h at 105° F, the EER would be

$$\begin{aligned} \text{EER} &= \text{Btu out} \div \text{power in} \\ &= 40{,}000 \div 4380 \\ &= 9.13. \end{aligned}$$

The SEER would then be

$$\begin{aligned} \text{SEER} &= \text{EER} \times 0.885 + 1.237 \\ &= 9.13 \times 0.885 + 1.237 \\ &= 8.08 + 1.237 \\ &= 9.32 \end{aligned}$$

where

$$\text{EER} = \text{energy efficient ratio}$$
$$0.885 = \text{a constant}$$
$$1.237 = \text{a constant.}$$

The heat pump, like any refrigeration system, will begin to lose capacity as the suction pressure drops. When the heat pump is operating in the heating mode, a capacity reduction occurs as the outdoor temperature drops. This also causes a reduction in the SEER and COP of the unit. This capacity reduction occurs because the vapor density of the refrigerant decreases as the suction pressure decreases. Therefore, the refrigerant vapor weighs less per cubic foot and the compressor pumps fewer pounds of refrigerant through the system. A refrigerant is capable of absorbing only a certain number of Btu per pound. When fewer pounds of refrigerant are circulated, the capacity is reduced accordingly.

Along with the decrease in the vapor density at lower suction pressures, the reexpansion stroke of the compressor piston is longer. Thus a shorter effective stroke causes a decrease in the refrigerant flow rate.

HEAT PUMP CAPACITY MEASUREMENT

Determining the capacity of a heat pump is a fairly simple task. Use the following steps:

1. Set the room thermostat to the heating or automatic position.
2. Set the room thermostat temperature selector to 90° F.
3. Turn off all electricity to the resistance heat strips. Only the heat pump should be operating during this test.
4. Measure the temperature rise through the indoor unit (Figure 8-40).

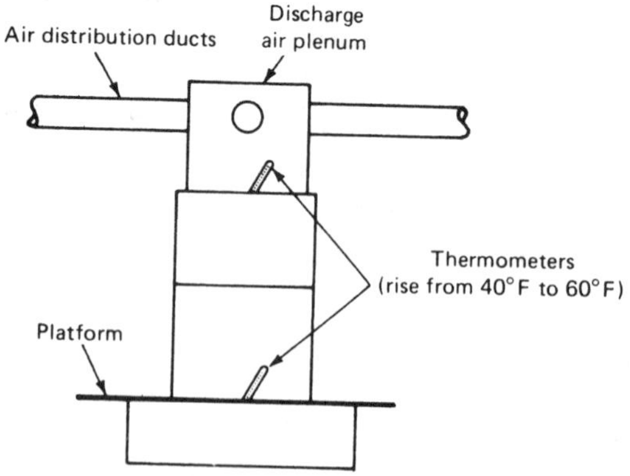
Discharge air plenum
Air distribution ducts
Thermometers
(rise from 40°F to 60°F)
Platform

FIGURE 8–40 Measuring temperature rise through a unit. (*Courtesy of Billy Langley, Heat Pump Technology Systems Design, Installation, and Troubleshooting, ©1989, p. 18; reprinted by permission of Prentice-Hall, Inc., Englewood Cliffs, N.J.*)

Use the following steps when measuring the temperature rise:

a. Use the same thermometer for determining the return and supply air temperatures.
b. Do not measure in the lined areas. True air temperature cannot be measured in areas affected by radiant heat.
c. Make the temperature measurements within 6 ft of the indoor unit. Measurements taken at the return and supply grilles are inaccurate.
d. Use the average temperature when more than one duct is connected to the plenum.
e. Make sure that the air temperature is stable before taking measurements.
f. Take measurements downstream from any mixed air.
g. Record the temperature difference of the return and supply air as ΔT.

5. Determine the Btu output by using the following formula:

$$\text{Btu} = \text{cfm} \times \Delta T \times 1.08$$

where

cfm = total measured air flow (cubic feet per minute)

ΔT = supply air temperature minus return air temperature

1.08 = specific heat of air constant.

Example:

The measured air flow of a unit is 1,600 cfm and the measured temperature rise is 30° F. The Btu capacity is

$$\text{Btu} = 1{,}600 \times 30 \times 1.08 = 51{,}840.$$

DUAL-FUEL HEAT PUMP SYSTEMS

The name *dual-fuel systems* comes from the fact that two fuel sources are used to power these systems. Normally these two fuels are either gas (either natural or LP) or fuel oil, and electricity. There are two reasons for choosing these types of systems: (1) They are used in many cases when the cooling part of an electric cooling-gas (oil) heating system fails and must be replaced while the furnace is still in good condition and the customer does not wish to replace the complete unit; or (2) they are used for economy in operation. These systems must be installed according to the equipment manufacturer's specifications to prevent damage to the unit and to obtain their designed operating efficiency.

The indoor coil of the heat pump is placed on top of the furnace just as in an ordinary gas heating-electric cooling application (Figure 8–41).

When such systems are installed, the economical heat pump unit is operated when the outdoor ambient temperatures are above the building balance point. Below the balance point, the furnace is used because gas or oil is more economical at these temperatures than operating both the heat pump and the required auxiliary heat strips. Also, the heat pump is less efficient at lower temperatures. As the out-

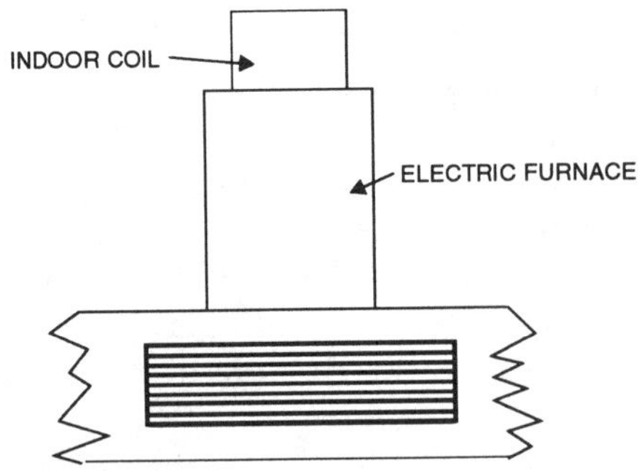
INDOOR COIL
ELECTRIC FURNACE

FIGURE 8–41
Location of indoor heat pump coil.

door temperature rises above the balance point, the heat pump will operate using the more economical electric energy. In this manner, as the outdoor ambient temperature rises above and falls below the building balance point, the most economical source of heating is used automatically.

The control circuit on these types of systems is designed to prevent operation of both the furnace and the heat pump at the same time. If both were to operate at the same time, the heat pump would act as a reheat device causing excessive compressor failure or permanent damage.

Basic Controls

The following is a description of the basic controls used on a dual-fuel heat pump system. (Be sure that the equipment manufacturer's recommended controls are used.)

Thermostat. A two-stage thermostat is used to control operation of the system. The first stage controls operation of the heat pump unit. The second stage energizes the furnace and deenergizes the heat pump. Because of the location of the indoor coil, it is necessary that the two systems do not operate at the same time.

Heat pump delay control. This is a thermostatic type of control that is mounted in the plenum. Its purpose is to sense the plenum temperature and prevent operation of the compressor until after the furnace has cycled and the discharge air temperature has dropped to a temperature that is below the control setting.

Defrost limit control. There is a thermostatic type of control that is mounted in the plenum. Its purpose is to stop operation of the furnace when the plenum temperature rises to the cut-off setting of the control.

Mild weather control. In some installations, the heat pump may be required to operate when the outdoor ambient temperatures are around 65° F to 70° F. When it is operated under these conditions, the compressor may occasionally be cut off by the high pressure control. The purpose of the mild weather control is to cycle the outdoor fan motor on and off, reducing the amount of heat taken into the system by the outdoor coil and reducing the head pressure.

Compressor monitor. Some manufacturers incorporate a compressor monitor thermostat in their unit. The purpose of this control is to stop operation of the heat pump when the outdoor ambient temperature drops below the setting of this thermostat. At this point the gas furnace is started to keep the building comfortable.

Operation

During mild temperatures, above the building balance point, the heat pump is capable of handling the building heat loss with relative ease and economy. However, when the outdoor ambient temperature falls below the building balance point temperature, the control system stops operation of the heat pump and starts operation of the furnace. The system will operate in this mode until the outdoor ambient temperature rises above the building balance point, at which time the control circuit will stop the gas furnace and start operation of the heat pump.

During the defrost cycle, the system switches from heating to cooling to melt the ice from the outdoor coil. The furnace is automatically started to prevent cool air from being blown into the building. When the indoor air temperature rises to about 90° F, the defrost limit control stops operation of the furnace. This is to protect the compressor from high compression ratios.

Equipment Selection

The Btu rating of the unit is determined in the normal manner. Then the unit is selected based on the model number, unit efficiency, cost, and any preference that the building owner may have. The approved indoor coil is selected from a chart to match the model heat pump used. The coil must be one that is authorized by the equipment manufacturer; otherwise the system may not operate as desired and damage to the compressor may be possible.

REVIEW QUESTIONS

1. Why is it possible for a heat pump system to produce heat during cold outdoor ambient temperatures?
2. How does a heat pump system move heat from one place to another?
3. How are heat pump units classified?
4. Is it easier for a heat pump to remove heat from lower or higher outdoor ambient temperatures?
5. What would be the advantage of having the reversing valve energized during the heating cycle of a heat pump system?
6. Name the operating cycles of a heat pump system.

7. What determines the condensing temperature of a refrigeration system?
8. What device causes the refrigerant to change directions in a heat pump system?
9. What is the most popular device used for controlling the refrigerant flow in a heat pump system?
10. What device is used to direct the refrigerant either around or through the flow control device on a heat pump system?
11. Where is the suction line accumulator located on a heat pump system?
12. What is the most popular method of defrosting a heat pump system?
13. What is probably the most popular method of defrost control used on heat pump systems?
14. Where are the supplementary heat strips located on a heat pump system?
15. Why are the supplementary heat strips placed in this location?
16. Name two reasons for the efficiency achieved by closed-loop, earth-coupled heat pump systems.
17. Name the two general types of earth-coupled heat pump systems.
18. What type of pipe is normally used in the earth-loop coil?
19. What should be done to prevent freezing of the earth-loop system?
20. What should be done to the closed-loop lake heat exchanger to maintain its efficiency?
21. What is the required water flow rate for most water source heat pump systems?
22. Of what is the COP of a heat pump system an indication?
23. Of what is the SEER of a heat pump system an indication?
24. Write the formula for determining the Btu output of a heat pump system.
25. On duel fuel heat pump installations, when does the secondary heat operate?

9

refrigerant recovery, recycling, and reclaim

Even though the ozone at ground level is a toxic pollutant and is responsible for smog and all kinds of health problems, the ozone in the stratosphere, 6 to 30 miles above us, is considered to be good ozone. While we are trying to eliminate ozone at our level, we want it to stay in the stratosphere.

STRATOSPHERIC OZONE

The stratospheric ozone layer is the main shield against ultraviolet radiation from the sun. A decrease in the stratospheric ozone will allow more of these ultraviolet rays to reach the earth. Scientists tell us that this radiation will cause an increase in skin cancer, cataracts (the leading cause of blindness in the United States), and, potentially, suppression of the immune system. When our immune systems are affected, all of us will be more susceptible to all types of diseases. Also, damage to the ozone layer presents a serious threat to our food supply by reducing crop yields. When the stratospheric ozone is depleted, all forms of life on land and in the sea are at great risk.

The problem is with chlorofluorocarbons (CFCs) and similar compounds, which are persistent and extremely stable chemicals that rise up into the atmosphere intact until they reach the stratosphere. The radiation from the sun causes them to break down, and the chlorine component is freed. The chlorine component then attacks and destroys the ozone.

In 1974, Drs. Sherwood Rowland and Mario Molina from the University of California published a paper demonstrating how CFCs destroy the stratospheric ozone. At this time there were no actual measurements of actual ozone loss; it was just a scientific theory. Because of this theory, the United States banned the use of CFCs in most aerosols in 1978.

Industry began to search for safe substitutes. Considerable progress was being made until the early 1980s, when the threat of further government regulation subsided. The search for substitutes came to a standstill, and worldwide use of CFCs continued to grow.

In 1985, scientists discovered a significant loss of the ozone layer over a portion of the southern hemisphere about the size of North America. Further measurements have revealed losses greater than 50% in the total stratospheric ozone column and greater than 95% at an altitude of 9 to 12 miles above the earth. When this hole over Antarctica was discovered, new importance was given to international efforts to understand and protect the ozone layer.

In September 1987, the Montreal Protocol on Substances that Deplete the Ozone Layer was negotiated and signed by more than two dozen nations. The United States was one of them. When the Protocol was enforced in January 1989, there were 68 nations which were parties to it. More nations have joined the Protocol since then.

However, only a short time after the signing of the Protocol, scientists observed and measured the ozone losses on a global scale and found that the destruction was not limited only to remote portions of Antarctica. Losses were detected over the United States, thus bringing the problem closer to home.

These measurements of actual ozone loss were significantly greater than computer models had predicted, thus raising serious questions about how adequate the control measures were that were set forth in the Montreal Protocol and Environmental Protection Agency (EPA) regulation. Even the Protocol amendments that were adopted in the June 1990 meeting of the signatories in London do not provide adequate protection. Four major areas that warrant further attention by national legislatures and by the parties to the Protocol are accelerating the CFC and methyl chloroform phase-out schedules; controlling and ultimately eliminating production and use of hydrofluorocarbons (HCFCs); eliminating the emissions of ozone-destroying compounds; and implementing effective trade sanctions. Each of these areas is covered by the Clean Air Act of 1990.

Natural chlorine in the stratosphere has a concentration of about 0.6 parts per billion. When the hole over Antarctica was discovered in 1985, the concentration was about 2.5 parts per billion. This was largely due to the emissions of CFCs, methyl chloroform, and similar chemicals. At the present time the concentration is greater than 3 parts per billion, a record high level, and is increasing.

The United States is the largest producer and user of ozone-depleting chemicals in the world. Even with the ban of CFCs in aerosol uses in 1978, the per capita use continues to be greater than that of Western Europe, Japan, China, the former Soviet Union, and India. CFCs are used as refrigerants in domestic refrigerators and freezers, automotive air conditioning units, and commercial refrigeration systems. They are used as blowing agents in the manufacture of furniture cushions and packaging materials. Methyl chloroform is used as a solvent for cleaning and degreasing metals and as a component of adhesives. The human race created these chemicals; we have a responsibility to lead the world in eliminating them and finding safe substitutes.

The EPA has estimated that the phase-out of CFCs and similar compounds scheduled in the Clean Air Act will benefit the entire U.S. population born before the year 2075, by eliminating almost 162 million cases of skin cancer, more than 3 million cancer deaths, and over 18 million cases of cataracts.

The EPA has also estimated that in addition to the aforementioned health problems, the economic and environmental benefits of the phase-out in the United States will be approximately \$58 billion through the year 2075. This figure includes \$41 billion due to the reduced damage to food crops.

There is yet another benefit. Most of us are familiar with the connection between CFCs and the destruction of the ozone layer. A less well known fact is the connection between CFCs and global climate change that is predicted to occur as a result of an intensified greenhouse effect.

The threat of uncontrolled global climate change is due to the accelerating accumulation of greenhouse gases in the atmosphere, primarily carbon dioxide, CFCs, and methane. These gases act as a thermal blanket, trapping the heat in the earth's atmosphere and causing the earth's surface temperature to rise.

Carbon dioxide emissions account for an estimated 50% of the predicted global warming. CFCs are estimated to account for a substantial part, about 15 to 20%.

Because each molecule of CFC has approximately 20,000 times more impact on the global climate than a single molecule of carbon dioxide, the control of CFCs is vital in preventing global warming.

Thus, there are two reasons to control the release of these ozone-destroying chemicals. First, it is essential if the destruction of the ozone layer is to be stopped. Second, it is the most effective single step that can be taken to curb the phenomenon known as the greenhouse effect.

THE CLEAN AIR ACT

Title VI of the Clean Air Act, "Protecting the Stratospheric Ozone," represents one more significant link in the global effort to safeguard the humans, plants, and animals of the planet from the sun's harmful ultraviolet radiation.

The Clean Air Act was enacted in 1990, and significant progress was made on the international level as well. The second meeting of the signatories of the original Montreal Protocol, in June of 1990, strengthened the Protocol. Because of these significant actions, 1990 will be remembered as the turning point in the political response to the threat of ozone depletion.

Efforts to protect the ozone layer have made great strides in a short period of time. After a decade of much debate and not much action, a growing international consensus began to develop in 1986 concerning the need to reduce the use of CFC refrigerants and other ozone-depleting substances. The ozone layer was originally regarded as a global concern which required international action to protect it.

Negotiations under the auspices of the United Nations Environment Program reached a breakthrough in September 1987, when the Montreal Protocol was signed by 23 nations, including the United States.

During the period since the signing of the Montreal Protocol, many things have happened. In an almost continuous stream of events, new scientific evidence established CFCs as the major cause of the Antarctic ozone hole, showed that ozone levels in the northern mid-latitudes (i.e., above the United States) had dropped more than anticipated, and revealed the potential for an Antarctic ozone hole.

The international political community, faced with even greater threats of ozone depletion, stepped up its efforts to gain widespread participation in the Protocol and to strengthen its requirements. As a result, by January 1991, 68 nations had joined in the agreement. Six months earlier, at the June 1990 meeting in London, amendments had been adopted to phase out CFCs and halons, to add and phase out other significant ozone-depleting chemicals, and to establish a landmark fund to support developing countries' participation in meeting the control requirements of the Montreal Protocol.

The new scientific evidence also increased the interest of Congress. With broad support and widespread interest at the grassroots level, Congress included in the Clean Air Act major provisions which expand broadly on previous legislation aimed at domestic efforts to protect the ozone layer.

Title VI is considered to be comprehensive. Like the Montreal Protocol, Title VI places limits on the production of ozone-depleting chemicals, but it goes further in that it also restricts their use, emissions, and disposal.

Title VI instructs companies about what ozone-depleting substances they must stop manufacturing. It also establishes a program to review their proposed substitute chemicals. This section of Title VI contains a phase-out schedule for both CFCs and HCFCs. HCFCs are a family of chemicals that will serve as transition substitute chemicals and will be used as replacements for some of the uses of CFCs. However, over a period of years, HCFCs must also be eliminated because they contain chlorine.

While the majority of the provisions directly affect the industrial sector of the world, consumers will also be affected by the provisions set forth for mandatory recycling of CFCs and HCFCs and by the warning labels that are required on many of the products they buy.

The following are some of the major provisions that are set forth in Title VI.

Phase-Out Requirements

Title VI sets out scheduled reductions leading to the phase-out of production of CFCs, halons, and carbon tetrachloride by the year 2000. It also freezes the production of HCFCs in the year 2015 and phases them out by the year 2030. Because these restrictions focus on production limitations, to the extent that these chemicals can be recovered, recycled, and reused, they may continue in commerce past the applicable phase-out date.

National Recycling And Emission-Reduction Program

This section called for EPA regulations by July 1, 1992 formally requiring emissions from all refrigerants (except mobile air conditioners) to be reduced to their "lowest achievable level." Regulations affecting emissions for all other uses of CFCs, halons, and methyl chloroform—and for all uses of HCFCs—are to take effect by November 1995. This section of the Clean Air Act also prohibited any person from knowingly venting any of the controlled substances, including HCFCs, during the servicing of refrigeration or air conditioning equipment (except for cars) beginning July 1, 1992, and required the safe disposal of these compounds by that date.

Warning Labels

To assist consumers in choosing among products and to aid service personnel in deciding when recycling and disposal are necessary, the Act establishes mandatory labeling requirements on all containers holding CFCs, other major ozone-depleting chemicals, and all products containing these chemicals (refrigerators, foam insulation, etc.). Under certain circumstances, warning labels may be required on products made with, but not containing, ozone-depleting chemicals (e.g., many electronics products and flexible foams) and, over a longer period of time, products made with or containing HCFCs.

Safe Alternatives

To ensure that the chemicals used as substitutes for ozone-depleting substances do not themselves create environmental problems, the EPA must be notified of, and evaluate, the overall environmental risks involved in using these substitutes.

Federal Tax

A tax will be assessed on each pound of ozone-depleting chemical sold or used by a manufacturer, producer, or importer. The amount of the tax shall be equal to (a) the base tax amount multiplied by (b) the ozone-depleting potential (ODP) for such chemical as stated in the Montreal Protocol and (c) the total pounds sold.

(a) Calendar Year	Base Tax Amount
1990 or 1991	$1.37
1992	1.67
1993 or 1994	2.65

The base tax amount shall increase by $0.45 each year after 1994.

(b) Chemical	ODP
CFC-11	1.00
CFC-12	1.00
CFC-113	0.8
CFC-114	1.0
CFC-115	0.6

Exceptions: No tax shall be imposed on the production, use of, or sale of any ozone-depleting chemicals:

- Which have been recycled for resale.
- Which are entirely consumed or transformed in the manufacture of any other chemical (companies can purchase these chemicals for transformation free of tax and need not apply for a refund or tax credit).
- Which are exported (however, an exemption will not exceed a company's percentage of the 1986 production that is exported, multiplied by the total tax).
- Which are produced with additional production allowances granted by the EPA under CFR Part 52 (Protection of Stratopheric Ozone).

SECTION 608 OF THE CLEAN AIR ACT

This section of the Act requires the EPA to develop regulations that limit emissions of ozone-depleting compounds during their use and disposal to the "lowest achievable level" and to maximize recycling. The Act also prohibited releasing refrigerant into the atmosphere during the maintenance, servicing, and disposal of refrigeration and air conditioning equipment beginning July 1, 1992.

As they now stand, the regulations require persons servicing or disposing of air conditioning and refrigeration equipment to observe certain service practices. As mentioned earlier, deliberate venting of refrigerants during service and disposal was prohibited beginning July 1, 1992. Under the proposal, refrigerant recovered and/or recycled could be returned to the same system or other systems owned by the same person without restriction. If the refrigerant changes ownership, however, that refrigerant would have to be reclaimed (cleaned to the ARI 700 standard of purity and chemically analyzed to verify that it meets this standard).

The Agency requires a certification program for recovery and recycling equipment. Under this program the EPA requires testing of the equipment to ensure that it minimizes refrigerant emissions during the recycling or recovery process. The Agency focuses on the recovery efficiency of the equipment, setting standards that would vary depending on the size and type of air conditioning or refrigeration equipment that was being serviced. Recovery equipment intended for small appliances, such as household refrigerators, household freezers, and water coolers, would be required to recover 80 to 90% of the refrigerant in the system. Recovery and recycling equipment intended for use with high pressure systems would have to achieve a vacuum of between 10 and 20 inches, depending on the size of the equipment being serviced; and equipment intended for use with low pressure systems would have to achieve a vacuum of 29 inches. The EPA requires that equipment purchased after January 1, 1993, be certified to meet these requirements.

Recovering and recycling refrigerants represent a critical component in stratospheric ozone protection. To encourage owners and users of refrigeration and air conditioning equipment to begin recovering and/or recycling CFCs as soon as possible, the EPA intends to propose grandfathering provisions for recycling and recovery equipment. Under such provisions, the use of equipment that could create at least a 4-inch vacuum (below atmosphere) for high pressure systems and a 25-inch vacuum for low pressure systems would not violate the prohibition on venting.

The EPA has developed regulations requiring the recycling of ozone depleting chemicals (both CFCs and HCFCs) during the servicing, repair, or disposal of refrigeration and air conditioning equipment. These regulations also require that refrigerant in appliances, machines and other goods be recovered from these items prior to their disposal.

TECHNICIAN CERTIFICATION

Technicians who service, install, and maintain air conditioning and refrigeration equipment must be certified by the EPA. The recovery and recycling equipment used during service operations must also be certified by the EPA. The technician certification is divided into four types: Type I (small appliances), Type II (high pressure), Type III (low pressure), and Type IV (universal certification). In the certification exam there will be 25 questions that are common to all 4 types of certification. These questions will focus on the environmental impact of CFCs and HCFCs, the regulations, and changing the outlook of the industry. There will also be questions concerning the filling and handling of refrigerant cylinders. There will also be questions that are related to the exposure levels allowed in equipment rooms. There will be 25 questions for each of Type I, Type II, and Type III certification which will also be specific for each sector covered by the particular certification. Type IV certification will have 75 questions including 25 from each of the Types I, II, and III, including safety, cylinder shipping, and refrigerant disposal.

Type I (Small Appliance) Certification

This test may be taken either on site or by mail. The required passing grade is 84%. The exam will include questions concerning recovery devices that are unique to the small appliance sector of the industry. Also included will be system-dependent recovery techniques.

Type II (High Pressure) Certification

The questions for this certification will deal with vacuum levels required, proper use of recovery equipment for removing both liquid and vapor refrigerant from the system, the purpose of system receivers, and the use of refrigerant monitors in equipment rooms. This is a closed-book, proctored test. The required passing grade is 70%.

Type III (Low Pressure) Certification

This certification covers any type of equipment that contains a low-pressure refrigerant, such as CFC-11. The questions for this type of certification will include evaporator leak testing methods and proper procedures for deep evacuating the system. This is a closed-book, proctored test. The required passing grade is 70%.

Type IV (Universal) Certification

The exam for this type of certification will have 75 questions. There will be 25 questions from each of the Types I, II, and III in addition to the 25 questions that are common to all 4 types of certification. The questions will also include shipping procedures, disposal and proper refrigerant cylinder handling. This is a closed-book, proctored test. The required passing grade is 70%.

The EPA has no regulations specifically for intermediate pressure and high pressure systems because they are so uncommon. However, for both Type II and Type III certification exams, there will be some general questions covering these types of systems.

Technicians have until November, 1994, to become certified. Then they can only work on the types of systems for which they are certified. When a technician is certified there will be no recertification required. However, EPA may require that technicians demonstrate proper techniques for recovery-recycle unit operation. If proper procedures are not followed the technician may lose certification.

CONTRACTOR SELF-CERTIFICATION

This is another type of certification that is required by the EPA. Those wishing this type of certification can obtain the proper form from either the EPA or OMB. The form is then filled out and mailed to the appropriate EPA regional office.

Those who service air conditioning and refrigeration systems (with the exception of automotive air conditioning systems) and those who dispose of appliances (with the exception of small appliances, room air conditioners, and automotive air conditioners) must self-certify that they either own or have leased certified recovery-recycle units. The deadline for self-certification was August 6, 1993.

Those persons who recover refrigerant from small appliances, room air conditioners, and automotive air conditioning systems before disposing of them must also be self-certified that the recovery/recycle units are EPA approved.

Self-certifications are not transferable to a new owner. When a business changes ownership, the new owner has 30 days to self-certify the equipment used.

EVACUATION STANDARDS

For systems containing a refrigerant charge of 200 lb or more of a high-pressure refrigerant such as HCFC-22 the system must be evacuated to 10 in. Hg during the recovery proceess and before the system is opened to the atmosphere for repairs, with the exceptions listed below. This requirement is because of the difficulty in reaching lower vacuums with these types of refrigerants.

Systems containing a charge of more than 200 lb of CFC-12 or CFC-502 must be evacuated to 15 in. Hg during the recovery process and before the system is opened to the atmosphere for repairs, with the exceptions as listed below.

Systems containing a charge of less than 200 lb of CFC-12 or CFC-502 must be evacuated to 10 in. Hg during the recovery process, with the exceptions listed below.

Low-pressure equipment must be evacuated to 29 in. Hg during the recovery process and before the system is opened to the atmosphere. This evacuation is because the equipment for recovering this type of refrigerant is capable of pumping this low pressure on these types of systems.

Evacuation Standard Exceptions

There are two exceptions that permit the service technician not to meet the above evacuation standards. They are:

1. "Non-major" repairs: These are repairs that, after completion, allow no evacuation of the refrigerant to the atmosphere; and
2. Leaks that will not allow the required evacuation to be reached.

It was decided that both of these instances would actually cause greater emissions when the required evacuation level was attempted than when less, or none at all, was used.

Non-major repairs. Non-major repairs mean that only a small opening is made for a short period of time—only a few minutes. This type of repair would allow only a small amount of refrigerant to escape and the amount of air and moisture that could enter the system is minimal. Also included in this category is the replacement of components such as safety and pressure switches, and filter-driers.

When making repairs of this nature the system can be evacuated to 0 psig for high-pressure equipment and pressurization to 0 psig for low-pressure systems when performing service operations that do not require the evacuation of refrigerant to the atmosphere before being placed back in operation.

Major repairs. This type of repair involves making large openings in the system. This includes such procedures as compressor replacement, condenser removal, evaporator removal, or an auxiliary heat exchanger removal. This class of repair can be categorized by the need to evacuate the system after repairs and before charging the system for operation. This is when any refrigerant left in the system is purged through the vacuum pump to the atmosphere during the evacuation process.

LEAK REPAIR REQUIREMENTS

Systems that contain a refrigerant charge of 50 lb or more must have any leak repaired within 30 days. Systems containing less than 50 lb are not covered by this requirement. This regulation became effective about June 21, 1993. It is the responsibility of the equipment owner to have the repair completed if the leak exceeds 15% or 35% of the total charge (explained later). The technician should inform the owner of the leak and of the requirements by the EPA. The owner cannot intentionally ignore any information which reveals that a leak exists. This is to prevent topping off the charge in a leaking system.

Flexibility

There are two types of system leaks mentioned above: 15% and 35% leaks. There is a certain amount of flexibility in the requirements concerning leak repairs. These are as follows.

Annual leak rate of 15% or higher. Air conditioning systems such as those used in commercial buildings and hotels must have the leaks repaired.

Annual leak rate of 35% or higher. Equipment that is used in an industrial process, commercial refrigeration, pharmaceutical systems, petrochemical systems, chemical systems, industrial ice machines and ice rinks must have all leaks repaired.

The technician can estimate the size of the leak by checking past invoices for service procedures requiring refrigerant to be charged into the system. There are probably other means of estimating the leak size if the unit is maintained by the service company or a maintenance department.

There is one exception to the above requirements, that being the instance when the owner decides to replace or retrofit a leaking system but must wait a couple of months before making the necessary changes. When this situation occurs, the owner must develop a detailed plan within 30 days of being informed, showing his intentions to replace or retrofit the equipment. This plan must be dated and kept on site and subject to EPA inspection. The repair or retrofit must be completed within one year from the date on which the plan was initiated.

COMPLIANCE DATES

The following are the compliance dates and what must be done to meet EPA guidelines:

June 14, 1993

On systems containing a refrigerant charge of 50 lb or more, such as commercial refrigeration or an industrial processing unit, if a leak of 35% or more per year is found, the system owner must be notified. The owner must then repair the unit within 30 days.

When a leak of 15% per year is found on any type of system containing a refrigerant charge of 50 lb or more, the owner is responsible for having it repaired within 30 days.

Exception. If the owner develops a replacement or retrofit plan within 30 days of notification, there is a one-year exemption from making the repairs. The plan must be kept on site and available for EPA inspection.

Purging. Only *de minimus* releases of refrigerant during the service procedure or equipment disposal are legal. Major releases are prohibited by law.

Recovery-recycle unit. To alter the design of a recovery-recycle unit in such a way as to affect its ability to meet EPA certification standards is illegal.

July 13, 1993

After this date to open a refrigeration system legally the service company must have at least one self-contained, EPA-certified recovery unit at the business location.

Before opening any high-pressure system with a refrigerant charge of 50 lb or more, it must be pumped down to a vacuum of 10 in. Hg before the system is opened if a major repair is to be performed. If a non-major repair is performed or if evacuation would substantially contaminate the recovered refrigerant, the system can be evacuated to 0 psig. The evacuation to 0 psig is also applicable if the system is not to be evacuated after the repair and before recharging the unit.

Before opening a low-pressure system the pressure inside the system must be increased to 0 psig before opening it. This is to prevent pulling moisture and air into the system and contaminating the system and refrigerant.

When evacuating small appliances—such as refrigerators, freezers, room units, packaged terminal air conditioning units, packaged terminal heat pump systems, dehumidifiers, vending machines, drinking water coolers, or units containing a refrigerant charge of 5 lb or less—the level of evacuation depends on whether or not the system compressor is operational.

If the system compressor is operational, both the self-contained and the system-dependent recovery units must produce a 90% evacuation level. When the system compressor is not operational the recovery procedure must remove 80% of the charge.

When disposing of a system containing refrigerant of any amount, an EPA approved recovery-recycle unit must be used to evacuate the system to the levels discussed above.

Any refrigerant removed from a system or a part of a system to be serviced must be removed with an EPA-approved recovery/recycle unit.

When disposing of a small appliance the last person to handle the appliance must recover any refrigerant or verify in writing that it has been recovered.

August 12, 1993

After this date it is illegal to open any refrigeration system for repair, maintenance, or disposal unless the technician has certified to the EPA the ownership of the EPA certified recovery/recycle equipment. The certification form is available from the EPA or OMB.

It is also illegal to sell any refrigerant to a new owner unless it has been reclaimed and certified to meet ARI Standard 700 purity. The used refrigerant can be used in another system with the same owner without prohibition.

Reclaimers must certify that the refrigerant has been cleaned to meet ARI 700 specifications and that not more than 1.5% of it will be vented during the reclamation process.

Reclaimers must report to EPA within 45 days from the end of the calendar year regarding the quantity of refrigerants received and reclaimed. They must also furnish the names of the persons who supplied the refrigerant to them.

November 15, 1993

After this date all recovery/recycle units must be certified by EPA to meet the approved evacuation levels.

After this date it is also illegal to sell anything but small appliances that are not equipped with a servicing aperture and a process tube so that the refrigerant can be recovered from the system.

The manufacturers of recovery/recycle units must have their units certified by an approved testing organization. The equipment must be able to meet the evacuation levels stated.

November 14, 1994

After this date a technician cannot legally open or dispose of refrigeration equipment unless proper technician certification has been attained.

Only training-testing organizations that are EPA approved can train and certify technicians after this date.

It is also illegal to sell refrigerant to anyone except EPA-certified technicians.

The recovery and recycling of refrigerants represents a critical step in protecting the stratospheric ozone layer. As an encouragement for owners and users of air conditioning and refrigeration equipment to start recovering and/or recycling CFCs as soon as possible, EPA will grandfather recycling and recovery equipment. However, these provisions require that the equipment certified before January 1, 1994 must pump a vacuum of 4 inches of mercury for high-pressure systems and a vacuum of 25 inches of mercury for low-pressure systems.

All individuals who service air conditioning and refrigeration systems must be certified. There are four classifications of certification: for servicing household appliances, for servicing high-pressure equipment with a refrigerant charge below 50 lb, for servicing high-pressure systems with a refrigerant charge above 50 lb, and for servicing low-pressure equipment. Certification will depend on the passing of a proctored, EPA-approved test. The agency will authorize organizations to administer the test.

There is also a provision requiring that contractors be certified. The contractors will be required to have sufficient recovery and recycling equipment to preform on-site recovery and recycling and will employ only certified technicians to service the equipment.

RECOVERY AND RECYCLING EQUIPMENT

Since heat pump systems are considered, generally, to be high pressure systems containing less than 50 pounds of refrigerant, we will discuss the accepted methods used for recovering refrigerant from these types of systems. One thing that must be remembered is never to use the system hermetic compressor-motor when attempting the refrigerant recovery process. To do so could have negative consequences. First, the oil could be pumped from the compressor crankcase, reducing the amount of oil available for use. Second, when it leaves the compressor the oil must pass through the compressor valves and may cause damage to them. Refrigeration compressors are vapor pumps and are not designed to move liquid. Third, the insulating effect of the compressor motor-winding insulation is greatly reduced when it is subjected to pressures below atmospheric, especially when no refrigerant is present. Thus, when the system pressure is lowered to the vacuum required by the EPA, the compressor-motor winding could short out, rendering the compressor inoperative and requiring replacement.

Recovery Equipment

There are three types of recovery units available. The one chosen should cover the majority of the systems that the technician will be working on. It must, however, be certified by ARI or be certifiable by EPA.

The recovery unit may have an internal system designed for various purposes. Recovery units may be equipped with an oil separator, an oil separator and multiple-pass filer, or an oil separator and single-pass filters (Figures 9–1 through 9–3).

Recovery Methods

There are two methods of recovering refrigerant from high pressure systems containing 50 pounds of refrigerant or less: liquid and vapor.

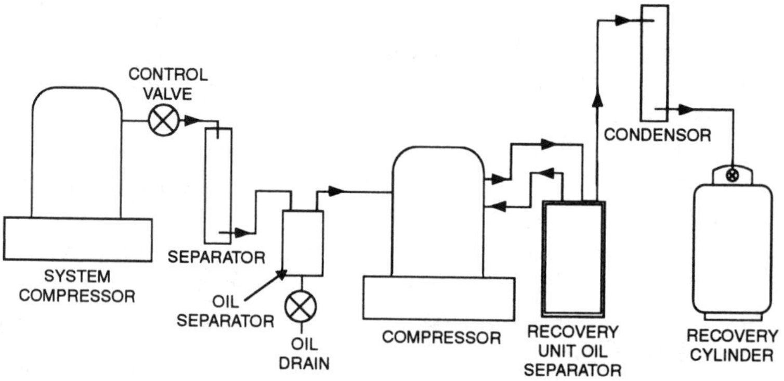
CONTROL
VALVE
CONDENSOR
SEPARATOR
SYSTEM
COMPRESSOR
OIL
SEPARATOR
OIL
DRAIN
COMPRESSOR
RECOVERY
UNIT OIL
SEPARATOR
RECOVERY
CYLINDER

FIGURE 9–1 Recovery/recycle unit schematic using oil separator.

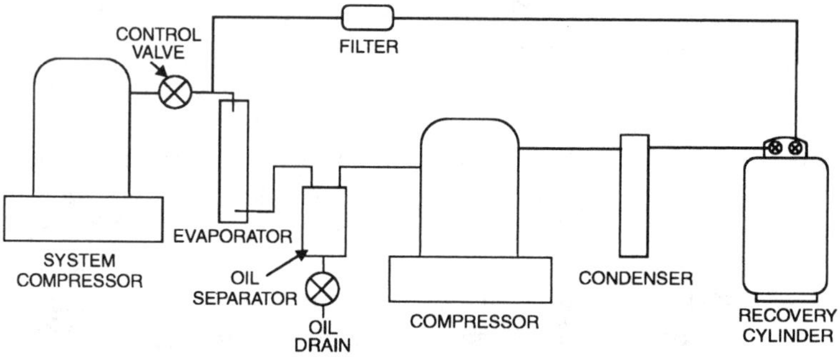
CONTROL
VALVE
FILTER
EVAPORATOR
SYSTEM
COMPRESSOR
OIL
SEPARATOR
OIL
DRAIN
COMPRESSOR
CONDENSER
RECOVERY
CYLINDER

FIGURE 9–2 Recovery/recycle unit schematic using multiple pass filters.

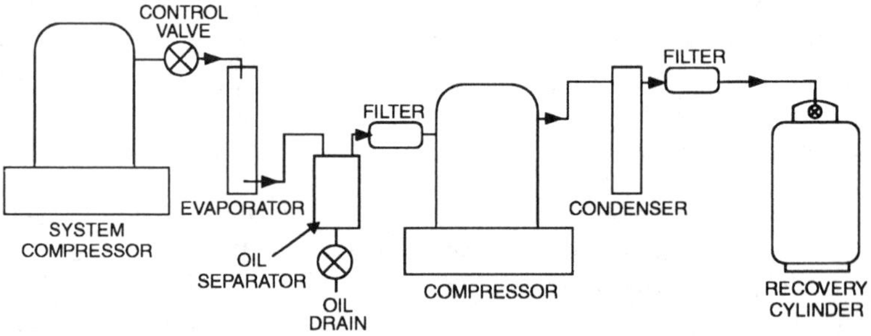
CONTROL
VALVE
FILTER
FILTER
EVAPORATOR
CONDENSER
SYSTEM
COMPRESSOR
OIL
SEPARATOR
OIL
DRAIN
COMPRESSOR
RECOVERY
CYLINDER

FIGURE 9–3 Recovery/recycle unit schematic using single pass filters.

Liquid recovery. The recovery process can be accomplished much faster if the refrigerant is removed in the liquid form. Thus, when the liquid recovery procedure is used much time, energy, and money will be saved. To use the liquid recovery method, connect the recovery unit to the system using the connections indicated in Figure 9–4. Caution should be used to prevent liquid refrigerant from

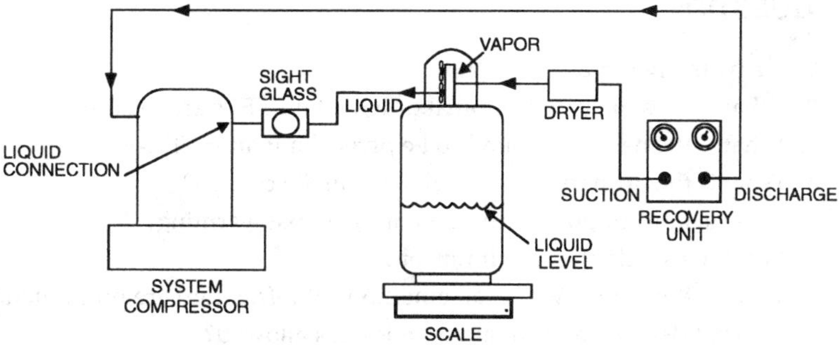
VAPOR
SIGHT GLASS
LIQUID
DRYER
LIQUID CONNECTION
SUCTION
DISCHARGE
RECOVERY UNIT
LIQUID LEVEL
SYSTEM COMPRESSOR
SCALE

FIGURE 9–4 Liquid recovery connections.

entering the recovery unit compressor when using this method. When this method is used, the liquid refrigerant is caused to flow into the cylinder because of the pressure difference between the cylinder and the system. The recovery unit suction removes the vapor from the top of the cylinder while the recovery unit discharge pressure causes an increase in the pressure over the liquid, causing it to flow into the recovery cylinder. A dryer placed in the vapor line to the recovery unit will help clean the refrigerant during the recovery process.

Vapor recovery. When the vapor recovery method is used, the recovery unit should be connected as shown in Figure 9–5. During this procedure, the vapor is drawn into the recovery unit by the recovery unit compressor. The vapor is then compressed, condensed, and forced into the refrigerant recovery cylinder. This method is best suited for systems containing small amounts of refrigerant and for those that prohibit a liquid connection. A dryer located in the vapor line will help clean the refrigerant during the recovery process. This method requires more time, but it is usually the only method that will work on all systems.

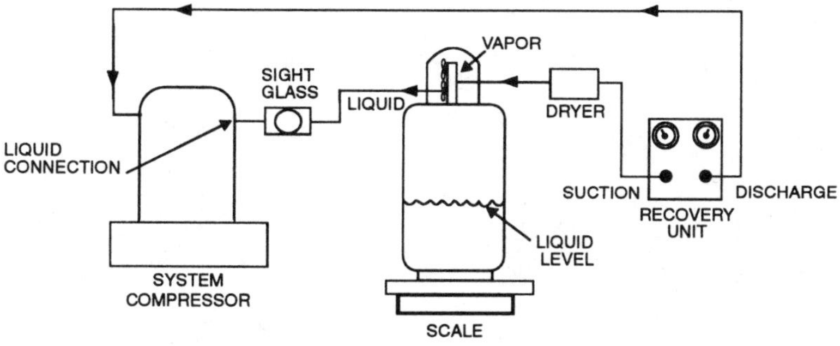
VAPOR
SIGHT GLASS
LIQUID
DRYER
LIQUID CONNECTION
SUCTION
DISCHARGE
RECOVERY UNIT
LIQUID LEVEL
SYSTEM COMPRESSOR
SCALE

FIGURE 9–5 Vapor recovery connections.

REVIEW QUESTIONS

1. Name the two types of ozone.
2. What causes the ozone to breakdown when CFCs are released?
3. What is the next compound to be phased out after CFCs?
4. What is the primary cause of global climate change?
5. What is most important in preventing global warming?
6. What does Title VI place limits on?
7. What is the date at which all emissions of refrigerant are prohibited?
8. To what degree are refrigerant emissions allowed?
9. Name the types of proposed technician certification.
10. Name the two recovery methods used on heat pump systems.

10

heating boilers and piping

The objectives of this chapter are:

- To acquaint you with boiler construction.
- To introduce you to the water flow through a boiler.
- To acquaint you with the different methods of boiler firing.
- To provide you with advantages and disadvantages of hydronic heating.
- To introduce you to the problems encountered in heat transfer in a boiler.
- To acquaint you with the theories of steam generation.
- To provide you with some solutions to problems encountered in boiler heat transfer.
- To acquaint you with the one-pipe system.
- To acquaint you with the two-pipe system.

Hydronic heating maintains comfort in the home by circulating hot water or steam to the secondary heat exchanger. This system basically consists of some or all of the following: a boiler, a pump, a secondary heat exchanger, and a series of piping. The principle of hydronic heating dates back to ancient times, when the Romans used it by circulating warm water through the walls and floors. Today, however, hydronics bears no resemblance to ancient methods. These systems are as modern as can be installed, with refinements such as zoning, which allows the temperature to vary in the different areas of the home.

HEAT TRANSFER

The transmission or flow of heat ranks in the field of engineering second only to problems involving the strength of materials. Unfortunately, most engineers are probably more familiar with the calculations applying the material strengths and stresses and can recognize the limitations of the laws and methods of solution. In their approach to a problem involving heat transfer, their possible blind acceptance of published elementary laws governing heat flow may lead to gross error in the solution.

The heat transfer problem in heating water, no matter what the design of the heater, is the transmission of the required amount of heat from hot combustion gases through a metal wall into the water. As noted, the chief resistance to the free flow of heat from the source to the point of use is the insulating effect of the film of stagnant gas that always tends to cling to a heat transfer surface. The resistance of the metal wall is insignificant. Likewise the slight resistance of a thin, stationary film of water on the other side is insignificant.

The relative importance of the three barriers to heat flow are dramatically illustrated in Figure 10–1. A combustion gas temperature of 820° F is assumed. In overcoming the resistance of the normal dead gas film, the temperature at the surface of the metal wall has dropped to 404° F. The slight resistance to conduction of heat through the steel tube wall or other heating surface causes a drop of only 3° F. The resistance of the water film is equivalent to a drop of 17° F. Of the total heat loss of 436° F from the hot gases to the water side of the tube, 416° F or 95% represents the direct effect of the stagnant gas film. There, obviously, is the point of attack.

For effective heat transfer and all that it means in the way of heating surface allowances, safe temperature differentials, economical operation, and life of equipment, the stagnant gas films must be effectively scrubbed off the heating surfaces. It has long been known that no method of combustion accomplishes this as well as a correctly engineered system of hydronic heating.

The scouring effect is produced mainly by the high velocity turbulent flow of hot gases through the firing area and the rapid expansion of the gases on burning that increases both velocity and turbulence.

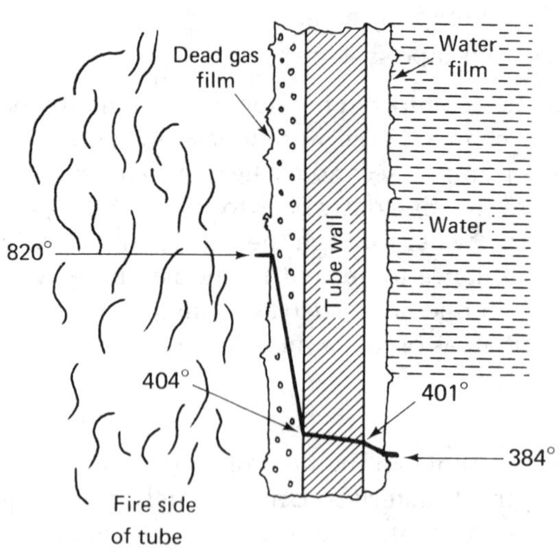
Dead gas film
Water film
Water
Tube wall
820°
404°
401°
384°
Fire side of tube

FIGURE 10–1
Heat transfer through a tube wall.
(Courtesy Sellers Engineering Co.)

The more efficient boilers are designed with a water back combustion area with water circulated all around the firebox. The crown sheet and side walls in the firebox provide a maximum amount of radiant heat transfer surface.

BOILERS

A boiler is defined as an enclosed vessel in which water may be heated to a given temperature or converted to steam at a given pressure and temperature, depending on the application. A fuel is used to heat the water and cause the desired effect. Some boilers use gas, some use electricity, and some use fuel oil.

The steam or hot water is distributed to the desired location through pipes. This steam or hot water is used in these types of applications to heat the structure to the desired comfort level.

There are two types of boilers used: the hot water boiler and the steam boiler. The hot water boiler has water throughout the complete system, including the distribution pipes. There is very little, if any, steam in this type of system. The steam boiler has only enough water to cover the heating surfaces as designed for proper, safe, and efficient operation. The steam is circulated through the piping arrangement by natural circulation. The type of boiler used will depend on the application requirements and the design engineer.

A hot water boiler maintains water temperatures between 120° F and 210° F. This water is pumped through piping to the secondary heat exchanger. The term *boiler* may bring to mind large heating plants such as those found in schools and other large buildings. These are commercial boilers. Actually, a modern cast iron boiler used for comfort heating is very compact. Most units are the size of an automatic washing machine, and some are as small as a suitcase and can be hung on the wall.

Basic Cast Iron Boiler Designs

There are four basic section designs for cast iron boilers.

1. Sections that have two equally sized horizontal push nipples on the vertical center line of the section, and with horizontal and equally sized tappings for supply and return (Figure 10–2).
2. Sections with two horizontal push nipples of different sizes off the vertical center of the section, and with a horizontal lower return tapping and horizontal or vertical upper supply tapping [Figure 10–3(a) and (b)].
3. Sections with three horizontal push nipples: two smaller nipples at the bottom and one larger nipple in the center at the top; one or two horizontal lower return tappings and one or more vertical top supply tappings [Figure 10–4(a) and (b)].

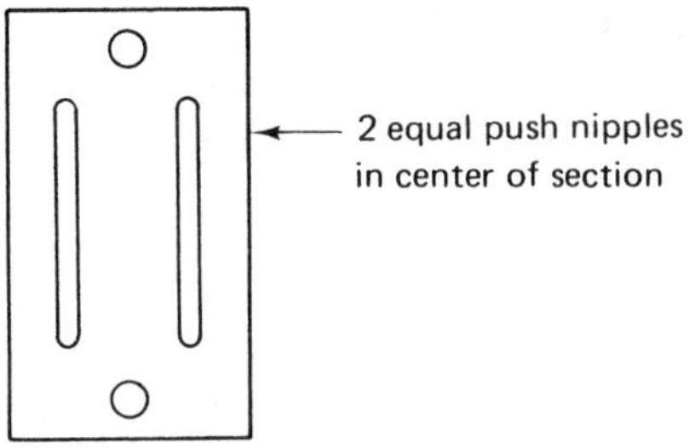
2 equal push nipples
in center of section

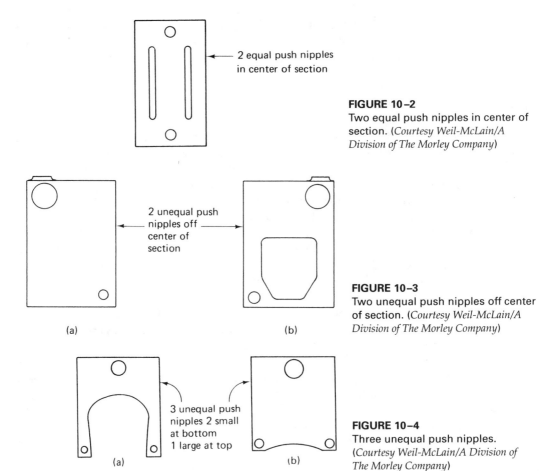

FIGURE 10–2
Two equal push nipples in center of section. *(Courtesy Weil-McLain/A Division of The Morley Company)*

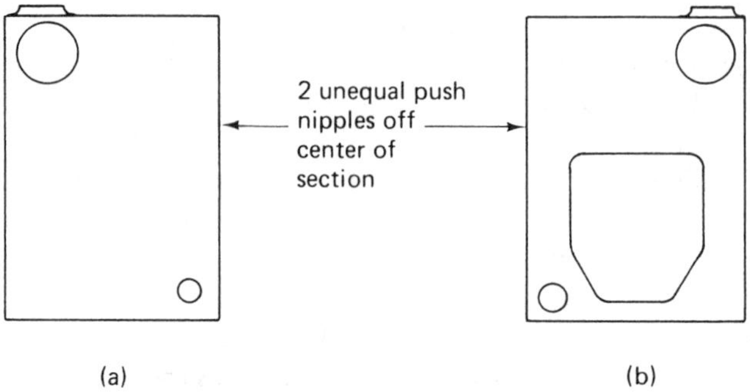
2 unequal push
nipples off
center of
section
(a)
(b)

FIGURE 10–3
Two unequal push nipples off center of section. *(Courtesy Weil-McLain/A Division of The Morley Company)*

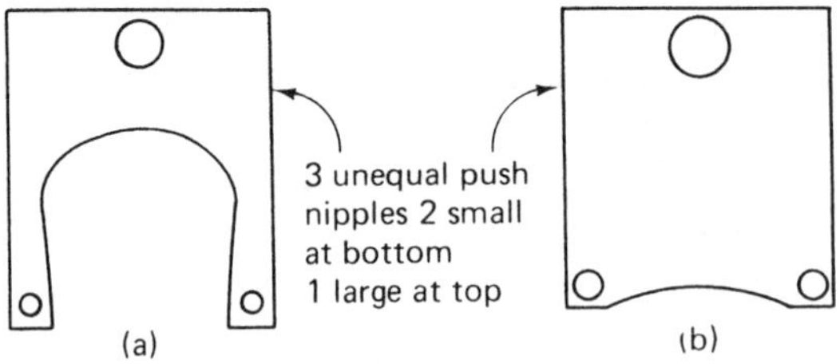
3 unequal push
nipples 2 small
at bottom
1 large at top
(a)
(b)

FIGURE 10–4
Three unequal push nipples. *(Courtesy Weil-McLain/A Division of The Morley Company)*

4. Split sections with horizontal or vertical lower screw nipples and horizontal upper screw nipples; two horizontal lower return tappings and one or more vertical top supply tappings (Figure 10–5).

The variety of nipple arrangements and the variety of supply and return locations make it evident that each type of boiler has a particular internal flow pattern best suited to its design, and that each type will react differently to a change in its flow pattern. All these boilers are basically designed for a normal upward flow.

Basic internal flow patterns. The four basic cast iron boiler designs have five basic flow patterns. Two of the basic flow patterns in the four basic boiler designs with normal upward flow are shown in Figures 10–6 and 10–7.

Figures 10–8, 10–9, and 10–10 illustrate a reverse flow and a combination of upward and reverse flow in the basic boiler designs. Keep in mind that these are internal flow patterns.

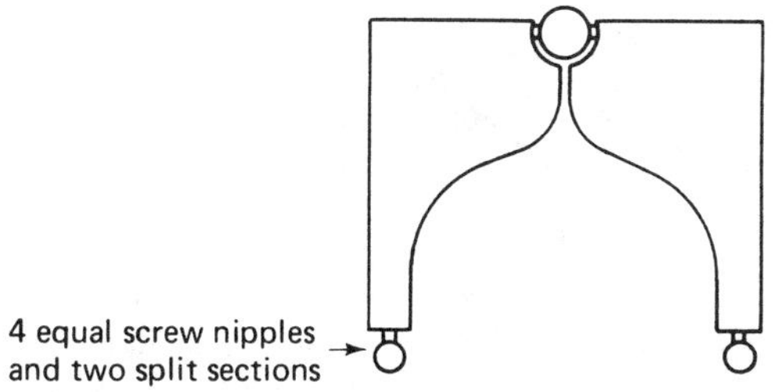
4 equal screw nipples
and two split sections

FIGURE 10–5
Split sections. (*Courtesy Weil-McLain/A Division of The Morley Company*)

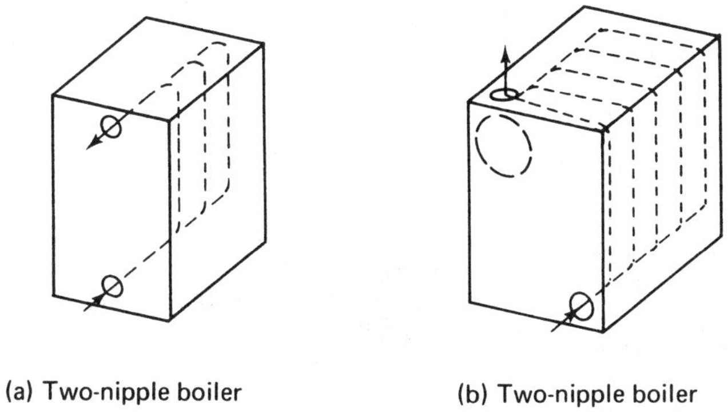
(a) Two-nipple boiler
(b) Two-nipple boiler

FIGURE 10–6
Upward U-shaped flow pattern in a two-nipple boiler. (*Courtesy Weil-McLain/A Division of The Morley Company*)

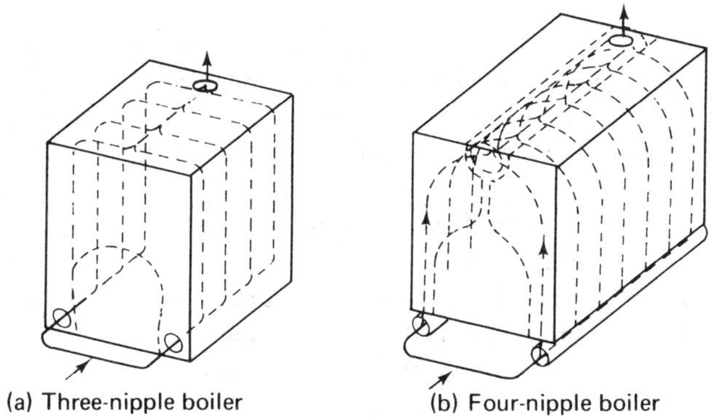
(a) Three-nipple boiler
(b) Four-nipple boiler

FIGURE 10–7
Upward cross flow in a three- and four-nipple boiler. (*Courtesy Weil-McLain/A Division of The Morley Company*)

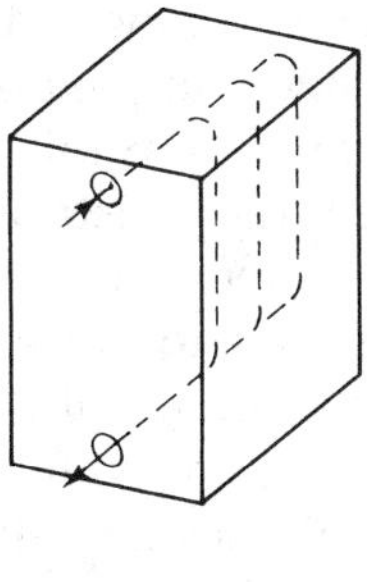

FIGURE 10–8
Reverse U-shaped flow pattern in a two-nipple boiler. (*Courtesy Weil-McLain/A Division of The Morley Company*)

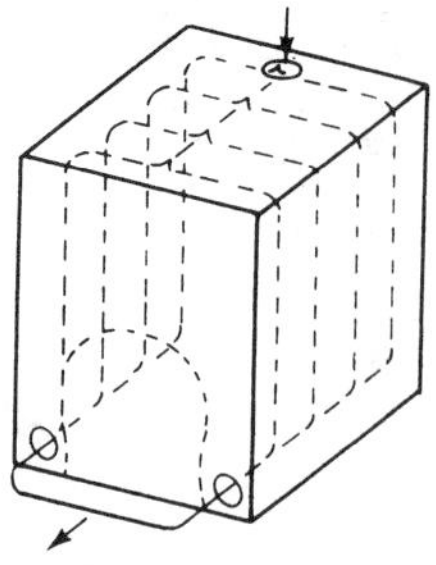

FIGURE 10–9
Reverse cross-flow pattern in a three- or four-nipple boiler. (*Courtesy Weil-McLain/A Division of The Morley Company*)

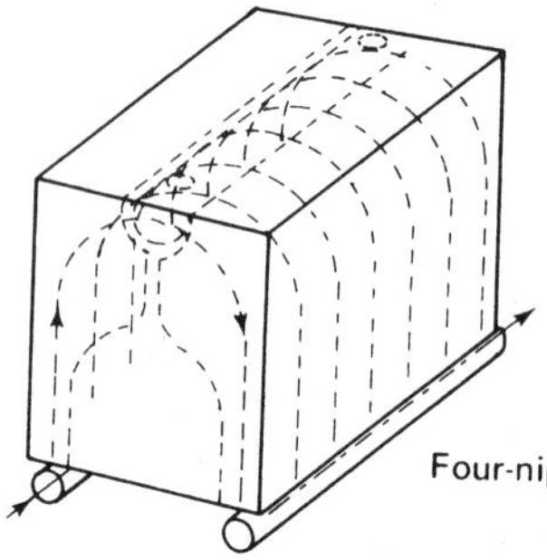
Four-nipple boiler

FIGURE 10–10
Combination upward and reverse cross-flow pattern. (*Courtesy Weil-McLain/A Division of The Morley Company*)

Almost all European cast iron boilers with two nipples in the center of the section have used the upward, U-shaped pattern [Figure 10–3(a)] with good results on gravity system for decades. In fact, many European gravity boilers of this design have supply and return tappings only in the back section, thus offering no choice but the upward, U-shaped flow.

Laboratory tests conducted on gas-fired boilers with two nipples, both off center of the section [Figure 10–3(b)] and on center of the section [Figure 10–3(a)], indicated a minimum temperature override and excellent temperature gradient with the upward, U-shaped flow for forced as well as gravity circulation. The key factors are that the return water is introduced into the axis of the lower nipple ports and that the stay arrangement in the waterways minimizes shortcuts.

This does not mean that the upward, U-shaped flow can be used in a sectional boiler of unlimited length. The limiting factor is the maximum allowable water velocity through the smallest-sized nipple port. It does mean, however, that the upward, U-shaped flow pattern, within this limitation, meets the requirements for internal flow—namely, equal water distribution, minimum temperature override, prevention of sudden great temperature change, and effective air elimination.

Similarly, field experience and laboratory tests have proven over the years that in boilers with three or more nipples per section (Figure 10–5), the upward cross flow meets the basic requirements for efficient internal flow better than other patterns. As in the two-nipple designs, the limiting factor for the length of the boiler is the maximum allowable water velocity consistent with good design.

Effect of the pump head. The proper placement of the pump and compression tank in relation to the boiler has been established as an important factor in the functioning of high pump head systems. The pump should be located at the supply water outlet side of the boiler (discharging away from the boiler) with the compression tank at the pump suction side. The purpose of including this subject with internal boiler water flow direction is to emphasize that the proper placement of the pump and compression tank, along with good system design in general, applies equally to systems where the boiler uses either an upward flow pattern, a reverse flow pattern, or a combination upward and reverse flow pattern. Regardless of the internal flow direction, the system design must always take into consideration the factors of pressure gradient, gpm, supply water temperature, temperature drop, and air control.

Regardless of the flow direction, a valve in the bypass line can provide a means of adjustment for the desired division of positive and negative pressure gradient in the system.

The proper placement of the pump and compression tank in relation to the boiler is the only complete answer to the elimination of undesirable effects of high positive pump head on the boiler and high negative pump head on other parts of the system. Reversing the internal boiler flow may help or hinder the pressure condition, depending on the location of the pump and compression tank, and may be of some advantage on existing systems with or without a divided pump head by changing the pressure on the boiler from a positive to a negative gradient. In new installations, the best choice would be a negative gradient. In new work, the best choice should be evident—begin with good system design.

It must be emphasized to the designer or installer of residential hot water systems that the location of the pump, either in the return or the supply line of a boiler, is not critical in heating systems with a low pump head (no more than 15 ft). The combination of the dynamic and static pressures in the system is, by definition, low enough to be disregarded.

The advantages of reverse flow are as follows:

1. Reverse flow may be useful in a complex piping layout, particularly on a replacement job, where a base-mounted pump may need a simplified connection to the boiler.
2. Reverse flow can be a compromise solution to a negative pump head, or to a positive pump head on a boiler in a system in which the head is divided.
3. It can correct an internal temperature problem in certain boiler designs. For example, it is known that steel boilers with a large water volume in one drum sometimes develop high temperature gradients within the shell. This condition has been minimized with downward (reverse) flow through the boiler. Recognizing this is only a temporary solution, several manufacturers of steel boilers have conducted extensive tests with various flow patterns. They found a better solution to the problem by using natural upward flow and by producing a U-shaped internal flow with a special internal elbow at the

return inlet. The supply outlet as well as the return inlet are located at the top of the boiler. A low temperature gradient for this type of boiler was achieved.

4. A reverse flow should only be used in boilers which are designed for it.

The question of whether reversing the flow in a cast iron boiler will increase the heat transfer from the flue gases in the boiler water may be answered in the affirmative, if the heat were transmitted by convection and/or conduction only and if a true counterflow between water and flue gases existed. However, it is a well-known fact that the flow of flue gases along the secondary heating surface of a cast iron boiler is not a true counterflow or parallel flow or cross flow, but a combination of the three. It is also known that a modern cast iron boiler has up to 60% primary heating surface where the heat is transmitted mainly by radiation and where the water flow direction has no bearing on the heat absorption. In other words, reversing the water flow through the boiler will not appreciably change the heat transfer from the flue gases to the boiler water.

The disadvantages of reverse flow include the following:

1. As air must rise against water flow direction, its elimination from the boiler becomes a problem. Air released from the heating of the boiler water can be trapped and can build air pockets. The pockets will often create hot spots and, in turn, steam bubbles. Steam bubbles also can be caused by an incorrectly sized or poorly operating firing device. On their way to the top of the boiler, steam bubbles will come in contact with the cool return water. This causes them to collapse, producing noise which is usually transmitted throughout the entire piping system. In severe cases steam bubbles can create vibration of the whole boiler.

   Air that will be drawn into the system can also cause noise and uneven heat distribution if the point of air elimination is not changed for reverse flow.
2. Since the water velocity within the boiler sections slows down considerably, no matter whether normal (upflow) or reversed (downflow), the reverse water flow direction within the boiler can be upset, particularly in a long boiler with many sections. In other words, it is possible that one part of the boiler can have reverse flow whereas in another part of the boiler the natural gravity flow is stronger than the reverse flow, thus upsetting the water circulation as well as the temperature gradients within the boiler. Actual tests have proven this fact. This condition will further aggravate air elimination and control functions.
3. Intermittent pump operation reverses the temperature pattern within the boiler each time the pump starts or stops and puts additional strain and stress on the boiler.
4. Controls in predetermined locations for normal upward flow generally do not provide satisfactory operation when used with reverse flow.
5. Reverse flow definitely decreases the output of a built-in tankless or storage heater because the cool return water comes in contact with the heater coil.

Boiler manufacturers are responsible for trouble-free operation of their product and, therefore, they must recommend the water flow best suited to each of their boiler designs.

Combustion chamber. Because the combustion area is entirely surrounded by water and because flame retention burners do not require hot refractory for combustion, a refractory combustion chamber is not required for this type of boiler. The advantages are obvious—the cost of the combustion chamber material and the labor to install it are saved, and there is no future replacement cost.

Rope seal. For forced draft firing, it is absolutely necessary that a boiler be completely gas-tight. A boiler with a separate base can be made gas-tight for a short time only. But when the firebox is surrounded with water, a separate base is not necessary; therefore, the seal between adjacent sections can be made permanently gas-tight [Figure 10–11(a)]. The seal between the platework and the boiler must also be permanently gas-tight.

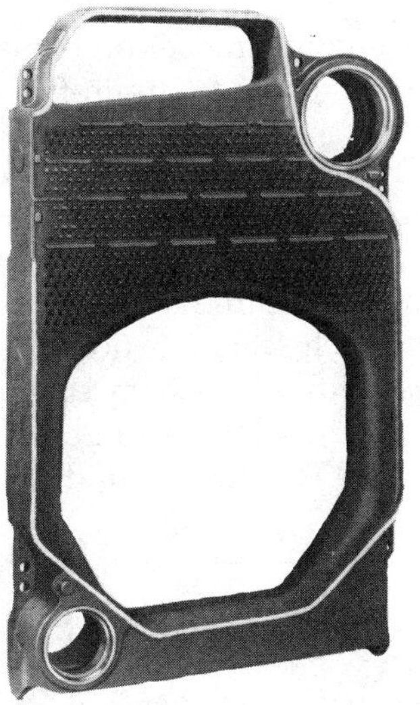

FIGURE 10–11 a
Boiler sections. (*Courtesy Weil-McLain/A Division of The Morley Company*)

The sections of these boilers have a grooved seal strip that receives the rope seal. When installed, the outer edge of the rope is accessible between sections, so that the boiler can be visually checked for tightness. The rope is compressible, allowing ample contraction and expansion of the boiler. Seal rope does not crack, lasts the life of the boiler, and assures a gas-tight assembly. The sections are not face ground, but retain the tough original skin which reduces the rate of rust growth. The rope seal is impervious to the intense heat in the firebox and to floor moisture.

Draw rods. Multiple short draw rods, instead of a single long rod and expansion washers, are desired to tie the sections together [see Figure 10–11(b)]. This

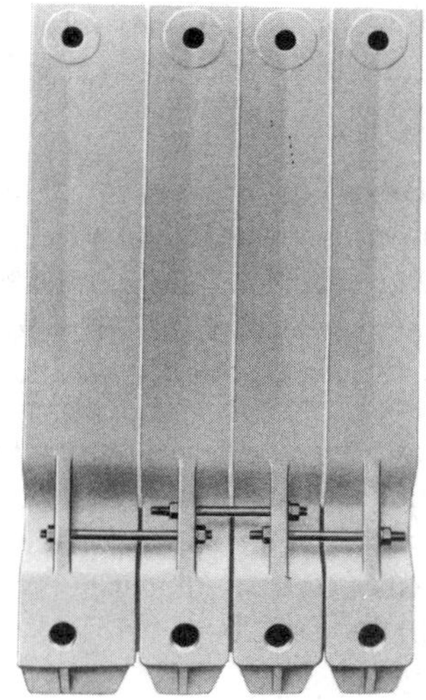

FIGURE 10–11 b
Boiler Sections. (*Courtesy Weil-McLain/A Division of The Morley Company.*)

feature is endorsed by leading insurance companies. The short draw rods permit faster, easier erection of the boiler and a strain-free assembly.

Flue gas travel. The flue gases rise from the combustion area into uptakes between each boiler section at the right and left side. The flue gases are directed through the two outside flue passages to the front of the boiler and then back through the center flueways to the flue connection (Figure 10–12).

The uptakes between each section expose more radiant heat-absorbing surface and divide the hot gas volume into many small gas streams. These gas streams are individually directed to the cooler sides of the firebox, creating a wiping action that increases heat absorption. The turn-around flue-gas travel at the upper part of

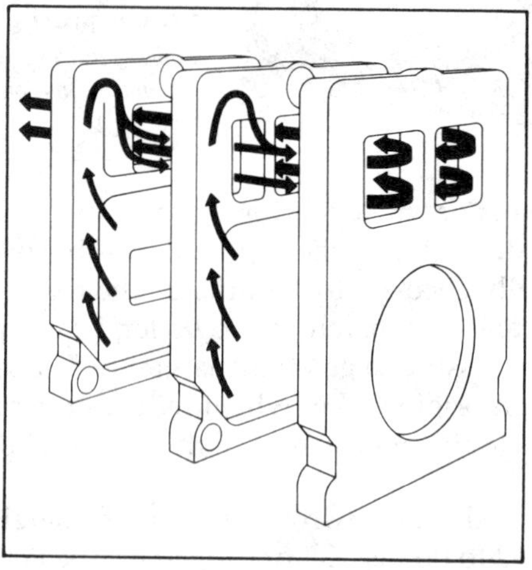

FIGURE 10–12
Flue gas travel. (*Courtesy Weil-McLain/A Division of The Morley Company*)

the uptakes uses the heating surface to the top of the sections. This feature aids in producing a dry steam in the steam boiler.

Multiple uptakes, combined with three-pass design that allows hot gases to wipe the entire flue area, assure balanced flue gas travel and prevent shortcuts to the chimney. Extra long flue gas travel and higher velocities increase heat absorption of the secondary heating surfaces.

Cleaning. Easy cleaning of the flue passages is a must. Most manufacturers provide a means of cleaning by removing the top jacket panel and the collector hood, which exposes all flue-ways for easy, straight-through cleaning with a steel brush, usually furnished with the boiler.

Fire Tube Boilers

The immersion (fire tube) boiler is also built to conform to the ASME code. Standard design pressures are 100 and 150 psi.

The simple design of the boiler shell lends itself to modern, precision fabrication procedures—full round cylinder rolling, production cutting, bevelling of heads, automatic welding of the head and shell seams, and automatic controlled tube rolling (Figure 10–13).

The firing tube. The firing tubes are copper clad on the water side. The 2-inch OD by number 13 gauge tubes (Figure 10–14) are formed by telescoping a 0.030-inch hard phosphorized copper tube over a 0.065-inch seamless steel tube and drawing over a mandrel and through a die for a tight bond. Long tube life is positively assured under the most aggressive conditions.

Straight-through, single-pass firing. There are no traps or pockets for gas accumulations (Figure 10–15) and no brick work or refractories anywhere. The identical firing tubes constitute individual combustion chambers—water backed

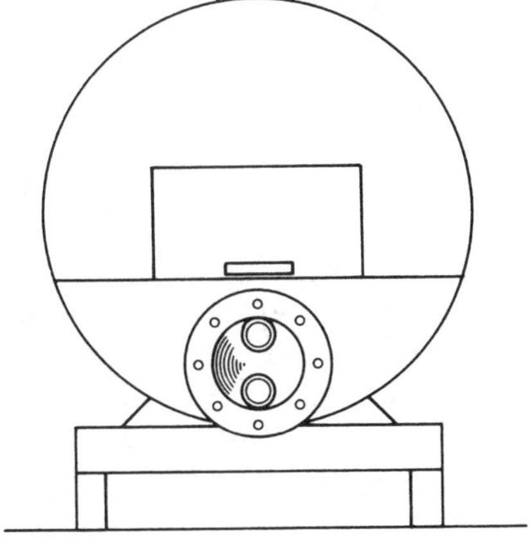

FIGURE 10–13
Immersion tube boiler. (*Courtesy Sellers Engineering Co.*)

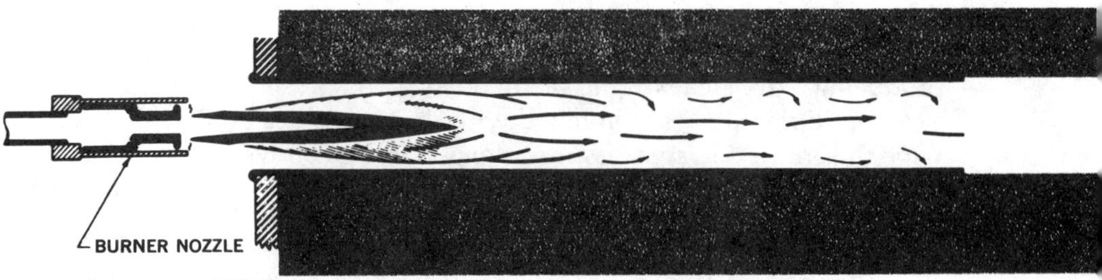
BURNER NOZZLE

FIGURE 10–14 Firing tubes. (*Courtesy Sellers Engineering Co.*)

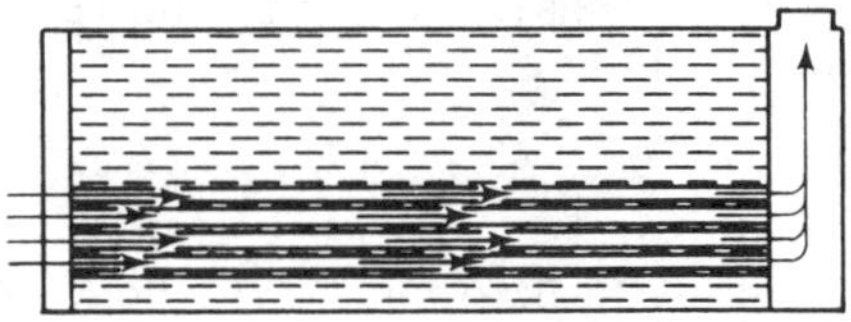

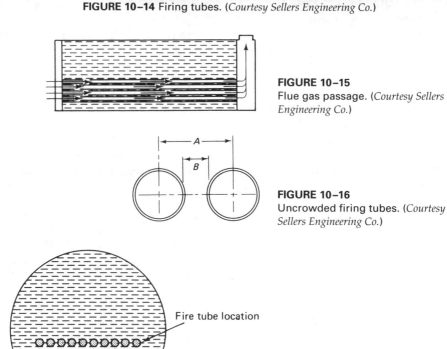

FIGURE 10–15
Flue gas passage. (*Courtesy Sellers Engineering Co.*)

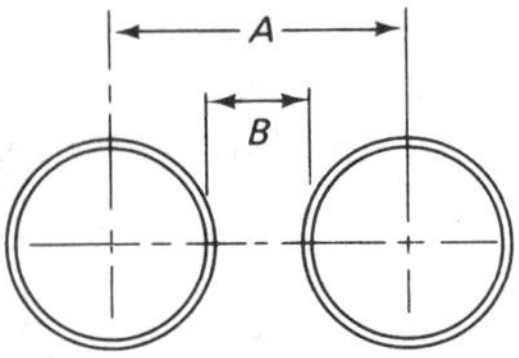
A
B

FIGURE 10–16
Uncrowded firing tubes. (*Courtesy Sellers Engineering Co.*)

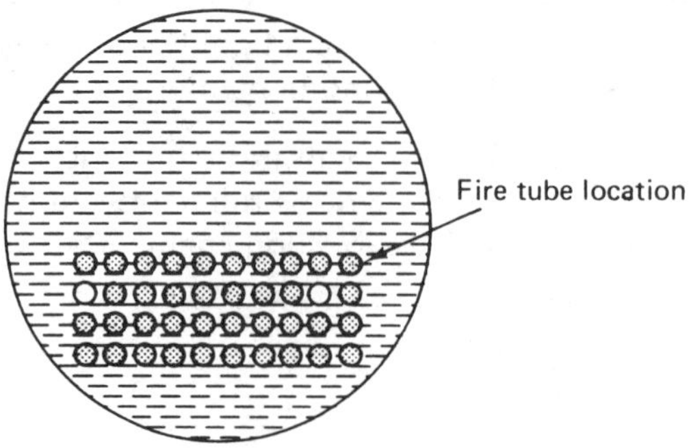
Fire tube location

FIGURE 10–17
Fire tube nest location. (*Courtesy Sellers Engineering Co.*)

all the way. There are no strains or stresses. The uncrowded shell permits a generous ligament, never less than 1 inch, between tubes (Figure 10–16). This removes the hazards of cracked heads and loosened tubes.

Only in immersion fired boilers can the total gas input be widely dispersed among numerous, small firing tubes (Figure 10–17). This eliminates hot spots and excessive scale deposits in hard water usage. It provides uniform expansion stresses with no likelihood of tube loosening. More importantly, every square foot of heating surface is put to work in a more uniform manner.

There is no simpler combustion system (in fact, no simpler construction) than the straight-through, single-pass design of immersion firing.

Only in immersion firing can heat be transferred to the water with so close an approach to uniformity. Immersion firing effectively scours away the insulating film of stagnant gases that tends to cling to heat transfer surfaces. This means longer life, less hard water scaling, and greater efficiency. Immersion firing design requires distribution of the gas input among numerous small-diameter firing tubes. The tubes are alike in all respects. They carry gases with equal temperature gradients. There are no strains set up. There is no loosening of tubes. There is no ignition of a large volume of gas at a single point, but instead, smooth and safe ignition of a series of small burners (Figure 10–18). There is no combustion chamber to be insulated against heat loss. Each burner fires into a separate, water-backed combustion chamber.

Electric Boilers

Electric boilers have been in use for many years. The first ones were crude in appearance, but they accomplished what they were designed to do. The lack of electrical power, especially in rural areas, has been the most outstanding hindrance in electric heating. In modern times, however, electric boilers are gaining popularity.

Electric hydronic boilers are designed specifically for hot water heating systems in homes and apartments. They are factory assembled and wired with circulator, compression tank, and controls; all parts are enclosed in a compact, clean-lined jacket (Figure 10–19).

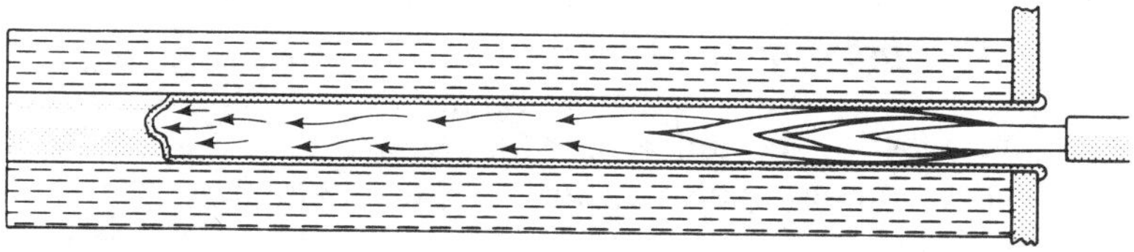

FIGURE 10–18 Fire tube and its burner. (*Courtesy Sellers Engineering Co.*)

WEIL-McLAIN

FIGURE 10–19
Electric hydronic boiler. (*Courtesy Weil-McLain/A Division of The Morley Company*)

The Weil-McLain Co., Inc. has designed a unique boiler control system which stages the heating elements on in 2 ½-minute intervals, and off in 10-second intervals. Thus, because of the time delay, full boiler capacity is not used to satisfy one zone of a multizone system.

These boilers are designed for fast, low-cost installation, for new buildings, or for replacement. The unit mounts on the wall, saving valuable living space, and no flue or vent is required. Standard components are used and electrical connections are required only for the power supply and thermostat. Piping connections are the same as for any hot water boiler.

Construction. The one-piece cast iron boiler sections are built to meet the requirements of ASME boiler and vessel code. Large water content (5.2 gallons) eliminates rapid internal temperature changes to assure better control response and a nearly constant supply water temperature.

The sections are insulated to reduce heat loss. A built-in air eliminator diverts air bubbles to an automatic air vent. No separate air-eliminating device is necessary. Internal separators in the section assure full flow of water over the heating elements.

Heating elements. The incoloy-sheathed, low density elements (approximately 55 W/in^2) resist the corrosive effects of all chemicals found in domestic water systems. If it is ever necessary to replace an element, a standard water heater element may be used.

Compression tank and circulator. Most boilers are furnished with a large-capacity compression tank that has a flexible diaphragm to prevent water from contacting the tank charge for positive system protection. The circulator is an industry standard, noted for its long life and dependability (Figure 10–19).

Internal fusing. Electric boilers are normally supplied with a separate fuse for each element leg plus a fuse for the circulator and control circuit.

Zoning. Because of the unique control systems employed in electric boilers, zoning may be easily accomplished with either zone valves or additional circulators. Indoor-outdoor controls and timing devices, necessary for many zoning applications, are not required with most electric boilers.

Contactors. Double line-break contactors close both legs of the power source on a call for heat, and open both legs of each element when the thermostat is satisfied.

STEAM-GENERATING THEORY

If a pan of water is set on a stove or some other source of heat, the heat will cause the temperature of the water to increase and, at the same time, expand in volume. When the temperature of the water reaches the *boiling point* (212° F, at sea level), a physical change will occur in the water, in that it starts to vaporize. If the temperature of the water is kept at the boiling point long enough, the water will continue to vaporize until it has all evaporated. The temperature of the water will not increase beyond the boiling point for a given pressure, even when more heat is added to the water after it starts boiling. The water will not get any hotter as long as it remains at the same pressure.

However, if a close-fitting lid is placed on the vessel of boiling water, the steam is prevented from escaping. This results in an increase in the pressure inside the vessel. If an opening is made in the lid, steam will escape through the opening at the same rate that it is being generated by the heat source. As long as water remains in the vessel, and as long as the pressure remains constant, the temperature of the water and steam will remain constant and equal.

The steam boiler operates on the same basic principle as the closed vessel of boiling water. It is as true with a boiler as with the closed vessel that the steam formed in the boiling process tends to push against the sides of the vessel and the surface of the water. Because of the downward force exerted on the surface of the water, a temperature greater than 212° F is required to boil the water. The higher temperature is possible simply by increasing the supply of heat. It must be remembered, therefore, that an increase in pressure means an increase in the boiling point temperature.

There are a number of technical terms that are used in steam generation. Some of the more common terms are as follows:

Degree: A degree is defined as a measure of heat intensity.

Dry saturated steam: Dry saturated steam is steam at the saturation temperature corresponding to a given pressure; it contains no water in suspension.

Heat: Technically speaking, heat is a form of energy that is measured in British thermal units (Btu). One Btu is the amount of heat required to raise one pound of water one degree Fahrenheit at sea level.

Latent heat: Latent heat is hidden heat that cannot be felt or measured with a thermometer. It is the amount of heat required to change the state of a substance with no change in its temperature.

Quality: The quality of steam is expressed in terms of percent. For instance, if a quantity of wet steam consists of 90% steam and 10% moisture, the quality of the steam is 90%.

Sensible heat: The term *sensible heat* refers to heat that can be measured with a thermometer or felt by the skin.

Steam: Simply stated, the term *steam* means water in the vapor phase.

Superheated steam: Superheated steam is steam at a higher temperature than its saturation temperature corresponding to a given pressure. For example, a boiler may operate on 415 pounds per square inch gauge. The corresponding saturation temperature for this pressure is 445° F. This will be the temperature of both the water in the boiler and the steam in the drum. (Charts are available from the ABMA, ASHRAE, and other such organizations for use in computing this pressure-temperature relationship.) This steam can be passed through a superheater, where the pressure will not be changed appreciably but the temperature will be increased to a higher number.

Temperature: Temperature may be defined as a measure in degrees of sensible heat. Heat that can be felt or measured.

Wet saturated steam: Wet saturated steam is steam at the saturation temperature corresponding to a given pressure; it contains water particles in suspension.

It should be obvious that all of the water in a vessel, if held at the boiling point long enough, will change to steam. What may not be as obvious is that there is a direct relationship between the pressure and the temperature. As long as the pressure is held constant, the temperature of the steam and the boiling water will remain the same.

Steam Boiler Design Requirements

It is essential that a steam boiler meet certain requirements before it is considered satisfactory for operation. Three important requirements are that the boiler be (1) safe to operate, (2) able to generate steam at the desired rate and pressure, and (3) economical to operate.

Everyone working on boilers should make it a point to be familiarized with the Boiler Code and other requirements that apply to the area in which the boilers are located.

The following are a few rules that were set up by the American Society of Mechanical Engineers (ASME). These show the general guidelines used by engineers when designing boilers.

For economy of operation, and to generate steam at the desired rate and pressure, the boiler must have:

1. Adequate water and steam capacity.
2. Rapid and positive water circulation.
3. A large steam-generating surface.
4. Heating surfaces which are easy to clean on both the water and gas sides.
5. Parts that are accessible for inspection.
6. A correct amount and proper arrangement of heating surface.
7. A firebox designed for the efficient combustion of fuel.

STEAM HEATING

The widespread use of steam for space heating today points up a long-recognized fact that steam, as a heating medium, has numerous basic characteristics which can be advantageously employed. Some of the most important advantages are as follows:

TABLE 10–1 Properties of saturated steam (approximate)

Absolute pressure	Gage reading at sea level	Temp. ° F	Heat in water Btu per lb.	Latent heat in steam (Vaporization) Btu per lb.	Volume of 1 lb steam ft^3	Wgt. of water lb. per ft^3
0.18	29.7	32	0.0	1076	3306	62.4
0.50	28.4	59	27.0	1061	1248	62.3
1.0	28.9	79	47.0	1049	653	62.2
2.0	28	101	69	1037	341	62.0
4.0	26	125	93	1023	179	61.7
6.0	24	141	109	1014	120	61.4
8.0	22	152	120	1007	93	61.1
10.0	20	161	129	1002	75	60.9
12.0	18	169	137	997	63	60.8
14.0	16	176	144	993	55	60.6
16.0	14	182	150	969	48	60.5
18.0	12	187	155	986	43	60.4
20.0	10	192	160	983	39	60.3
22.0	8	197	165	980	36	60.2
24.0	6	201	169	977	33	60.1
26.0	4	205	173	975	31	60.0
28.0	2	209	177	972	29	59.9
29.0	1	210	178	971	28	59.9
30.0	0	212	180	970	27	59.8
14.7	0	212	180	970	27	59.8
15.7	1	216	184	968	25	59.8
16.7	2	219	187	966	24	59.7
17.7	3	222	190	964	22	59.6
18.7	4	225	193	962	21	59.5
19.7	5	227	195	960	20	59.4
20.7	6	230	196	958	19	59.4
21.7	7	232	200	957	19	59.3
22.7	8	235	203	955	18	59.2
23.7	9	237	205	954	17	59.2
25	10	240	208	952	16	59.2
30	15	250	219	945	14	58.8
35	20	259	228	939	12	58.5
40	25	267	236	934	10	58.3
45	30	274	243	929	9	58.1
50	35	281	250	924	8	57.9
55	40	287	256	920	8	57.7
60	45	293	262	915	7	57.5
65	50	298	268	912	7	57.4
70	55	303	273	908	6	57.2
75	60	308	277	905	6	57.0
85	70	316	286	898	5	56.8
95	80	324	294	892	5	56.5
105	90	332	302	886	4	56.3
115	100	338	309	881	4	56.0
140	125	353	325	868	3	55.5

Steam's Ability to Give Off Heat

Properties of saturated steam may be found in steam tables (Table 10–1), which give much information regarding the temperature and heat contained in 1 pound of steam for any pressure. For example, to change 1 pound of water to steam at 212° F at an atmospheric pressure of 14.7 psia requires a heat content of 1150.4 Btu, which is made up of 180.1 Btu of sensible heat (the heat required to raise 1 pound of water from 32° F to 212° F) and 970.3 Btu of latent heat. Latent heat is the heat added to change the 1 pound of water at 212° F into steam at 212° F. This stored-up heat is required to transform the water into steam, and it reappears as heat when the process is reversed to condense the steam into water.

Because of this fact, the high latent heat of vaporization of 1 pound of steam permits a large quantity of heat to be transmitted efficiently from the boiler to the heating unit with little change in temperature.

Steam Promotes Its Own Circulation

For example, steam will flow naturally from a higher pressure (as generated in the boiler) to a lower pressure (existing in the steam lines). Circulation or flow is caused by the lowering of the steam pressure along the steam supply mains and the heating units due to the pipe friction and to the condensing process of the steam as it gives up heat to the space being heated (Figure 10–20).

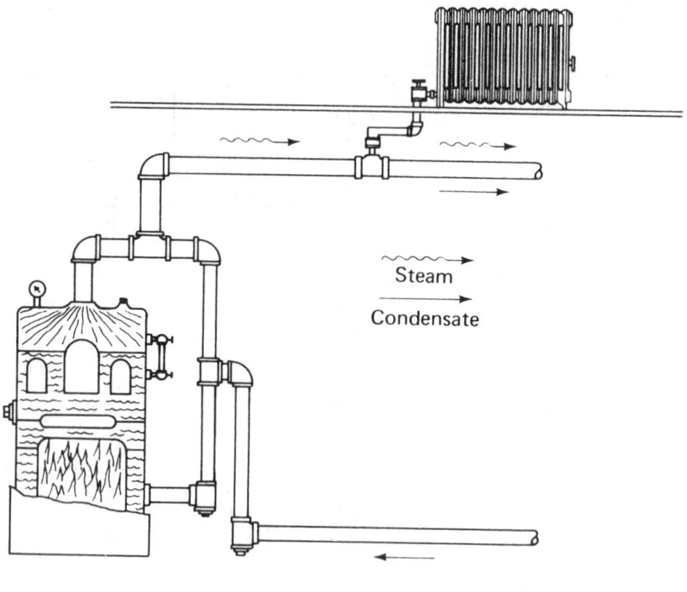
Steam
Condensate

FIGURE 10–20
Steam and condensate flow in lines. *(Courtesy ITT Hoffman Specialty)*

Because of this fact, the natural flow of steam does not require a pump, such as that needed for hot water heating, or a fan, as employed in warm air heating.

Steam Heats More Readily

Steam circulates through a heating system faster than other fluid mediums. This can be important where fast pick-up of the space temperature is desired. It will also cool down more readily when circulation is stopped. This is an important consideration in spring and fall when comfort conditions can be adversely affected by long heating-up or slow cooling-down periods (Figure 10–21).

Steam Heating Is More Flexible

Other advantages in using steam as a heating medium can be found in its easy adaptability to meet unusual conditions of heat requirements with a minimum of attention and maintenance. Here are some examples:

1. Temporary heat during construction is easily provided without undue risks and danger of freeze-ups.
2. Additional heating units can be added to the existing system without making basic changes to the system design.
3. Increased heat output from heating units can be easily accomplished by increasing the steam pressure the proper amount.
4. Steam heating systems are not prone to leak; however, leaks that may occur in the system piping, pipe fittings, or equipment cause less damage than leaks

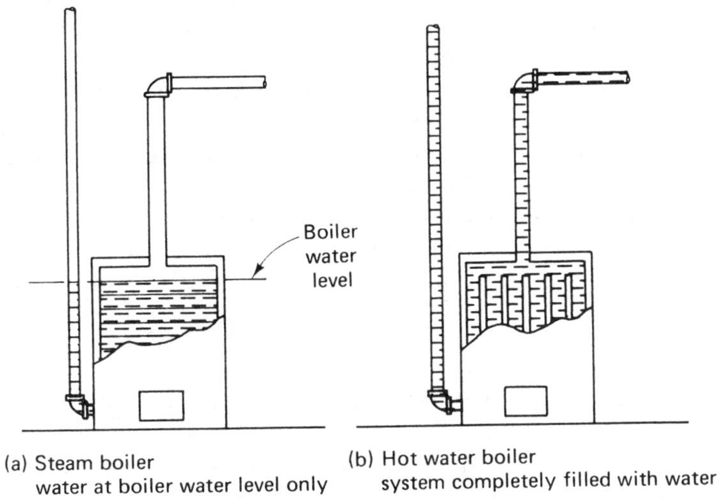
Boiler
water
level
(a) Steam boiler
water at boiler water level only
(b) Hot water boiler
system completely filled with water

FIGURE 10–21
Boiler water comparison. (*Courtesy ITT Hoffman Specialty*)

in systems using hot water. One cubic foot of steam condenses into a relatively small quantity of water. In many cases, a small leak does not cause any accumulation of water at the location of the leak; instead it evaporates into the air and causes no damage.

5. Repair or replacement of system components such as valves, traps, heating units, and similar equipment can be made by simply closing off the steam supply. It is not necessary to drain the system and to spend additional time to reestablish circulation. There is less need to worry abut freezing since the water in a steam heating system is mainly in the boiler (Figure 10–21). The boiler water can easily be protected from freezing during shut-downs, during new construction, and during repairs or replacement of parts by installing an aquastat below the boiler water level to control water temperature.
6. Steam is a flexible medium when used in combination processes and heating applications. These often require different pressures that are easily obtained. In addition, exhaust steam, when available, can be used to the fullest advantage.
7. Steam heating systems are considered to be lifetime investments. Many highly efficient systems are in operation today after more than 50 years of service.

Steam Is Easy to Distribute and Control in a Heating System

The distribution of steam to heating units is easy to accomplish with distribution orifices located at the steam inlet to the unit. Metering orifices can be used along with proper controls to maintain steam pressure in accordance with the flow characteristic of the metering orifices. These orifices can be either a fixed-type or a variable-type. The controls for steam systems are simple and effective. They include those used to control space temperature by the application of "on-off" valves. Modulating controls can also be applied which respond to indoor-outdoor temperature conditions to control the quantity of steam flowing to orificed radiators.

Steam heating systems fall into two basic classifications—one-pipe systems and two-pipe systems. These names are descriptive of the piping arrangement used to supply steam to the heating unit and to return condensate from the unit. It is a one-pipe system when the heating unit has a single pipe connection through which steam flows to it and condensate returns from it at the same time. In a like manner, it is a two-pipe system when the heating unit has two separate pipe connections—one for the steam supply and the other for the condensate return.

One-Pipe System

Modern automatically fired boilers promote rapid steaming and assure quick pick-up of the space temperature from a cold start. The natural circulation of steam in the system, in combination with the simplicity of piping and air venting, makes this type of system the most economical as well as a desirable method of heating.

A one-pipe system (Figure 10–22), properly designed for gravity return of the condensate to the boiler with open type air vents on the ends of mains on each heating unit, requires a minimum of mechanical equipment. The result is a low initial cost for a very dependable system.

The modern one-pipe system is a simple up-feed system. The basic equipment used is described as follows.

Steam boiler. It is automatically fired and equipped with suitable controls to maintain system pressure. It also has the required safety devices for proper burning of fuel and should be equipped with an automatic water feeder and low water cut-off (Figure 10–23).

Heating units. One-pipe systems can be designed to use convectors, wall fin-tube, and similar heat output units (Figure 10–24).

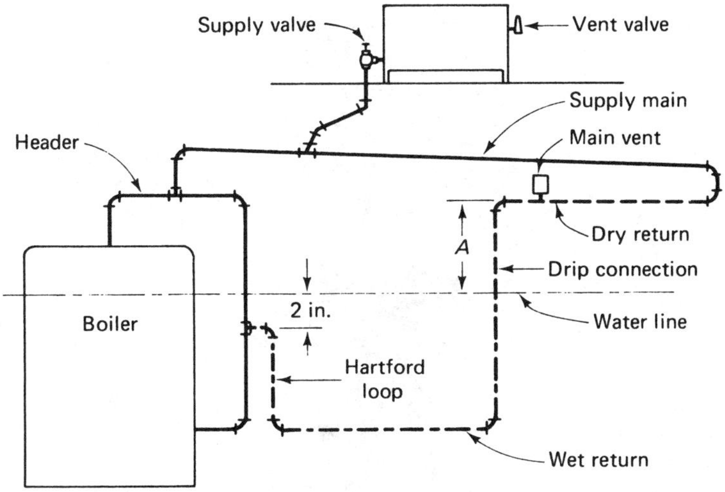
Supply valve
Vent valve
Supply main
Main vent
Header
Dry return
A
Drip connection
2 in.
Water line
Boiler
Hartford
loop
Wet return

FIGURE 10–22
Basic one-pipe up-feed system. (*Courtesy ITT Hoffman Specialty*)

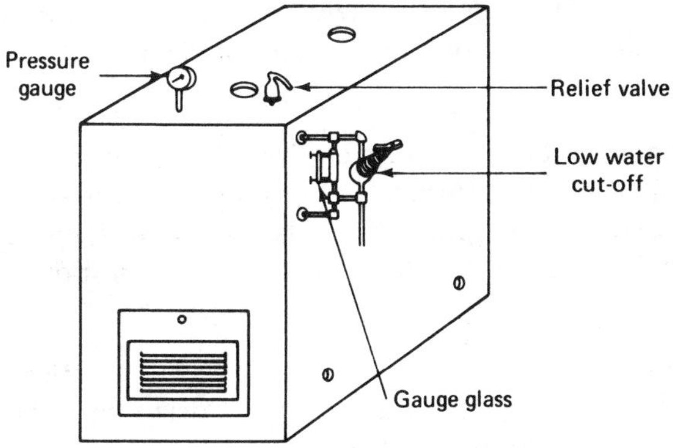
Pressure
gauge
Relief valve
Low water
cut-off
Gauge glass

FIGURE 10–23
A steam boiler. (*Courtesy ITT Hoffman Specialty*)

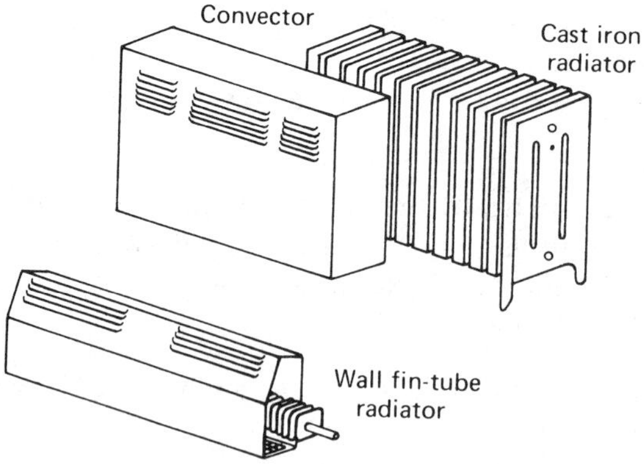
Convector
Cast iron radiator
Wall fin-tube radiator

FIGURE 10–24
Heating units. (*Courtesy ITT Hoffman Specialty*)

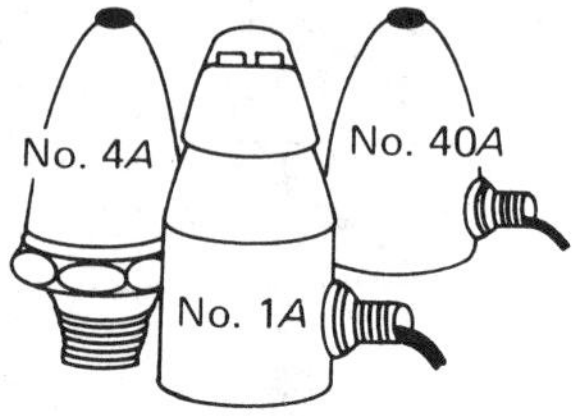
No. 4A
No. 40A
No. 1A

FIGURE 10–25
Air vents. (*Courtesy ITT Hoffman Specialty*)

FIGURE 10–26
Radiator supply valve. (*Courtesy ITT Hoffman Specialty*)

Air vents. Steam cannot circulate or radiate heat until all the air has been vented from the system. Each heating unit and the end of each steam main must be equipped with an air vent valve (Figure 10–25).

Radiator valves. Each radiator must be equipped with an angle pattern, radiator supply valve installed at the bottom inlet tapping (Figure 10–26).

The typical system illustrated in Figure 10–22 also shows the piping components of a one-pipe system. The names and functions of these components are as follows.

Header. Boilers, depending on their size, have one or more outlet tappings. The vertical steam piping from the tapped outlet joins a horizontal pipe called a *header.* The steam supply mains are connected to this header.

Steam supply main. The supply main carries steam from the boiler to the radiators connected along its length. It also carries condensate accumulation from these units back to the drip connection (Figure 10–22). When the condensate flow in the supply main is the same direction as the steam flow, the system is called *a parallel flow system.*

Drip connection. The drip connection is the vertical length of pipe connecting the remote end of the steam supply main to the wet return (Figure 10–22).

Wet return. The return piping that carries the condensate accumulation back to the boiler and is installed below the level of the boiler water line is called *a wet return* (Figure 10–22). It is completely filled with water and does not carry air or steam. When the system is first filled with water or is cold, the pressure throughout the system is the same, or balances. Therefore, the water is at the same level in the boiler and drip connection as indicated by the boiler water line.

Dry return. The dry return is that portion of the return line located above the boiler water line. In addition to carrying condensate, it also carries steam and air. The end of the dry return must be located at the proper height to maintain the minimum required distance above the boiler water line (Figure 10–22).

The illustration shown in Fig. 10–27 will be used to describe important operating principles of a one-pipe system. Steam is generated in the boiler when fuel is burned and heat added to the boiler water. This causes an increase in steam to the heating units. As steam flows through the steam supply main, there is a loss in steam pressure due to the resistance to flow caused by pipe friction. An additional pressure loss is caused by the condensing process in the steam main, other piping, and radiators. The sum of these pressure losses is the pressure drop, which results in a lower pressure on the surface of the water in the drip connection at the end of the steam supply main as compared to the higher pressure acting on the surface of the water in the boiler. This difference in pressure causes the water to rise in the drip connection. The measured difference between these two levels is the pressure drop of the system. The end of the steam supply main must be a minimum distance above the boiler water line for any gravity return one-pipe system. Two other

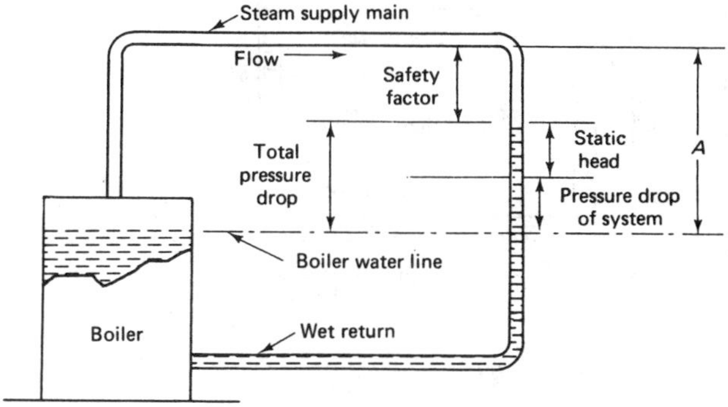
Steam supply main
Flow
Safety factor
Total pressure drop
Static head
Pressure drop of system
A
Boiler water line
Boiler
Wet return

FIGURE 10–27.
Pressure drop in a one-pipe steam system. (*Courtesy ITT Hoffman Specialty*)

factors must be considered, and the distance they represent must be added to the system pressure drop to obtain the proper distance. The static head represents the height of water required to return the condensate to the boiler. The safety factor represents an additional height to allow for unusual heating up conditions.

The Hartford loop. The Hartford loop is a special arrangement of return piping at the boiler. Its purpose is to reduce the likelihood of an insignificant quantity of water creating a low water condition that can cause damage to a steam heating boiler. It came into general use in 1919 and was primarily designed for use with heating systems having gravity wet returns.

In effect, it consists of two loops of pipe forming two U-tubes (Figure 10–28). The first loop is around the boiler and the second loop is composed of the *drip connection, the wet return,* and a short riser called the *loop riser.* In the first loop, an equalizer line runs from the boiler header down the side of the boiler to the boiler return connection. A pressure balance is maintained in this loop because the steam pressure on the top of the water in the equalizer pipe is the same as the pressure on top of the water in the boiler.

In the second loop, the short riser is connected to the equalizer pipe with a close nipple, the center of which must be not less than 2 inches below the boiler water line. Here again, there is a balance between the pressure on top of the water in the equalizer pipe, in the boiler, and in the drip.

Should a leak occur in the wet return or should bad operating conditions be experienced, the boiler could drain down only until the water line fell to the bottom of the nipple connection between the loop riser and the equalizer pipe. Suffi-

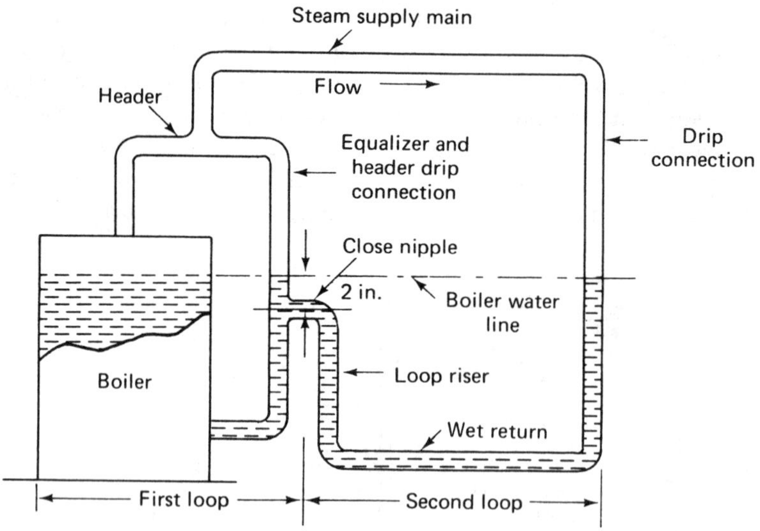
Steam supply main
Flow
Header
Equalizer and
header drip
connection
Drip
connection
Close nipple
2 in.
Boiler water
line
Boiler
Loop riser
Wet return
First loop
Second loop

FIGURE 10–28
The Hartford loop. (*Courtesy ITT Hoffman Specialty*)

cient water will remain in the boiler to cover the crown sheet or section and prevent damage to the boiler.

It is important to use a close nipple to construct the Hartford loop. If a long nipple is used at this point and the water line of the boiler becomes low, water hammer noise will occur. Some designers prefer using a γ fitting that, if used, also must be 2 inches below the boiler water line.

The use of a Hartford loop is not recommended where the condensate is returned to the boiler by a condensate pump. It can be a source of noise resulting from the introduction of relatively cold water to boiler at the hottest boiler water temperature.

Two-Pipe Steam Heating Systems

Although a two-pipe system has fundamental differences from a one-pipe system, many components and piping installation practices are common to both systems. Also, the two-pipe system employs many advantages of using steam as a heating medium. Two-pipe systems are applicable to a variety of structures from small residences to large commercial buildings, office buildings, apartment buildings, and industrial complexes.

Two-pipe systems are designed to operate at pressures ranging from subatmospheric (vacuum) to high pressure. Although they use many practical piping arrangements to provide upflow or downflow systems, they are conveniently classified by the method of condensate return to the boiler. Condensate can be returned to the boiler by gravity or by use of any one of several mechanical means.

By definition, a system is a two-pipe system when the heating unit has two separate pipe connections—one used for the steam supply and the other used for the condensate return.

Basic equipment for two-pipe systems is used in various combinations depending on the type of system or its design. Some of these same components are used for one-pipe systems and were described for these systems. However, all equipment used for two-pipe systems will be described and discussed as follows.

A steam boiler. Boilers are manufactured as either a sectional cast iron boiler or as a steel boiler. The modern boiler is automatically fired using electricity, oil, or gas as the fuel. Steam boilers are provided with suitable pressure controls, safety firing devices, and protection against damage caused by low water conditions (Figure 10–29).

Heating units. Two-pipe systems use a variety of heat output units such as cast iron radiators, convectors, unit ventilators, and heaters (Figure 10–30).

Thermostatic traps. Thermostatic traps are the most common of all types used in two-pipe steam heating systems. They are used because they (1) are simple in construction; (2) are small in design and weight; and (3) have adequate capacity for usual heating system pressure. They are designed to open in response to pres-

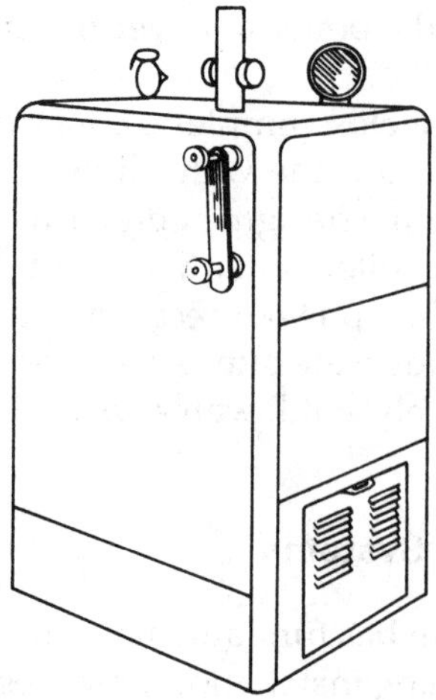

FIGURE 10–29
Steam boiler. (*Courtesy ITT Hoffman Specialty*)

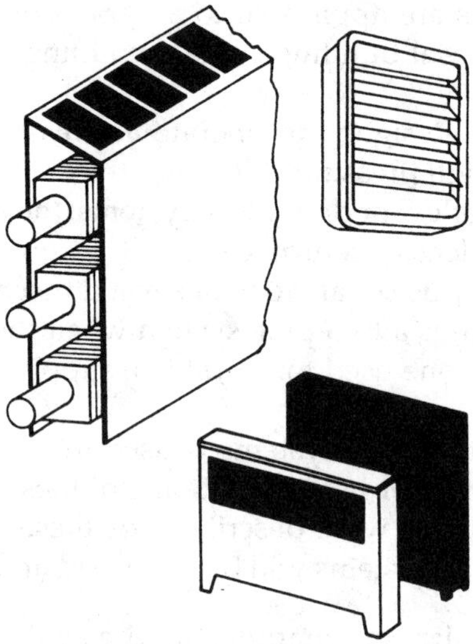

FIGURE 10–30
Heating units. (*Courtesy ITT Hoffman Specialty*)

sure and temperature to discharge air and condensate and to close against the passage of steam. The temperature at which a thermostatic trap will open is variable but is at the required number of degrees below the saturated temperature for the existing steam temperature; the trap opening temperature is called *temperature drop.*

Balanced pressure-type thermostatic traps are those which have been described. They employ thermal elements made from metal bellows, a series of diaphragms, or special cells made from diaphragms (Figure 10–31). There are some types that employ special shapes of bimetal for the thermal elements.

The radiator return connection is equipped with a thermostatic trap. Radiator returns are also used as drip traps and to handle condensate from fan-coil units such as a unit heater. For these and similar applications, a cooling leg (an adequate length

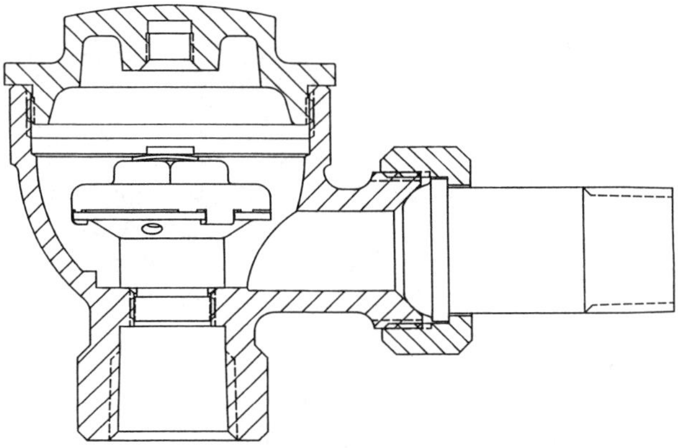

FIGURE 10–31
Thermostatic trap with Dura-Stat ® module. (*(Courtesy ITT Hoffman Specialty)*

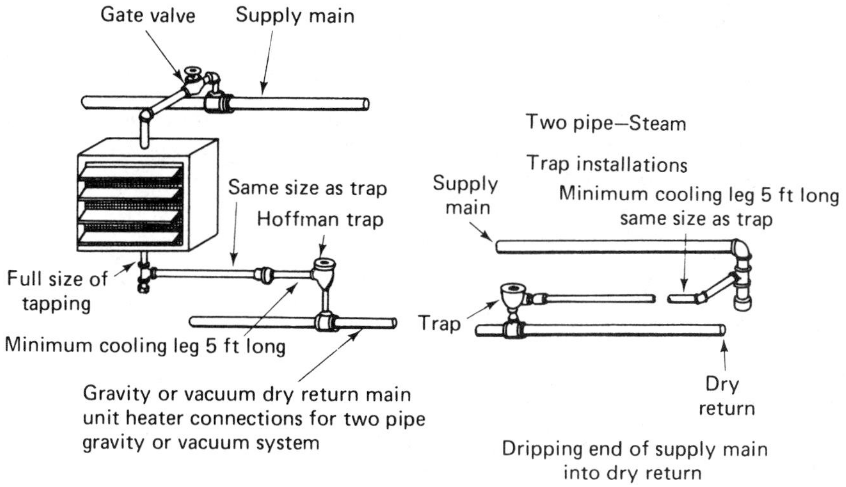
Gate valve
Supply main
Same size as trap
Hoffman trap
Full size of tapping
Minimum cooling leg 5 ft long
Gravity or vacuum dry return main
unit heater connections for two pipe
gravity or vacuum system
Two pipe—Steam
Trap installations
Supply main
Minimum cooling leg 5 ft long
same size as trap
Trap
Dry return
Dripping end of supply main
into dry return

FIGURE 10–32
Cooling leg location. *(Courtesy ITT Hoffman Specialty)*

of pipe) must be installed between the equipment or drip and the thermostatic trap (Figure 10–32). The cooling leg permits the condensate accumulation to cool sufficiently to open the trap in order to discharge the condensate, thus preventing flooding of the equipment.

Thermostatic traps are made in angle, straightway, swivel, and vertical patterns and can be used over a wide range of pressures from subatmospheric (vacuum) to high pressure steam.

Mechanical steam traps. There are several different mechanical traps used in two-pipe heating systems. They include float and thermostatic traps, float traps, inverted bucket traps, and open or upright bucket traps. All of these have different operating characteristics that make them applicable to a variety of heat output units, steam mainlines, or riser drips.

1. *Float and thermostatic traps.* A float and thermostatic trap is often called an F and T trap (Figure 10–33). This trap opens and closes in response to the rising and lowering of the float caused by changes in the level of condensate entering the trap body. The discharge of condensate is continuous due to the throttling action of the pin or valve in the seat port. The thermostatic air bypass remains open when air or condensate is present at a temperature below its designed closing temperature. When steam enters the trap body, the thermostatic air bypass is closed. An F and T trap will discharge condensate at any temperature up to a temperature very close to the saturated steam temperature corresponding to the pressure at the trap inlet. For this reason, cooling legs are not required and condensate accumulations are kept free from steam lines or from equipment being served. An F and T trap is a first-choice selection for dripping the end of a steam main, the heels of upfeed steam risers, and the bottom of downfeed steam risers (Figure 10–34). It is also an excellent choice for handling the condensate from fan-coil units such as unit heaters, unit ventilators, and ventilating coils.

2. *Float trap.* A float trap does not have an internal thermostatic air bypass. There are applications for which air is not a problem, and a float trap can therefore be used to handle condensate only. There are specifications in which float traps are called for, using an external thermostatic air bypass. The external bypass is a thermostatic trap piped around the inlet and outlet of the trap body.

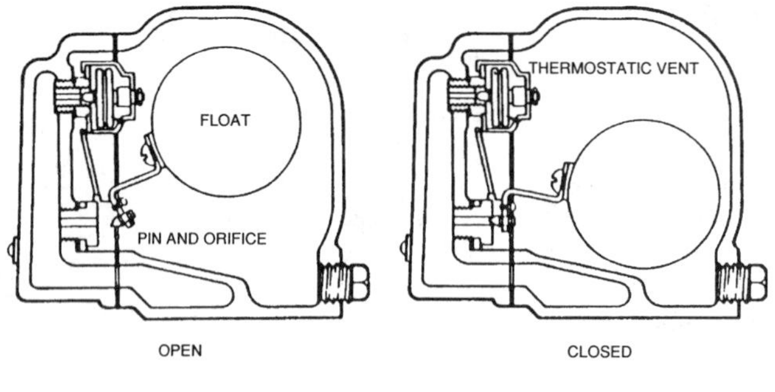
FLOAT
PIN AND ORIFICE
THERMOSTATIC VENT
OPEN
CLOSED

FIGURE 10–33
A float and thermostatic trap. (*Courtesy ITT Hoffman Specialty*)

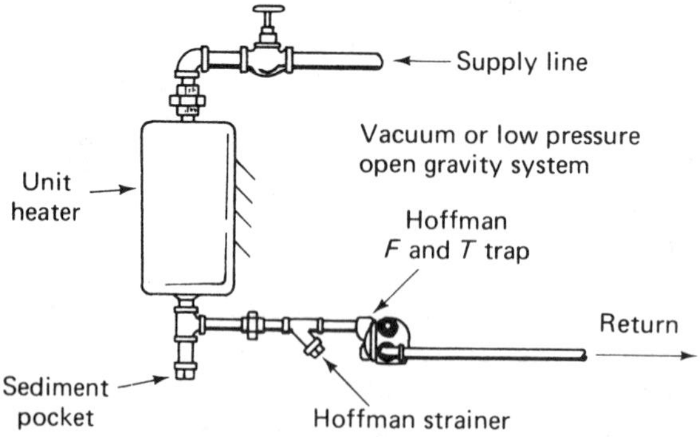
Supply line
Vacuum or low pressure
open gravity system
Unit
heater
Hoffman
F and T trap
Return
Sediment
pocket
Hoffman strainer

FIGURE 10–34
F and T trap application. (*Courtesy ITT Hoffman Specialty*)

3. *Inverted bucket trap.* An inverted bucket trap is so named because the float is an inverted bucket that operates the leverage mechanism to open and close the trap (Figure 10–35). This trap will operate to discharge condensate at any temperature up to the saturated temperature corresponding to the steam pressure at the trap inlet. It operates in cycles at a frequency that depends on the condensate load being handled. This inverted bucket trap is simply constructed and can handle condensate for many industrial types of requirements. It will handle the condensate from heating fan-coil units that must be lifted to discharge to the return mains located above the equipment (Figure 10–36). Before an inverted bucket trap is put into operation, it must be *primed* or filled with water. It operates best at or near full load conditions and when loads do not vary over a wide range.

4. *Upright bucket trap.* This trap operates on the principle of condensate entering the trap body to "float" the upright bucket (Figure 10–37). This action closes

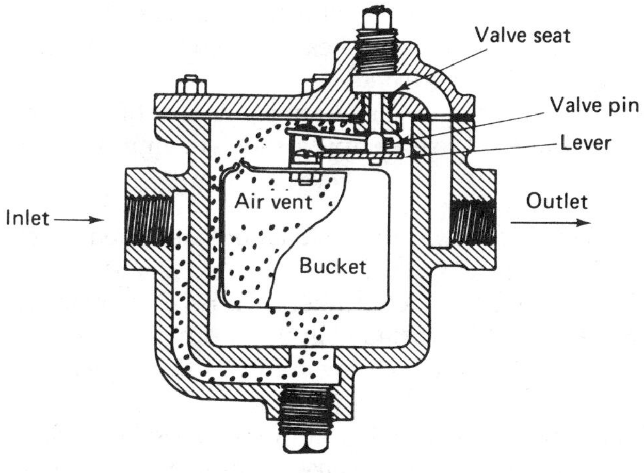
Valve seat
Valve pin
Lever
Inlet
Air vent
Outlet
Bucket

FIGURE 10–35
Inverted bucket trap. (*Courtesy ITT Hoffman Specialty*)

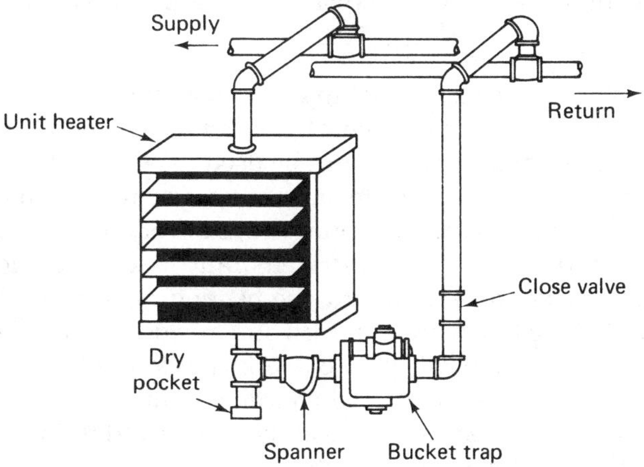
Supply
Return
Unit heater
Close valve
Dry pocket
Spanner
Bucket trap

FIGURE 10–36
Installation of a bucket trap. (*Courtesy ITT Hoffman Specialty*)

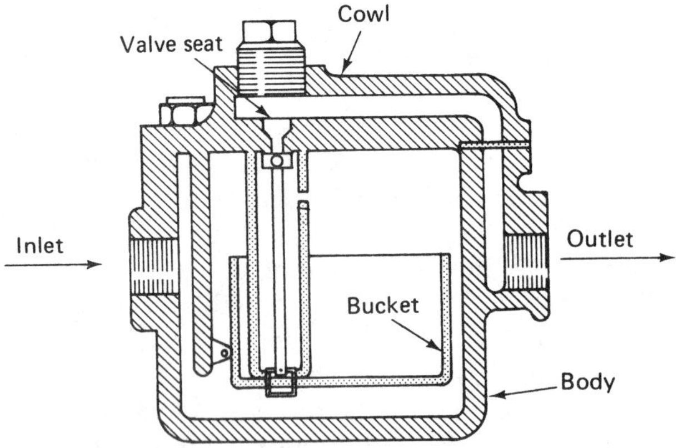
Cowl
Valve seat
Inlet
Outlet
Bucket
Body

FIGURE 10–37
Upright bucket trap. (*Courtesy ITT Hoffman Specialty*)

the discharge port. As condensate continues to enter the trap body, it rises to fill the space surrounding the bucket until it overflows into the bucket. When this occurs, the bucket sinks and the discharge port is opened. Steam pressure then discharges the condensate and the operating cycle is repeated.

Although the construction of this trap is not so simple as that of an inverted bucket trap, the upright bucket trap will handle applications having wide variations of load or pressure.

Main vent air valves. For small two-pipe installations designed to return condensate directly to the boiler by gravity, air must be vented from the system. A main vent air valve must be used at the proper location. These vents are the same as those used for one-pipe systems and are installed at the end of the return main ahead of the point where it drops below the boiler water line to become a wet return.

The main vent valve is designed to permit air accumulations to be discharged (vented) to the atmosphere and to close against the passage of steam or water (Figure 10–38). The proper installation of a main vent is shown in Figure 10–39. The incorrect method is also shown.

Condensate pump. When systems increase in size and higher steam pressures are required to circulate the steam, the condensate cannot be returned to the boiler by gravity. Some type of mechanical means must be used to perform this return function. The accepted method for modern steam heating systems is to use a condensate pump (Figure 10–40). The condensate pump consists of a receiver on which is mounted single or multiple electric motor-driven pump assemblies, float switches, and other electrical controls as required by the heating system. A condensate pump must be located at the low point of return and as close to the boiler as possible. The return main must be uniformly pitched $\frac{1}{4}$ inches to 10 ft to the receiver inlet so condensate can flow by gravity into the receiver. This is important so that condensate and air can be separated as it flows in the piping and the air can be vented to the atmosphere through the vent connection.

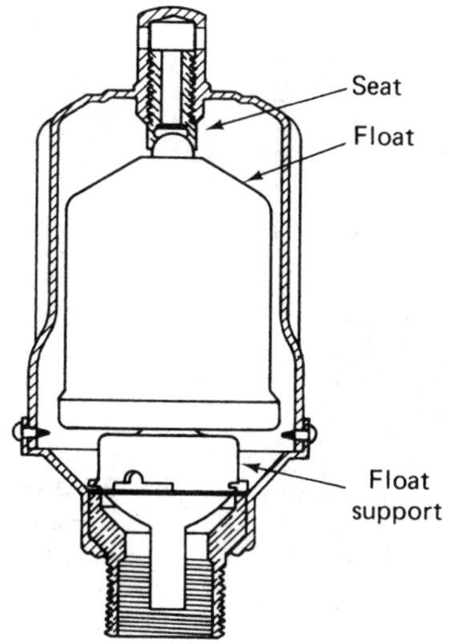
Seat
Float
Float support

FIGURE 10–38
Main vent valve. (*Courtesy ITT Hoffman Specialty*)

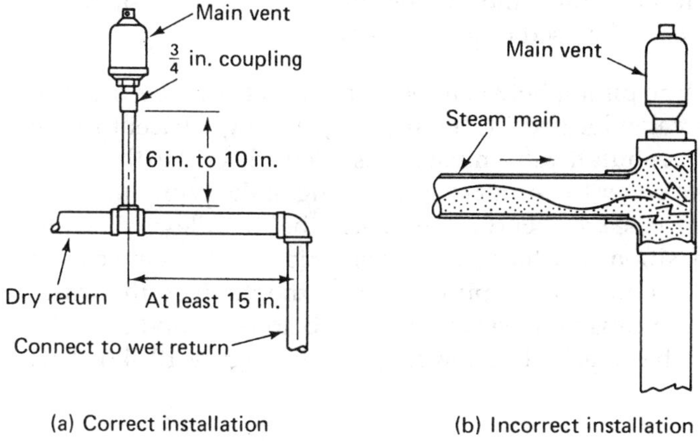
Main vent
$\frac{3}{4}$ in. coupling
6 in. to 10 in.
Dry return
At least 15 in.
Connect to wet return
(a) Correct installation
Main vent
Steam main
(b) Incorrect installation

FIGURE 10–39
Correct and incorrect method of installing a main vent. (*Courtesy ITT Hoffman Specialty*)

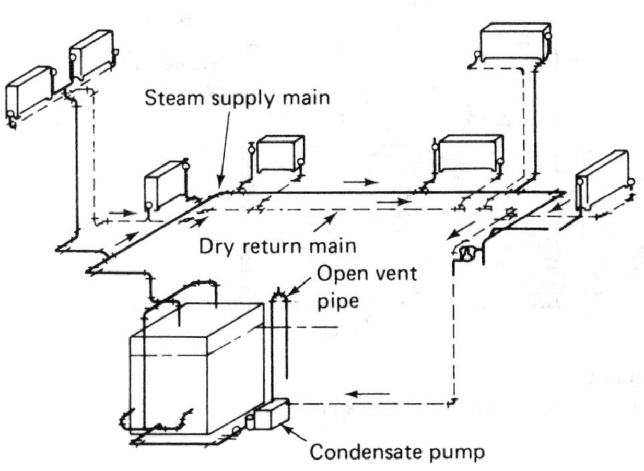
Steam supply main
Dry return main
Open vent pipe
Condensate pump

FIGURE 10–40
Mechanical condensate return. (*Courtesy ITT Hoffman Specialty*)

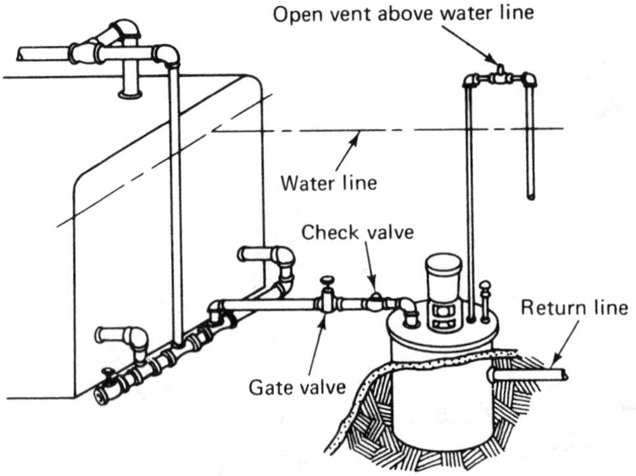
Open vent above water line
Water line
Check valve
Return line
Gate valve

FIGURE 10–41
Underground condensate pump.
(*Courtesy ITT Hoffman Specialty*)

When it is necessary to locate the condensate pump below the floor level to obtain uniform pitch to the receiver inlet, the condensate pump can be installed in a pit or an underground type can be used (Figure 10–41).

The condensate discharge piping between the pump and the boiler must be equipped with a discharge valve located in the discharge piping adjacent to the pump discharge connection. A shutoff valve must be installed adjacent to the check valve and between it and the boiler to permit easy servicing of the pump unit.

As condensate flows into the receiver (Figure 10–42), the water level rises until the float reaches its top position, at which time the electrical contacts are closed to start the pump motor. As condensate is pumped from the receiver, the water level is lowered sufficiently to cause the electrical contacts to open and stop the pump motor when the float has reached its lowest position (Figure 10–43). This

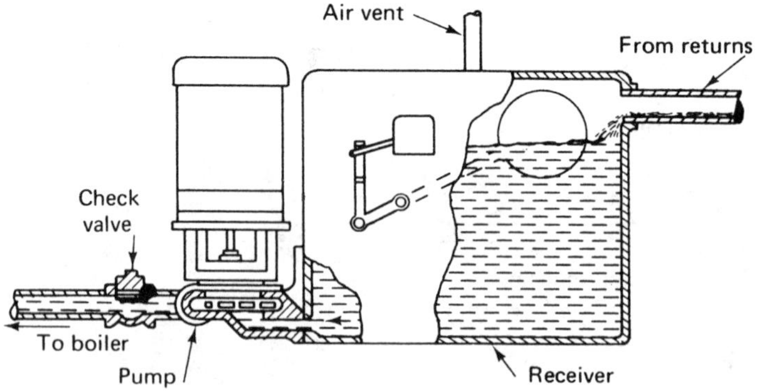
Air vent
From returns
Check valve
To boiler
Pump
Receiver

FIGURE 10–42
Condensate tank full. (*Courtesy ITT Hoffman Specialty*)

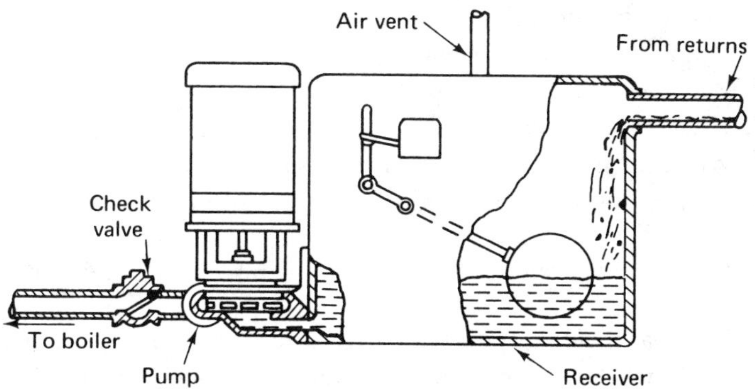
Air vent
From returns
Check
valve
To boiler
Pump
Receiver

FIGURE 10–43
Condensate tank empty. (*Courtesy ITT Hoffman Specialty*)

cycle is repeated as often as necessary to handle the condensate from the system and maintain boiler water level at the proper height.

The float switch for a condensate pump is adjustable so that the quantity of water being discharged for each operating cycle can be set to satisfy the water level condition required for the boiler. Duplex condensate pumps are often equipped with mechanical alternators that will cause the lead pump to operate every other time. The alternator is designed to cause both pumps to operate together to provide double capacity when the system condensate load demands such operation.

REVIEW QUESTIONS

1. What type of refinement is possible with hydronic heating systems?
2. Name the types of boilers used in hydronic comfort heating systems.
3. What most retards the flow of heat from the source to the point of use?
4. What is the range of water temperatures most popular for hot water boilers?
5. In what direction should the pump discharge in a hot water boiler system?
6. In a hot water boiler system, where should the compression tank be located?
7. When should reversed water flow be used in a hot water heating system?
8. What direction of water flow when using an intermittent pump operation causes stress on the boiler?
9. What are the standard design pressures of fire tube boilers?
10. In an immersion-type fire tube boiler, what is inside the tubes?
11. In an immersion-type fire tube boiler, what happens to the insulating gas film?
12. In a compression tank, what is used to prevent water from contacting the tank charge?
13. In a steam boiler, at a given pressure what happens to the temperature of the water?
14. Why is a temperature greater than 212° F required to boil water in a closed vessel?
15. Why must a steam boiler have the correct amount and proper arrangement of heating surface?

16. Define latent heat.
17. Define dry saturated steam.
18. Why does steam flow naturally?
19. What part of a steam heating system does not carry steam or air?
20. Where must a Hartford loop be connected to a boiler?
21. What major advantage does a two-pipe steam heating system have over a one-pipe system?
22. In a two-pipe stem heating system, one pipe carries steam. What does the other pipe carry?
23. In a steam heating system, what is the radiator return connection equipped with?
24. Under what conditions does an inverted bucket trap operate best?
25. How is a condensate return line installed?

11

comfort heating furnaces

The objectives of this chapter are:

- To bring to your attention the air requirements for gas furnaces.
- To introduce you to the furnace components and their sequence of operation.
- To acquaint you with the combustion flow through a gas furnace.
- To acquaint you with the air flow path through a furnace.
- To describe to you the temperature rise through a furnace and its use in heating systems.
- To instruct you in the insulation requirements of furnaces.

Heating furnaces and equipment are probably the most abused and misused machinery in the modern home. Normally, they are purchased and installed in too small a space, and that is the last thought given to them until they stop performing.

VENTILATION AIR REQUIREMENTS

Natural draft gas furnaces, like people, are designed to function in a sea of air. Take away the air supply and problems occur—heating units that operate erratically or not at all.

In the "good old days," warm air furnaces were installed in the basement, surrounded by plenty of air to breathe. Modern-day furnaces, however, are crammed into a space in the modern home the size of a broom closet. The furnace would still be all right, though, if someone would only remember that it takes lots of air.

For every 1,000 Btu of rated input, a natural draft gas unit requires a total of 45 ft^3 of replacement air (i.e., 15 ft^3 for complete combustion; 15 ft^3 dilution air at the draft hood; 15 ft^3 ventilation air for the control of the temperature within the furnace room.

If we take a warm air natural draft furnace, for example, with a rated input of 100,000 Btu, we would see that it needs 4,500 ft^3; this means that to satisfy proper operating conditions for the furnace requires 45 complete roomsful of air each hour. A good task for a crack under the door!

Modern architectural design created this problem by putting equipment in rooms much too small. Because replacement air and venting are interdependent, the responsibility for correcting this situation rests with the installer or, later, with the service technician. Fortunately, this situation can easily be corrected.

Following are some easy-to-use instructions for replacement air supply.

Divide the combined rated appliance inputs by 1,000 to get the number of $in.^2$ of free area required for each of two grills (Figure 11–1). Install both grills, one grill low and the other high above the draft hood opening, to connect the appliance room to the large interior room as shown.

Divide the combined rated appliance by 2,000 (if openings are located in an outside wall, divide by 4,000) to get the number of $in.^2$ of free area required for each of two grills (Figure 11–2). Install both grills, one high and one low, to connect the appliance room to the outdoors as shown.

Divide the combined rate appliance input by 4,000 to get the number of $in.^2$ of free area required for each of two grills (Figure 11–3). Install both grills, one high and one low, to connect the appliance room to ventilated attic as shown.

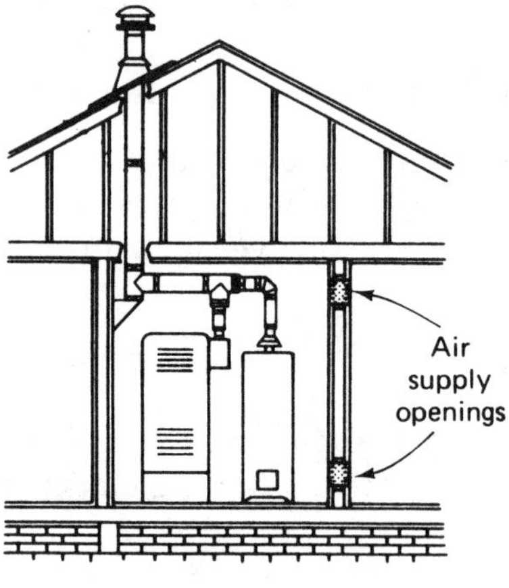
Air
supply
openings

All air from inside building

$$\text{Free area of each grill} = \frac{\text{Total input}}{1000}$$

(Use 2 grills facing into large interior room)

FIGURE 11–1
All air from inside the building. (*Courtesy William Wallace Division, Wallace-Murray Corp.*)

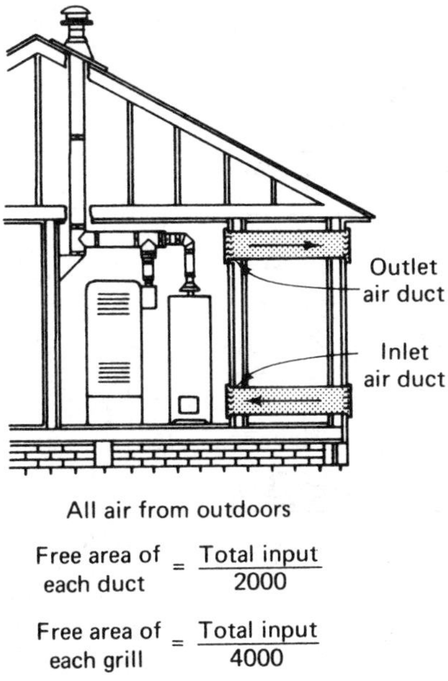
Outlet
air duct
Inlet
air duct
All air from outdoors
Free area of each duct = Total input / 2000
Free area of each grill = Total input / 4000

FIGURE 11–2
All air from outdoors. (*Courtesy William Wallace Division, Wallace-Murray Corp.*)

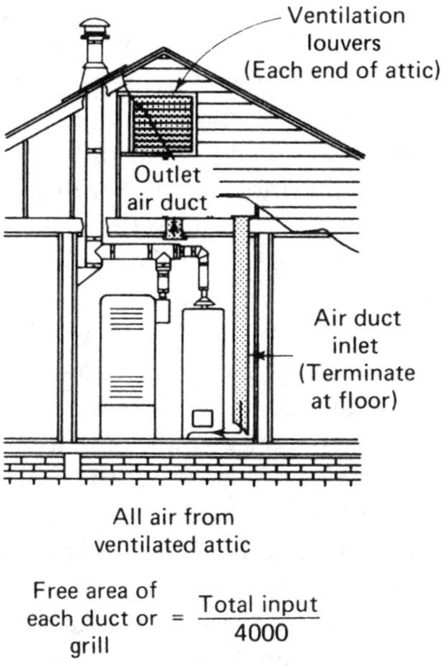
Ventilation
louvers
(Each end of attic)
Outlet
air duct
Air duct
inlet
(Terminate
at floor)
All air from
ventilated attic
Free area of each duct or grill = Total input / 4000

FIGURE 11–3
All air from outdoors through ventilated attic. (*Courtesy of William Wallace Division, Wallace-Murray Corp.*)

Divide the combined rate appliance input by 4,000 to get the number of in.^{2} of free area required for each of two grills (Figure 11–4). Install both grills, one high and one low, to connect the appliance room to ventilated spaces as shown.

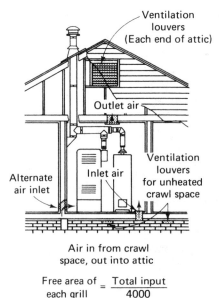

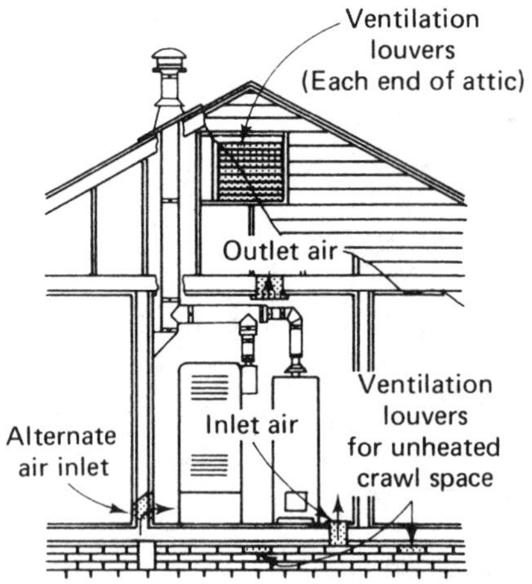
Ventilation
louvers
(Each end of attic)
Outlet air
Ventilation
louvers
for unheated
crawl space
Inlet air
Alternate
air inlet
Air in from crawl
space, out into attic
Free area of each grill = Total input / 4000

FIGURE 11–4
All air from outdoors-inlet from ventilated crawl space and outlet to ventilated attic. (*Courtesy William Wallace Division, Wallace-Murray Corp.*)

Sequence of Operation (Gas Furnace)

It should be evident that when the furnace is not in operation there are no air requirements. However, when the thermostat demands heating, demands are made of several components of the furnace and related equipment.

In operation, while the furnace is at rest the only functioning part is the pilot burner, if one is used. When the thermostat calls for heat, gas is admitted from the gas distribution system to the furnace. Before the gas can enter the combustion area, it must pass through a gas pressure regulator, where the pressure is reduced to approximately 3 $^1/_2$ inches of water column for natural gas. From the pressure regulator, the gas must go through a main gas valve, which functions on demand from the thermostat. These two components are usually combined into one housing and termed a combination gas control (Figure 11–5).

After the main gas valve has opened, the gas is admitted to the gas manifold pipe and through the main burner orifices. From the main burner orifices, the gas goes into the main burner, where it is mixed with primary air for combustion. As the gas flows from the main burner head, it is ignited by the pilot burner flame and a tremendous amount of heat is released.

At this point, secondary air is mixed with the flame to ensure complete combustion. The gas has now been changed to vent gases that leave the combustion area (firebox) and release additional heat while passing through the heat exchanger (Figure 11–6). On leaving the heat exchanger, the combustion gases enter the venting system and flow to the outside atmosphere.

As the temperature within the circulating air passages of the heat exchanger is raised, the fan control completes an electrical circuit to the indoor fan motor, thus

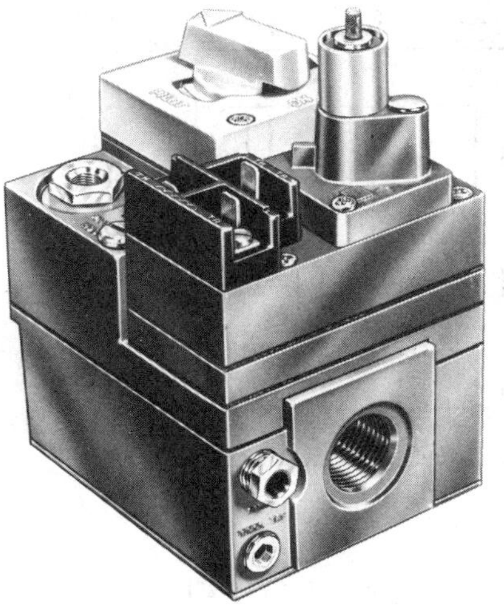

FIGURE 11–5
Combination gas control. (*Courtesy Honeywell Inc.*)

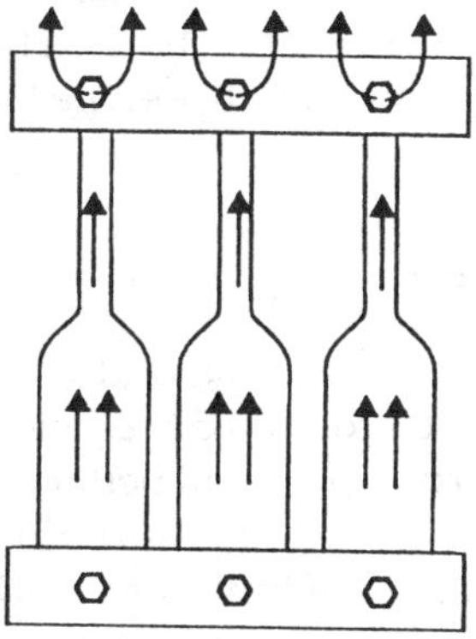

FIGURE 11–6
The vent gas path through a heat exchanger.

causing the air to be forced through the air filter, the furnace heat exchanger, and into the duct system, where it is distributed to the individual rooms (Figure 11–7).

The heating system is now in normal operation. Should the temperature within the furnace become unsafe, the limit control will interrupt the control circuit, causing the supply of gas to stop.

As the thermostat becomes satisfied, the sequence of operation is reversed. The thermostat contacts open, allowing the main gas valve to close, thus stopping the flow of gas to the burners. The fan will continue to circulate the air until the furnace temperature has fallen sufficiently. When the furnace temperature reaches the lower set point, the fan control contacts open and the fan quits circulating the air. The heating system is now at rest and will remain in this condition until the thermostat demands additional heat, at which time the sequence of operation will be repeated.

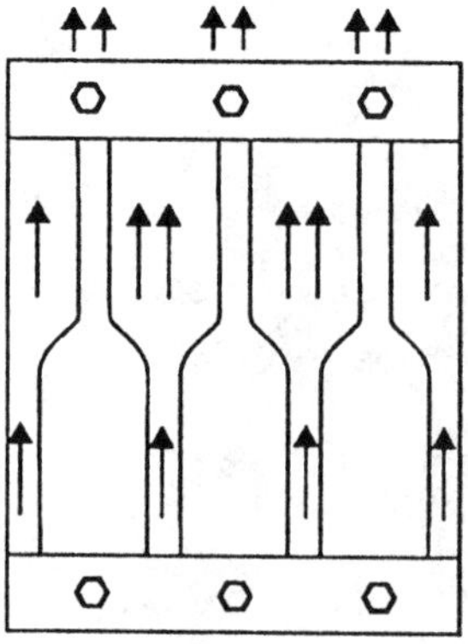

FIGURE 11–7
Circulating air path through a furnace.

Basically, there are three types of gas furnaces: the upflow, the counterflow, and the horizontal furnace. The direction of air flow through the furnace is the determining factor for its name. It should be noted, however, that these are single-purpose, or single-mounting, furnaces. That is, a horizontal furnace cannot be used in any other position. To do so would upset the burner and venting action.

HIGH EFFICIENCY GAS FURNACES

The so-called standard furnaces are rated at 80% efficiency. However, in reality, most of them are somewhere around 65% efficient. The need to conserve energy has caused most builders to use more and better insulation, to include better vapor barriers around buildings, to use weather stripping, to use better construction techniques to improve the tightness of buildings, and to take other energy conservation measures. Buildings, as a result, are more energy efficient, due in part to less air infiltration. This reduction in air infiltration has resulted in the furnaces and other gas appliances being starved for combustion air, causing a need for a more positive combustion air supply. The high efficiency gas furnaces provide this necessary air with a combustion air fan or inducer.

The desire to conserve energy has also resulted in more efficient gas furnace designs. Gas furnaces that are designed to produce 90% and upward efficiency are known as *condensing gas furnaces.* The higher efficiency is obtained by condensing the water vapor from the flue gases and transferring the latent heat that is given up to the circulating air for the inside of the building. The water vapor is condensed in a device called a secondary heat exchanger.

Secondary Heat Exchangers

There are several methods used to provide a secondary heat exchanger, including a finned coil and additional heat exchangers.

Finned coils. The finned coil is located in the air stream ahead of the regular heat exchanger (Figure 11–8). The products of combustion first enter the primary heat exchanger from the combustion zone, and heat is transferred to the circulating air the same as in a standard gas heating furnace. The products of combustion then pass through a tail pipe, or other type of design, into the secondary heat exchanger. The cooler air passes over the finned heat exchanger first and causes the water vapor to condense, giving up the latent heat of condensation to the circulating air. The circulating air then flows over the warmer primary heat exchanger where it absorbs more heat.

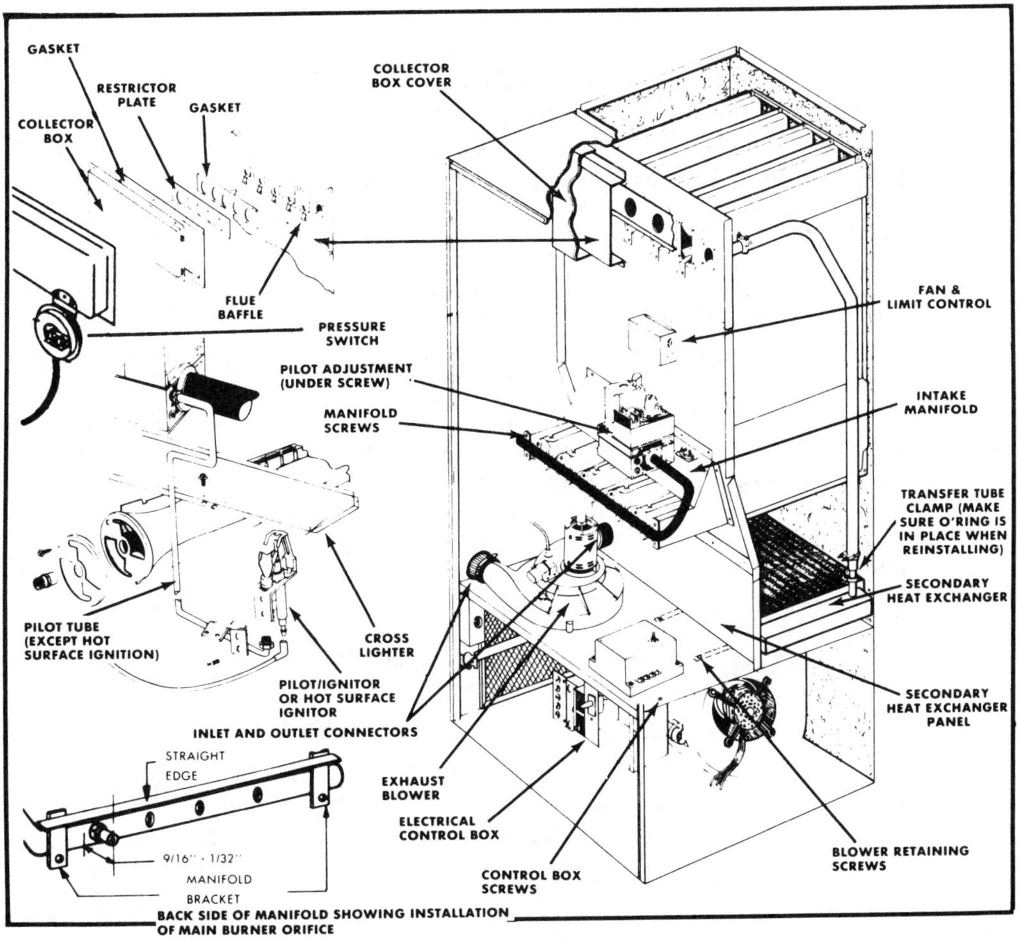
GASKET
RESTRICTOR PLATE
GASKET
COLLECTOR BOX
COLLECTOR BOX COVER
FLUE BAFFLE
PRESSURE SWITCH
PILOT ADJUSTMENT (UNDER SCREW)
MANIFOLD SCREWS
FAN & LIMIT CONTROL
INTAKE MANIFOLD
TRANSFER TUBE CLAMP (MAKE SURE O'RING IS IN PLACE WHEN REINSTALLING)
SECONDARY HEAT EXCHANGER
PILOT TUBE (EXCEPT HOT SURFACE IGNITION)
CROSS LIGHTER
SECONDARY HEAT EXCHANGER PANEL
PILOT/IGNITOR OR HOT SURFACE IGNITOR
INLET AND OUTLET CONNECTORS
STRAIGHT EDGE
EXHAUST BLOWER
ELECTRICAL CONTROL BOX
9/16" · 1/32"
MANIFOLD
BRACKET
CONTROL BOX SCREWS
BLOWER RETAINING SCREWS
BACK SIDE OF MANIFOLD SHOWING INSTALLATION OF MAIN BURNER ORIFICE

FIGURE 11–8
Finned coil secondary heat exchanger location. [*Courtesy of Inter-City Products, Inc.* (*USA*).]

Additional heat exchangers. When an additional heat exchanger is used, the products of combustion pass from the primary heat exchanger to the secondary heat exchanger. Some of the circulating air passes over the primary heat exchanger, and some passes over the secondary heat exchanger (Figure 11–9). The circulating air does not pass through both the primary and the secondary heat exchangers on each pass through the furnace. Some manufacturers use two secondary heat exchangers. The first one increases the efficiency to the 80 to 85% efficient range and the second increases the efficiency to the 95 to 98% efficient range. When only one secondary heat exchanger of this type is used, the furnace generally is not considered to be a condensing gas furnace.

Primary Heat Exchangers

The primary heat exchangers are basically the same as those used on standard gas furnaces. They generally include the combustion zone and primary heat transfer surfaces. The heat exchangers used in condensing gas furnaces are made of a type of metal that has a high resistance to corrosion. This is necessary because of the corrosive nature of the condensed products of combustion.

Condensate Drain

The secondary heat exchangers are equipped with a condensate drain connection that allows the condensate to be drained into the city drain system. This corrosive condensate should not be allowed to drain onto the ground or other places where damage can be caused. Plastic pipe is normally used as the drain line because of the corrosive nature of the condensate. Be sure to follow the proper procedure for installing the drain. This is generally covered in detail in the manufacturers' installation literature. These instructions should be followed unless there is a conflict with the local ordinance, in which case the local ordinance should be used.

Vent Pipe

Plastic pipe of the proper type, such as schedule-40 PVC, PVC-DWV, SDR-21 PVC, SDR-26 PVC, or ABS-DWV pipe, or any other recommended type, is used. The size

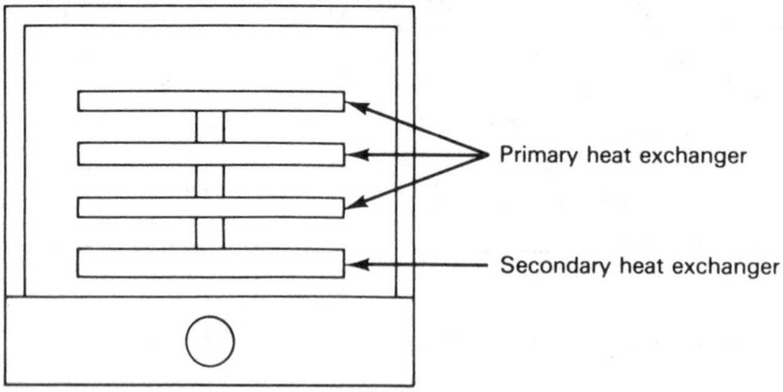
Primary heat exchanger
Secondary heat exchanger

FIGURE 11–9
Additional heat exchanger for increased efficiency. (*Courtesy of Billy Langley,* High-Efficiency Gas Furnace Troubleshooting Handbook, *© 1991, p. 3; reprinted by permission of Prentice-Hall, Inc., Englewood Cliffs, N.J.*)

is determined by the manufacturer of the furnace. Plastic pipe may be used because the flue gas temperature has been reduced below the safe operating temperature of the vent pipe material. The size is generally smaller than the type B vent for the same size lower efficiency furnace because the water vapor has been removed and the reduced temperature causes a reduction in the volumetric area of the flue gases. The installation of the vent pipe should conform to the code that is in force at the time of installation at the job location.

Combustion Air

The combustion air is supplied to high efficiency gas furnaces by a motor-driven blower. There are two types of combustion systems in use: forced draft and induced draft.

Forced draft system. The forced draft system forces the air into the combustion zone under pressure (Figure 11–10). In this type of system, the combustion zone and the vent system have a slight pressure inside. The combustion air adjustments must be made according to the manufacturer's specifications for the furnace in question. The blower supplies all of the air needed for proper combustion and provides the force required to move the products of combustion through the vent system and into the atmosphere. Anytime a leak occurs anywhere in the system, the combustion process will be upset and proper, if any, operation will not be possible.

Induced draft. The induced draft system draws, or pulls, the air into the combustion zone through the heat exchanger and discharges it into the vent system.

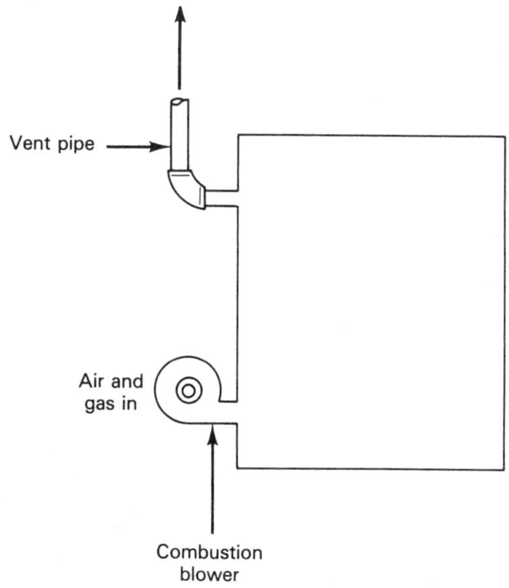
Vent pipe
Air and gas in
Combustion blower

FIGURE 11–10
Forced draft heat exchanger. (*Courtesy of Billy Langley,* High-Efficiency Gas Furnace Troubleshooting Handbook, *© 1991, p. 4; reprinted by permission of Prentice-Hall, Inc., Englewood Cliffs, N.J.*)

(Figure 11–11). The combustion zone and the heat exchanger are under a negative pressure. The vent system may have either atmospheric pressure or a slightly positive pressure inside it. The combustion air adjustments must be made according to the manufacturer's specifications for the furnace in question. The blower supplies all of the air needed for proper and complete combustion and provides the force required to move the products of combustion through the vent system and into the atmosphere. Anytime a leak occurs anywhere in the system, the combustion process will be upset and proper, if any, operation will not be possible.

Combustion and Ventilation Air Requirements

The flow of combustion and ventilation air to the furnace must not be restricted in any way. The air openings placed in the furnace cabinet during construction must be kept free of any obstructions that would in any way restrict the flow of air to the furnace. Such obstructions would affect the efficiency as well as the safe operation of the furnace. Furnaces must have a sufficient quantity of air for proper and safe operation and performance. This must be kept in mind during installation and service procedures.

For complete combustion and ventilation, the furnace requires approximately 20 ft^3 of air for every 1,000 Btu of natural gas burned. Thus, for each 1,000 Btu of natural gas consumed, a total of 20 ft^3 of air must be supplied to the furnace room. *Warning:* The air used for combustion and ventilation purposes must not come from a source containing corrosive contaminants. Therefore, when installing a condensing gas furnace in a commercial building, a workshop, or a beauty shop it may be necessary to provide outside air for combustion and ventilation purposes.

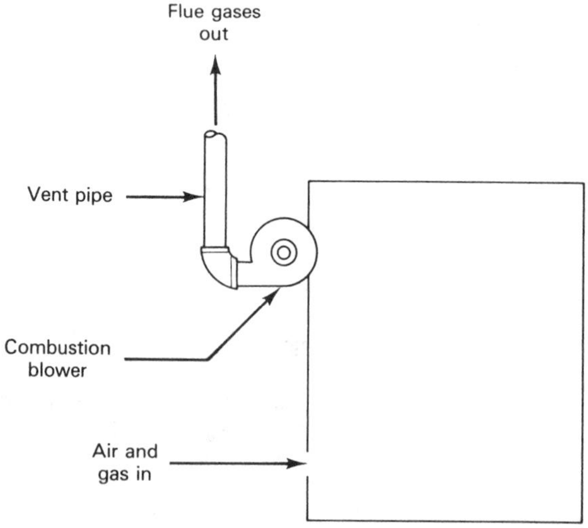
Flue gases out
Vent pipe
Combustion blower
Air and gas in

FIGURE 11–11
Induced draft heat exchanger. (*Courtesy of Billy Langley,* High-Efficiency Gas Furnace Troubleshooting Handbook, *© 1991, p. 5; reprinted by permission of Prentice-Hall, Inc., Englewood Cliffs, N.J.*)

The combustion air must be free of all acid-forming chemicals, such as fluorine, chlorine, and sulfur. These elements are generally found in aerosol sprays, detergents, bleaches, cleaning solvents, air fresheners, paint and varnish removers, refrigerants, and many other commercial and household cleaning products. When the vapors from these products are burned in a gas flame, acid compounds are formed. These acid compounds then increase the dew point temperature of the flue gas products and are highly corrosive after they are condensed.

Be sure to follow the furnace manufacturer's recommendations or the local code requirements when installing a high efficiency gas furnace. This will assure proper and safe performance of the furnace.

Vent Pipe Installation

Proper vent pipe installation is critical to the safe operation of the furnace. Improper installation leads to danger of property damage, bodily injury, or death. Carefully read and follow the instructions provided here and those that apply to the furnace in question.

Condensing gas furnaces remove both sensible and latent heat from the combustion flue gases. Removal of the latent heat results in the condensation of flue gas water vapor. The condensed water vapor drains from the secondary heat exchanger into the combustion blower and out of the unit into a PVC drain trap (Figures 11–12 and 11–13).

The furnace must be vented to the outdoors using the proper connections and the proper size vent pipe, which is made from the required type of material in accordance to the local codes and ordinances.

When a substitute piping is used, it must be connected to the furnace with the proper fittings for the type of pipe being used. All joints, fittings, and so on must be cemented, sealed, or mechanically connected to prevent leakage of the flue gases.

All of the instructions, guidelines, and limitations outlined in this section for PVC piping must be followed unless they are in conflict with the type of material being used, local codes, or the requirements of the furnace manufacturer.

The vent must be installed in compliance with Part 7, Venting of Equipment, of the National Fuel Code NFPA 54/ANSI Z 223.1, 1984; local codes or ordinances; these instructions; and good trade practices.

Each vent must serve only one furnace. Do not connect to an existing vent or chimney.

Vertical venting is preferred because there will be some moisture in the flue gases that may condense in the vent pipe (see instructions for horizontal vents). The vent must exit the furnace at the designated point. If necessary, a drain trap assembly must be installed (such as when a side vent exit is used) to provide the necessary 5-inch water column against the vent pressure. The drain trap must be constructed with the proper parts. Make sure that all parts fit properly and are correctly oriented before beginning any solvent cementing. Observe the following guidelines and limitations when constructing the vent assembly.

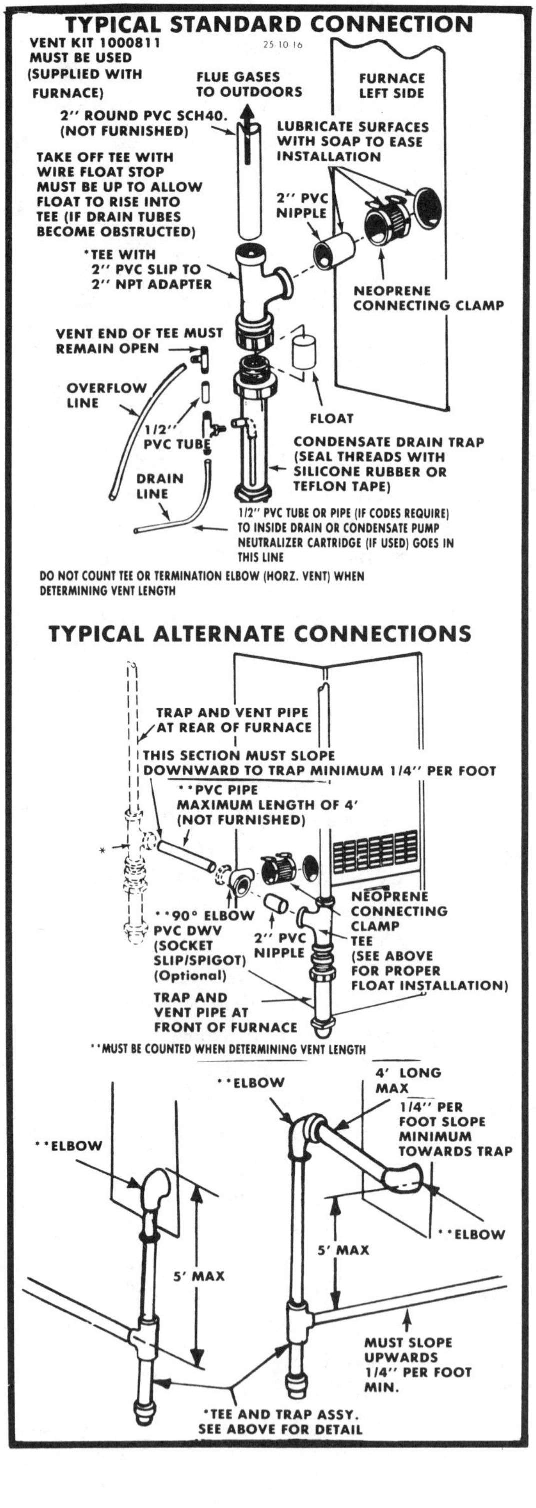

TYPICAL STANDARD CONNECTION
VENT KIT 1000811 MUST BE USED (SUPPLIED WITH FURNACE)
25 10 16
FLUE GASES TO OUTDOORS
FURNACE LEFT SIDE
2'' ROUND PVC SCH40. (NOT FURNISHED)
LUBRICATE SURFACES WITH SOAP TO EASE INSTALLATION
TAKE OFF TEE WITH WIRE FLOAT STOP MUST BE UP TO ALLOW FLOAT TO RISE INTO TEE (IF DRAIN TUBES BECOME OBSTRUCTED)
2'' PVC NIPPLE
*TEE WITH 2'' PVC SLIP TO 2'' NPT ADAPTER
NEOPRENE CONNECTING CLAMP
VENT END OF TEE MUST REMAIN OPEN
OVERFLOW LINE
1/2'' PVC TUBE
FLOAT
CONDENSATE DRAIN TRAP (SEAL THREADS WITH SILICONE RUBBER OR TEFLON TAPE)
DRAIN LINE
1/2'' PVC TUBE OR PIPE (IF CODES REQUIRE) TO INSIDE DRAIN OR CONDENSATE PUMP NEUTRALIZER CARTRIDGE (IF USED) GOES IN THIS LINE
DO NOT COUNT TEE OR TERMINATION ELBOW (HORZ. VENT) WHEN DETERMINING VENT LENGTH
TYPICAL ALTERNATE CONNECTIONS
TRAP AND VENT PIPE AT REAR OF FURNACE
THIS SECTION MUST SLOPE DOWNWARD TO TRAP MINIMUM 1/4'' PER FOOT
**PVC PIPE MAXIMUM LENGTH OF 4' (NOT FURNISHED)
*
NEOPRENE CONNECTING CLAMP
**90° ELBOW PVC DWV (SOCKET SLIP/SPIGOT) (Optional)
2'' PVC NIPPLE
TEE (SEE ABOVE FOR PROPER FLOAT INSTALLATION)
TRAP AND VENT PIPE AT FRONT OF FURNACE
**MUST BE COUNTED WHEN DETERMINING VENT LENGTH
**ELBOW
4' LONG MAX
1/4'' PER FOOT SLOPE MINIMUM TOWARDS TRAP
**ELBOW
**ELBOW
5' MAX
5' MAX
MUST SLOPE UPWARDS 1/4'' PER FOOT MIN.
*TEE AND TRAP ASSY. SEE ABOVE FOR DETAIL

FIGURE 11–12
Vent trap and furnace connection. [*Courtesy of Inter-City Products, Inc. (USA).*]

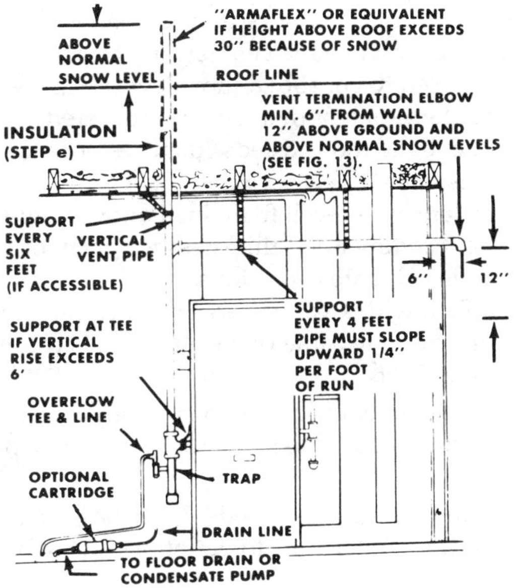
"ARMAFLEX" OR EQUIVALENT IF HEIGHT ABOVE ROOF EXCEEDS 30" BECAUSE OF SNOW
ABOVE NORMAL SNOW LEVEL
ROOF LINE
VENT TERMINATION ELBOW MIN. 6" FROM WALL 12" ABOVE GROUND AND ABOVE NORMAL SNOW LEVELS (SEE FIG. 13).
INSULATION (STEP e)
SUPPORT EVERY SIX FEET (IF ACCESSIBLE)
VERTICAL VENT PIPE
6"
12"
SUPPORT EVERY 4 FEET PIPE MUST SLOPE UPWARD 1/4" PER FOOT OF RUN
SUPPORT AT TEE IF VERTICAL RISE EXCEEDS 6'
OVERFLOW TEE & LINE
OPTIONAL CARTRIDGE
TRAP
DRAIN LINE
TO FLOOR DRAIN OR CONDENSATE PUMP

FIGURE 11–13
Vent installation. [*Courtesy of Inter-City Products, Inc. (USA).*]

correctly oriented before beginning any solvent cementing. Observe the following guidelines and limitations when constructing the vent assembly.

1. The vent diameter must not be reduced.
2. The drain trap assembly, when used, must be within 4 ft horizontally and 5 ft vertically (lower only) of the furnace vent connector. Some typical examples are shown in Figure 11–12. All vent piping from the furnace to the trap must slope downward a minimum of $^1/_4$ inch per foot of run. All vent piping from the trap to the vent termination must slope upward a minimum of $^1/_4$ inch per foot of run.

   The drain trap assembly may not be installed in any unconditioned space if there is any chance of condensate freezing inside the trap or the drain lines. The drain trap must be reasonably accessible so that homeowners can check it.

   Any elbows used to change from vertical to a horizontal run should be of the type DWV to provide the correct slope in the horizontal run. If other types of elbows are used, then two 45° elbows should be used, in place of one 90°, with the elbows slightly misaligned to provide a slope in the horizontal runs.
3. All horizontal pipe runs must be supported at least every 4 ft with metal pipe strapping. No sags or dips are permitted.
4. All vertical pipe runs must be supported every 6 ft where accessible.
5. The vent pipe must be insulated if there is any chance of condensate freezing inside the pipe. This can occur if the vent pipe passes through an

unconditioned space, such as an attic, crawl space, uninsulated chase, or a masonry chimney. It can also occur when the vent terminates above the roof or if an exterior vertical riser is used to get above snow levels (Figure 11–14). The local climatic conditions and the vent length must be considered.

If the vent height above the roof exceeds 30 inches because of snow accumulations, it must be insulated.

Armaflex® or equivalent closed-cell foam insulation should be used for interiors or exteriors. The recommended thickness is 1 inch, or use multiple layers if required for extreme climate conditions.

Fiberglass or equivalent with a vapor barrier is recommended for exterior use only. The recommended R value of 7 up to 10 ft is recommended—R-11 if the exposure exceeds 10 ft.

6. If it is necessary to insulate the vent pipe and a chimney is used as a chase, the top of the chimney must be sealed flush, or crowned, so that only the vent pipe protrudes.
7. When the vent height above the roof exceeds 30 inches, or if an exterior vertical riser is used on a horizontal vent to get above the snow levels, the

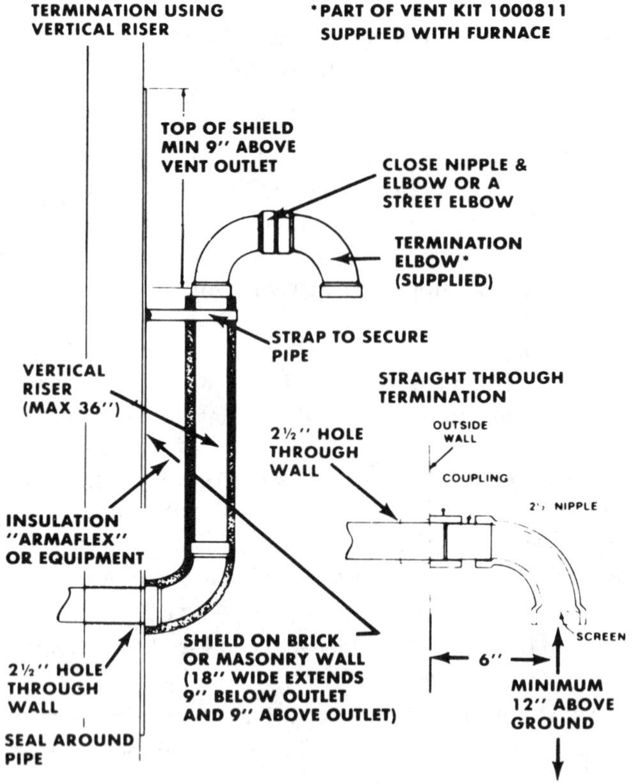
TERMINATION USING
VERTICAL RISER
*PART OF VENT KIT 1000811
SUPPLIED WITH FURNACE
TOP OF SHIELD
MIN 9" ABOVE
VENT OUTLET
CLOSE NIPPLE &
ELBOW OR A
STREET ELBOW
TERMINATION
ELBOW*
(SUPPLIED)
STRAP TO SECURE
PIPE
VERTICAL
RISER
(MAX 36")
STRAIGHT THROUGH
TERMINATION
2½" HOLE
THROUGH
WALL
OUTSIDE
WALL
COUPLING
INSULATION
"ARMAFLEX"
OR EQUIPMENT
SHIELD ON BRICK
OR MASONRY WALL
(18" WIDE EXTENDS
9" BELOW OUTLET
AND 9" ABOVE OUTLET)
SCREEN
6"
MINIMUM
12" ABOVE
GROUND
2½" HOLE
THROUGH
WALL
SEAL AROUND
PIPE

FIGURE 11–14
Vent termination. [*Courtesy of Inter-City Products, Inc.* (*USA*).]

exterior portion must be insulated. Use only moisture-resistant insulation, such as Armaflex.

8. The maximum vent length is 60 total equivalent feet with each 45° elbow (maximum of 8) counting as 2 1/2 ft and each 90° (maximum of 4) counting as 4 ft. Do not count the take-off tee (Figure 11–12) or the vent termination elbow on a horizontal vent. For example, a 40-ft vent pipe with four 90° elbows (20 ft) equals 60 equivalent feet; or 40 ft of vent pipe with two 45° elbows (5 ft) and three 90° elbows (15 ft) equals 60 equivalent feet.
9. The minimum vent length is 5 ft.
10. Do not install the vent pipe in the same chase with a vent from another gas or fuel-burning appliance.
11. Do not install the vent pipe within 6 inches of a vent pipe from another gas or other fuel-burning appliance.
12. The vent pipe can run in the same chase or adjacent to a supply or vent pipe from water supply or waste plumbing.

    The vent pipe must be fastened to the furnace using the manufacturer's recommended procedure. Do not attempt to cement the pipe to the blower housing.

    If a vertical riser exceeds 6 ft, secure the tee and trap assembly to the furnace with a mounting strap to remove excessive weight from the clamp connection.

    An optional 90° elbow (PVC DWV socket × slip/spigot) may be used and fastened to the combination blower outlet coupling using a 2 × 2-inch PVC nipple. See Figure 11–12 for the drain trap assembly.

Joining pipe and fittings. All pipe, fittings, solvent cement, and procedures must conform to the American Institute and the American Society for Testing and Material (ANSI/ASTM) standards. Use ASTM D1785, D2466, and D2665 PVC primer and solvent cement, and ASTM D2564 Procedure for Cementing Joints, Reference ASTM D2855.

Improper installation leads to danger of fire or bodily injury. PVC solvent cements and primers are highly flammable. Provide adequate ventilation and do not assemble near a heat source or open flame. Do not smoke. Avoid skin or eye contact. Observe all cautions and warnings printed on material containers.

All joints in the PVC vent must be properly sealed using the following materials and procedures.

For proper installation, do not use solvent cement that has become curdled, lumpy, or thickened. Do not thin. Observe shelf precautions printed on the containers. For applications below 32° F, use only low temperature solvent. Use PVC cleaner-primer and PVC medium body solvent cement.

1. Cut the pipe and square the end; remove ragged edges and burrs. Chamfer the end of the pipe; then clean the fitting socket and pipe joint area of all dirt, grease, or moisture.

2. After checking the pipe and socket for proper fit, wipe the socket and pipe with cleaner-primer. Apply a liberal coat of primer to the inside surface of the socket and the outside of the pipe. Do not allow the primer to dry before applying the cement.
3. Apply a thin coat of cement evenly to the socket. Then quickly apply a heavy coat of cement to the pipe and insert the pipe into the fitting with a slight twisting motion until it bottoms out. The cement must be fluid; if not, recoat.
4. Hold the pipe in the fitting for 30 seconds to prevent the tapered socket from pushing the pipe out of the fitting.
5. Wipe all excess cement from the joint with a rag. Allow 15 minutes before handling. Cure time varies according to the fit, temperature, and humidity. Stir the solvent cement frequently while using it. Use a natural bristle brush or the dauber supplied with the can. The proper brush size is 1 inch.

Condensate drain/neutralizer. The drain line and the overflow line can be $1/2$-inch PVC flex tube or schedule 40 with a disconnect union so the trap can be removed. The trap assembly provides 5 inches water column, so no additional trap is required. Drains must terminate at an outside drain.

Do not run an outside drain. Freezing of the condensate could cause property damage.

If a condensate pump is used, or if local codes require one, install a condensate neutralizer cartridge in the drain line. Be sure to install the cartridge in a horizontal position only. Install an overflow line in routing to a floor drain or a sump pump (Figure 11–12).

If no inside floor drain is available, a condensate pump or sump pump must be used. The condensate neutralizer must be used with either type of pump.

The condensate pump must have an auxiliary safety switch to prevent operation of the furnace and resulting overflow of condensate in the event of pump failure. The safety switch must be wired through the R circuit *only* (low voltage) to provide operation in either of the heating or cooling modes.

Horizontal vents. The furnace may be vented horizontally through an outside wall, using all of the applicable instructions discussed under "Vent Pipe Installation," with these additional requirements. The requirements and limitations for horizontal venting are strict. *All horizontal vent installations must be made in accordance with these instructions.*

Vent location. The vent location must meet the requirements listed in the following instructions or applicable codes, whichever specifies the most clearance or strictest limitation. The combustion products and moisture in the flue gases may condense as they leave the terminal elbow. The condensate may freeze on the exterior wall, under the eaves, and on surrounding objects. Some discoloration to the exterior of the building may occur.

The vent must be installed with the following clearances and requirements (Figure 11–15).

1. Twelve inches above ground level, above normal snow levels (when practicable), and 6 inches out from the wall. Ice or snow may cause the furnace to shut down if the vent becomes obstructed. If required, use a vertical riser to shield the vent or prevent blockage from drifting snow (Figure 11–16).

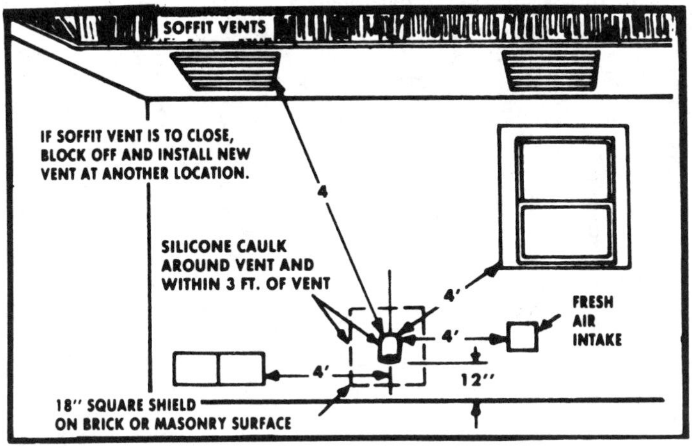
SOFFIT VENTS
IF SOFFIT VENT IS TO CLOSE, BLOCK OFF AND INSTALL NEW VENT AT ANOTHER LOCATION.
4
SILICONE CAULK AROUND VENT AND WITHIN 3 FT. OF VENT
4'
FRESH AIR INTAKE
4'
4'
12"
18" SQUARE SHIELD ON BRICK OR MASONRY SURFACE

FIGURE 11–15
Minimum clearances. [*Courtesy of Inter-City Products, Inc.* (*USA*).]

2. Not above any walkway or area that may create a hazard nuisance or be detrimental to the operation of any other equipment.
3. Four feet from and not above or below any door, window, gravity inlet, or forced air inlet for the building.
4. At least 4 ft from any soffit or under an eave vent.
5. Do not vent under any kind of patio or deck.
6. Locate the vent on the side of the building away from prevailing winter winds when practical but taking into consideration other limitations to determine the best overall location. If installed on a side with prevailing winds, consider the possible effects of moisture damage from freezing on the walls or overhangs (under eaves) and use protective measures such as shielding (step 7 below) and/or sealing cracks, seams, and joints (step 1 above), but extend the area of sealing to a minimum of 6 ft.
7. On brick or masonry surfaces, use a rust-resistant shield (18 in^2) behind the vent. If a vertical rise is used, the shield must extend 9 inches above and 9 inches below, as shown in Figure 11°–16. The shield can be wood, plastic, sheet metal, etc.
8. Do not locate the vent too close to shrubs, because condensate may stunt or kill them.
9. Calk all cracks, seams, and joints within 3 ft of the vent.

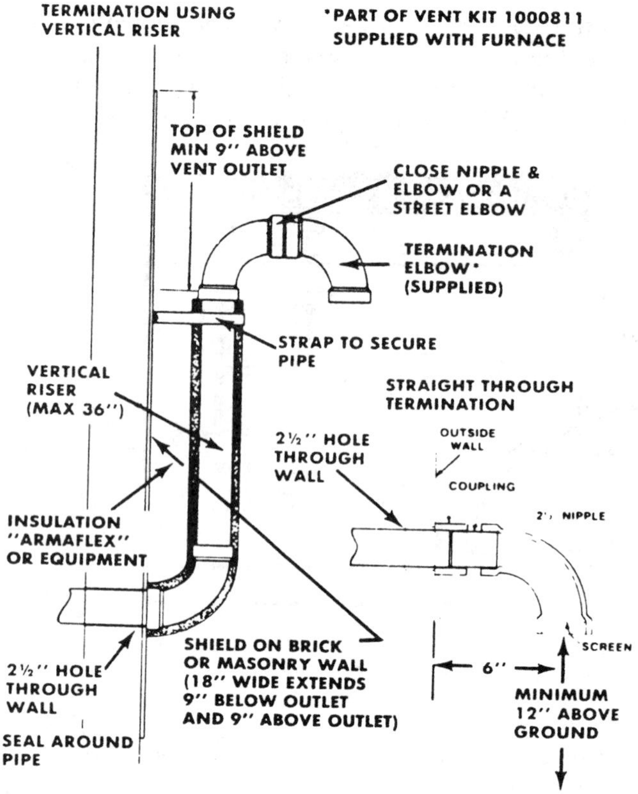
TERMINATION USING VERTICAL RISER
*PART OF VENT KIT 1000811 SUPPLIED WITH FURNACE
TOP OF SHIELD MIN 9" ABOVE VENT OUTLET
CLOSE NIPPLE & ELBOW OR A STREET ELBOW
TERMINATION ELBOW* (SUPPLIED)
STRAP TO SECURE PIPE
VERTICAL RISER (MAX 36")
STRAIGHT THROUGH TERMINATION
2½" HOLE THROUGH WALL
OUTSIDE WALL
COUPLING
2½ NIPPLE
INSULATION "ARMAFLEX" OR EQUIPMENT
SCREEN
SHIELD ON BRICK OR MASONRY WALL (18" WIDE EXTENDS 9" BELOW OUTLET AND 9" ABOVE OUTLET)
6"
2½" HOLE THROUGH WALL
MINIMUM 12" ABOVE GROUND
SEAL AROUND PIPE

FIGURE 11–16
Typical vent piping. [*Courtesy of Inter-City Products, Inc.* (*USA*).]

Vent termination. The vent termination elbow must be installed as shown in Figure 11–16.

1. Cut a 1 ½-inch-diameter hole through the exterior wall. Do not make the hole oversized, or it will be necessary to add a sheet of metal or plywood plate on the outside with the correct size hole in it. Check the hole size by making sure it is smaller than the coupling or elbow that will be installed on the outside. The coupling or elbow must prevent the pipe from being pushed back through the wall.
2. Extend the vent pipe through the wall ¾ to 1 inch and seal the area between the pipe and the wall.

STRAIGHT-THROUGH TERMINATION (NO VERTICAL RISER)

3. Install the coupling, 2 ½-inch-long nipple, and termination elbow as shown in Figure 11–16.

TERMINATION USING EXTERIOR RISER

4. Install the elbows and vent pipe (maximum 36 inches long) to form a rise as shown in Figure 11–16.

5. Secure the bent pipe to the wall with a galvanized strap or other rust-resistant material to restrain the pipe from moving.
6. Insulate the pipe with Armaflex or equivalent moisture-resistant, closed-cell foam insulation. If situations require that the pipe be run on the exterior of the wall to reach a suitable termination location, it must be properly insulated. It must be boxed in and sealed against moisture if fiberglass insulation is used.

Gas Supply and Piping

The American Gas Association (AGA) rating plate is stamped with the model number, type of gas, and gas input rating. Incorrect installation leads to danger of property damage, bodily injury, or death. Make sure that the furnace is equipped to operate on the type of gas available. Models designed as natural gas are to be used with natural gas only.

Furnaces designated for use with liquified petroleum (LP) gas have orifices sized for commercially pure propane gas. They must not be used with butane or a mixture of butane and propane unless properly sized orifices are installed by a licensed LP installer.

Gas supply. The recommended gas supply pressures are 7 inches water column pressure for natural gas and 11 inches water column pressure for LP gas. A maximum gas supply pressure of 14 inches water column should not be exceeded in either gas. A minimum gas supply pressure of 4 $^1/_4$ inches water column for natural gas and 11 inches water column for LP is required for the purpose of input adjustment, and it should not be allowed to vary downward, because this will decrease the input to the unit.

The gas input to the burners must not exceed the rated input shown on the rating plate. On natural gas the manifold pressure should be 3 $^1/_2$ inches water column. The manifold pressure should be 10 inches water column for LP gas. For operation at altitudes above 2,000 ft, an orifice change or a manifold pressure adjustment may be required to suit the gas being supplied. Check with the gas supplier.

For elevations over 2,000 ft, the furnace should be derated based on the standard input rating. Do not base on the alternate input rating.

Orifice sizes. Make certain that the unit is equipped with the correct main burner and pilot burner orifices.

Factory-sized orifices for natural gas and LP gas are generally listed in the manufacturer's literature.

Gas piping. The gas pipe supplying the furnace must be properly sized to handle the combined appliance loads, or it must run directly from the gas meter (natural gas) or LP gas second-stage regulator and supply only the furnace. It must be the correct size for the length of run and furnace rating. The length of pipe or tubing should be measured from the gas meter for natural gas or the LP gas second-stage regulator, which is usually just outside the building wall.

Determine the minimum pipe size from the proper table, basing the length of run from the main line, or source to the furnace (Tables 11–1 and 11–2). Use the correct size of pipe. Piping that is too small will not allow enough gas to reach the furnace and will reduce the heat output of the furnace.

Check the gas line installation for compliance with the local codes.

TABLE 11–1 Gas pipe sizes/capacity, natural gas [*Courtesy of Inter-City Products, Inc. (USA).*]

Length of Pipe - Ft	Capacity-Btuh Per Hour Input		
	1/2″	3/4″	1″
20′	92,000	190,000	350,000
40	63,000	130,000	245,000
60	50,000	·105,000	195,000

TABLE 11–2 Gas tubing and pipe sizes, LP gas [*Courtesy of Inter-City Products, Inc. (USA).*]

Length In Feet	Capacity-Btuh Per Hour Input			
	1/2″**	3/4″**	1/2″	3/4″
20′	62,000	216,000	189,000	393,000
40′	41,000	145,000	129,000	267,000
60′	35,000	121,000	103,000	217,000

Connecting the gas piping. Refer to Figure 11–17 for the general layout of the furnace. It shows the basic fittings you will need. The following rules apply:

1. Use black iron or steel pipe and fittings or other piping that is approved by local codes.

    a. If a gas connector is used, it must be acceptable to the local authority. The connector may not be used inside the furnace or be secured or supported by the furnace or ductwork. The connectors should comply with one of the following standards or a superceding standard: ANSI Z 21.24a, 1983, Metal Connectors for Gas Appliances; or ANSI Z 21.45b, 1983, Flexible Connectors of Other Than All-Metal Construction for Gas Appliances.

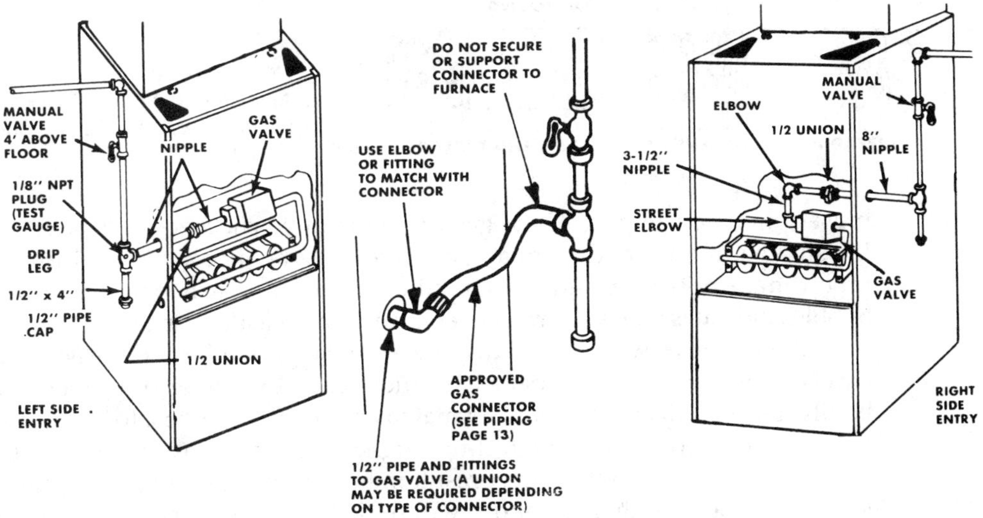
MANUAL VALVE 4' ABOVE FLOOR
1/8" NPT PLUG (TEST GAUGE)
DRIP LEG
1/2" x 4"
1/2" PIPE CAP
NIPPLE
GAS VALVE
1/2 UNION
LEFT SIDE ENTRY
DO NOT SECURE OR SUPPORT CONNECTOR TO FURNACE
USE ELBOW OR FITTING TO MATCH WITH CONNECTOR
APPROVED GAS CONNECTOR (SEE PIPING PAGE 13)
1/2" PIPE AND FITTINGS TO GAS VALVE (A UNION MAY BE REQUIRED DEPENDING ON TYPE OF CONNECTOR)
MANUAL VALVE
ELBOW
1/2 UNION
8" NIPPLE
3-1/2" NIPPLE
STREET ELBOW
GAS VALVE
RIGHT SIDE ENTRY

FIGURE 11–17
Proper piping practice. [*Courtesy of Inter-City Products, Inc. (USA).*]

2. Use pipe joint compound on the male threads only. The pipe joint compound must be resistant to the action of the LP gases (Figure 11–18).
3. Use ground joint unions.
4. Install a drip leg to trap dirt and moisture before it can enter the gas valve. The drip leg must be a minimum of 3 inches long.
5. Use two pipe wrenches when making the connections to the valve to keep it from turning.
6. Provide a $\frac{1}{8}$-inch National Pipe Thread (NPT) plug for a test gauge connection immediately upstream of the gas supply connection for the furnaces.
7. Install a manual shutoff valve.
8. Tighten all joints securely.

ADDITIONAL LP GAS REQUIREMENTS

9. All connections made at the storage tank should be made by a licensed LP gas dealer.
10. An LP gas dealer should check all the lines and connections from the storage tank to the heating unit when the unit is connected to the storage tank.
11. Two-stage regulators should be used by the LP installer.
12. All gas piping should be checked out by the LP installer.

Checking the gas piping. Test all piping for leaks. When checking gas piping to the furnace, shut off the manual gas valve for the furnace. The gas pressure

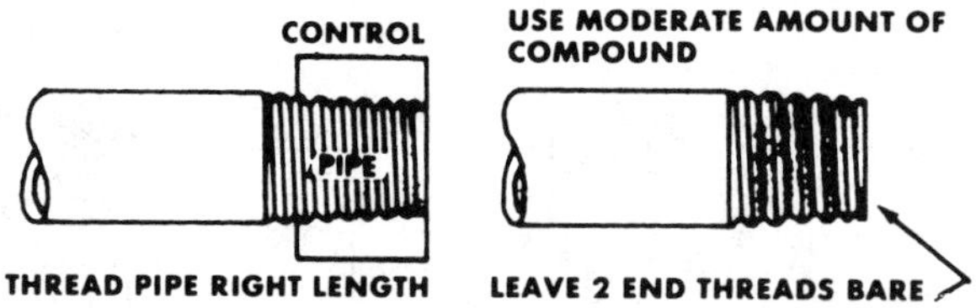
CONTROL
PIPE
THREAD PIPE RIGHT LENGTH
USE MODERATE AMOUNT OF COMPOUND
LEAVE 2 END THREADS BARE

FIGURE 11–18
Application of pipe joint compound. [*Courtesy of Inter-City Products, Inc. (USA).*]

must not exceed ½ psig. If the gas piping is to be checked with pressure above ½ psig, the furnace and manual gas valve must be disconnected during the test procedure. Apply soap suds (or a liquid detergent) to each joint. The formation of bubbles indicates a leak. Correct even a very small leak at once.

If the orifices were changed, make sure the pilot tube and burner orifices are checked for leakage. Improper installation leads to danger of property damage, bodily injury, or death. Never use a match or open flame to test for leaks. Never exceed the specified pressure for testing. Higher pressures may damage the gas valve and cause overfiring, which may result in heat exchanger failure. Liquified petroleum gas is heavier than air and it will settle in any low area, including open depressions, and will remain there unless the area is properly ventilated.

Never attempt to start up the unit before thoroughly ventilating the area.

Electrical Wiring

Improper installation leads to danger of bodily injury or death. Turn off the electric power at the fuse box or service panel before making any electrical connections. A grounded connection must be completed before making the line voltage connections. All line voltage connections must be made inside the furnace junction box. All electrical work must conform to the National Electrical Code ANSI/NFPA, No. 70, 1984, or current edition.

Grounding. A green wire pigtail is generally installed for the ground connection. Use an insulated copper conductor (No. 14 AWG) from the unit to a ground connection in the electric service panel or a properly driven and electrically grounded ground rod.

Electric power supply. The line voltage section is completely factory wired. It is only necessary to run No. 14 AWG hot, neutral, and ground wires from the power supply circuit (15 A) through a disconnect switch (if required by codes) to furnish power to the unit. Do not connect to existing lighting or other circuits.

Do not complete the line voltage connections until the unit is permanently grounded. All line voltage connections and the ground connections must be made with copper wire (Figure 11–19).

Optional equipment wiring. All wiring (except the thermostat) from the furnace to the optional equipment, such as humidifiers or electronic air cleaners, or between optional equipment, must conform to the temperature limitations for type T wire and be installed in accordance with the manufacturer's instructions supplied with the equipment.

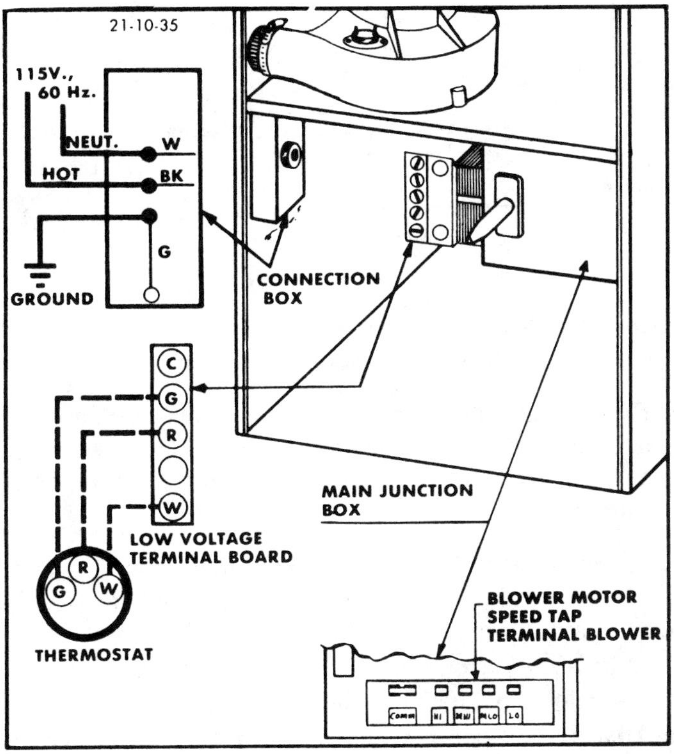
21-10-35
115V., 60 Hz.
NEUT.
W
HOT
BK
G
GROUND
CONNECTION BOX
C
G
R
W
MAIN JUNCTION BOX
LOW VOLTAGE TERMINAL BOARD
R
G
W
THERMOSTAT
BLOWER MOTOR SPEED TAP TERMINAL BLOWER

FIGURE 11–19
Electrical connections. [*Courtesy of Inter-City Products, Inc.* (*USA*).]

Humidifier/electronic air cleaner. The power connection for a humidifier or electronic air cleaner must be made through a sail switch, installed in the ductwork, if the furnace has a single-pole, double throw (SPOT) fan relay with only three terminals.

If the manufacturer does not supply a sail switch, consult the place of purchase.

If the furnace has a double-pole, double throw (DPDT) fan relay with six terminals, the power connections can be made to the furnace fan relay (Figure 11–20).

With these connections and those shown in Figure 11–19, the humidifier will be powered when the furnace is fired and the circulating blower comes on. The

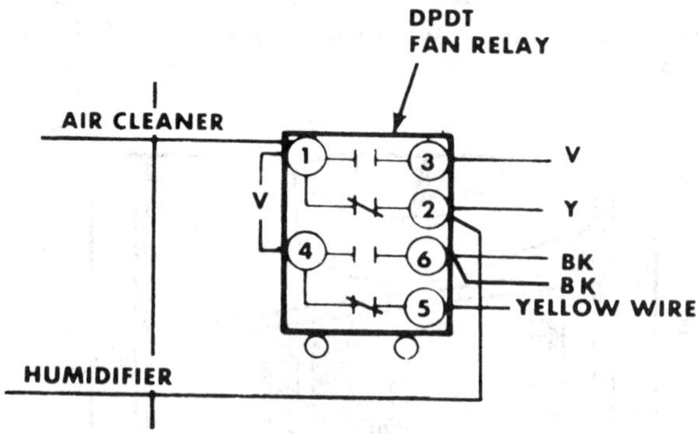
DPDT
FAN RELAY
AIR CLEANER
V
1
3
2
4
6
5
V
Y
BK
BK
YELLOW WIRE
HUMIDIFIER

FIGURE 11–20
Humidifier and air cleaner connections. [*Courtesy of Inter-City Products, Inc. (USA).*]

electronic air cleaner will be powered anytime the circulating air blower is on, whether for heating, cooling, or air circulation.

Blower speeds. When it is necessary to change the blower speed, the manufacturer's specifications for the particular furnace in question should be consulted. This procedure will save time and money and perhaps prevent damage to the equipment.

Thermostat. The location of the thermostat has an important effect on the operation of the unit. Follow the instructions that are included with the thermostat for correct mounting and wiring.

Heat anticipator. Set the heat anticipator in accordance with the thermostat instructions to the values shown on the gas valve or the manufacturer's data.

Thermostat connection, heating only. Connect the two wires from the thermostat to terminals R and W on the transformer low voltage terminal board. If the thermostat has a fan on switch, it will connect to terminal G (Figure 11–19).

Adding air conditioning. Use the following general steps when adding air conditioning to the unit:

1. Obtain a heating-cooling thermostat and four-wire thermostat cable. Replace the existing thermostat and cable. Connect the wires to Y, W, G, and R on the low voltage terminal board and to Y, W, G, and R on the thermostat.
2. The condensing unit will have a contactor in it. Connect its 24-V coil to terminals Y and C on the low voltage terminal board. It may be necessary to consult the manufacturer's data for connections other than these.
3. Follow all of the instructions packed with the condensing unit and the evaporator coil.

The furnace fan relay will now change speeds automatically as the thermostat is switched to heat or cool.

Ductwork and Filter

Improper installation leads to danger of bodily injury or death. The return air must not be drawn from inside the furnace closet or utility room. The return air duct must be sealed to the furnace casing.

Cool air from an evaporator passing over the heat exchanger may cause condensation to form inside the heat exchanger, causing failure to the heat exchanger.

The air distribution system should be designed and installed in conformation with manuals published by American Society of Heating, Refrigerating and Air-conditioning Engineers (ASHRAE) or other approved methods in conformance with local codes and good accepted trade practices.

When a furnace is installed so that the supply air ducts carry air circulated by the furnace to an area outside the space containing the furnace, the return air must also be handled by a duct or ducts sealed to the furnace casing and terminating outside the space containing the furnace. This is to prevent drawing possibly hazardous combustion products into the circulated air.

When air conditioning is installed with the furnace, the air conditioning cooling coil (evaporator) must be on the outlet side of the furnace, or any evaporator and blower can be separate from the furnace. This means the same duct would be used but the air will go around the furnace during the cooling cycle. With a separate blower and evaporator, the dampers must seal properly for good air flow control. Chilled air going through the furnace could cause condensation and shorten the furnace life. The dampers can be either automatic or manually operated. If manually operated, they must be equipped with a means to prevent operation of either unit unless the damper is in the full heat or cool position. Purchase them locally.

Ductwork sizing. The existing or new ductwork must be sized to handle the correct amount of air flow for either heating only or heating and cooling.

Refer to the manufacturer's specifications for the equipment in question for the proper air flow characteristics.

Ductwork installation. Ductwork that is installed in attics or other areas that are exposed to the outside temperatures should be insulated with a minimum of 2 inches of insulation and have an indoor-type vapor barrier. Ductwork in other indoor, unconditioned areas should have a minimum of 1 inch insulation with an indoor-type vapor barrier.

Ductwork connections. The return air can enter through either both sides or through the bottom of most makes of furnaces.

By using an optional return air cabinet, the back side may also be used on some makes.

For side connections, cut out an area that is large enough to handle the filter size without making too large a hole in the casing. The procedure is generally outlined in the manufacturer's data.

Filter. The size and type of filter supplied with the furnace will handle the air flow through it if central air conditioning is used with the furnace.

If external filter grilles are used, filters that comply with the specifications in Table 11–3 should be used. The filter size and type must be adequate to handle the cfm requirements based on heating only or heating/cooling application. See the manufacturer's instructions for the cfm data. If the filters provided are suitable for heating applications only, be sure to advise the homeowner so that he or she is aware that the filter size will have to be increased if air conditioning is added.

TABLE 11–3 Remote filter sizes. [*Courtesy of Inter-City Products, Inc.* (*USA*).]

CFM AIRFLOW	DISPOSABLE TYPE FILTER LOW VELOCITY/300 FPM		CLEANABLE TYPE FILTER HIGH VELOCITY/500 FPM	
	Minimum Surface Area (Sq. In.)	Recommended Nominal Size	Minimum Surface Area (Sq. In.)	Recommended Nominal Size
800	384	20 X 25	231	14 X 20
900	432	20 X 25	260	15 X 20
1000	480	20 X 30	288	14 X 25
1100	528	20 X 30	317	15 X 25
1200	576	14 X 25 (2)	346	16 X 25
1300	624	14 X 25 (2)	375	20 X 25
1400	672	16 X 25 (2)	404	20 X 25
1500	720	16 X 25 (2)	432	20 X 25
1600	768	20 X 25 (2)	461	20 X 25
1700	816	20 X 25 (2)	490	20 X 30
1800	864	20 X 25 (2)	519	20 X 30
1900	912	20 X 30 (2)	548	24 X 25
2000	960	20 X 30 (2)	576	24 X 25

ELECTRIC FURNACES

Even though electric heating is not new, forced-air electric furnaces (Figure 11–21) have been available in large quantities for only a few years. Electric furnaces, although similar in appearance to gas furnaces, are quite different. They do not require gas piping or venting. They do, however, require that high capacity electric circuits be installed to the furnace (see Table 11–4). Electric furnaces are more flexible than gas furnaces because there is no specific position in which to install them. The only requirement is that all components be accessible for installation and service.

Sequence of Operation (Electric Furnace)

Although there are few differences in the operation of an electric furnace and a gas furnace, we will go through the sequence of operation for clarity. As the temperature in the conditioned area falls, the thermostat contacts close the electric control circuit. The control circuit energizes some relays, which completes the line voltage electric circuit to the heating elements. When sufficient time has passed or

FIGURE 11–21
Forced air electric furnace. (*Courtesy Weatherking Heating-Cooling.*)

TABLE 11–4 Ratings and specifications

Model Number	Btu	kW	Electrical Rating	Amps	Min. Wire Size	Number Circuits
EFS-04	13660	4	240/1/60	16.6	#8	1
EFS-05	17065	5	240/1/60	20.8	#8	1
EFS-08	27300	8	240/1/60	33.3	#6	1
EFS-10	34130	10	240/1/60	41.7	#6	2
EFS-12	40960	12	240/1/60	50.0	#6	2
EFS-15	51200	15	240/1/60	62.5	#6	2
ERS-16	54610	16	240/1/60	66.7	#6	2
EFS-20	68260	20	240/1/60	83.3	#4	2
EFS-24	81920	24	240/1/60	100.0	#4	3
EFS-25	85325	25	240/1/60	104.1	#4	3
EFS-30	102390	30	240/1/60	12.0	#4	3

(*Courtesy Dearborn Stove Co.*)

the temperature has risen, the fan control turns on the blower to circulate the air as with the gas furnace. In some installations the fan is energized when the first elements are energized. The limit controls on electric furnaces are on each individual heating element and respond to that temperature.

When the thermostat is satisfied, the control circuit is interrupted and the heating relays remove the elements from the line voltage circuit. After the required time has lapsed, or the temperature has fallen enough, the fan motor will stop. Thus, the heating unit is at rest.

Electric furnace diagram: There are many different ways to wire an electric furnace for the desired operation. Figures 11–22 and 16–29 show an example of one manufacturer's diagram.

BTU COMPARISON

Gas Furnaces

The Btu rating of gas furnaces is certified by the American Gas Association. The input rating per hour is placed on the name plate, a permanent part of the furnace. The input rating is always higher than the output or bonnet capacity, which is usually about 80% of the input rating of the furnace. This 20%, known as waste gas, is used to operate the venting system. Vent gas is not actually wasted; however, it is not used to heat the conditioned area and is, therefore, termed *waste heat.*

The input rating of gas furnace is the maximum amount of Btu input that can safely be allowed to that furnace. Most manufacturers design their equipment to operate close to these ratings. However, these furnaces can be operated safely on about 80% of the input rating. Any input less than this is dangerous because there will not be enough heat to operate the vent system properly. If the input Btu is increased above the name plate rating, two hazards exist: (1) The vent system cannot remove the extra gas properly; and (2) the extra heat at the burner head will tend to overheat it along with the heat exchanger, thus resulting in possible burn-out of both the burners and the heat exchanger. Both conditions are costly and hazardous.

Electric Furnaces

The Btu rating of electric furnaces is certified by the National Electrical Manufacturers Association (NEMA). Electric furnaces are rated by the kilowatt (kW) input per hour. The kilowatts are the converted to obtain the Btu rating (Table 11–5). Electric furnaces are virtually 100% efficient. That is, for each Btu input to the furnace, almost 1 Btu of heat is delivered to the conditioned area. No heat is used for vent operation. The Btu rating of an electric furnace can be changed by adding or removing heating elements to meet the requirements. These elements may be added as long as there is a space available to install them.

TEMPERATURE RISE

The temperature rise through a furnace is the rise in temperature of the air as it passes through the heating elements. The Btu input and the air delivery of the blower are the major factors determining the temperature rise.

The temperature rise in gas furnaces will normally be rated from 60° to 100° F, depending on the unit design conditions. It is usually stated on the furnace name plate. Thus, the discharge air temperature of a heating furnace warming an 80° F area will be 80° F + 60° F = 140° F minimum.

WIRING DIAGRAM

MODELS 020H & 023H

WARNING

Pull out furnace safety disconnect from unit control panel door before servicing the furnace.
Not suitable for use on systems exceeding 120 volts to ground.
Ne convient pas aux installations de plus de 120 volts a la terre.

NOTES:

1. See Furnace Data Label for recommended wire sizes.
2. Single (1) supply circuit required for Canadian installations (see installation instructions)
3. Thermostat anticipator: (0.40 amps).
4. Use Class "K" or "RK" replacement fuses only (60 amp required).
5. Secure 24V pigtails connections (to thermostat) with wire nuts and tape with approved electrical tape.
6. 24V pigtails are not used with relay control. See wiring diagram on Relay for 24V connections.
7. Connect Blower Plug to receptacle on Relay Control (if installed). Refer to wiring diagram on Relay.
8. Type 105C thermoplastic or equivalent must be used if replacing any original unit wiring.

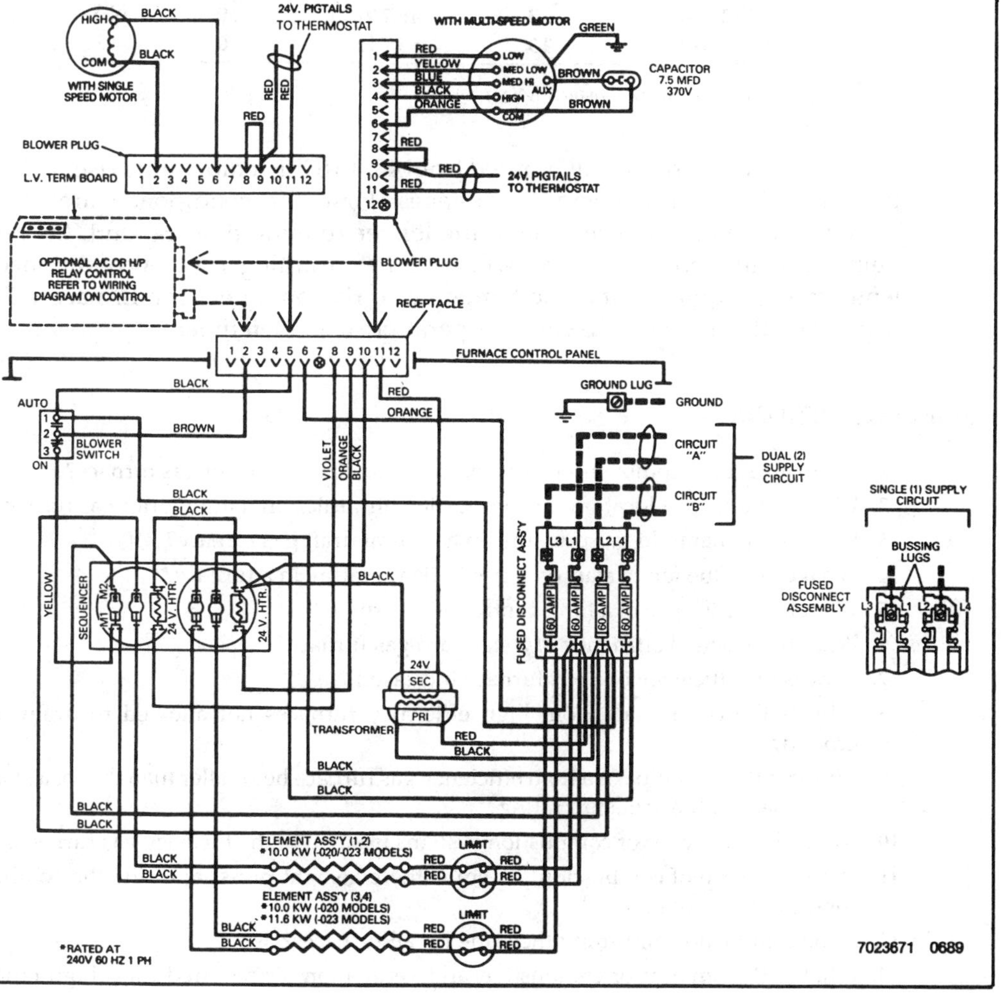

24V. PIGTAILS TO THERMOSTAT
WITH MULTI-SPEED MOTOR
WITH SINGLE SPEED MOTOR
HIGH
COM
BLACK
RED
YELLOW
BLUE
ORANGE
BROWN
GREEN
LOW
MED LOW
MED HI
AUX
COM
CAPACITOR 7.5 MFD 370V
BLOWER PLUG
L.V. TERM BOARD
24V. PIGTAILS TO THERMOSTAT
OPTIONAL A/C OR H/P RELAY CONTROL REFER TO WIRING DIAGRAM ON CONTROL
RECEPTACLE
FURNACE CONTROL PANEL
GROUND LUG
GROUND
AUTO
ON
BLOWER SWITCH
VIOLET
CIRCUIT "A"
CIRCUIT "B"
DUAL (2) SUPPLY CIRCUIT
SINGLE (1) SUPPLY CIRCUIT
BUSSING LUGS
FUSED DISCONNECT ASSEMBLY
FUSED DISCONNECT ASS'Y
60 AMP
YELLOW
SEQUENCER
24 V. HTR.
24V
SEC
PRI
TRANSFORMER
ELEMENT ASS'Y (1,2)
*10.0 KW (-020/-023 MODELS)
LIMIT
ELEMENT ASS'Y (3,4)
*10.0 KW (-020 MODELS)
*11.6 KW (-023 MODELS)
*RATED AT 240V 60 HZ 1 PH
7023671 0689

FIGURE 11–22
Electric furnace wiring diagram. (*Courtesy NORDINE.*)

TABLE 11–5 Btu/kW rating

Model Number	kW @ 240 V Nom.	Btu @ 240 V Nom.	Total 1-Ph Amp.	Max 3-Ph Amp.
PT-2-1-03	3	10,681	16	NA
PT-2-1-04	4	14,244	22	NA
PT-2-1-05	5	17,805	26	NA
PT-2-1-06	6	21,365	29	NA
PT-2-1-08	8	28,487	39	NA
PT-2-1-10	10	35,610	48	40
PT-2-1-12	12	42,730	54	44
PT-2-1-13	13	46,292	60	NA
PT-2-1-15	15	53,415	70	40

(Courtesy Electric Products Mfg. Co.)

An electric furnace will be rated from 40° to 80° F temperature rise. This lower temperature rise has several advantages. First, the conditioned area temperature will be more uniform due to the longer running time. Second, the relative humidity of the conditioned area will be higher, resulting in lower conditioned area temperature requirements. The temperature rise on an electric furnace will also depend on the number of elements operating at a given time.

REVIEW QUESTIONS

1. How much total combustion air is required for a natural draft gas furnace?
2. What is the correct location for combustion air grilles on natural draft gas furnaces?
3. What is the manifold gas pressure on a natural draft gas furnace?
4. What causes the fan to start on a rise in temperature in a furnace?
5. What is the purpose of a limit control on a furnace?
6. What is the actual efficiency of a standard gas furnace?
7. How is the efficiency of a gas furnace increased?
8. Why is the condensate from high efficiency furnaces not allowed to drain on the ground?
9. Why can the vent pipe of a high efficiency gas furnace be smaller than that of a standard gas furnace of the same Btu rating?
10. Name the two types of combustion systems used on high efficiency gas furnaces.
11. On which type of combustion system is there a slight pressure inside the combustion zone and the vent system?
12. Is the combustion air setting the same for all furnaces?
13. What is the amount of combustion and ventilation air required for a high efficiency furnace?

14. What must the combustion air be free of?
15. Why are condensing gas furnaces so efficient?
16. Where must the drain trap assembly on a condensing gas furnace be installed during freezing weather?
17. What precaution must be observed when installing a condensate pump on a condensing gas furnace?
18. What is the minimum gas supply pressure to a natural gas furnace?
19. What is the purpose of a drip leg in gas piping?
20. How should a thermostat heat anticipator be set?
21. To what does the limit control respond on an electric furnace?
22. Who certifies the Btu rating of electric furnaces?
23. What does the input rating of a gas furnace indicate?
24. What are the major factors that determine the temperature rise through a furnace?
25. What is the normal temperature rise of an electric furnace?

12

infrared heating

The objectives of this chapter are:

- To instruct you in the theory of infrared heating.
- To acquaint you with different methods used in infrared heating.
- To bring to your attention the desired location of infrared heating units.
- To introduce you to how space heating is accomplished with low air temperatures.

Large open buildings and areas exposed to strong drafts, such as shipping docks, can create many problems for the heating contractors responsible for installing the equipment used for human comfort and for the protection of materials and equipment.

When conditions such as these exist, infrared heating will often provide the desired solution. Infrared heating is a form of radiation that resembles and behaves like light rays. Its invisible rays race through space at the speed of light to warm solid objects such as floors, equipment, and people, before the air is heated.

Like the sun, heat generated by an infrared energy source is absorbed rapidly by solid objects, which in turn release heat to the air around them.

INFRARED THEORY

Heat may be transmitted from one body to another without altering the temperature of the intervening medium. That this is true may be proven by a simple, everyday experience. If you stand before a fire or a hot radiator, you experience a sensation of warmth that is not due to the temperature of the air. This is evident if a screen is placed between you and the fire: The sensation immediately disappears. Such would not be the case if the air had a higher temperature.

This phenomenon, which is called thermal radiation, is but one of the many forms of radiant heat energy that is continuously being emitted and absorbed in various degrees by all bodies. All radiant energy may be regarded as a form of wave motion known as *electromagnetic phenomena.* This type of wave motion should not be confused with sound waves, spring vibrations, and other elastic mechanical waves that may occur in solids, liquids, and gases only. Radiant energy waves may be transmitted even through a vacuum.

Infrared radiation is a form of energy that is propagated as an electromagnetic wave, like radio waves, light waves, and X-rays. The energy is transmitted at a velocity of 186,272 mps; the common source of infrared radiation is a hot radiating body. While being transmitted this way, energy is called *radiant energy* and is not heat. The energy will be instantaneously converted to heat on being absorbed by an absorbing medium.

A given body under a fixed set of conditions will emit radiation of various wavelengths. The intensity of the radiation in the various wavelengths is different. The type of radiation emitted is characterized by the band of wavelengths having the greatest intensity. The curves in Figure 12–1 show the distribution of intensity of radiation with wavelengths for a *black body.* (A black body is one that emits the maximum possible radiation at a given temperature. The adjective has nothing to do with the color of the body.) Table 12–1 provides the same information in table form. An examination of this table or of the curves shown in Figure 12–1 would result in the following conclusions:

1. An increase in temperature causes a decrease in the wavelength at which maximum energy emission occurs.
2. An increase in temperature causes a rapid increase in energy emission at any given wavelength and the total energy emission.
3. The total rate of energy emission at any given temperature and for any range of wavelengths is given by the area under the curve for that temperature taken over the wavelength range being considered.

The differentiation among various types of radiation, such as light radiation, thermal radiation, etc., is rather indefinite since radiation of all wavelengths will heat bodies to some extent.

In the use of high intensity gas-fired infrared generators, we will be primarily concerned with the infrared portion of the spectrum (Figure 12–2). A continuous interchange of energy among bodies results from the reciprocal process of radiation and absorption. Thus, if two bodies are at different temperatures within an enclosure, the hotter body receives from the colder body less energy than it radiates. Consequently, its temperature decreases. The colder body receives more energy than it radiates, and its temperature increases.

This interchange of energy continues even after thermal equilibrium is reached, except that both bodies then receive as much energy as they radiate. According to this theory, which agrees well with observations, a body would

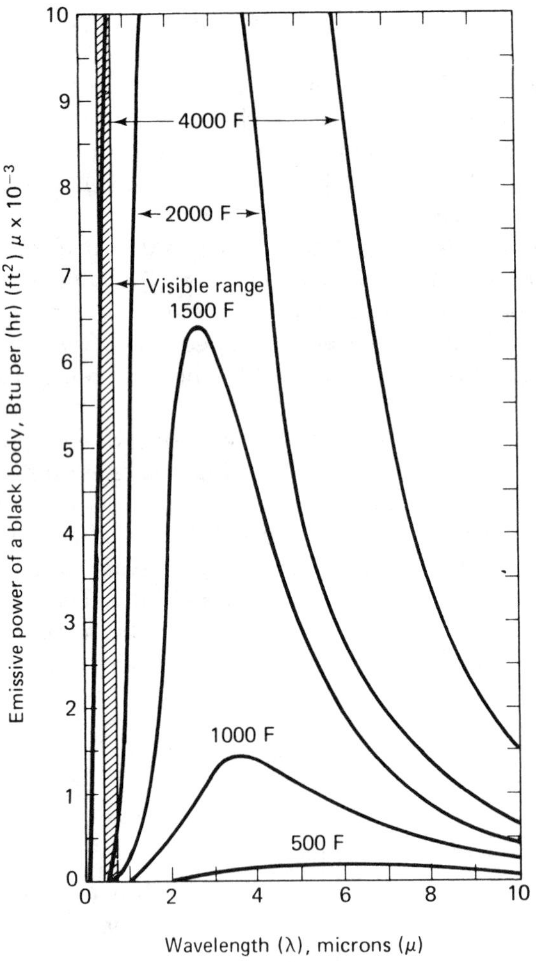

10
9
8
7
6
5
4
3
2
1
0
0
2
4
6
8
10
Emissive power of a black body, Btu per (hr) (ft²) μ x 10⁻³
Wavelength (λ), microns (μ)
4000 F
2000 F
Visible range
1500 F
1000 F
500 F

FIGURE 12–1
The emissive power generated by a black body. *(Courtesy Thermal Engineering Corp.)*

cease to emit thermal radiation only when its temperature had been reduced to absolute zero.

The higher the temperature of an emitting body, the faster it radiates infrared energy. Obviously, the amount of energy radiated per unit of time will also be proportional to the amount of exposed surface.

Relationships involving thermal radiation are expressed by a number of radiation formulas. Some have been deduced theoretically and have been proven true by experiment, whereas others are purely empirical. Only the equation that relates to the emission of energy from infrared heaters will be discussed here. However, to understand these formulas, their uses, and their limitations, it is necessary to understand some of the definitions and phenomena of radiation theory.

TABLE 12–1 Black body thermal radiation—Btu/hr-ft^2
(1) In various wavelength bands (quality); (2) Total thermal radiation (quantity); (3) Wavelength at maximum radiation

Source Temp °F	(1) Wave Length Band—Microns 0–0.7	0.7–1	1–2	2–3	3–4	4–5	5–6	6–8	8–10	10+	(2) Total Thermal Rad.	(3) Wave Length At Max. Rad. Microns
500			1	27	99	155	174	308	215	482	1460	5.43
600			3	70	201	278	285	442	300	590	2170	4.92
700			12	162	376	445	414	620	376	705	3110	4.50
800			32	325	620	672	575	809	468	840	4340	4.14
900			76	576	963	952	752	1035	565	960	5880	3.84
1000			174	960	1400	1305	975	1250	660	1095	7820	3.57
1100			316	1525	1982	1628	1245	1510	765	1220	10,190	3.34
1200		1	556	2250	2673	2053	1518	1763	876	1370	13,060	3.14
1300		3	973	3200	3424	2550	1830	2010	1025	1483	16,500	2.96
1400		8	1564	4404	4380	3090	2059	2340	1095	1650	20,590	2.80
1500		18	2388	5960	5335	3600	2415	2640	1245	1780	25,380	2.66
1600		34	3557	7720	6410	4340	2715	2910	1365	1920	30,970	2.53
1700		71	5059	9840	7570	5050	3035	3215	1540	2060	37,440	2.41
1800		126	7040	21,310		5525	3450	3590	1620	2200	44,870	2.31
1900	5	225	9550	25,440		6300	3840	3890	1760	2350	53,360	2.21
2000	12	13,090		29,935		11,210		4220	1965	2455	62,990	2.12
2500	157	43,125		58,650		17,300		6015	2500	3155	131,200	1.76
3000	1281	109,660		96,850		23,900		7640	3205	3940	246,400	1.51
4000	20,415	405,560		194,620		38,110			21,780		680,500	1.17
5000	129,900	1,005,400		311,750		51,950			29,935		1,528,000	0.96

(*Courtesy Thermal Engineering Corp.*)

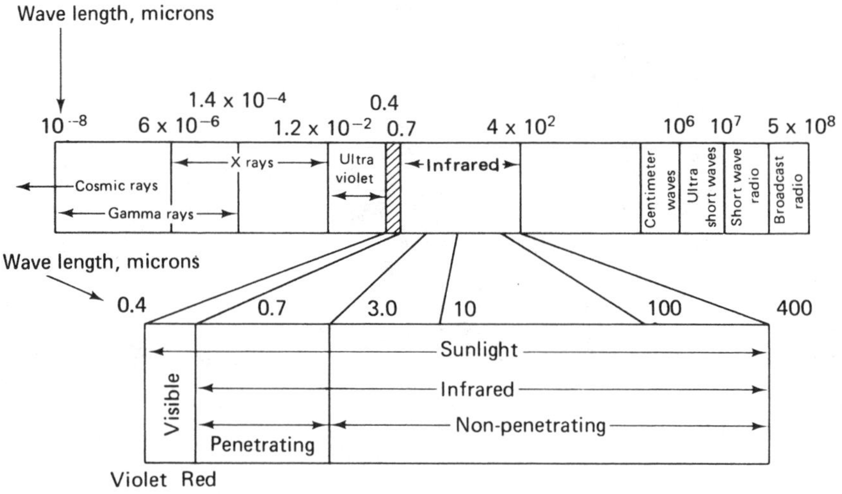
Wave length, microns
1.4 x 10−4
0.4
10−8
6 x 10−6
1.2 x 10−2
0.7
4 x 102
106
107
5 x 108
X rays
Ultra violet
Infrared
Cosmic rays
Gamma rays
Centimeter waves
Ultra short waves
Short wave radio
Broadcast radio
Wave length, microns
0.4
0.7
3.0
10
100
400
Sunlight
Infrared
Non-penetrating
Penetrating
Visible
Violet Red

THERMAL RADIATION, which includes both visible and infrared waves, is but a part of the spectrum of the family of electromagnetic waves. The quantity and quality of thermal radiation depends solely upon the temperature of the surface emitting the radiation. All electromagnetic waves are transmitted at the speed of light. All heated solids and liquids, and some gases emit thermal radiation. Conversely, these materials absorb a part of the visible and infrared waves and convert them to heat. At surface temperatures greater than 1000 F, a part of the radiation is visible

FIGURE 12–2
Thermal radiation. *(Courtesy Thermal Engineering Corp.)*

Absorptivity and Emissivity

Absorptivity is defined as the ratio of radiant energy by an actual surface at a given temperature to that absorbed by a black body at the same temperature. A black body can also be defined as one that will absorb all incidental radiation. In nature, there is no truly black body for thermal radiation; however, the black body concept is useful in comparing the absorption and emission of radiant energy by different materials.

The *emissivity* of a surface is the ratio of the energy it emits to that emitted by a black body at the same temperature. The emissivity of most materials increases with the temperature of the material.

In nature, absorptivity and emissivity for most materials are approximately equal and have a value of less than 1.0. This is true for most materials at temperatures normally encountered. However, absorptivity depends on the quality of the incident radiation as well as the surface temperature. If the incident radiation is from a high source such as the sun, the emissivity and absorptivity may differ markedly. In space heating applications, the quality of the radiation is an important factor. For this reason high intensity gas-fired heaters operate at a surface temperature that allows the energy to be emitted at wavelengths that will be readily absorbed by most materials encountered in space, spot, or process heating applications.

Total Energy Emission

Infrared energy emitted by a black body depends on its absolute temperature only and can be simply computed by use of the Stefan-Boltzman law, which states:

$$E_{bb} = aT^4$$

where

E_{bb} = total thermal radiation emitted by a black body

a = Stefan-Boltzman constant

= 0.1713×10^{-8}

T = absolute temperature °R.

Column 2 of Table 12–1 provides E_{bb} for a range of temperatures from 500° to 5,000° F (960° to 5460° R).

The foregoing has all been related to a perfect radiator. Since all infrared devices are less than perfect, some modification of the black body laws is needed to provide good engineering estimates of the thermal radiation emitted by the device.

By including the emissivity of the actual emitter and its effective area in the Stefan-Boltzman law, a good approximation of the thermal radiation emitted can be made. The formula then becomes:

$$E_{bb} = AeaT^4$$

where

E_{ac} = total thermal radiation emitted by the actual emitter surface

A = effective or solid area of the emitter surface

e = emissivity of the emitter surface

a = Stefan-Boltzman constant of 0.1713×10^{-8}

T = absolute temperature of the emitter surface °R.

The Thermal Engineering Corp. has developed a radiation calculator (Table 12–2), which can be used instead of the foregoing calculation. The dashed heavy line shows the total thermal radiation that can be expected from the TEC B45 or PB45 ceramic surface (Figure 12–3).

The burner used in these or any other models is designed to produce a ceramic surface temperature of 1,739° F.

The overall area of a TEC B45 burner is 156 in^2. However, the slots occupy 29% of the total area; therefore, the effective area or solid area is 0.71×156, or 111 in^2. A scale for the solid area percentage is included on the Thermal Engineering Corp.'s *Thermal Radiation Calculator* (Table 12–2). Emissivity of the Thermal Engineering Corp. ceramic is approximately 0.90.

TABLE 12–2 Thermal radiation calculator

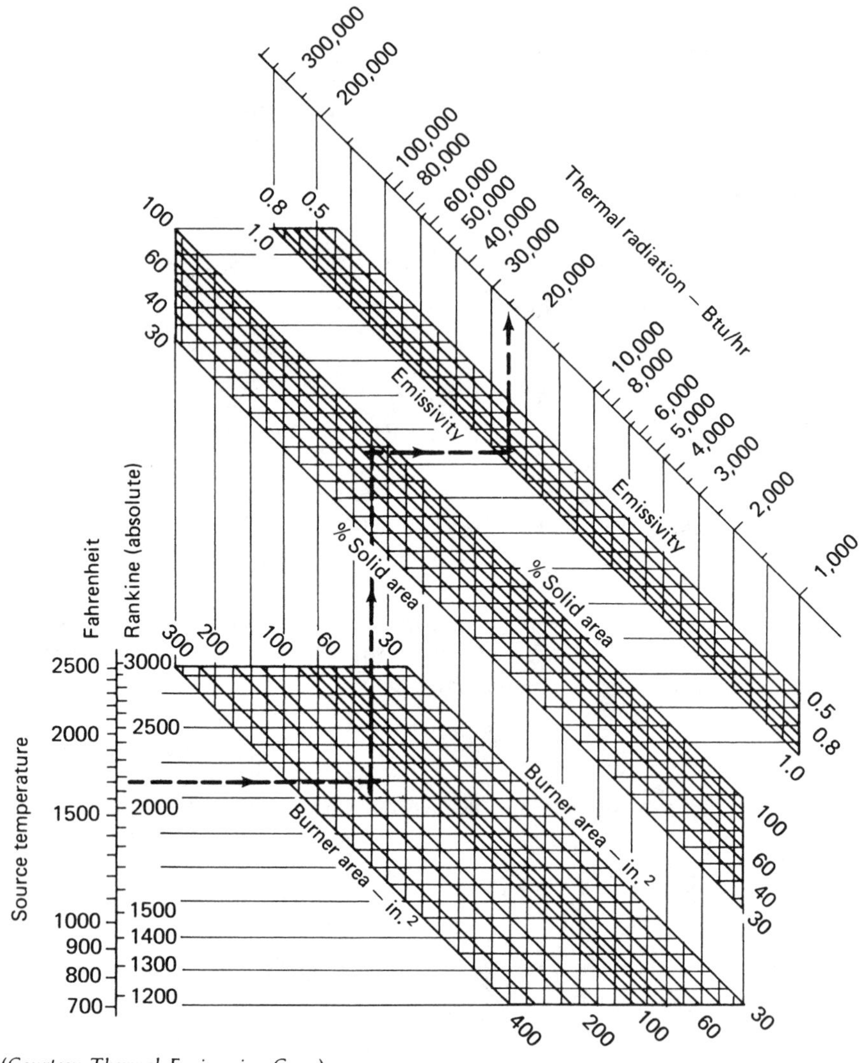
Thermal radiation – Btu/hr
300,000
200,000
100,000
80,000
60,000
50,000
40,000
30,000
20,000
10,000
8,000
6,000
5,000
4,000
3,000
2,000
1,000
Emissivity
0.8
0.5
1.0
Emissivity
0.5
0.8
1.0
% Solid area
100
60
40
30
% Solid area
100
60
40
30
Burner area – in.2
300
200
100
60
30
Burner area – in.2
400
200
100
60
30
Source temperature
Fahrenheit
Rankine (absolute)
2500
2000
1500
1000
900
800
700
3000
2500
2000
1500
1400
1300
1200

(Courtesy Thermal Engineering Corp.)

If the surface temperature of the actual emitter has been taken as the direct reading of an optical pyrometer, no correction need be made for emissivity.

(Note: An optical pyrometer reading must be corrected for the emissivity of the surface being measured. Therefore, if the direct reading is taken, the emissivity can be taken as 1.0 and the actual emissivity need not be known.)

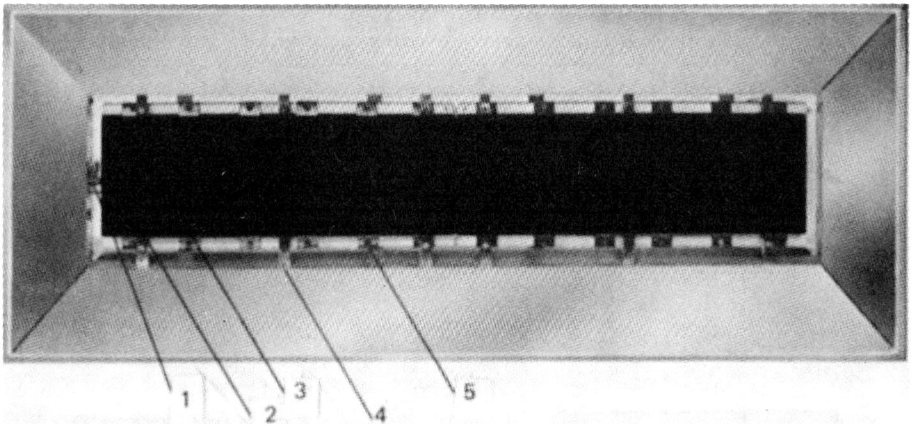
1
2
3
4
5

FIGURE 12–3
Infrared burner unit. 1. Pilot burner assembly. 2. Ceramic burner surface. 3. Full nichrome screen. 4. Screen mounting bracket. 5. Ceramic assembly mounting clip. (*Courtesy Thermal Engineering Corp.*)

Infrared Energy Exchange

To determine the actual energy transferred between two objects, additional factors must be considered. One is the emissivities of both objects and the other is the portion of energy leaving one object that strikes the other. Because of the difficulty in obtaining these factors, no attempt will be made here to describe their determination and use.

The Thermal Engineering Corp. has used these principles in determining energy deliveries of each heater, and these data appear in Tables 12–3 and 12–4.

SPACE HEATING WITH GAS-FIRED INFRARED HEATERS

An intervening heat transfer medium is not required in the transfer of energy by radiation. This phenomenon allows high intensity radiant heaters to be employed for space heating very effectively. Three methods exist by which heat may be transferred: conduction, convection, and radiation.

In the conventional methods of space heating, air is used as the intervening medium to transfer the energy from its source to the people or objects to be heated. In such a system, convection and, to some lesser degree, radiation are employed. The energy is transmitted from the flame by conduction, radiation, and convection through a heat exchanger, generally to air or water. The energy is then distributed to the area to be heated by the use of convection.

Air currents passing over an individual subject at temperatures less than body temperatures will cause a cooling sensation to be experienced by the individual subject. Consequently, when convection means are employed to heat a

TABLE 12–3 Energy transfer between objects
Btu hr/ft² delivery of B30 (30° angle) to a vertical surface 1 ft × 1 ft (midpoint 3 ft above floor).

h FT	*d* FT 2	4	6	8	10	12	14	16	18	20
7	125	119	81	52	35	25	18	14	12	9
9	44	61	54	42	31	23	18	14	12	9
11	19	28	33	30	25	20	16	13	10	9
13	10	18	21	21	19	17	14	12	10	9
15	5	11	14	15	15	13	12	10	9	8
17	4	7	10	11	11	10	10	9	8	7
19	3	4	6	8	9	9	8	7	6	6
21	2	3	4	6	6	6	6	6	6	5
23	1	3	3	4	5	5	5	5	5	4

(Courtesy Thermal Engineering Corp.)

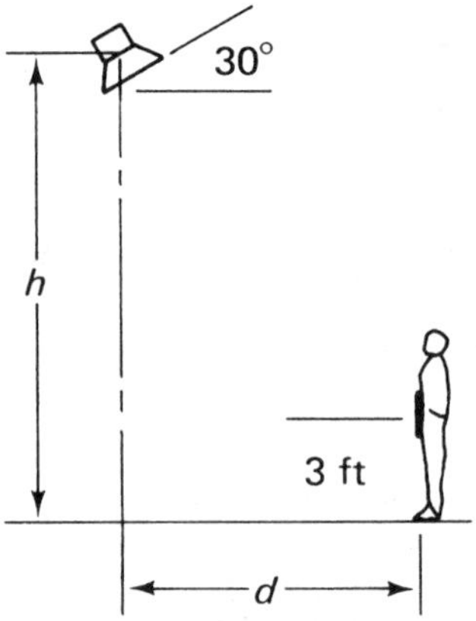
30°
h
3 ft
d

space, the air leaving the heat exchanger must be at a relatively high temperature in comparison to the body temperature. This is essential to prevent a chilling effect on the occupants of the building.

The temperature required to create this condition makes the air buoyant or lighter than the surrounding air and causes it to rise toward the ceiling. In most applications when convection heat is used, the temperatures at the top of the building are generally higher than those at floor level because of the buoyant effect of the warm air.

In gas-fired infrared heaters, two heat exchangers are employed in the distribution of energy. The emitting surface of the burner is a radiant heat exchanger that transmits the infrared energy to the various absorbing mediums in the surrounding area below the heater. The energy is instantaneously converted to heat on absorption of the infrared energy by these absorbing mediums.

TABLE 12–4 Energy transfer between objects
Btu/hr/ft² delivery of PB 30 (30° angle) to a vertical surface 1 ft × 1 ft (midpoint 3 ft above floor).

h FT	*d* FT 2	4	6	8	10	12	14	16	18	20
7	405	242	110	57	27	13	7	3		
9	104	194	109	64	38	25	15	10	6	4
11	37	91	92	61	40	27	19	15	9	7
13	18	48	74	55	39	28	20	15	12	10
15	8	26	46	49	37	27	21	16	12	11
17	6	15	26	39	32	25	21	16	13	10
19	4	9	16	27	30	25	20	14	13	9
21	3	6	10	16	22	20	17	13	12	9
23	2	4	7	12	17	18	15	13	12	9

(*Courtesy Thermal Engineering Corp.*)

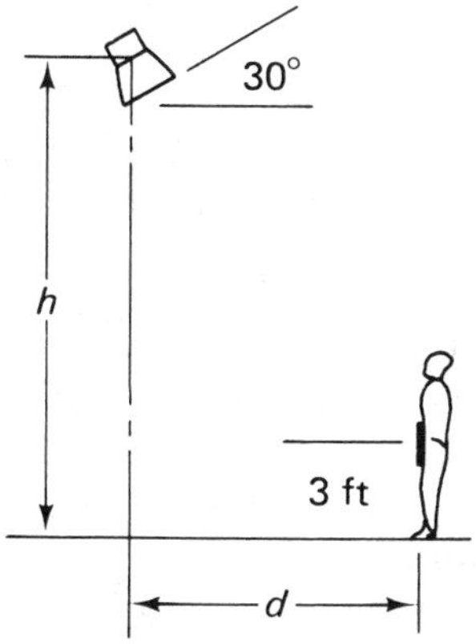
30°
h
3 ft
d

MODEL NO.	1 UNIT	4 OR MORE UNITS IN LINE** 10 FT APART	14 FT APART
PB30	1.0	2.16	1.52
PB45	1.5*	3.24	2.28
PC55	1.8*	3.89	2.74
PB60	2.0*	4.32	3.04
PB75	2.5*	5.40	3.80
PB90	3.0*	6.48	4.56
PC110	3.6*	7.78	5.47

*For section through short axis of unit. Units deliver elliptical pattern. Delivery is greater to sides.
** For average delivery

For section through short axis of unit. Units deliver ellliptical pattern. Delivery is greater to sides.
**For average delivery.

The air circulating in close proximity to the various materials and objects that have absorbed the infrared energy is heated. Thus, in effect, the floor and the other equipment and structures in the building become a very large heat exchanger, reflector, and radiator for the infrared energy.

One important difference in the operation of an infrared system is that the floor and other objects that become the heat exchanger are heated to a much lower temperature than is the heat exchanger in a convection heater. The air circulating in close proximity to the floor and other objects that have absorbed the energy cannot be heated any higher than the temperature of the objects themselves. Consequently, the air heated in an infrared system does not have the buoyant effect found in a conventional convection system. Because the air is not buoyant and because it cannot be heated any warmer than the heat exchanger from which it receives its heat, the heat must be concentrated in the working area of the building instead of rising to the roof.

Since the air is warmest closest to the floor and other absorbing mediums, it must decrease in temperature as it attempts to rise. Consequently, the heat loss through the walls and ceilings of the building is decreased because of the lower ambient temperatures prevailing on the upper portion of the walls and ceiling.

Since the energy has been distributed by a radiant method of heat transfer, no high velocity air circulation is required that would otherwise create uncomfortable conditions because of the relatively low air temperature.

Naturally, the exhaust gases escaping from the infrared heaters are buoyant. They will rise toward the ceiling and will help offset the ceiling or roof losses. The mounting height of the infrared heaters will determine the temperature of the ceiling. Table 12–5 accounts for variations in installation height as related to ceiling height. The use of this table will be explained as follows:

Obviously, lower ceiling temperatures will result when the heaters can be installed at low levels in relationship to the ceiling height. When the heaters are installed at a low level, the surrounding air will have an opportunity to dilute the exhaust gases and lower their temperature before the gases progress to the level of the ceiling. Under such conditions, the energy of the exhaust gases becomes contained in a greater volume of air resulting in relatively low ceiling temperatures.

The radiant energy accounting for more than one half of the total energy input to the heater is available to offset heat losses of the floor, lower walls, and infiltration. Stated in another manner, infrared energy emitted from high bay heaters does not heat the air first. This results in heat conservation. To a lesser degree, additional heat is conserved since the wall and roof internal surface coefficients are decreased because of reduced air movement resulting from use of infrared heaters.

This decrease lowers the *U* values or, in other words, provides a measure of insulation without changing construction.

Table 12–5 considers the aforementioned factors and allows the conventional heat loss to be corrected when a conservation of energy can be realized with the installation of an infrared heating system.

TABLE 12–5 Installation height as related to ceiling height

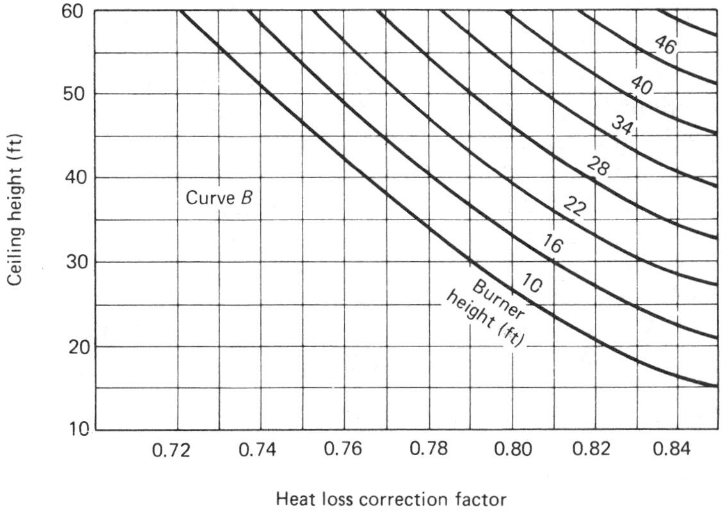
60
50
40
30
20
10
Ceiling height (ft)
Curve B
46
40
34
28
22
16
10
Burner height (ft)
0.72
0.74
0.76
0.78
0.80
0.82
0.84
Heat loss correction factor

(*Courtesy Thermal Engineering Corp.*)

It has been shown that when the occupants of an area are exposed to direct radiation from the heaters and/or the reduction of radiation losses to warmer floors and surrounding materials, lower ambient air temperatures can be used. In fact, in most cases it is insisted on. This further reduces energy consumption and, if anticipated, can result in lower initial system costs.

SPACE HEATING AT LOW AIR TEMPERATURES

The previous section on space heating dealt only with the heating of a building when air temperatures are maintained at or slightly below traditional levels.

Another approach to heating an entire space is to provide a thermally comfortable environment and yet allow the inside air temperature to fall well below traditional levels.

When heating an entire space at lower than normal air temperature, the comfort of the space is governed by two factors: (1) The direction of the radiation received by the subject from every heater and surface that his or her body "sees," and (2) the warming of the air in the space that results from convection heat transfer from the floor and other surfaces that are heated. A balance must be achieved between the two heating effects within which the body can adjust to attain comfort. When infiltration rates are high, direct radiation can be depended on to minimize the cost of heating the infiltrating air. At very high air change rates caused by industrial ventilation, consideration must be given to the use of make-up air systems to temper the incoming air required to maintain a suitable pressure balance.

When infiltration is minimal, the building is fairly well insulated and has relatively low ceilings, and the air temperature closely approximates normal room temperatures, direct radiation is of minor consequence. However, when these conditions are absent, the direct radiation becomes a major factor in the total comfort.

As heat loss increases per ft² of floor space, the benefits of direct radiation absorptivity by the occupants become of greater importance in establishing comfortable conditions. Table 12–6 is intended to correct normal heat losses when the absorption of direct radiation is beneficial to the extent that ambient temperatures on the inside of the building can be lowered, resulting in lower heat losses. Also, less energy would be lost due to air changes when lower ambient temperatures exist inside the building. Table 12–6 also corrects for energy conserved in the air changes due to the lower inside temperatures.

Table 12–7 corrects for the energy conserved due to lower ceiling temperatures when the heaters are installed at a lower level in relation to the ceiling.

Determination of Energy Input

1. Calculate the conventional heat loss of the building using the normal inside air temperatures.
2. Determine the heat loss per ft² by dividing the total heat loss by the total floor space of the building.
3. Use Table 12–6 to determine the correction factor *A*.
4. Use Table 12–7 to determine the correction factor *D*.
5. Multiply the correction factor *A* by *D* and by the total heat loss to obtain the proper energy input for an infrared heating system.

TABLE 12–6 Normal heat loss correction

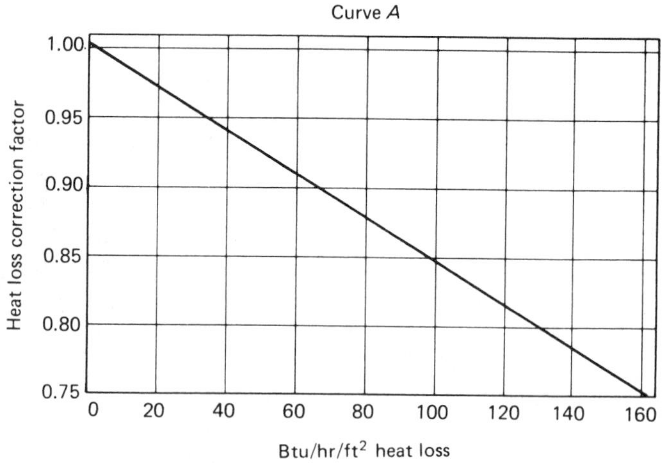
Curve A
1.00
0.95
0.90
0.85
0.80
0.75
Heat loss correction factor
0
20
40
60
80
100
120
140
160
Btu/hr/ft² heat loss

(*Courtesy Thermal Engineering Corp.*)

TABLE 12–7 Lower ceiling temperature correction

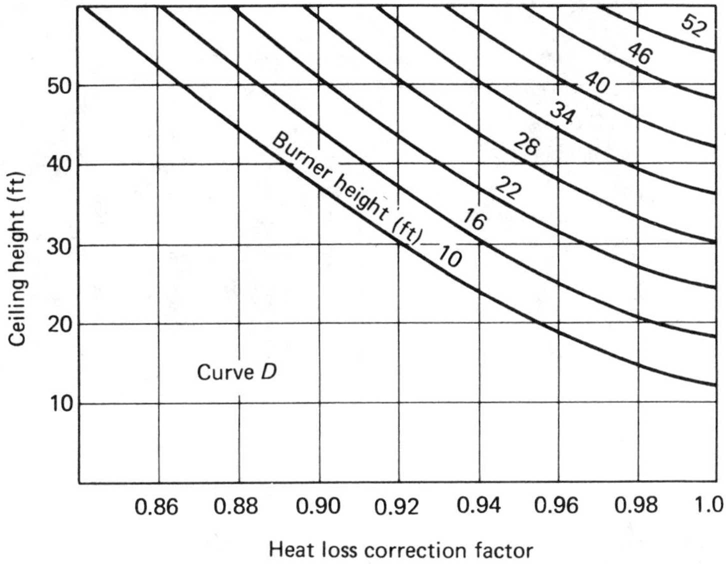
Ceiling height (ft)
50
40
30
20
10
Burner height (ft)
52
46
40
34
28
22
16
10
Curve D
0.86 0.88 0.90 0.92 0.94 0.96 0.98 1.0
Heat loss correction factor

(Courtesy Thermal Engineering Corp.)

6. Be sure that the infiltration rates for ventilation are sufficient to dilute the CO_2 to 5,000 parts per mil (ppm) or less.
7. Check for condensation possibilities.

LOCATION OF INFRARED HEATING UNITS

Because comfort depends on the radiation being received, the heaters must be placed so that all areas are covered. This differs from the perimeter system, in which the heat loss of the building is the prime consideration. Areas adjoining exposed walls should have greater energy to overcome body losses to cold surfaces (Figure 12–4).

Spot and Area Heating

When the occupancy of a building is low or work stations are static and the entire area does not require general heating, spot heating with infrared heaters will provide the most economical solution, from the standpoint both of first cost and of operational cost (Figure 12–5).

The major purpose is to heat the user. A person loses heat through radiation losses to surrounding surfaces that are lower in temperature than his or her clothing. Clothing also loses heat by convection to the surrounding air temperature. The prime factor is air velocity.

FIGURE 12–4
Perimeter heating system. *(Courtesy Modine Manufacturing Co.)*

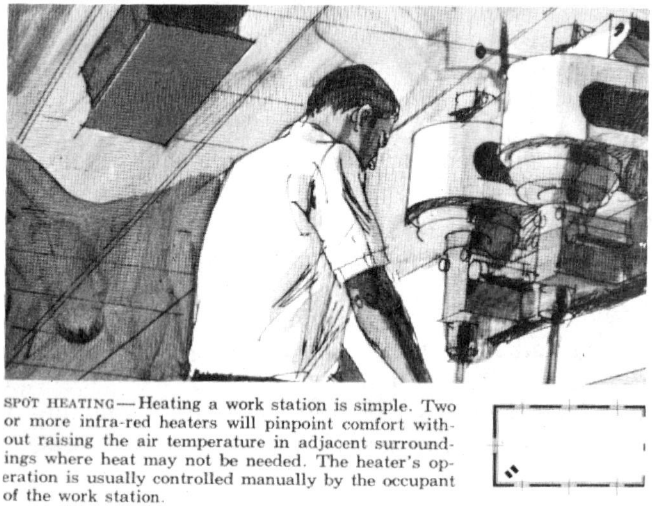
SPOT HEATING—Heating a work station is simple. Two or more infra-red heaters will pinpoint comfort without raising the air temperature in adjacent surroundings where heat may not be needed. The heater's operation is usually controlled manually by the occupant of the work station.

FIGURE 12–5
Spot heating. *(Courtesy Modine Manufacturing Co.)*

Other factors that influence a person's heat losses are the amount of clothing he or she wears and the heat lost as a result of the person's degree of activity (internal heat gains).

Design Procedure

The Thermal Engineering Corp. has developed a surface heat loss chart (see Table 12–8) that provides a reasonably accurate estimate of a person's surface losses. The procedure is summarized as follows:

1. Determine the design air, the surrounding temperature, and the clothing to be worn.
2. Determine the design air velocity.
3. Enter at the top or bottom horizontal scales both the air temperature and the clothing to be worn.
4. Follow the vertical line up or down until the air velocity line is crossed.

TABLE 12–8 Surface heat losses (Btu/hr/ft²), radiation, and convection losses for person doing light bench work

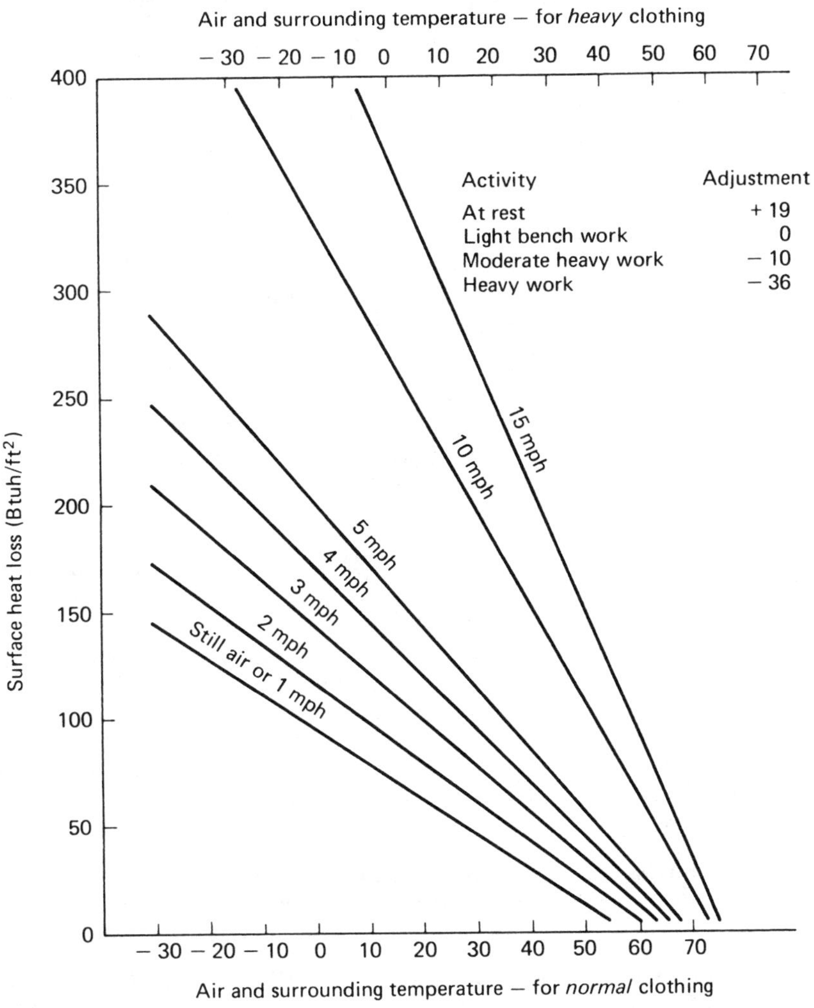

Air and surrounding temperature — for *heavy* clothing
– 30 – 20 – 10 0 10 20 30 40 50 60 70
Surface heat loss (Btuh/ft²)
400
350
300
250
200
150
100
50
0
Activity
Adjustment
At rest + 19
Light bench work 0
Moderate heavy work – 10
Heavy work – 36
15 mph
10 mph
5 mph
4 mph
3 mph
2 mph
Still air or 1 mph
– 30 – 20 – 10 0 10 20 30 40 50 60 70
Air and surrounding temperature — for *normal* clothing

(*Courtesy Thermal Engineering Corp.*)

5. At this point, extend a horizontal line to the left and read the surface loss from the vertical scale.
6. Adjust the reading (if necessary) according to the adjustment table on the chart.

The amount in step 6 gives the surface heat loss of the person in Btu/hr/ft^2. This amount of heat must be delivered to the person's body or clothing surface in order that he or she remain in thermal comfort.

Tables 12–9 and 12–10 give the delivery of heaters at varying heights and differences from the person. Several solutions can be found, but the most economical should be used.

For larger areas when spot heating is desired, the same design procedure is followed but the multipliers of Table 12–10 are used. Higher design temperatures for the air and surroundings may be possible in these applications because the greater number of units may increase these temperatures. These factors must be considered and the requirements adjusted.

For optimum design, the person or persons being heated should be heated by infrared energy from at least two sides.

ROBERTS-GORDON CO-RAY-VAC

The Roberts-Gordon CO-RAY-VAC is a compact, highly efficient, self-vented, infrared gas heating system working on a patented vacuum-firing principle. The

TABLE 12–9 *U* Factors for various fuels and types of equipment

Fuel	Equipment	Unit	*U* Factor
Natural Gas	CO-RAY-VAC	ft^3	0.295
Natural Gas	Unit Heater	ft^3	0.430
Natural Gas	Boiler	ft^3	0.482
LP Gas	Unit Heater	lb	0.020
LP Gas	Boiler	lb	0.023
LP Gas	CO-RAY-VAC	lb	0.014
Fuel Oil	Boiler	gal	0.00437
Coal	Boiler	lb	0.0592

(Courtesy Roberts-Gordon Appliance Corp.)

TABLE 12–10 Correction factor for outdoor design temperatures

Outdoor design temperature F	−20	−10	0	+10	+20
Correction factor	0.778	0.875	1.00	1.167	1.400

(Courtesy Roberts-Gordon Appliance Corp.)

CO-RAY-VAC basically consists of small overhead gas burners (each having a gas input of 40,000 Btu/hr) connected from one combustion chamber to the next by a standard 2 1/2-inch steel pipe (Figure 12–6), usually suspended from 10 to 15 ft above the area to be heated and directly under a bright-surface metal reflector. Generally, the combustion chambers are spread 15 to 21 ft apart, while the reflectors direct the "draftless" infrared rays emitted by the steel pipes downward to blanket the entire work area. The system uniformly heats a span, not a spot. No heat is wasted to the ceiling! And there are no air blasts to stir up dust, dirt, and germs, as with conventional heating systems. Greatly improved employee productivity is also reported by users (Figure 12–7).

Vacuum Combustion Principle

The products of combustion from the connected burners are totally exhausted to the outdoors by means of a vacuum pump or exhauster. Each burner depends on vacuum to introduce a filtered air-gas mixture to complete the combustion process. The system fires and operates with extreme fuel economy under a unique, highly efficient vacuum principle (Figure 12–8).

Combustion efficiency is approximately 93%. Almost every Btu is extracted from the combustion products. Exhaust temperatures are extremely low, proving CO-RAY-VAC's unmatched heat transfer efficiency.

No Venting or Condensation Problems

The vacuum pump in a CO-RAY-VAC system discharges all flue products directly to the outdoors through a single vent. No make-up air for flue product dilution is required. Moisture from the flue products, which may cause condensation damage, is eliminated. The problem of negative air pressures in the building, which may create toxic fumes or other hazardous conditions with gravity-vented heating units, is not encountered with CO-RAY-VAC's method of venting. Normally, air for com-

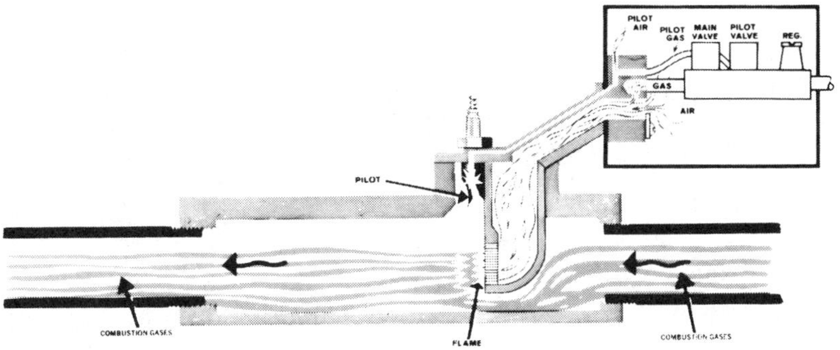
PILOT AIR
PILOT GAS
MAIN VALVE
PILOT VALVE
REG.
GAS
AIR
PILOT
COMBUSTION GASES
FLAME
COMBUSTION GASES

FIGURE 12–6
CO-RAY-VAC combustion chamber with burner and controls assembly.
(Courtesy Roberts-Gordon Appliance Corp.)

FIGURE 12–7
Typical CO-RAY-VAC installations. *(Courtesy Roberts-Gordon Appliance Corp.)*

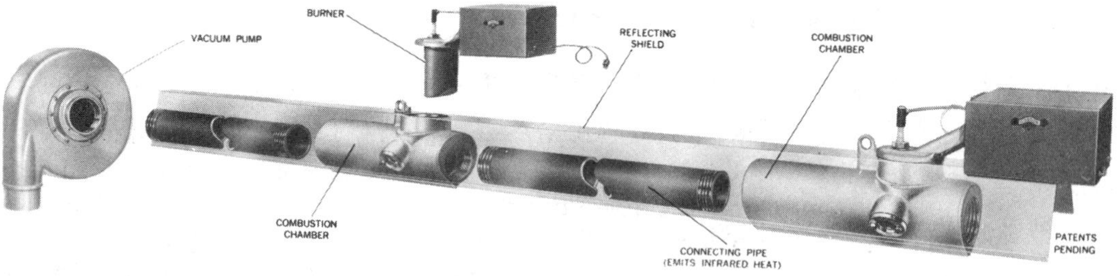
BURNER
VACUUM PUMP
REFLECTING SHIELD
COMBUSTION CHAMBER
COMBUSTION CHAMBER
CONNECTING PIPE (EMITS INFRARED HEAT)
PATENTS PENDING

FIGURE 12–8
CO-RAY-VAC vacuum system. *(Courtesy Roberts-Gordon Appliance Corp.)*

bustion is taken from the enclosed space, but in special cases the equipment is designed to provide for outside air to be ducted to the burners.

High Operating Safety

The blue flame in each burner is visible through a small glass-sealed port. The flame is completely enclosed and protected from sudden drafts. Moreover, the vacuum principle guards against combustion products or gas leaking into the building. A vacuum failure would produce an automatic positive mechanical interlock to provide a 100% gas shut-off.

The maximum operation temperature of the CO-RAY-VAC is about 900° F. The infrared rays invisibly emitted are below glowing temperature and are readily and safely absorbed by people and most common materials.

Rugged Durability

Each combustion chamber (containing the firing device) is a heavy-duty alloy casting which, together with the standard 2 1/2-inch steel pipe, constitutes an exceptionally rugged heat exchanger of extremely long life. The metal reflectors will last indefinitely.

Infrared heating has wide use. It can be found in all kinds of buildings—from sprawling factories to large indoor sports arenas.

The heat can be directed with pinpoint accuracy into a concentrated work area to solve spot heating problems. Cold doorways or hallways, for example, are made more comfortable by an adequate number of infrared heaters properly located. Heaters located along the outside walls of a building where heat loss is usually greatest will provide comfortable working conditions.

Versatile infrared heaters may be installed in groups for zone control to maintain the temperature demands of certain areas. They may be controlled automatically or manually.

REVIEW QUESTIONS

1. When infrared heating is used, is the air between objects heated?
2. What is the form of wave motion known as in infrared heating?
3. Is this form of motion the same as sound waves?
4. Is the energy transmitted by an infrared heater known as heat?
5. Are wave form lengths the same for all bodies and conditions?
6. What is a *black body*?
7. How is the emissivity of a body affected by a change in temperature?
8. What is the average value of the absorptivity and emissivity of a body?
9. Does the distance between bodies affect their emissivity and absorptivity?
10. For what purpose is the Stefan-Boltzman law used?
11. What is the ceramic surface temperature of the TEC infrared heater?
12. With what is a direct reading for emissivity taken?
13. What two factors must be considered when determining the infrared energy exchange?
14. What are the three methods of heat transfer?
15. How many heat exchangers are used in infrared heating?
16. Why is the heat loss less with infrared heating than with conventional furnace heating methods?
17. Does the installation of an infrared heater affect the ceiling temperature?
18. What two factors govern the comfort of a space when heating with lower than normal air temperatures?
19. When is the direct absorption of radiation by occupants most important?
20. When would the perimeter system best be used?
21. When would spot heating best be used?
22. What principle does the Roberts-Gordon CO-RAY-VAC unit use?

23. What is used for a connection between the CO-RAY-VAC units?
24. Why are there no condensation problems with the CO-RAY-VAC system?
25. What is the maximum operating temperature of the CO-RAY-VAC units?
26. What percentage of the building heat loss is used in determining the number of CO-RAY-VAC units to install?

13

humidification

The objectives of this chapter are:

- To familiarize you with the terms used in humidification.
- To introduce you to the advantages of proper humidification.
- To give you the desired humidity requirements for comfort conditioning.
- To provide you with the different types of humidifier operation.

Humidification is, unfortunately, one of the least understood, most important aspects of comfort conditioning. It is understandable that it is misunderstood because humidity is intangible: You cannot see it, you cannot touch it, it has no color, no odor, and no sound.

WHAT IS HUMIDITY?

To make sure we are starting out on common ground, let us define some of the words used in discussing humidity.

Humidity is the water vapor within a given space. *Absolute humidity* is the weight of water vapor per unit volume. *Percentage humidity* is the ratio of the weight of water vapor per pound of dry air to the weight of water vapor per pound of dry air saturated at the same temperature. *Relative humidity* is the ratio of the mol fraction of water vapor present in the air to the mol fraction of water vapor present in saturated air at the same temperature and barometric pressure. Approximately, it equals the ratio of the partial pressure or density of the water vapor in the air to the saturation pressure or density, respectively, of the water vapor at the same temperature.

Although there is a difference between percentage humidity and relative humidity, it is only slight and is practically negligible at normal room temperatures. So, for our purposes, we can say that relative humidity indicates the amount of water vapor actually in the air expressed as a percentage of the maximum amount that the air could hold under the same conditions.

WHY WE NEED HUMIDIFICATION

The amount of moisture in the air has a direct bearing on personal comfort or discomfort. Just as extremely high relative humidity in the summer gives one a soggy feeling, low humidity in the winter in heated homes gives one a dried-out feeling.

When cold air enters the house during the winter, it is heated; thus its moisture holding capacity is increased. If moisture is not added to this air, indoor relative humidity drops below the minimum range for personal comfort. In most homes without humidification equipment, heated indoor air is much too dry for the well-being of its occupants, their pets, plants, and furnishings. Proper humidity control is simple with a humidifier. Simply dial the desired humidity on the humidistat.

Properly humidified air enhances personal comfort and well-being by helping to prevent throat irritations, nasal discomfort, bronchial aggravations, and itchy dry skin caused by hot, dry indoor air. Floors, doors, frames, and wood furniture will have a minimum of drying out, cracking, or warping. Draperies and upholstering stay fresh and wear longer. Annoying minor shocks caused by static electricity are greatly reduced. House plants stay fresher and prettier. A home can be kept more comfortable at lower temperatures with properly humidified air. This helps lower heating costs.

As stated before, the warmer the air, the more moisture it can hold. Air in a home heated to 70° F can hold about 8 grains of moisture per ft^3. That is 100% relative humidity. If there are only 2 grains/ft^3 in the home, this is one quarter of the capacity of the air to hold moisture. Therefore, the relative humidity is also one quarter or 25%. The air can hold four times as much water.

However, the important thing to remember is what happens to air when it is heated. The outdoor-indoor relative humidity conversion chart (Table 13–1) illustrates this.

Because of this capability of warm air to hold more water than cold air, a substantial reduction of relative humidity is taking place in every unhumidified or underhumidified home where winter heating is prevalent.

To solve this problem, we add moisture artificially so that more water is available for this thirsty air to hold. That is, we humidify because of benefits as important as heating to overall indoor comfort and well-being during the heating season. And these benefits are actually what must be maintained.

These benefits can be grouped into three general classifications:

TABLE 13–1 Outdoor-indoor relative humidity conversion chart

Outdoor relative humidity														
100%	2%	3%	4%	6%	7%	9%	11%	14%	17%	21%	26%	31%	38%	46%
95%	2%	3%	4%	5%	7%	8%	10%	13%	16%	20%	24%	30%	36%	44%
90%	2%	2%	4%	5%	6%	8%	10%	12%	15%	19%	23%	28%	34%	41%
85%	2%	2%	4%	5%	6%	8%	9%	12%	15%	18%	22%	27%	32%	39%
80%	2%	2%	4%	5%	6%	7%	9%	11%	14%	17%	20%	25%	30%	37%
75%	2%	2%	3%	4%	5%	7%	8%	10%	13%	16%	19%	23%	28%	36%
70%	1%	2%	3%	4%	5%	**6%**	8%	10%	12%	15%	18%	22%	26%	32%
65%	1%	2%	3%	4%	5%	6%	7%	8%	11%	14%	17%	20%	25%	30%
60%	1%	2%	3%	3%	4%	5%	7%	8%	10%	13%	15%	19%	23%	28%
55%	1%	1%	2%	3%	4%	5%	6%	8%	9%	12%	14%	17%	21%	25%
50%	1%	1%	2%	3%	4%	4%	6%	7%	9%	10%	13%	16%	19%	23%
45%	1%	1%	2%	3%	3%	4%	5%	6%	8%	9%	12%	14%	17%	21%
40%	1%	1%	2%	2%	3%	4%	4%	6%	7%	8%	10%	12%	15%	18%
35%	1%	1%	2%	2%	3%	3%	4%	5%	6%	7%	9%	11%	13%	16%
30%	1%	1%	1%	2%	2%	3%	3%	4%	5%	6%	8%	9%	11%	14%
25%	1%	1%	1%	1%	2%	2%	3%	3%	4%	5%	6%	8%	10%	12%
20%	+%	1%	1%	1%	1%	2%	2%	3%	3%	4%	5%	6%	8%	10%
15%	+%	+%	1%	1%	1%	1%	2%	2%	3%	3%	4%	5%	6%	7%
10%	+%	+%	+%	1%	1%	1%	1%	1%	2%	2%	3%	3%	4%	5%
5%	+%	+%	+%	+%	+%	+%	1%	1%	1%	1%	1%	1%	2%	2%
0%	0%	0%	0%	0%	0%	0%	0%	0%	0%	0%	0%	0%	0%	0%
Outdoor temperature	−20°	−10°	−5°	0°	+5°	**+10°**	+15°	+20°	+25°	+30°	+35°	+40°	+45°	+50°

(Courtesy Research Products Corp.)

1. Comfort.
2. Preservation.
3. Health.

Benefit Number One: Comfort

Did you ever step out of the shower, start shaving, and notice how warm it is in the bathroom? It is actually muggy. It will probably be about 75° F in the bathroom and the relative humidity will probably be about 75–80%, because of the water vapor added to the air while showering. Now the phone rings, and you have to step out into the hall to answer it. What happens? You freeze. Although the temperature is probably about 70° F, just 5° F cooler than in the bathroom, you shiver. This is because you have just become an evaporative cooler. The air out in the hall is dry. The relative humidity is probably about 10 to 15%. You are wet and this thirsty air goes to work on your skin. As it evaporates the water, your skin is cooled. This same thing continues day after day, every winter, in millions of homes. People turn their thermostats up to 75° F and more in order to feel warm. Even then, it feels drafty and chilly because the evaporative cooling process is going on. Proper relative humidity levels make you feel more comfortable at lower thermostat settings.

But this cold feeling is not the only discomfort caused by too dry air. Static electricity, usually an indication of low relative humidity levels, is a condition that is constantly annoying. Proper relative humidity will alleviate this discomfort.

Benefit Number Two: Preservation

The addition or reduction of moisture drastically affects the qualities, the dimensions, and the weight of hygroscopic materials.

Wood, leather, paper, and cloth, although they feel dry to the touch, contain water—not a fixed amount of water, but an amount that will vary greatly with the relative humidity level of the surrounding air. Take, for example, 1 ft^3 of wood with a bone-dry weight of 30 pounds. At 60% relative humidity, the wood will hold over 3 pints of water. Now, if the relative humidity is lowered to 10%, the water held by the wood will not fill even a 1-pint bottle. Thus, in effect, we have withdrawn $2\,^1/_2$ pints of water from the wood by lowering the relative humidity from 60% to 10% (1 pint of water = approximately 1 pound).

This type of action continues, not only with wood, but with every material in the home that is capable of absorbing moisture. Paper, plaster, fibers, leather, glue, hair, skin, . . . practically everything in the home. These materials shrink as they lose water and swell as they absorb water. When the water loss is rapid, warping and cracking take place. Also, as the relative humidity changes, the condition and dimensions of the materials change. This is why humidity must be added. This is why proper relative humidity is important.

What are the effects of this constantly changing or constantly low moisture content of the air? They are damaging. Furniture construction is affected. Glue dries out, joints separate, rungs fall out of chairs, and cracks appear. The plaster walls dry out and crack. Joints and wall studs shrink, causing the room to be out of square. The wood paneling separates and cracks. The boards in the floor separate. Musical instruments lose their tone. Pieces of art, books, and documents dry out and either break or crack. The rugs wear out quickly, simply because a dry fiber will break while a moist fiber will bend.

Now that the problems of too little humidity have been discussed, let us discuss the problem of too much humidity and the effect of vapor pressure.

Perhaps you have noticed windows that fog during the winter—maybe a small amount of fog on the lower corners or a whole window fogging or completely frosting over. These conditions are an indication of too high indoor relative humidity.

This condensation is due to the effect of vapor pressure. Dalton's law explains vapor pressure: *In a gaseous mixture, the molecules of each gas are evenly dispersed throughout the entire volume of the mixture.* Taking the house as the volume involved, water vapor molecules move throughout the entire home. Because of the tendency of these molecules to disperse evenly, or to mix, the moisture in the humidified air moves toward the drier air. In other words, in a house the moist indoor air attempts to reach the drier outside air. It moves toward the windows where there is a lower temperature and, therefore, causes an increase in relative humidity to a point at which the water vapor will condense out on the cold surface of the window. This is the dew point, and it occurs at various temperatures, depending on the type of windows in the home.

TABLE 13–2 Condensation temperatures

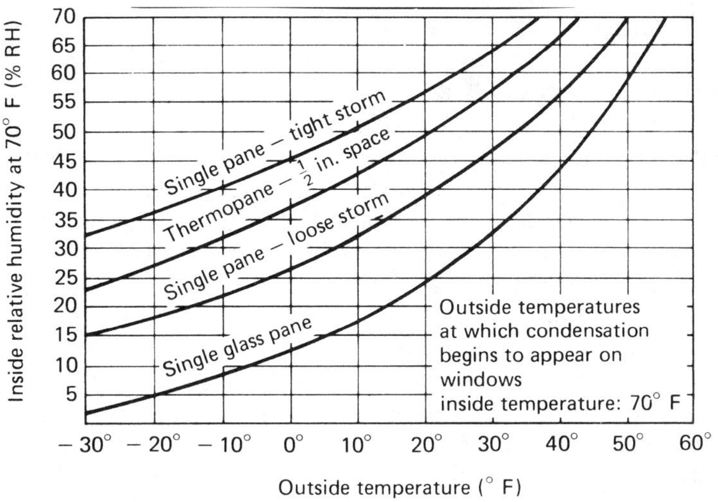
Inside relative humidity at 70° F (% RH)
70
65
60
55
50
45
40
35
30
25
20
15
10
5
Single pane – tight storm
Thermopane – 1/2 in. space
Single pane – loose storm
Single glass pane
Outside temperatures at which condensation begins to appear on windows
inside temperature: 70° F
– 30° – 20° – 10° 0° 10° 20° 30° 40° 50° 60°
Outside temperature (° F)

(Courtesy Research Products Corp.)

With an indoor temperature of 70° F and an outdoor temperature of 20° F, for example (Table 13–2), condensation begins on a single glass plane at about 24% relative humidity, at 38% for a single pane with a loose storm sash, at 50% for a thermopane with 1/2-inch space between the panes, and at 58% for a single pane with a tight storm sash.

Usually condensation on inside windows is a type of measurement of the allowable relative humidity inside a home. We can furthermore assume that if this

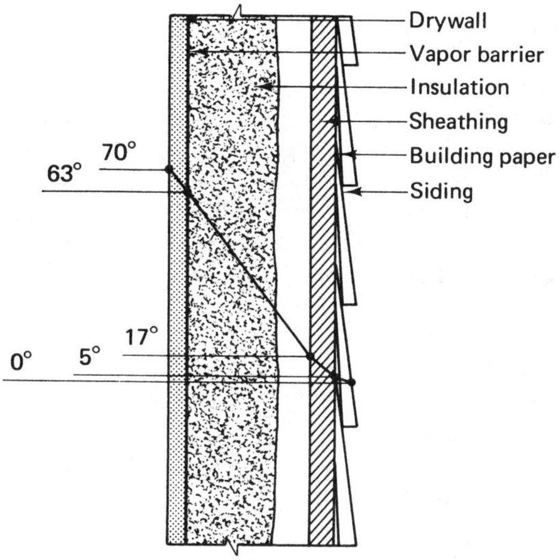
Drywall
Vapor barrier
Insulation
Sheathing
Building paper
Siding
70°
63°
17°
0°
5°

FIGURE 13–1
Typical outside frame wall construction. (*Courtesy Research Products Corp.*)

condensation activity is taking place on windows, it may also be taking place within the walls if there is no vapor barrier.

A vapor barrier, as the name implies, is a material that restricts the movement of water vapor molecules (Figure 13–1). Examples of a typical vapor barrier are aluminum foil, polyethylene film, plastic wall coverings, plastic tiles, and some types of paint and varnish. Actually, practically every home has a vapor barrier of some type that at least retards the movement of the water molecules from a high vapor pressure area (inside) to a low vapor pressure area (outside).

The typical outside wall has a dry wall, or plaster, a vapor barrier (on the warm side of the insulation), the insulation, air space, sheathing, building paper, and siding. With an indoor temperature of 70° F, a relative humidity of 35%, and an outside temperature of 0° F, what happens to the temperature of the air passing through the wall? It drops to about 60° F at the vapor barrier, down to 17° F at the sheathing, and on down to 0° F outside. And, if we checked a psychrometric chart, we would find that with an indoor temperature of 70° F and an indoor relative humidity of 35%, the dew point would be 41° F. This temperature occurs right in the middle of the insulation in the wall. This is where we have condensation, and this is where we would have trouble without a vapor barrier and without controlled humidification.

The important aspect is, then, properly controlled relative humidity to avoid the damaging effects of too dry air and, equally as important, to avoid the damaging effects of too high relative humidity.

Benefit Number Three: Health

Dr. Arthur W. Proetz, an eye, nose, and throat specialist, says, in the annals of *Otology, Rhinology and Laryngology:*

> In the struggle between the nose and the machinery in the basement, sometimes the heater wins and sometimes the cooler, but seldom the nose. The nasal mucus contains some 96% water. To begin with, it is more viscous than mucus elsewhere in the body and even slight drying increases the viscosity enough to interfere with the work of the cilia. Demands on the nasal glands are great even under usual conditions and they cannot cope with extreme dryness indoors in winter. Experience has shown that with approaching winter, the first wave of dry-nose patients appears in the office when the relative humidity indoors falls to 25%. It would seem, therefore, that 35% would be required as a passing grade but 40% something to shoot for. It boils down to this: a pint of water is a lot of water for a small nose to turn out. In disease or old age, it simply doesn't deliver and drainage stops and the germs take over.

The right-hand column in Table 13–3 is the number of cases of respiratory disease per 1,000 population, taken from the U.S. Public Health statistics. The left-hand column is a typical indoor relative humidity in Madison, Wisconsin during these same months. Note that there is an obvious and definite correlation between good health and high relative humidities and between poor health and low relative humidities. All the facts point toward a positive connection between humidity and health.

TABLE 13–3 Respiratory disease per 1,000 population

MONTH	AVERAGE INDOOR RELATIVE HUMIDITY			TOTAL NUMBER OF CASES OF RESPIRATORY DISEASES PER 1,000 POPULATION
July	59%	High Humidity	Good Health	17
August	56%			20
September	51%			23
October	40%			29
November	20%	Low Humidity	Poor Health	40
December	12%			58
January	12%			92
February	14%			102
March	18%			89
April	25%			55
May	35%	High Humidity	Good Health	35
June	57%			22

(Courtesy Research Products Corp.)

THE CORRECT INDOOR RELATIVE HUMIDITY

While some humidity conditions may be ideal for health and comfort, they are, in many cases, less ideal for other reasons. An indoor relative humidity of 60% may fulfill all the requirements for health and comfort, but it can result in damage to walls, furnishings, etc. The fogging of windows is usually an indication of too high relative humidity, and it must be remembered that this same condensation is taking place inside walls and other places vulnerable to damage by excessive moisture (Table 13–4).

It is, therefore, necessary to set safe limits of indoor relative humidity levels to receive the maximum benefits from correct humidity, without making the structure itself susceptible to damage. It is recommended that Table 13–5 be followed to ensure these benefits.

EFFECT OF WATER CHARACTERISTICS

Only distilled water or rain water caught before it reaches the ground is free from minerals. Water from wells, lakes, and rivers all contain varying amounts of minerals in solution. These minerals are picked up as the water moves through or across water soluble portions of the earth's surface. In many cases, the level of these minerals is sufficiently high to make water-conditioning equipment necessary to remove the objectionable minerals for normal domestic use.

TABLE 13–4 Condensation on windows

Inside Temp. 70° F.	Outside Temperature at Which Condensation Will Probably Occur*	
Inside Relative Humidity	Single Glass	Double Glass Thermopane
50%	43°	18°
45%	38°	11°
40%	34°	2°
35%	28°	−8°
30%	22°	−20°
25%	15°	−30°
20%	8°	
15%	0°	
10%	−11°	

(Courtesy Research Products Corp.)
*With constant air circulation over the windows, a higher inside relative humidity can be maintained. Heavy drapes, closed blinds, etc. will have an adverse effect.

TABLE 13–5 Temperature-humidity table

Outside Temperature	Recommended R.H.
+20° and above	35%
+10°	30%
0°	25%
−10°	20%
−20°	15%

(Courtesy Research Products Corp.)

It is common knowledge that water evaporated from a tea kettle leaves a residue known as lime. Since evaporation of water is the only way to create and distribute water vapor into the air present in homes, it is apparent that mineral residue resulting from evaporation presents a problem.

Water hardness varies in different localities. Drinking water contains some hardness, consisting primarily of calcium carbonate and/or magnesium carbonate. This hardness is expressed in grains per gallon (Table 13–6).

If 1 gallon of average water hardness is evaporated, a residue of 25 grains remains. If 100 gallons of water are evaporated to provide humidity, 2,500 grains or 5.7 ounces of solids will build up on the evaporating surface.

TABLE 13–6 Water hardness table

Class of Hardness	Grains Hardness per Gallon	% Figure in U.S.
Low	3–10	30
Average	10–25	55
High	25–50	15

(Courtesy Research Products Corp.)

HOW RELATIVE HUMIDITY IS MEASURED

There are two types of instruments normally used in measuring relative humidity: hygrometers and psychrometers.

Hygrometers

Hygrometers are the instruments most commonly seen in homes and offices. They are manufactured in a variety of models. Most of them use a device that changes in dimensions as the relative humidity changes. This dimension change actuates a dial from which the relative humidity can be read. This type of instrument, when properly calibrated, will provide reasonably accurate relative humidity readings (Figure 13–2). The humidistat supplied with humidifiers is a type of hygrometer. The accuracy of these instruments can be checked by the second type of measuring device, the psychrometer. For convenience and maximum accuracy, the humidistat should be wall-mounted in the living area.

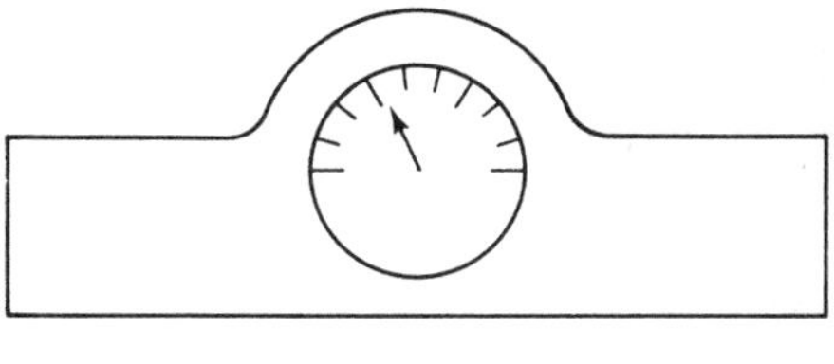

FIGURE 13–2
Desk hygrometer.

Sling-Type Psychrometer

This instrument is used by humidifier installers and others to determine the exact relative humidity readings necessary for calculations and recommendations (Figure 13–3).

For accurate results, correct sling psychrometer techniques should be utilized. The following hints are helpful to ensure accurate readings. Use a clean wick and distilled water. Be sure the wet and dry bulb thermometers are matched, with no mercury separation. Rotate the psychrometer at approximately 600 feet per minute (fpm).

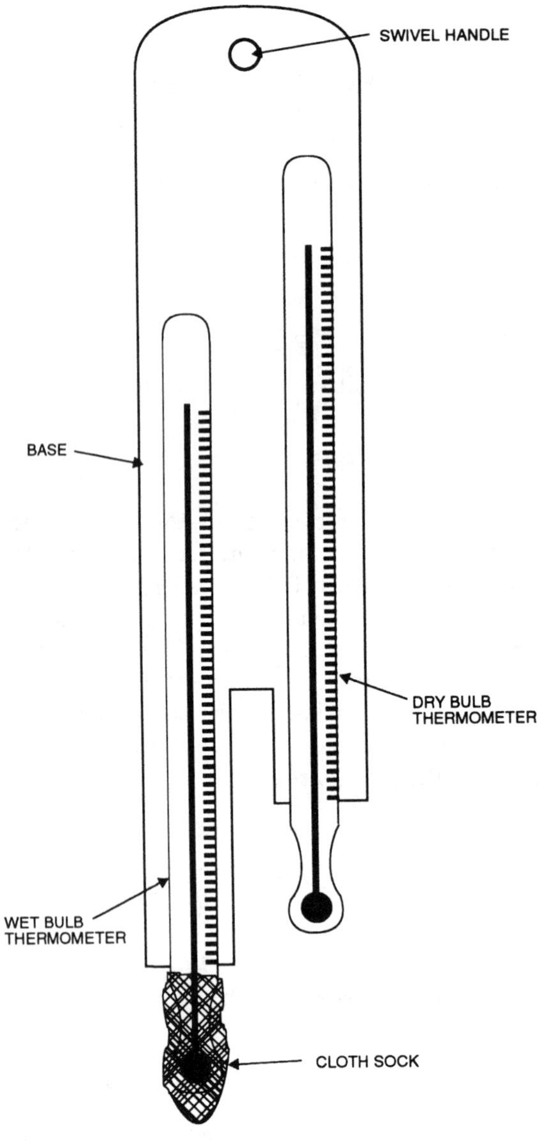
SWIVEL HANDLE
BASE
DRY BULB
THERMOMETER
WET BULB
THERMOMETER
CLOTH SOCK

FIGURE 13–3
Sling psychrometer.

How to Take Readings

1. Dip the wick on the wet bulb thermometer in distilled water, preferably (only one dipping per determination of relative humidity, but never between readings). The progressive evaporation of the moisture in the wick, until it reaches equilibrium with the moisture in the air, is the determining factor of the wet bulb reading.

2. Whirl the sling psychrometer for 30 seconds. Take the readings quickly on the wet bulb thermometer first, then on the dry bulb, and jot them down. Whirl the psychrometer again, taking readings at 30-second intervals for five successive readings, and jot down the temperatures each time, or until the lowest reading shows a leveling off or a return curve (two or more nearly identical successive readings).
3. Use a psychrometric chart or table to obtain the relative humidity.

THE PSYCHROMETRIC CHART

The psychrometric chart simplifies the determination of air properties and eliminates many tedious calculations. We are concerned with five specific psychrometric terms that can be found on the chart: (1) the dry bulb temperature; (2) the wet bulb temperature; (3) the relative humidity; (4) the absolute humidity; and (5) the dew point temperature. If we know any two of these air properties, we can, from the chart, determine the other three.

The first term, *dry bulb temperature,* is the temperature measure of an ordinary thermometer. This temperature scale runs horizontally across the bottom of the chart (Chart 13–1). The dry bulb temperature lines are the straight vertical lines.

The second air property found on the psychrometric chart is the *wet bulb temperature,* or the temperature resulting when water is evaporated off a cloth covering an ordinary thermometer. As you can see, the wet bulb scale is measured along the curve portion of the psychrometric chart (Chart 13–2), from the lower left to the upper right. The wet bulb lines run diagonally across the chart.

Next, we find the *relative humidity.* On a complete psychrometric chart, relative humidity lines are the only curved lines on the chart (see Chart 13–3). The various relative humidities are indicated on the lines. There is no coordinate scale as with the other air properties.

The fourth component of a psychrometric chart is the *absolute humidity,* or the actual weight of water vapor in the air. The scale for absolute humidity is a verti-

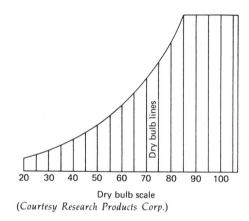

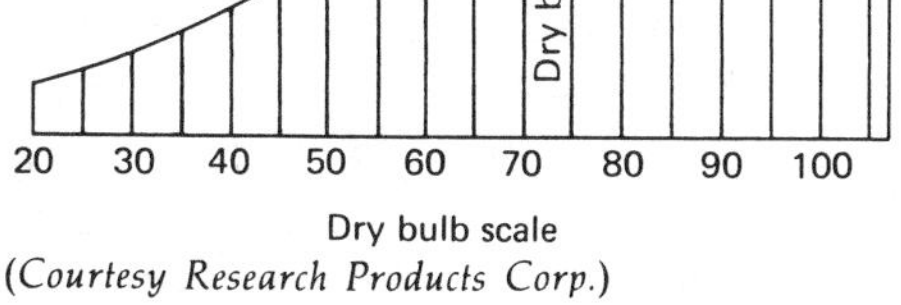
Dry bulb lines
20
30
40
50
60
70
80
90
100
Dry bulb scale
(Courtesy Research Products Corp.)

CHART 13–1
Dry bulb temperature lines. (*Courtesy Research Products Corp.*)

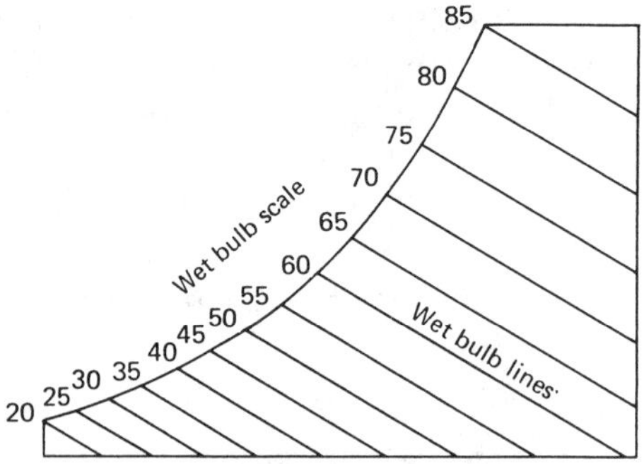
85
80
75
70
65
60
55
50
45
40
35
30
25
20
Wet bulb scale
Wet bulb lines

(Courtesy Research Products Corp.)

CHART 13–2
Wet bulb temperature lines. *(Courtesy Research Products Corp.)*

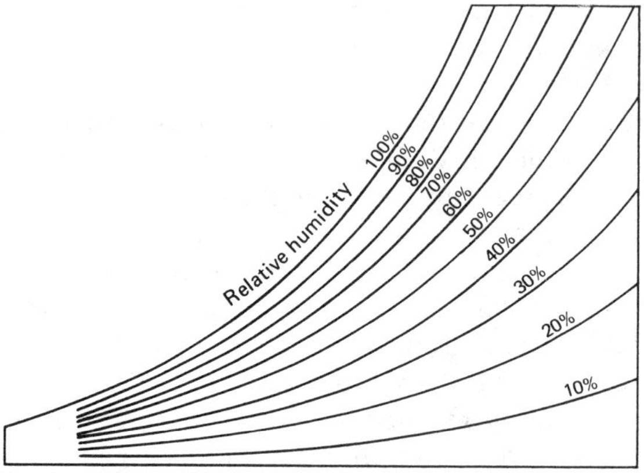
Relative humidity
100%
90%
80%
70%
60%
50%
40%
30%
20%
10%

(Courtesy Research Products Corp.)

CHART 13–3
Relative humidity lines. (*Courtesy Research Products Corp.*)

cal scale on the right side of the psychrometric chart (Chart 13–4). The absolute humidity lines run horizontally from this scale.

The *dew point temperature* is the fifth air property that is included on a psychrometric chart. This is the temperature at which moisture will condense on a surface. The scale for the dew point temperature is identical to the scale for the wet bulb temperature (Chart 13–5). However, the dew point lines run horizontally across the chart, not diagonally as is the case with the wet bulb temperature lines.

When we put together the five charts we have just covered, we will have a complete psychrometric chart (Chart 13–6).

Let us take an example to explain the working of the chart. Let us assume that we have used a sling psychrometer and have taken readings of a dry bulb temperature of 72° F and a wet bulb temperature of 56° F. We know two factors. The 72° F dry bulb temperature is found on the bottom scale. The wet bulb scale is on the curved outside line of the chart on the left. The wet bulb temperature line is 56° F. Extending the two lines, we find that they intersect. From this point we can determine any other information we need. The relative humidity is, for example, 35%.

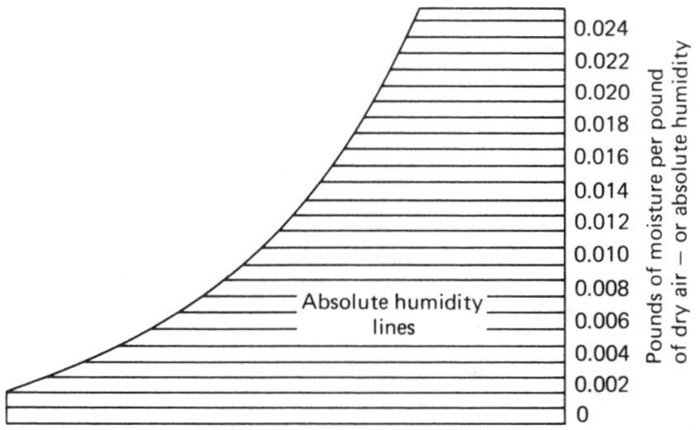
0.024
0.022
0.020
0.018
0.016
0.014
0.012
0.010
0.008
0.006
0.004
0.002
0
Absolute humidity lines
Pounds of moisture per pound of dry air – or absolute humidity

(*Courtesy Research Products Corp.*)

CHART 13–4
Absolute humidity lines. (*Courtesy Research Products Corp.*)

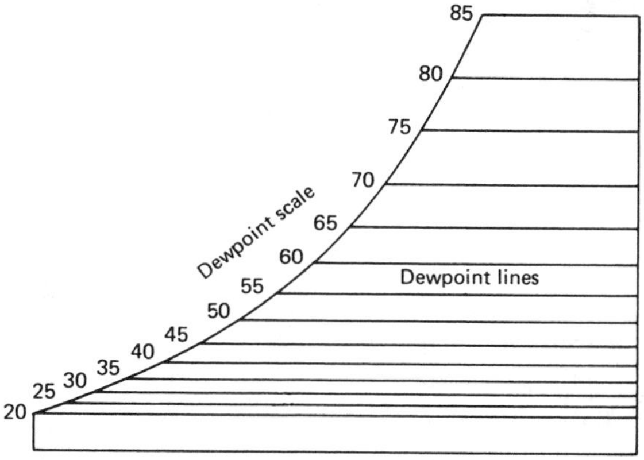
85
80
75
70
65
60
55
50
45
40
35
30
25
20
Dewpoint scale
Dewpoint lines

(*Courtesy Research Products Corp.*)

CHART 13–5
Dew point temperature lines. (*Courtesy Research Products Corp.*)

The absolute humidity is 0.0058 pounds of water per pound of air. The dew point temperature is 42° F.

HUMIDIFIERS

There are many, many humidifiers available. They vary in price, they vary in capacity, and they vary in principle of operation.

For classification purposes, it is simpler, and more logical, to consider humidifiers in three general types:

1. Pan-type units.
2. Atomizing-type units.
3. Wetted-element-type units.

The pan-type unit is the simplest type of humidifier (Figure 13–4). Its capacity is low. On a hot radiator, it might evaporate 0.0083 gallons of water per hour. In

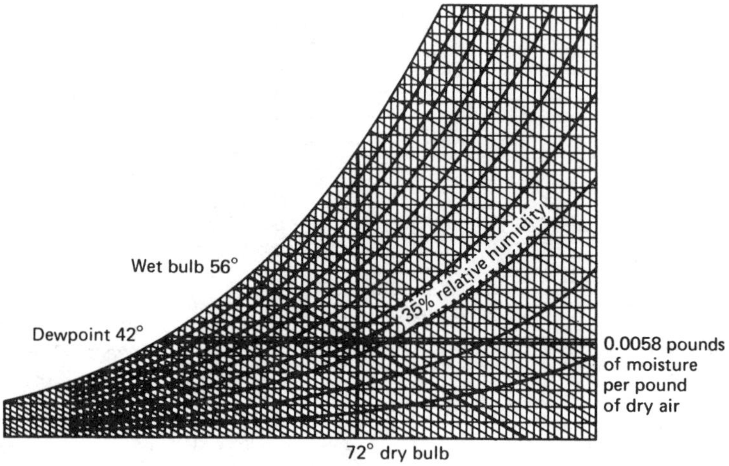
Wet bulb 56°
Dewpoint 42°
35% relative humidity
0.0058 pounds of moisture per pound of dry air
72° dry bulb

(*Courtesy Research Products Corp.*)

CHART 13–6
Psychrometric chart. (*Courtesy Research Products Corp.*)

the warm air plenum of a furnace, it would evaporate approximately 0.18 gallons of water per hour. To increase the capacity, the air-to-water surfaces can be increased by placing water wicking plates in the pan. The capacity increases as the air temperature in the furnace plenum increases.

Greater capacity is also possible through the use of steam, hot water, or an electric heating element immersed in the water. A 1,200-W heating element, for ex-

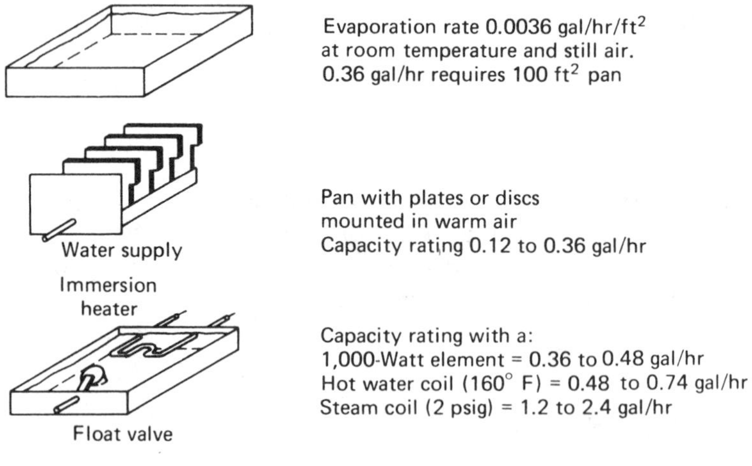
Evaporation rate 0.0036 gal/hr/ft²
at room temperature and still air.
0.36 gal/hr requires 100 ft² pan
Water supply
Pan with plates or discs
mounted in warm air
Capacity rating 0.12 to 0.36 gal/hr
Immersion
heater
Capacity rating with a:
1,000-Watt element = 0.36 to 0.48 gal/hr
Hot water coil (160° F) = 0.48 to 0.74 gal/hr
Steam coil (2 psig) = 1.2 to 2.4 gal/hr
Float valve

FIGURE 13–4
Pan-type humidifier. *(Courtesy Research Products Corp.)*

ample, in a container with water supplied by a float valve could produce 0.48 gallons of moisture per hour.

The second type of humidifier is the atomizing type (Figure 13–5). This device atomizes the water by throwing it from the surface at a rapidly revolving disc. It is generally a portable or console unit, although it can be installed so that water particles will be directed into a ducted central system.

The third type is the wetted-element-type humidifier. In its simplest form, it operates in the manner of an evaporative cooler. Air is either pushed or pulled through a wetted pad material or filter and evaporative cooling takes place (Figure 13–6). By increasing the air flow or by supplying additional heat, the evaporation rate of the humidifier can be increased. The heat source for evaporation can be free from an increase in water temperature or an increase in air temperature.

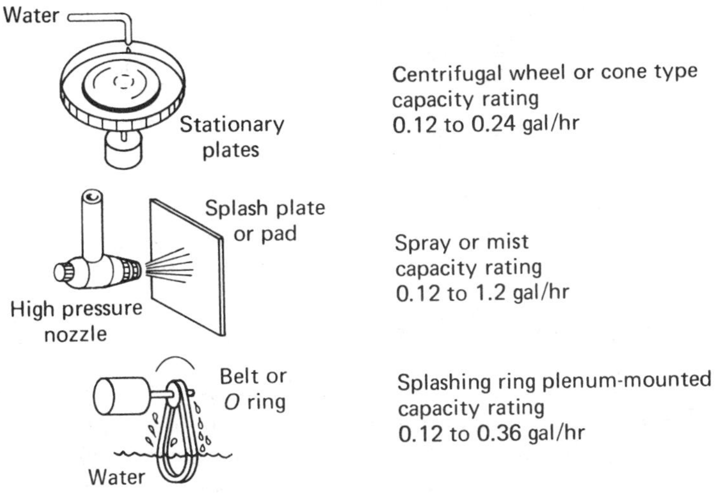
Water
Stationary plates
Centrifugal wheel or cone type
capacity rating
0.12 to 0.24 gal/hr
Splash plate or pad
High pressure nozzle
Spray or mist
capacity rating
0.12 to 1.2 gal/hr
Belt or O ring
Water
Splashing ring plenum-mounted
capacity rating
0.12 to 0.36 gal/hr

FIGURE 13–5
Atomizing-type humidifier. (*Courtesy Research Products Corp.*)

The air for evaporation can be taken from the heated air of the furnace plenum and directed through the humidifier by the humidifier fan, or it can be drawn through the wetted element by the air pressure differential of the furnace blower system.

Furnace-mounted humidifiers, usually of the wetted type, can be constructed so that they produce 1.2 or more gallons of moisture per hour. Because of their capacity, this type usually has a humidistat or control that will actuate a relay or water valve and start a fan that operates until the control is satisfied. Normally more water is supplied to the unit than is evaporated, and this flushing action washes a large portion of the hardness salts from the evaporative element to a floor drain to eliminate them from the humidifying system.

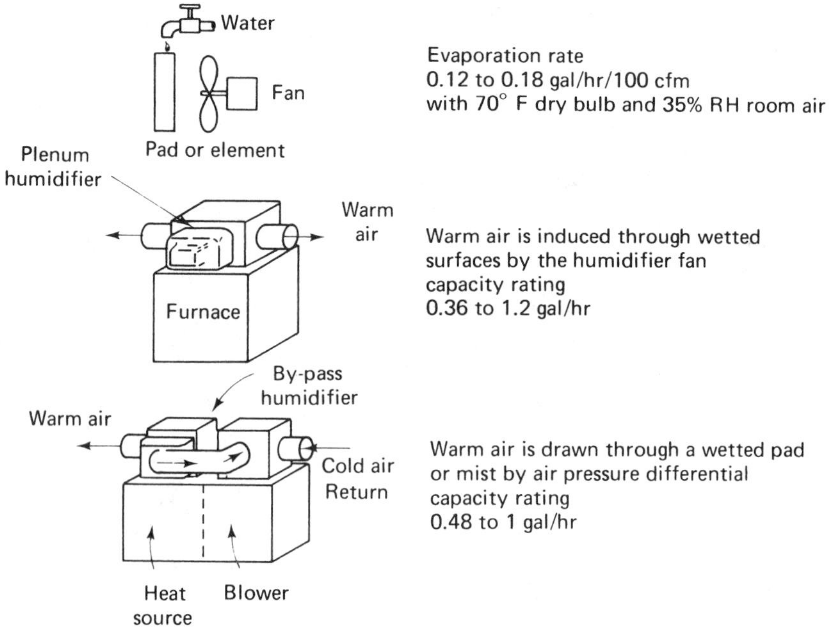
Water
Fan
Pad or element
Plenum humidifier
Warm air
Furnace
By-pass humidifier
Warm air
Cold air Return
Heat source
Blower
Evaporation rate
0.12 to 0.18 gal/hr/100 cfm
with 70° F dry bulb and 35% RH room air
Warm air is induced through wetted surfaces by the humidifier fan
capacity rating
0.36 to 1.2 gal/hr
Warm air is drawn through a wetted pad or mist by air pressure differential
capacity rating
0.48 to 1 gal/hr

FIGURE 13–6
Wetted-element-type humidifier. (*Courtesy Research Products Corp.*)

Humidifying with a Heat Pump

Heat pump systems are sophisticated systems that require more than the usual attention to airflow, sizing, and installation techniques. Proper humidification is just as important for homes that use heat pump systems as with any other type of heating system. Because heat pump systems operate with a lower discharge air temperature than conventional systems, the water supply to the humidifier is taken from the hot water line from the water heater. The flushing water that carries most of the minerals from the humidifier down the drain after the humidifying process is completed has been cooled to the touch. These systems may use either a forced-air-type humidifier or a bypass-type humidifier. For bypass installations, see Figure 13–7. Table 13–7 shows the specifications for the Aprilaire® Models 440/445.

HUMIDIFIER SIZING

Sizing for humidification is similar to sizing a system for heating and cooling. The humidifier capacity required will be determined by various factors: (1) the volume of the area being humidified; (2) the air change rate (infiltration or ventilation); (3) the inside and outside design conditions; and (4) other sources of humidity. Each factor is discussed as follows:

TABLE 13–7 Specifications for Aprilaire ® Models 440/445. (*Courtesy Research Products Corp.*)

SPECIFICATIONS	
STATIC PRESSURE DROP ACROSS SUPPLY AND RETURN PLENUM	.2"
CAPACITY—GALS. PER HOUR (at 120° plenum temperature)	0.7
VOLUME OF HOME* (CU. FT.)	
LOOSE HOUSE	8,000
AVERAGE HOUSE	16,000
TIGHT HOUSE	32,000
PLENUM OPENING	13¾" WIDE x 11¼" HIGH
ELECTRICAL DATA	
Model 440	24V—60Hz—.7 amp.
Model 445	120V—60Hz—.5 amp.

WATER FEED (Model 440) (Orifice-metered)	brown orifice, .024" diameter opening (6 gph)
ALSO INCLUDED	
Model 440	Humidistat, Saddle Valve, Transformer
Model 445	Humidistat, Saddle Valve
UNIT SIZE	
Model 440	16½" W. x 15" H. x 8" D.
Model 445	16½" W. x 18" H. x 8" D.
DISCHARGE DUCT OPENING	10" x 3¼" (Will accept std. 10" x 3¼" to 6" dia. round fitting)
SHIPPING WEIGHTS	
Model 440	15 lbs.
Model 445	19 lbs.

* Calculations based on (1) 20°F. and 70% R.H. Outside Air. (2) 70°F. and 35% R.H. Room Air. (3) Internal Moisture Gain of 1 lb./hr. (4) 70% Blower Operation. On heat pumps the capacity of the Model 440 using 140°F. service hot water is 0.6 gph.

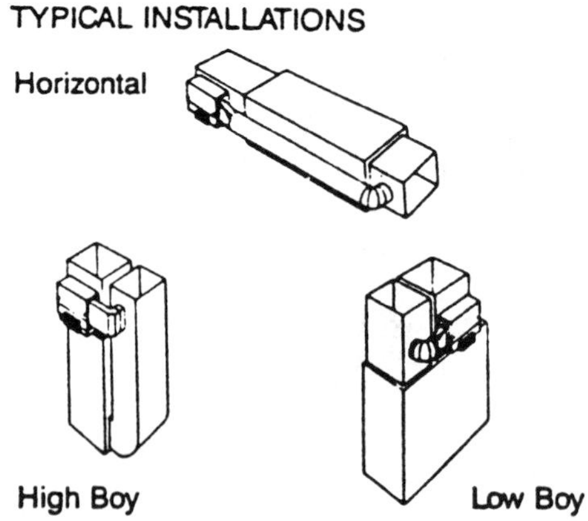
TYPICAL INSTALLATIONS
Horizontal
High Boy
Low Boy

FIGURE 13–7
Typical bypass humidifier installations. (*Courtesy Research Products Corp.*)

1. The volume of the home being humidified: The volume can be determined from a floor plan or from measurements taken within the home. If the basement is heated and ventilated, its volume should be included.
2. Air change rate: The amount of infiltration was probably calculated when computing the heating and cooling load.

 The following method may also be used:

 An average house will normally have about one air change per hour. A tight house may have as little as one-half air change per hour and a loose house may have as high as two air changes per hour. For definition purposes, an average house is assumed to have insulation in the walls and ceilings, vapor barriers, loose storm doors and windows, and may or may not have a fireplace. If it has a fireplace, however, it will be dampered. The tight house will be well insulated, have vapor barriers, tight storm doors and windows with weather stripping, and its fireplace will be dampered. The loose house will probably be one constructed before 1930, have little or no insulation, no storm doors or windows, no weather stripping, no vapor barriers, and often will have a fireplace without an effective damper.
3. Design conditions: The principal factors involved are (a) the desired indoor temperature and relative humidity; and (b) the prevailing outdoor temperature and relative humidity.

Comfort is usually the prime requirement in residential humidification. The human body is comfortable within a broad range of relative humidity. Humidities of 35% or more are desirable from the health standpoint. However, in the winter these high humidities can present a condensation problem. There is, however, a compromise between the humidities preferable from a health standpoint and the humidities desirable from a construction standpoint (Table 13–8). Because of the condensation problem, the humidity maintained within a home should be lowered as the outdoor temperature drops.

TABLE 13–8 Recommended indoor relative humidity levels

OUTDOOR TEMP., °F	RECOMMENDED HUMIDITY, %
20° and above	35%
10°	30%
0°	25%
−10°	20%
−20°	15%

(Courtesy Research Products Corp.)

However, if the house is designed to withstand higher humidities or if for medical reasons a higher humidity must be maintained, then higher values can be used.

4. Other sources of humidity: The principal sources of humidity in residential buildings are cooking; bathing, especially taking showers; washing clothes; and drying clothes in an unvented clothes drier. When excessive humidity conditions are encountered from these sources, some means to reduce the humidity, such as venting the clothes dryer, or installing a vent-a-hood over the cook stove, should be done.

CALCULATIONS

When calculating the amount of humidity required, the following formula can be used:

$$\mathrm{H} = \frac{VR(W_{\mathrm{i}} \times W_{\mathrm{o}})}{13.5 \times 8.3}$$

where

H = gallons of moisture per hour required to maintain indoor design conditions

V = number of changes of air per hour

R = number of changes of air per hour

W_i = pound of moisture per pound of dry air at the desired indoor conditions (from the psychrometric chart)

W_o = pound of moisture per pound of dry air at outdoor conditions (from the psychrometric chart).

The value of 13.5 ft^3 of air per pound of air is an average that is suitable for calculations where extreme accuracy is not required. The value of 8.3 is the number of pounds of moisture in a gallon.

HUMIDIFIER OPERATION

Sequence of Operation

Most humidifiers are wired into the control circuit so that they will operate only when the indoor fan is in operation. If the humidifier were allowed to operate without the indoor fan, the moisture would accumulate mostly in the discharge air plenum of the heating unit and not be distributed into the heated space. Excess moisture could also cause the heat exchanger and cabinet parts to rust and become inoperative, or dangerous to operate.

When the heating unit is operating, outdoor air is pulled into the heated space through cracks around doors, windows, and other structural cracks. This infiltrated air is then pulled through the heating unit by the indoor fan. When this air is heated it expands, becomes drier, and the relative humidity is reduced proportionately. The purpose of the humidifier is to replace this moisture and raise the level of humidity to the comfort level.

The heated discharge air from the heating unit is passed through the humidifier and moisture is added to it. The air will absorb more moisture at this time when it is the warmest and driest. The warm, moist air from the humidifier is then mixed with the remainder of the discharge air and is directed into the living space to raise the relative humidity.

Humidifiers require both 120V ac and 24V ac electricity for operation (Figure 13–8). Bypass-type humidifiers are wired differently from the forced-air-

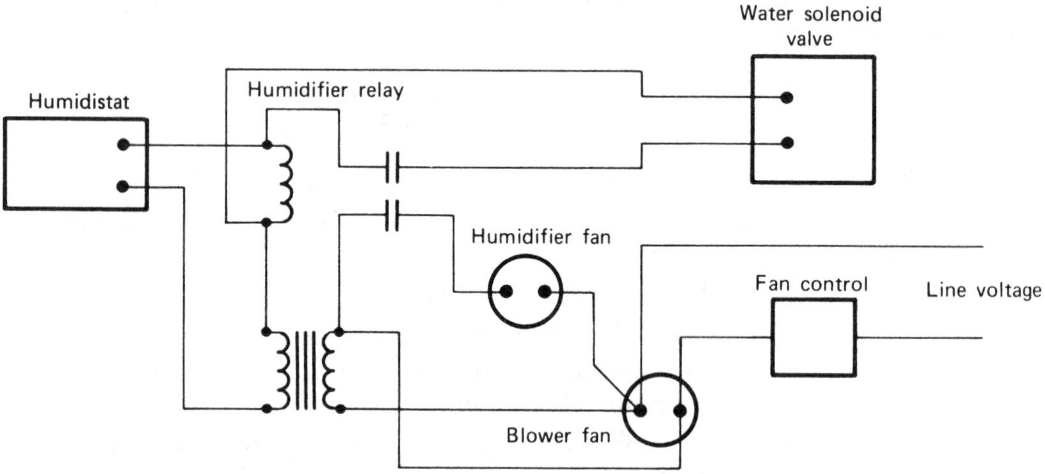
Water solenoid valve
Humidistat
Humidifier relay
Humidifier fan
Fan control
Line voltage
Blower fan

FIGURE 13–8
Typical humidifier wiring diagram. (*Courtesy of Billy C. Langley,* Electric Controls for Refrigeration and Air Conditioning, *2E, ©1988, p. 106; reprinted by permission of Prentice-Hall, Inc., Englewood Cliffs, NJ.*)

type humidifiers (Figure 13–9). The 120V circuit is powered only when the indoor fan motor is in operation. The 24V circuit is supplied by a transformer that is powered from the 120V line from the indoor fan motor. When wired in this manner, the humidifier will operate only when the indoor fan motor is in operation.

The 24V control circuit is made up of the humidistat and the electromagnetic relay coil. The 120V power circuit is made up of the transformer primary coil, the water valve coil, and the humidifier fan motor (on forced-air models). The humidifier fan motor draws the heated air from the discharge air plenum of the heating unit and passes it through the wetted element of the humidifier, where it picks up moisture and then pushes it back into the discharge air plenum. There it is mixed with the discharge air from the heating unit. Some installations have the water valve coil and the humidifier fan motor wired into a separate 120V circuit, as shown in Figure 13–8.

Bypass-type humidifiers have no fan motor and, therefore, do not require the separate 120V fan circuit (Figure 13–9).

When the heating unit indoor fan motor is energized, 120V power is applied to the primary side of the transformer and the control circuit. The water valve coil is energized through the contacts of the humidifier relay when the humidistat demands more humidity. On a demand for more humidity, the 24V coil of the water solenoid is energized, allowing water to flow over the wetted pad of the humidifier. If the humidifier is a forced-air type, the humidifier fan motor is also energized by these same relay contacts. In the forced-air-type humidifier, the air is pulled into the humidifier by the fan motor, passed over the wetted element where humidity is absorbed by the air, and is then forced into the heating unit plenum (Figure 13–10).

When a bypass humidifier is used, the air is forced into and through the humidifier by the pressure differential between the discharge air plenum and the return air plenum (Figure 13–11). As the heated space is warmed and humidified, either the thermostat or the humidistat can stop operation of the humidifier. Both must be demanding before the humidifier will operate, but either can stop operation of it.

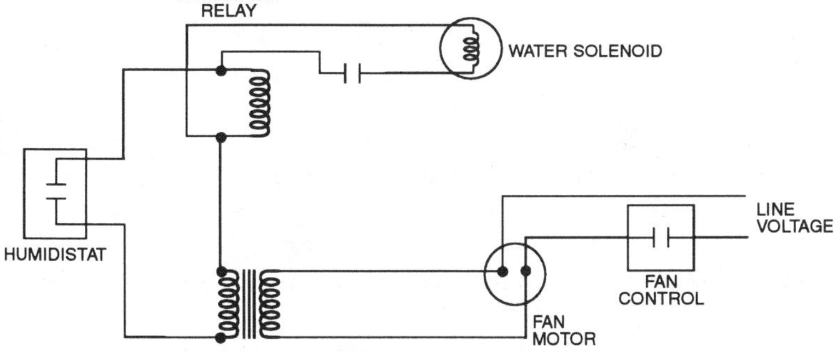
HUMIDIFIER
RELAY
WATER SOLENOID
HUMIDISTAT
LINE
VOLTAGE
FAN
CONTROL
FAN
MOTOR

FIGURE 13–9
Typical wiring diagram for bypass humidifier.

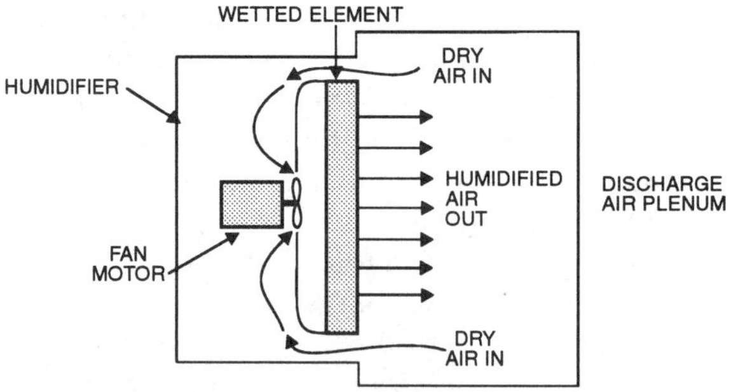
WETTED ELEMENT
HUMIDIFIER
DRY AIR IN
HUMIDIFIED AIR OUT
DISCHARGE AIR PLENUM
FAN MOTOR
DRY AIR IN

FIGURE 13–10
Air flow through a forced air type humidifier (top view).

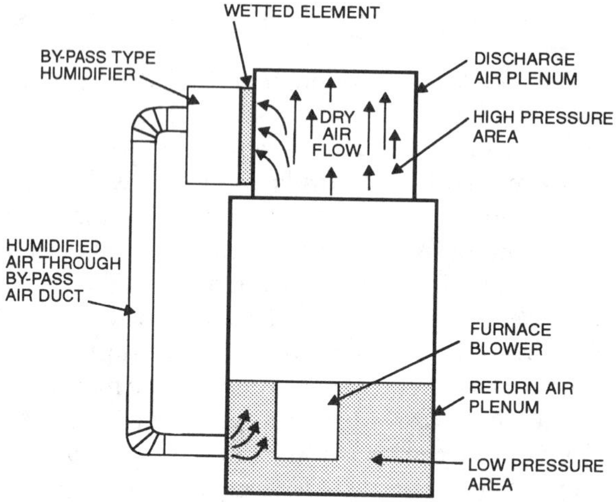
WETTED ELEMENT
BY-PASS TYPE HUMIDIFIER
DISCHARGE AIR PLENUM
HIGH PRESSURE AREA
DRY AIR FLOW
HUMIDIFIED AIR THROUGH BY-PASS AIR DUCT
FURNACE BLOWER
RETURN AIR PLENUM
LOW PRESSURE AREA

FIGURE 13–11
Air flow through a bypass type humidifier.

The Model 50 Electronic Relay

The model 50 electronic relay has been developed by Research Products Corporation to specifically interface with Aprilaire® Humidifiers with most heating and cooling systems. It is more reliable and less expensive than other interfacing equipment such as sail switches, temperature switches, and other relays (Figure 13–12). This control is ideally functional for use with multispeed blower motors and with 240V ac heat pump systems or electric furnaces.

The model 50 is easily installed around the common lead of the indoor blower motor and introduced into the circuitry (Figures 13–13 and 13–14).

An electrical conduit is not required because only low-voltage (24V) wiring is used. The model 50 does not measure voltage—it senses current. It is an electronic relay which will operate with up to 30-ampere current draw.

Important: The wire lead under the bracket must carry a minimum of 4 amps for proper operation. If less than 4 amps is available, simply wrap the lead wire around the bracket so that the same wire passes between the bracket and the relay

ACTUAL SIZE

FIGURE 13–12
Model 50 electronic relay. (*Courtesy Research Products Corp.*)

housing two or more times, as required for the proper current. The 24V ac wire leads can carry a maximum of $\frac{1}{2}$ amp.

Aprilaire® Humidifiers

The operating principle of all Aprilaire® humidifiers is basically the same, and all use nature's own process (the introduction of humidity in the form of water vapor), with the refinements necessary to provide positive, accurate control.

The humidistat, preferably located in the living area, is conveniently set at the desired humidity level and activates the unit whenever the humidity falls below the setting. Water is supplied to the distribution pan (Figure 13–15), from where it flows evenly across the water panel evaporator. Thirsty, dry air is forced through the wetted panel and the now humidified air carrying water as a vapor is distributed throughout the living area of the home.

Figure 13–16 illustrates the amount of mineral residue from the evaporation of just 1 gallon of 20 hardness water (that is high hardness). About 360 gallons of water must be evaporated monthly in a typical 13,000 ft^3 home (1,625 ft^2) to maintain proper relative humidity.

In some humidifiers, this imposing amount of mineral deposit build-up can cause a malfunction requiring service. And, in some, the minerals can be distributed throughout the home in the form of white dust.

Control

A precision-made, accurately calibrated humidistat is required for maximum performance (Figure 13–17). For maximum convenience, it is usually located in the living area. All that is required is to set the dial (and reset as necessary) for the desired humidity.

Models 110, 112, 445

1. Follow the step-by-step procedure noted on the installation template for mounting the unit. DO NOT connect the electrical, water supply or drain at this time.
2. See the special wiring instructions below:
   a. Attach the electronic relay around the common lead of the furnace blower motor (see detail Z). The electronic relay must be located at least 3" from any transformer. The metal bracket on the relay must not touch any metal.

   b. Connect one lead from the electronic relay to one of the 24 VAC wires from the humidifier with wire nut (A). (Additional wire length may be required.)

   c. Connect the remaining lead from the electronic relay to the humidistat wire with a wire nut (B).

   d. To complete the circuit, connect the second lead from the humidistat to the second 24 VAC lead from the humidifier, with a wire nut (C).

   e. The 120 Volt power cord from the humidifier must connect to a 120 VAC source INDEPENDENT of the furnace blower motor.

   CAUTION: THE HUMIDIFIER WILL MALFUNCTION IF THE 120 VAC POWER IS CONNECTED TO THE FURNACE MOTOR CIRCUIT.
3. Complete the installation following the instructions on the template supplied with the humidifier.

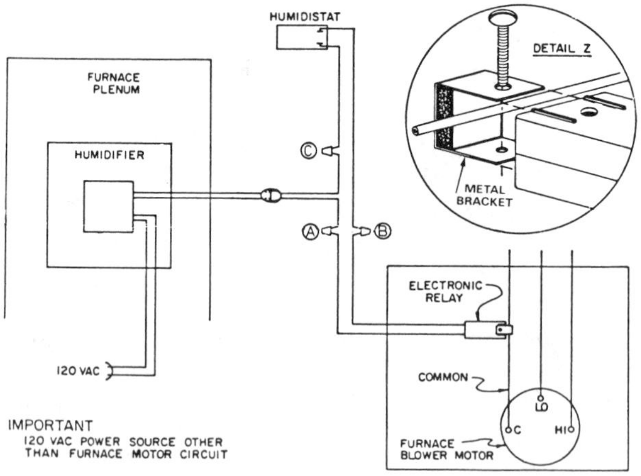
HUMIDISTAT
DETAIL Z
FURNACE PLENUM
HUMIDIFIER
C
A
B
METAL BRACKET
ELECTRONIC RELAY
COMMON
LO
C
HI
FURNACE BLOWER MOTOR
120 VAC
IMPORTANT
120 VAC POWER SOURCE OTHER THAN FURNACE MOTOR CIRCUIT

FIGURE 13–13
Model 50 electronic relay with models 110, 112, and 445 Aprilaire® humidifiers. (*Courtesy Research Products Corp.*)

Models 220, 224, 440

1. Follow the step-by-step procedure noted on the installation template for mounting the unit. NOT connect the electrical, water supply or drain at this time.
2. See the special wiring instructions below:
   a. Attach the electronic relay around the common lead of the furnace blower motor (see detail Z). The electronic relay must be located at least 3" from any transformer. The metal bracket on the relay must not touch any metal.

   b. Connect one lead from the water solenoid valve to the 24 VAC transformer (A).

   c. Connect the other lead from the water solenoid to the humidistat (B).

   d. Connect one lead from the electronic relay to the second terminal of the 24 VAC transformer (C). (Additional wire length may be required.)

   e. To complete the circuit, connect the remaining electronic relay lead to the humidistat (D).

   f. The 120 VAC/ 24 VAC step down transformer supplied with the humidifier MUST be connected to a 120 VAC source INDEPENDENT of the furnace blower motor.

   CAUTION: THE HUMIDIFIER WILL MALFUNCTION IF THE 120 VAC POWER IS CONNECTED TO THE FURNACE MOTOR CIRCUIT.
3. Complete the installation following the instructions on the template supplied with the humidifier.

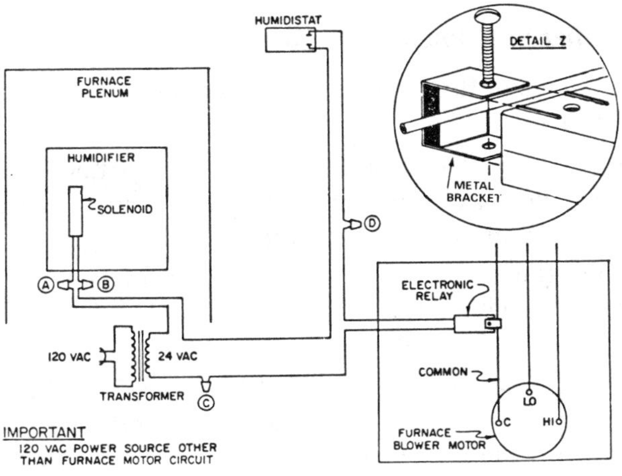
HUMIDISTAT
DETAIL Z
FURNACE PLENUM
HUMIDIFIER
SOLENOID
METAL BRACKET
Ⓓ
Ⓐ
Ⓑ
ELECTRONIC RELAY
120 VAC
24 VAC
TRANSFORMER
Ⓒ
COMMON
LO
C
HI
FURNACE BLOWER MOTOR
IMPORTANT
120 VAC POWER SOURCE OTHER THAN FURNACE MOTOR CIRCUIT

FIGURE 13–14
Model 50 electronic relay with models 220, 224, and 440 Aprilaire® humidifiers. (*Courtesy Research Products Corp.*)

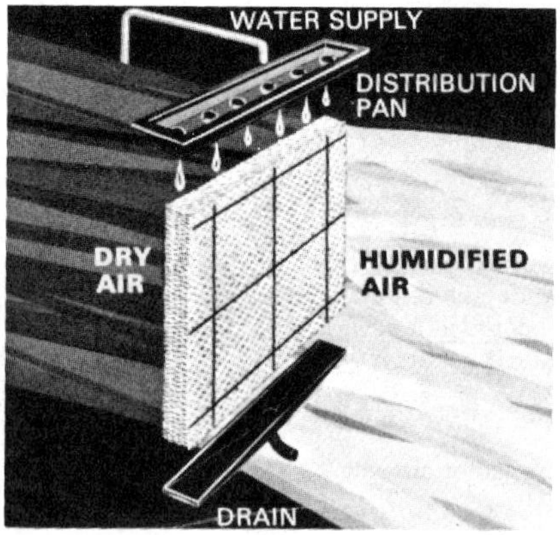
WATER SUPPLY
DISTRIBUTION PAN
DRY AIR
HUMIDIFIED AIR
DRAIN

FIGURE 13–15
Aprilaire® humidifier operation. (*Courtesy Research Products Corp.*)

FIGURE 13–16
Mineral residue. *(Courtesy Research Products Corp.)*

15 20
RP

FIGURE 13–17
Humidistat. (*Courtesy Research Products Corp.*)

Important Aprilaire Humidifier Principles

1. Drain (Figure 13–18). The water flushes a significant amount of the minerals completely out of the system.
2. Distribution pan (Figure 13–19). The scientifically designed pattern of circular inlets permits water to flow, by gravity, uniformly across the water panel evaporator for more efficient, higher capacity humidification.
3. Water reservoir and liner (Figure 13–20). The partitioned reservoir controls scale and separates fresh and circulated water. A removable liner simplifies cleaning on some models.
4. Water panel evaporator and scale control insert (Figure 13–21). The water panel is absorbent, with high saturation ability, to ensure uniform evaporation, optimum efficiency, and high capacity. It traps trouble-causing minerals.

FIGURE 13–18
Aprilaire® humidifier drain. (*Courtesy Research Products Corp.*)

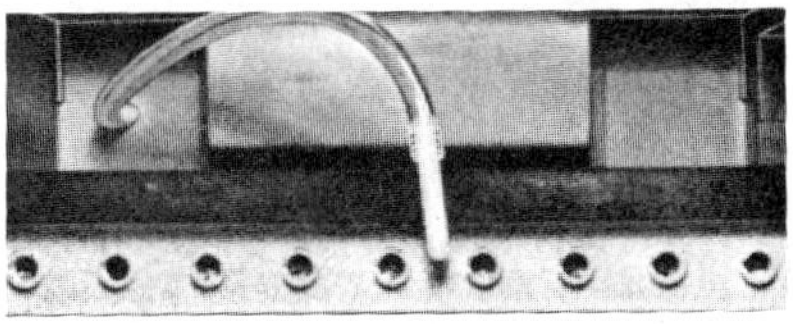

FIGURE 13–19
Aprilaire® distribution pan. (*Courtesy Research Products Corp.*)

FIGURE 13–20
Aprilaire® partitioned reservoir. (*Courtesy Research Products Corp.*)

FIGURE 13–21
Aprilaire® water panel evaporated and scale control insert. (*Courtesy Research Products Corp.*)

The scale control insert localizes excess mineral deposits—another aid in reducing maintenance.

5. Solenoid valve (Figure 13–22). This valve is electrically controlled for positive water supply and shutoff. A Teflon orifice assures a positive, metered flow of water.

A complete line of Aprilaire® humidifiers offers a choice of models to fulfill the requirements of installations from homes to commercial establishments.

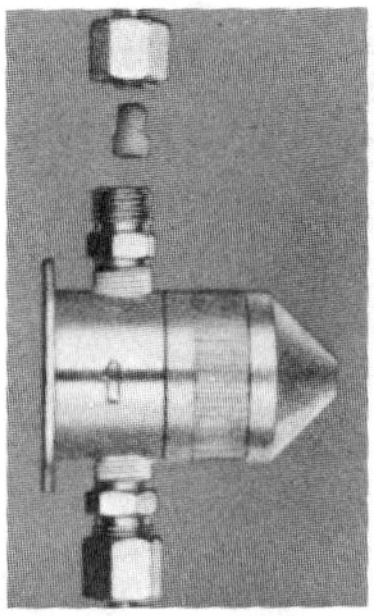

FIGURE 13–22
Aprilaire® solenoid valve. (*Courtesy Research Products Corp.*)

The models 110 and 112 (Figure 13–23) are forced-air furnace-attached units that incorporate a fan and a drain and represent the ultimate in controlled humidification for residential applications.

The model 330 (Figure 13–24) is a self-contained unit designed especially for homes heated by hot water or steam systems, and equally suitable for electric or any other type of heating system. It can be installed in the basement, utility room, or heated crawl space.

Aprilaire
Model 110

FIGURE 13–23
Aprilaire® models 110 and 112 humidifiers. (*Courtesy Research Products Corp.*)

The Aprilaire® models 440 and 445 (Figure 13–25) and the Chippewa models 220 and 224 (Figure 13–26) are bypass-type humidifiers that can be installed on either the supply or the return air plenum of a forced-air furnace. They are ideal for installations where plenum space is limited. All of these are equipped with a drain except the model 445, which has a water circulating system and is designed for installations where drain facilities are not available, or where high water hardness is not a problem.

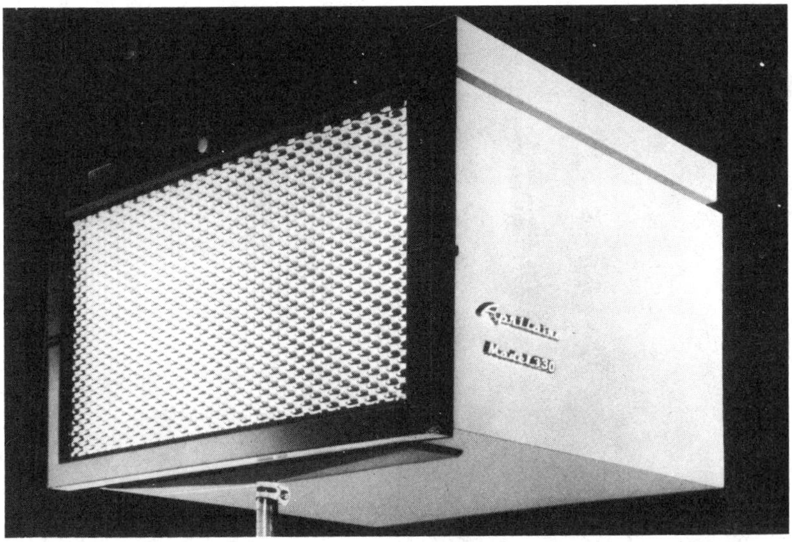

FIGURE 13–24
Aprilaire® model 330 humidifiers. (*Courtesy Research Products Corp.*)

FIGURE 13–25
Aprilaire® models 440 and 445 humidifiers. (*Courtesy Research Products Corp.*)

Lau Humidifiers

The Lau Vapor-Air residential humidifiers operate on the proven vapor-wheel principle (Figure 13–27). By taking dry air and forcing it through a saturated, rotating media on the vapor-wheel, the dry air becomes comfortably humidified. The moisture-laden air produced by the humidifier is then carried throughout the home (Figure 13–28).

Chippewa
Model 224

FIGURE 13–26
Aprilaire® Chippewa humidifier. (*Courtesy Research Products Corp.*)

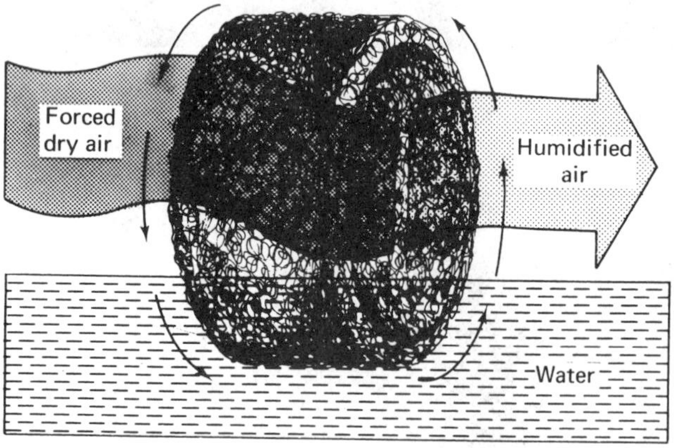
Forced dry air
Humidified air
Water

FIGURE 13–27
Lau Vapor-Wheel. (*Courtesy Lau, Inc.*)

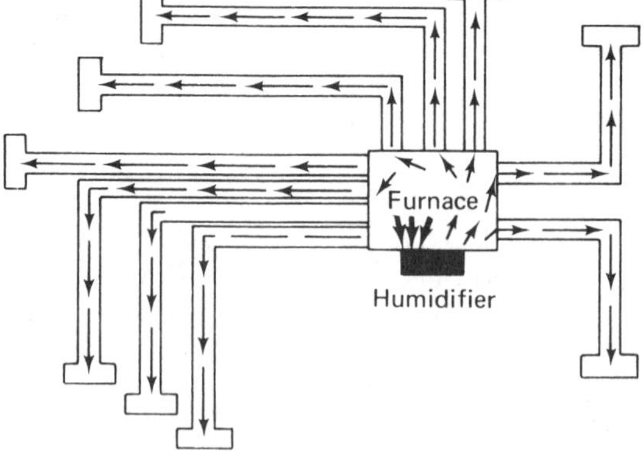
Furnace
Humidifier

FIGURE 13–28
Moisture distribution throughout a home. (*Courtesy Research Products Corp.*)

The Lau Vapor-Wheel principle is an extremely economical method of operation. A Lau Vapor-Air Humidifier costs no more to operate per day than a 60-watt light bulb.

Applications: The Lau Vapor-Air 2 is a highly versatile, easy to install, automatic central system humidifier (Figure 13–29). It is designed for use on all residential and commercial forced warm air heating systems. It is capable of delivering up to 38 gallons of water per 24 hours of operation.

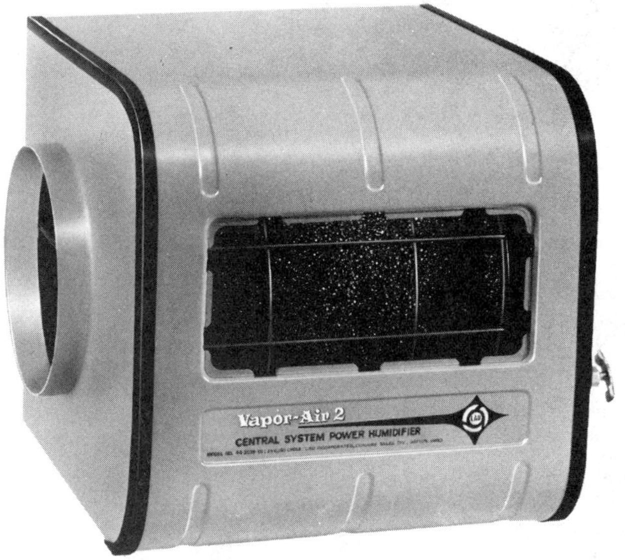
Vapor-Air 2
CENTRAL SYSTEM POWER HUMIDIFIER

FIGURE 13–29
Lau Vapor-Air 2 humidifier. (*Courtesy Lau, Inc.*)

During operation, air is drawn into the unit through the flexible ducting by the static pressure differential between the warm air and cold air ducts (Figure 13–30). The unit is designed so that the air drawn into the unit passes through the saturated media pad, where it becomes properly humidified, and is forced into the heating system.

The Vapor-Air 2 can be installed on any forced warm air heating system. It may be mounted on either the warm air or cold air plenum. The unit is completely reversible so that, depending on the installation, the flexible duct may be mounted on either the right or left side of the humidifier. It is extremely easy to service. The lid raises for maximum access, the media wheel is easily removable, and the reservoir slides out (Figure 13–30). A water shutoff valve on the side of the unit eliminates the need to turn off the water at the saddle valve.

The Vapor-Air 3 is a trim, appliance-styled, completely automatic central system power humidifier designed for average to large homes requiring up to 20 gallons of moisture per day (Figure 13–31).

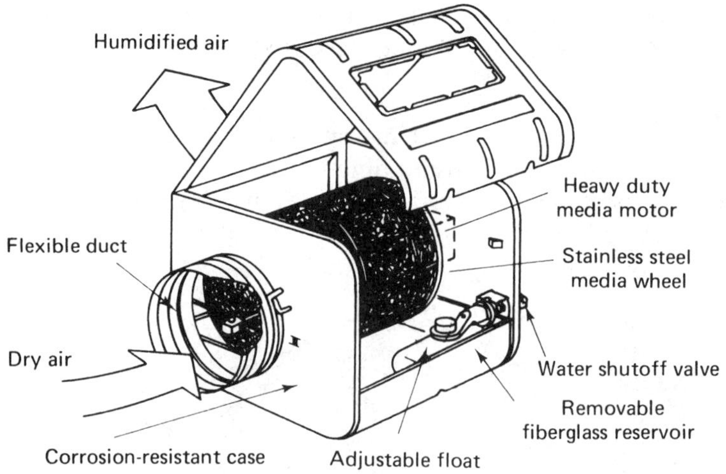
Humidified air
Heavy duty media motor
Flexible duct
Stainless steel media wheel
Dry air
Water shutoff valve
Removable fiberglass reservoir
Corrosion-resistant case
Adjustable float

FIGURE 13–30
Lau Vapor-Air 2 humidifier air flow. (*Courtesy Lau, Inc.*)

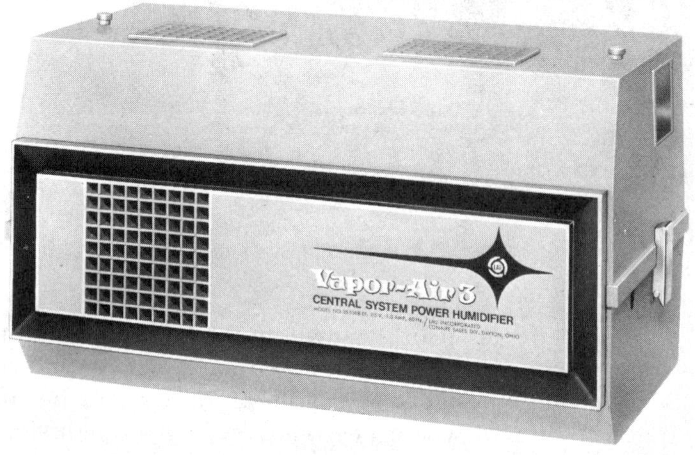
Vapor-Air 3
CENTRAL SYSTEM POWER HUMIDIFIER

FIGURE 13–31
Lau Vapor-Air 3 humidifier. (*Courtesy Lau, Inc.*)

The specifications for the Vapor-Air 3 are shown in Table 13–9.

For a central system, bypass-type humidifier, the Vapor-Air 4 is designed for smaller homes with forced warm air heating systems (Figure 13–32). This unit requires no electrical connections and has a capacity of up to 14 gallons per day.

In operation, the Vapor-Air 4 uses the Lau Vapor-Wheel principle coupled with the Lau Dyna-Drive Assembly (Figure 13–33). A multi-blade propeller is

TABLE 13–9. Vapor-Air specifications. (*Courtesy Lau, Inc.*)

Duct Temp. Fahrenheit	Gal/24 Hr	Lb/Hr
180°	19.50	6.8
160°	16.95	5.9
*140°**	*14.75**	*5.1**
120°	11.75	4.1
100°	9.29	3.2
80°	6.88	2.4

(*Courtesy Lau, Inc.*)

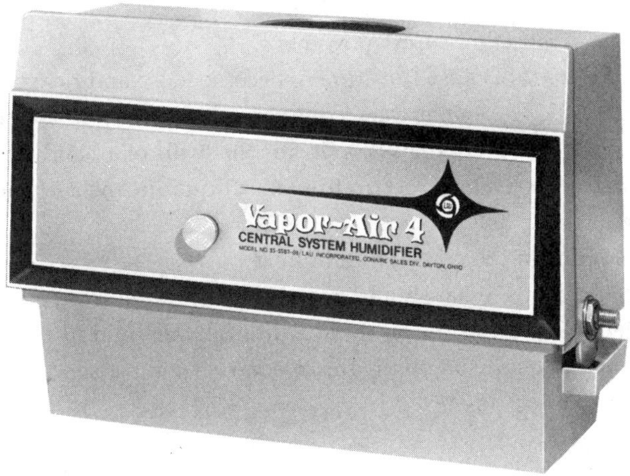
Vapor-Air 4
CENTRAL SYSTEM HUMIDIFIER

FIGURE 13–32
Lau Vapor-Air 4 humidifier. (*Courtesy Lau, Inc.*)

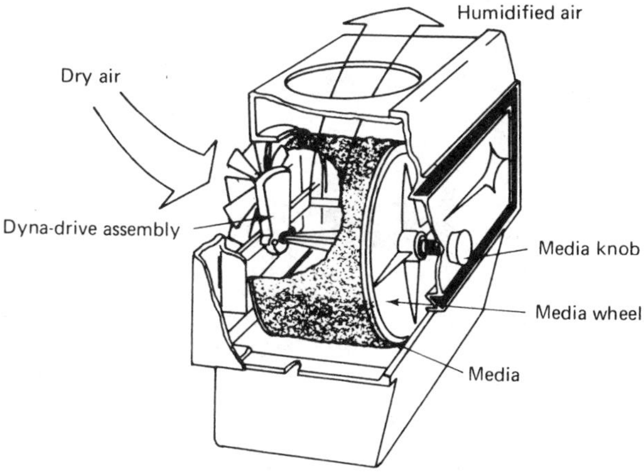
Humidified air
Dry air
Dyna-drive assembly
Media knob
Media wheel
Media

FIGURE 13–33
Operating principle of Lau Vapor-Air 4. (*Courtesy Lau, Inc.*)

driven off the pressure differential air velocity that takes place between the warm and cold air plenum in a forced-air heating system. The air-driven propeller drives the reduction gear that rotates the media wheel at 2 rpm with an 85 cubic feet per minute (cfm) bypass at .2 static pressure for the designated capacity.

REVIEW QUESTIONS

1. What is the water vapor within a space known as?
2. What is the weight of water vapor per unit volume known as?
3. Why is the air more muggy in the bathroom after taking a shower than in the rest of the house?
4. Will moisture move more easily to a higher or a lower temperature area?
5. Will moisture tend to be more evenly distributed in a building with the same temperature throughout?
6. What is used in construction to restrict moisture movement?
7. What type of humidity conditions will cause furniture to become loose and rickety?
8. With an ambient temperature of 35° F, what is the recommended relative humidity?
9. What is the best indication that there is condensation inside the walls of a building?
10. How many test readings are required to get an accurate relative humidity measurement?
11. Which type of humidifier works like an evaporative cooler?
12. Why is more water than is needed supplied to a humidifier?
13. As the outdoor air temperature drops, what should be done to the humidistat?
14. On a bypass-type humidifier, what causes the air to flow through the humidifier?
15. Why should the humidifier operate only when the furnace fan is running?

14

electric and electronic ignition systems

Both electric and electronic ignition systems are part of the control systems used on modern furnaces. The ignition system may light the pilot burner, or the main burner directly, depending on the manufacturer's system design. Each control system manufacturer has a slightly different wiring and operating sequence for the control systems that it manufactures. There are several types of each; therefore, it would be almost impossible to cover all of them in a manual of this type. However, an awareness of these types can be helpful in servicing, installing, and maintaining any of the other types encountered.

The main types are the glow coil, hot surface ignitor, and the direct spark ignitor. Some state and local laws require the use of automatic ignitors and have done away with the standard standing pilot on new equipment.

ELECTRIC IGNITION

Electric ignition systems generally use 2.5, 24, or 120 V AC as the source for ignition. In most instances, the electrical circuit is used to cause a temperature high enough to ignite the gas as it escapes from the burner.

Glow Coil Ignitors

Glow coil ignitors use a coil of resistance wire placed in the pilot gas stream. When the electricity is applied to the glow coil, it gets red hot and ignites the pilot gas (Figure 14–1). A single-pole, double-throw thermopilot relay is used in conjunction with the glow coil (Figure 14–2). When the pilot is not burning, the relay

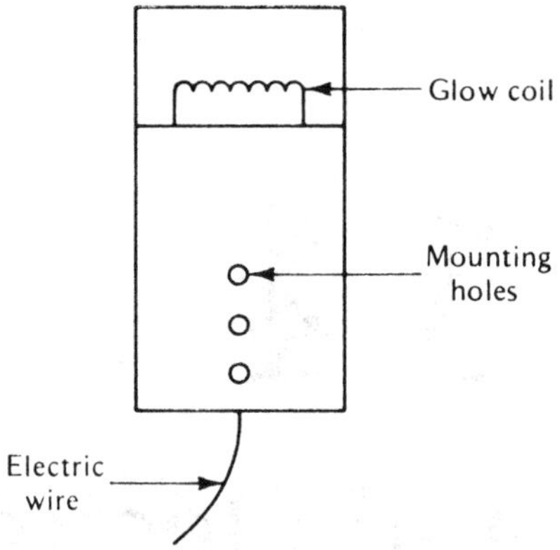
Glow coil
Mounting holes
Electric wire

FIGURE 14–1
Glow coil ignitor. (*Courtesy of Billy Langley,* Heating, Ventilating, Air Conditioning, and Refrigeration, *©1990, p. 549; reprinted by permission of Prentice-Hall, Inc., Englewood Cliffs, N.J.*)

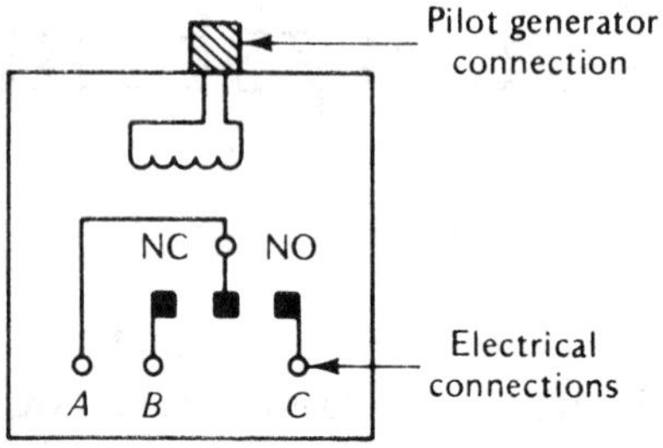
Pilot generator connection
NC
NO
A
B
C
Electrical connections

FIGURE 14–2
Thermopilot relay in deenergized position. (*Courtesy of Billy Langley,* Heating, Ventilating, Air Conditioning, and Refrigeration, *©1990, p. 549; reprinted by permission of Prentice-Hall, Inc., Englewood Cliffs, N.J.*)

contacts are in a position to supply 2.5 V AC electric current to the glow coil. After the pilot is ignited and the thermocouple is heated sufficiently, the movable contact moves to complete the 24-V AC circuit to the temperature control circuit (Figure 14–3). The complete control system must be wired properly to provide the desired results (Figure 14–4). This type of ignitor is used to light or relight pilot burners in the case of flame failure. These units are assembled in a draft protected holder, which provides a means for either horizontal or vertical mounting to a pilot burner (Figure 14–5).

Glow coil ignitors are connected to an electrical power source of 2.5 volts. The glow coil is heated because of electrical resistance. After the pilot is lighted, the electrical power to the glow coil is interrupted. This step is necessary because continuous electrical power to the glow coil will cause it to burn out and become inoperative. Use of glow coil ignitors is limited to the lighting of pilot burners only; they should not be used for main burner ignition.

ELECTRONIC IGNITION SYSTEMS

Electronic ignition systems are designed to conserve energy by shutting off the pilot burner gas when there is no call for heat from the thermostat. These controls save from 3 to 5% of the gas normally used in standing pilot installations and are known as intermittent ignition devices (IIDs).

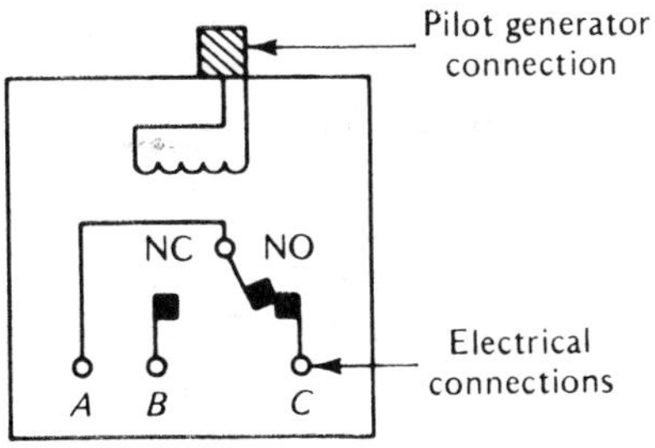
Pilot generator
connection
NC
NO
Electrical
connections
A
B
C

FIGURE 14–3
Thermopilot relay in energized position. (*Courtesy of Billy Langley,* Heating, Ventilating, Air Conditioning, and Refrigeration, *©1990, p. 549; reprinted by permission of Prentice-Hall, Inc., Englewood Cliffs, N.J.*)

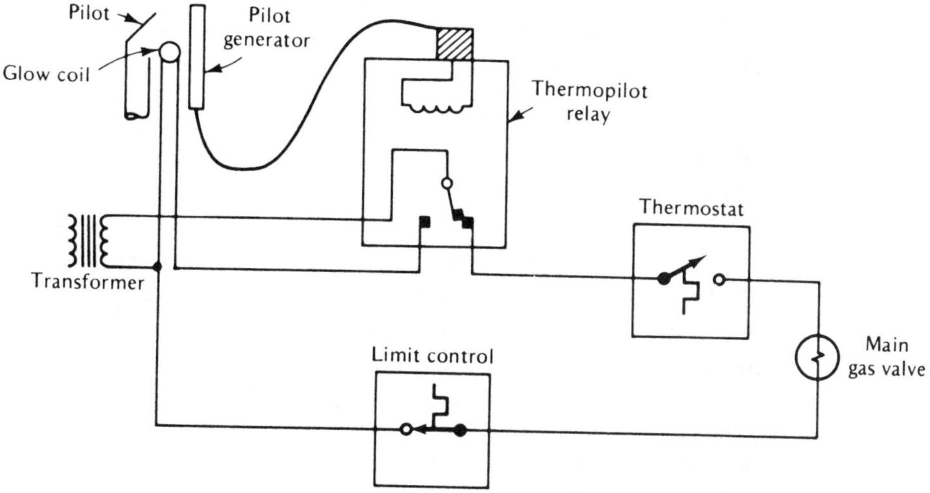
Pilot
Pilot
generator
Glow coil
Thermopilot
relay
Thermostat
Transformer
Limit control
Main
gas valve

FIGURE 14–4
24-V temperature control circuit with automatic relight. (*Courtesy of Billy Langley,* Heating, Ventilating, Air Conditioning, and Refrigeration, *©1990, p. 550; reprinted by permission of Prentice-Hall, Inc., Englewood Cliffs, N.J.*)

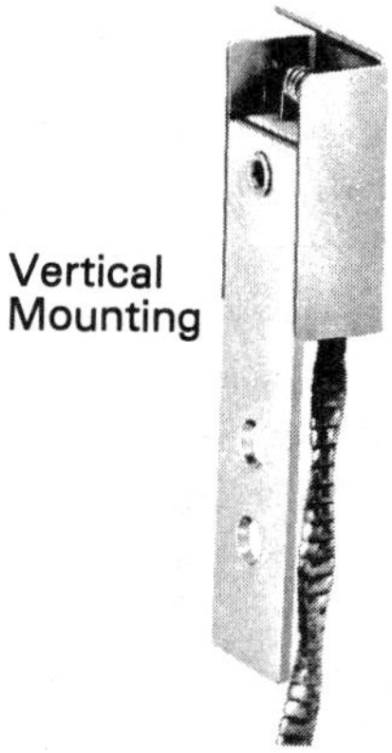
Vertical
Mounting

FIGURE 14–5
Glow coil. (*Courtesy of Billy Langley,* Heating, Ventilating, Air Conditioning, and Refrigeration, *©1990, p. 548; reprinted by permission of Prentice-Hall, Inc., Englewood Cliffs, N.J.*)

Components and Parts Description

An IID includes the following components:

1. Ignition control: This component houses the electronic circuitry used to control the safety, sequencing, and operation of the heating system.
2. Gas valve: The gas valve electronically controls both the pilot and the main burner gas supply. It may also include a gas pressure regulator and a manual shutoff valve.
3. Electrode pilot burner: The electrode pilot burner incorporates an electrode that lights the pilot gas on a call for heat from the temperature control.
4. Flame sensor: The flame sensor detects the presence of the pilot flame and acts as a key part of the sensing circuit.

Sequence of Operation

The following is the sequence of operation of electronic ignition systems:

1. When the temperature controller calls for heat, the spark transformer in the ignition control and the pilot gas valve are automatically energized.
2. The spark lights the pilot gas on each operating cycle.
3. The flame sensor proves the presence of the flame. The ignition control then shuts off the spark transformer. At the same time, the main burner gas valve is energized. (Some models permit the spark to continue for a short period of time after the main burner gas is ignited.) On 100% lockout models, a shutdown of the entire system will occur if the pilot gas does not light within some fixed period of time (usually 30 seconds).
4. The main burner gas lights and the system continues the normal operating cycle.
5. When the temperature control is satisfied, the main burner and pilot gas valves are deenergized, shutting off all gas flow to the burners.

Principles of Operation

Basically there are two different types of electronic ignition systems in use today. One type uses a spark to ignite the burner gas, and the other type uses a hot surface to ignite the burner gas.

The spark ignition systems can be further divided into two types: (1) the intermittent ignition device, and (2) the direct spark ignition (DSI) system. The IID is the most popular spark ignition system used on residential systems. It is used to light the pilot burner gas, which, in turn, lights the main burner gas. In these installations, the pilot flame must be proven before the main burner valve is energized.

The direct spark ignition system uses a spark to light the main burner gas directly without the use of a pilot burner. These systems are also popular on residential units and in commercial applications that require energy savings and automatic ignition for the main burner gas. Some manufacturers use a spark plug similar to the one used in an automobile. Others make use of a spark electrode directly over the main burner.

Sensing Methods

When a standing pilot is used, heat is used to operate a thermocouple to energize the pilot safety circuit. In installations using the IID, flame conduction or rectification is used. To aid in understanding the principles of flame conduction, we must first have a thorough understanding of the structure of a gas flame (Figure 14–6).

When the flame is properly adjusted with the proper air-gas ratio, there are three cones or zones present:

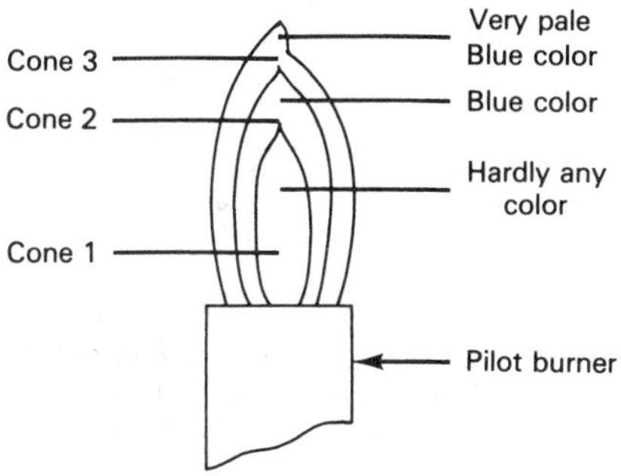
Cone 3
Cone 2
Cone 1
Very pale
Blue color
Blue color
Hardly any
color
Pilot burner

FIGURE 14–6
Structure of a gas flame. (*Courtesy of Billy Langley,* High-Efficiency Gas Furnace Troubleshooting Handbook, *©1991, p. 228; reprinted by permission of Prentice-Hall, Inc., Englewood Cliffs, N.J.*)

1. In cone 1, or the inner cone, the flame will not burn because there is an excess of fuel to air.
2. Cone 2, or the intermediate cone, surrounds the inner cone. This area has a slightly different color of blue and is where the proper combustion occurs. This cone takes part of its combustion air from the surrounding air, or secondary air.
3. Cone 3, also known as the outer cone or envelope, contains an excessive amount of air, which it gets from the surrounding secondary air. This envelope is also a different color of blue from the other two cones.

Cone 2, or the intermediate cone, is the place that we are most concerned with. Since this is the place where the best combustion occurs, it is a prime location for the sensing probe of the electronic ignition device.

In actuality, a flame is but a series of small, controlled, explosions that cause the immediate area to become ionized. The ionization of this area causes the area to become electrically conductive (Figure 14–7). This conductive flame is often thought of as being an electrical switch located between the flame sensor and the pilot burner tip. If there is not a flame between the sensing probe and the pilot burner tip, the switch is open. When the flame makes contact with both the pilot

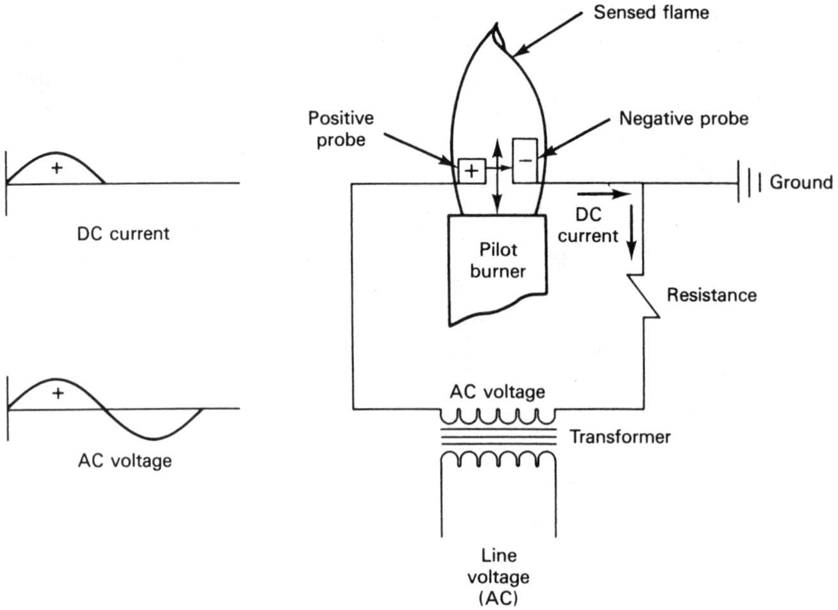

Sensed flame
Positive probe
Negative probe
Ground
DC current
DC current
Pilot burner
Resistance
AC voltage
AC voltage
Transformer
AC voltage
Line voltage (AC)

FIGURE 14–7
Principle of flame rectification. (*Courtesy of Billy Langley,* High-Efficiency Gas Furnace Troubleshooting Handbook, *©1991, p. 229; reprinted by permission of Prentice-Hall, Inc., Englewood Cliffs, N.J.*)

burner tip and the sensing probe, the switch is closed. In Figure 14–8 the flame is used to conduct an AC signal. In this case, both of the probes have approximately the same amount of area exposed to the flame. However, this is not a safety device signal because it does not identify the type of current conducted by the flame. This type of system could mistake an electrical short and allow the main gas valve to be opened without a means of ignition. To prevent this, a difference in type of flame must be noted. This difference is known as flame rectification (Figure 14–9).

Flame Rectification

When flame rectification is used, the flame and probes are used in a similar manner, with one important exception: The area of one of the probes exposed to the flame must have a greater area than the other one.

In Figure 14–9 the flame is again used to conduct an AC signal. Both of the probes are in contact with the flame. In this case the probe with the greatest area attracts the most free electrons and becomes electrically negative. In this condition, the direction of flow of electrons and, therefore, the current is from the positive probe to the negative probe. The AC voltage sine wave has not been changed, but the negative portion of the current sine wave is not present. Thus, the AC current has been changed to pulsating DC current. The result is flame rectification.

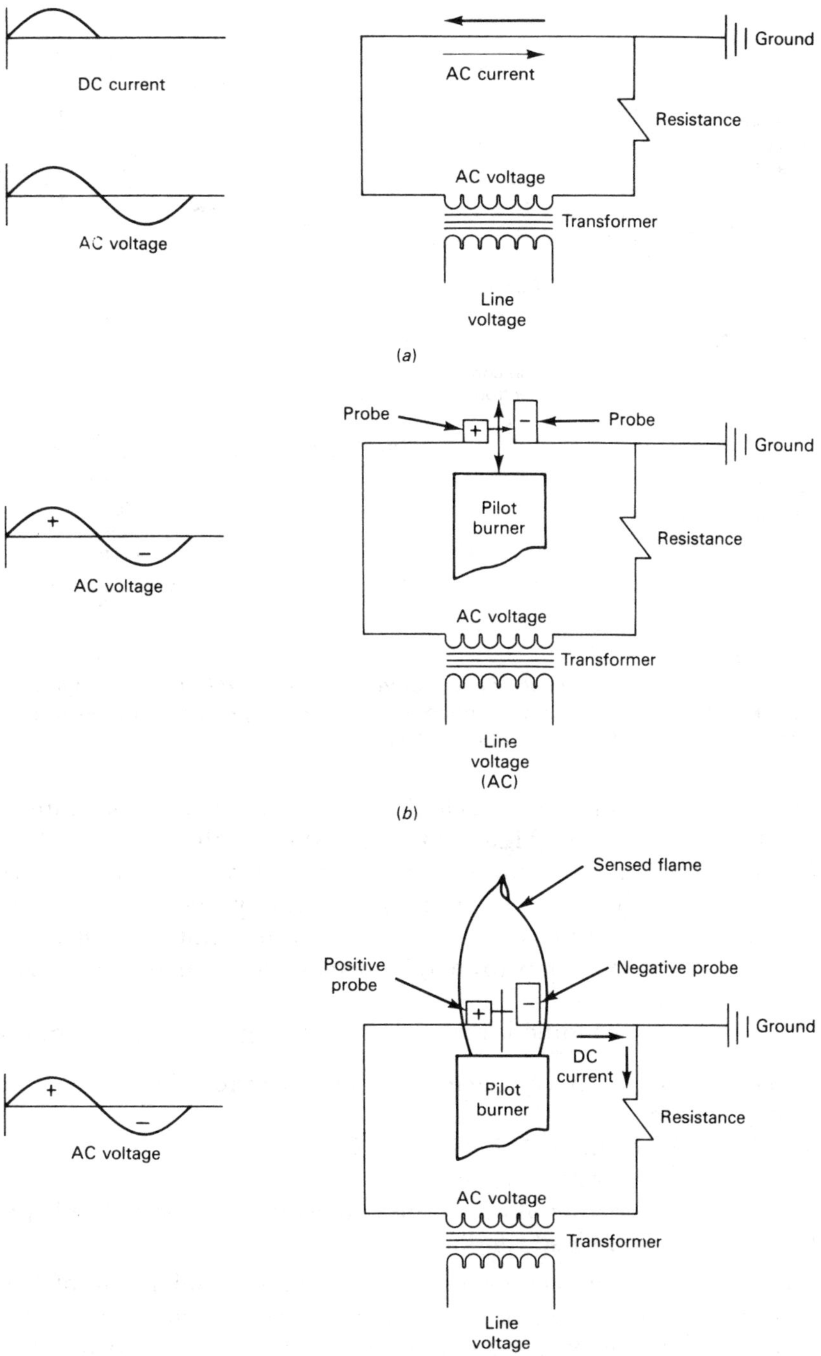

DC current
AC voltage
AC current
Ground
Resistance
AC voltage
Transformer
Line
voltage
(a)
Probe
Probe
Ground
Pilot
burner
Resistance
AC voltage
+
−
AC voltage
Transformer
Line
voltage
(AC)
(b)
Sensed flame
Positive
probe
Negative probe
Ground
DC
current
Pilot
burner
Resistance
+
−
AC voltage
AC voltage
Transformer
Line
voltage

FIGURE 14–8
(a) Normal AC circuit; (b) flameout, switch is open, no current is flowing; (c) pilot is lit, flame is sensed, and current is flowing. (*Courtesy of Billy Langley,* High-Efficiency Gas Furnace Troubleshooting Handbook, *©1991, p. 230; reprinted by permission of Prentice-Hall, Inc., Englewood Cliffs, N.J.*)

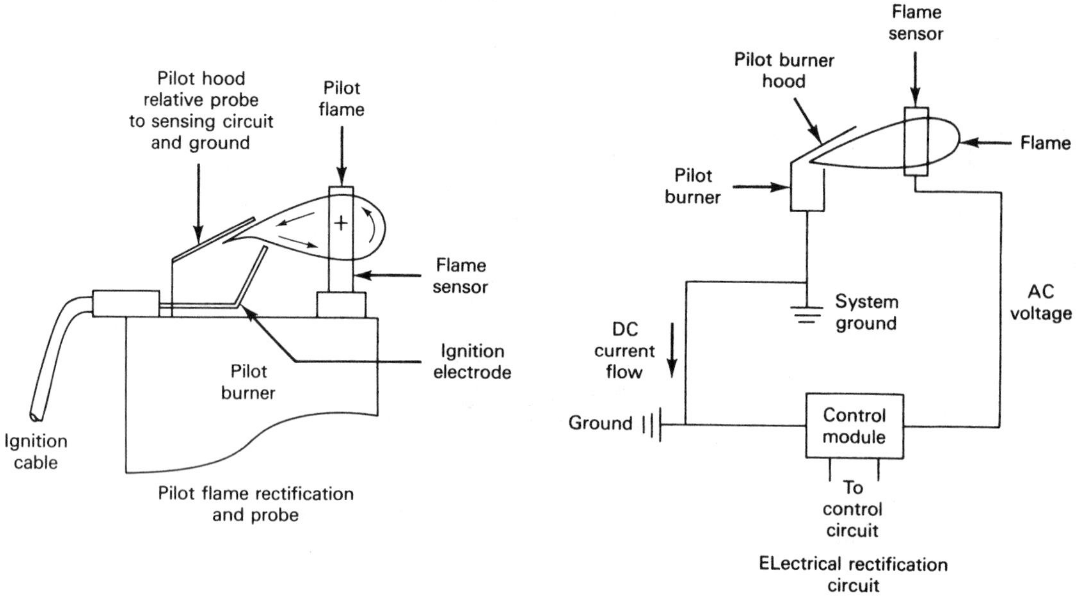
Pilot hood relative probe to sensing circuit and ground
Pilot flame
Flame sensor
Ignition electrode
Pilot burner
Ignition cable
Pilot flame rectification and probe
Flame sensor
Pilot burner hood
Flame
Pilot burner
System ground
AC voltage
DC current flow
Ground
Control module
To control circuit
ELectrical rectification circuit

FIGURE 14–9
Electrical rectification by use of pilot flame. (*Courtesy of Billy Langley,* High-Efficiency Gas Furnace Troubleshooting Handbook, *©1991, p. 231; reprinted by permission of Prentice-Hall, Inc., Englewood Cliffs, N.J.*)

To make the principle useful in IIDs, a pilot and flame sensor must be used in place of the two probes [Figure 14–8(b)]. After ignition of the pilot, a micro-ampere DC current flow is conducted through the flame, from the flame sensor (the positive probe) to the pilot burner tip (the negative probe). Acting as the negative probe, the pilot burner tip completes the electrical circuit to ground. The IID system uses this DC current flow to energize a relay, which, in turn, energizes the main burner gas valve.

The following conditions have a direct bearing on every IID application.

Voltage: The supply voltage to the ignition controls should be within the following ranges:

120-V AC controls: 102 to 132 V AC
24-V AC controls: 21 to 26.5 V AC

24-V AC systems should use transformers that will provide adequate power under maximum load conditions.

Gas pressure: The inlet gas pressure should be a minimum of 1 inch water column above the equipment manufacturer's recommended gas manifold pressure. The inlet gas pressure must never be allowed to fall below the equipment manufacturer's recommended minimum inlet gas pressure. The maximum inlet gas pressure for natural gas should be limited to 10.5 inches water column. On LP applications, the inlet gas pressure should be limited to a maximum of 14 inches water column.

Temperature: Electronic ignition controls should never be exposed to temperatures greater than 150° F or less than -40° F.

Pilot application: The pilot and flame sensor application is the most critical aspect of the IID application.

The pilot flame must make contact with the pilot burner tip and completely surround the sensor probe. A microammeter is necessary to verify that the proper amount of current is maintained through the pilot flame. If the proper amount of current is not maintained for the equipment in use, the unit will not operate as it was designed. This could be indicated by rapid short cycling of the main burner flame or no main burner flame. Flame rectification ignition systems respond in less than 0.8 seconds to a loss of flame signal. Thus, any deflection of the pilot flame away from the flame sensor, or the pilot burner tip, could result in rapid cycling (chattering) of the main burner gas valve, or prevent the main burner from coming on.

Other conditions that could cause the failure of the main burner to come on or rapid chattering of the main burner valve are (1) a pilot flame that is too small, or (2) gas pressure that is too low for proper pilot flame impingement on the flame sensor (the main burner gas valve will not be energized). It is also possible for drafts or unusual air currents to deflect the pilot flame away from the flame sensor. Deflection of the pilot flame may also be caused by main burner ignition concussion or rollout of the main burner flame.

Another consideration is the condition of the pilot flame. If the pilot flame is hard and blowing, the grounding area of the pilot is reduced to a point so that necessary current flow is not being maintained, and a shutdown of the system will result.

The positioning of the flame sensor is also critical in the pilot application. Positioning of the flame sensor should be such that it will be in contact with the second cone or combustion area of the flame. Passing the flame sensor through the inner cone of the pilot flame is not a recommended procedure. For this reason, a short flame sensor may provide a superior signal over a longer one. The final determination of the sensor location (length) is best determined by the use of a microammeter.

Checking Out Intermittent Pilot Systems

This section is presented with permission of Honeywell Incorporated.

Check out the gas control system at the following times:

1. At the initial installation of the appliance.
2. As a part of regular maintenance procedures. Maintenance intervals are determined by the application, as indicated in the following section.
3. As a first step in troubleshooting.
4. Anytime work is done on the system.

Applications

Electronic ignition systems are used on a wide variety of central heating equipment and on heating applications such as stoves, agricultural equipment, industrial heating equipment, and outdoor pool heaters. Two important reasons for choosing electronic ignition are:

1. To save fuel.
2. To avoid nuisance shutdowns of equipment mounted where the pilot can be easily blown out or fouled.

Fuel savings result from eliminating the standing pilot. This is the motivation behind building codes in some areas that prohibit standing pilots in new construction.

Furnaces located in crawl spaces, in attics, or on rooftops are excellent candidates for electronic ignition because the pilot is more apt to be blown out and because it is often hard to reach. Furnaces and heating appliances in locations where the air is dusty or greasy often use electronic ignition to avoid problems associated with clogging the pilot orifice.

Some of these applications may make heavy demands on electronic ignition systems because of moisture, corrosive chemicals, dust, or excessive heat in the environment. In these situations, special steps may be required to prevent nuisance shutdowns and premature control failure. With the exception of intermittent pilot retrofit, do not apply electronic ignition systems in the field.

Improper installation can cause a fire or explosion, and property damage, injury, or loss of life.

1. If you smell gas or suspect a gas leak, turn off the gas at the normal service valve and evacuate the house. Do not try to light any appliance. Do not touch any electrical switch or the telephone in the building until you are sure that no spilled gas remains.
2. A gas leak test, described next in steps 1 and 5, must be done on the initial installation and anytime work is done involving the gas piping.

   Step 1: Perform a visual inspection.

   1. With the electric power off, make sure that all wiring connections are clean and tight.
   2. Turn on the electric power to the appliance and the ignition module.
   3. Open the manual shutoff valve in the gas line to the appliance.
   4. Do a gas leak test ahead of the gas control if the gas piping has been disturbed.

   Gas Leak Test: Paint the pipe joints with a rich soap-and-water solution. Bubbles indicate a gas leak. Tighten the joints to stop the leak.

   Step 2: Review the normal operating sequence and timing summary (Figures 14–10 and 14–11 and Table 14–1).

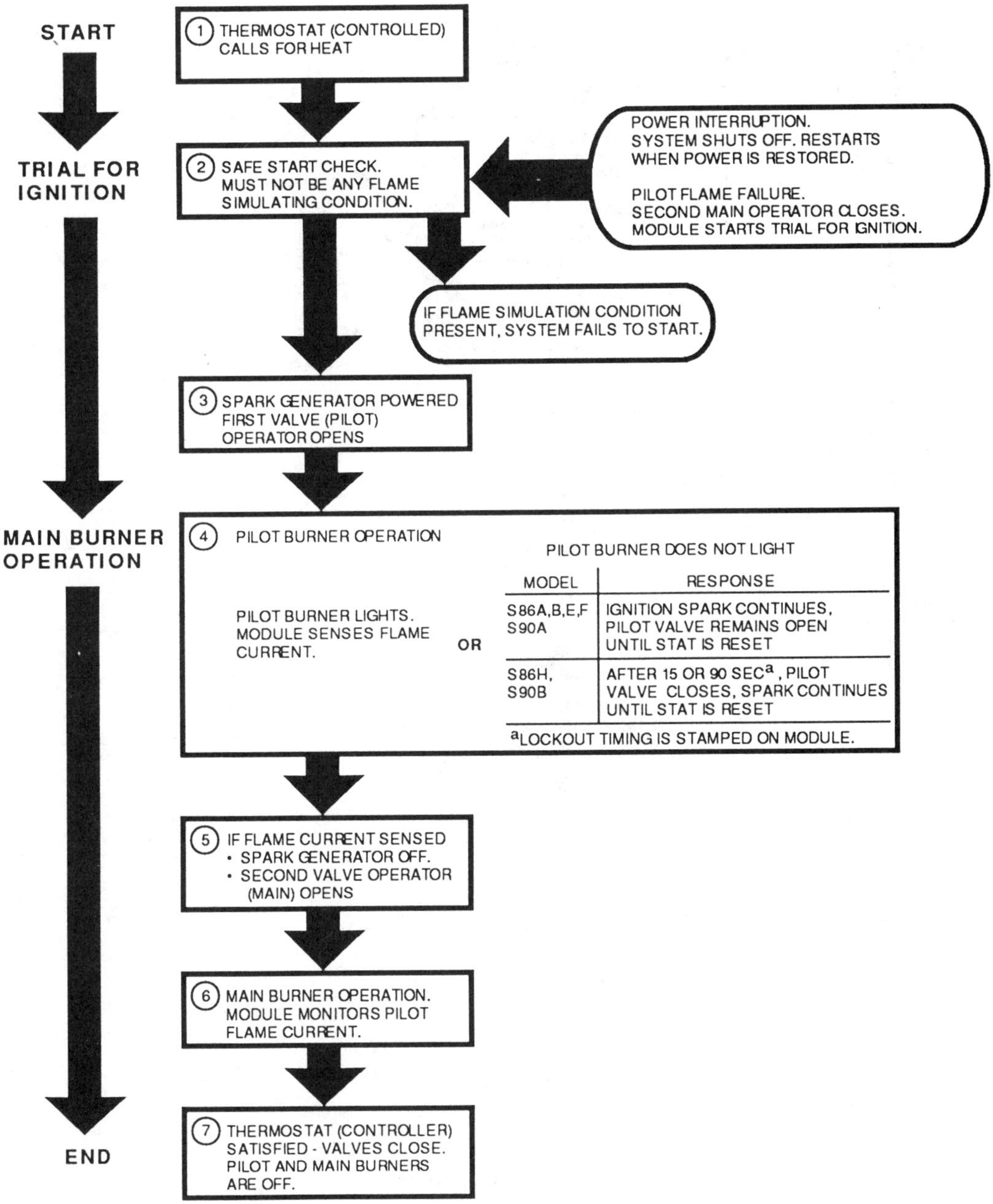
START
TRIAL FOR IGNITION
MAIN BURNER OPERATION
END
1 THERMOSTAT (CONTROLLED) CALLS FOR HEAT
2 SAFE START CHECK. MUST NOT BE ANY FLAME SIMULATING CONDITION.
POWER INTERRUPTION. SYSTEM SHUTS OFF. RESTARTS WHEN POWER IS RESTORED.
PILOT FLAME FAILURE. SECOND MAIN OPERATOR CLOSES. MODULE STARTS TRIAL FOR IGNITION.
IF FLAME SIMULATION CONDITION PRESENT, SYSTEM FAILS TO START.
3 SPARK GENERATOR POWERED FIRST VALVE (PILOT) OPERATOR OPENS
4 PILOT BURNER OPERATION
PILOT BURNER LIGHTS. MODULE SENSES FLAME CURRENT.
OR
PILOT BURNER DOES NOT LIGHT
MODEL
RESPONSE
S86A,B,E,F S90A
IGNITION SPARK CONTINUES, PILOT VALVE REMAINS OPEN UNTIL STAT IS RESET
S86H, S90B
AFTER 15 OR 90 SEC[a], PILOT VALVE CLOSES, SPARK CONTINUES UNTIL STAT IS RESET
[a]LOCKOUT TIMING IS STAMPED ON MODULE.
5 IF FLAME CURRENT SENSED
• SPARK GENERATOR OFF.
• SECOND VALVE OPERATOR (MAIN) OPENS
6 MAIN BURNER OPERATION. MODULE MONITORS PILOT FLAME CURRENT.
7 THERMOSTAT (CONTROLLER) SATISFIED - VALVES CLOSE. PILOT AND MAIN BURNERS ARE OFF.

FIGURE 14–10
S86 and S90 normal operating sequence. (*Courtesy of Honeywell.*)

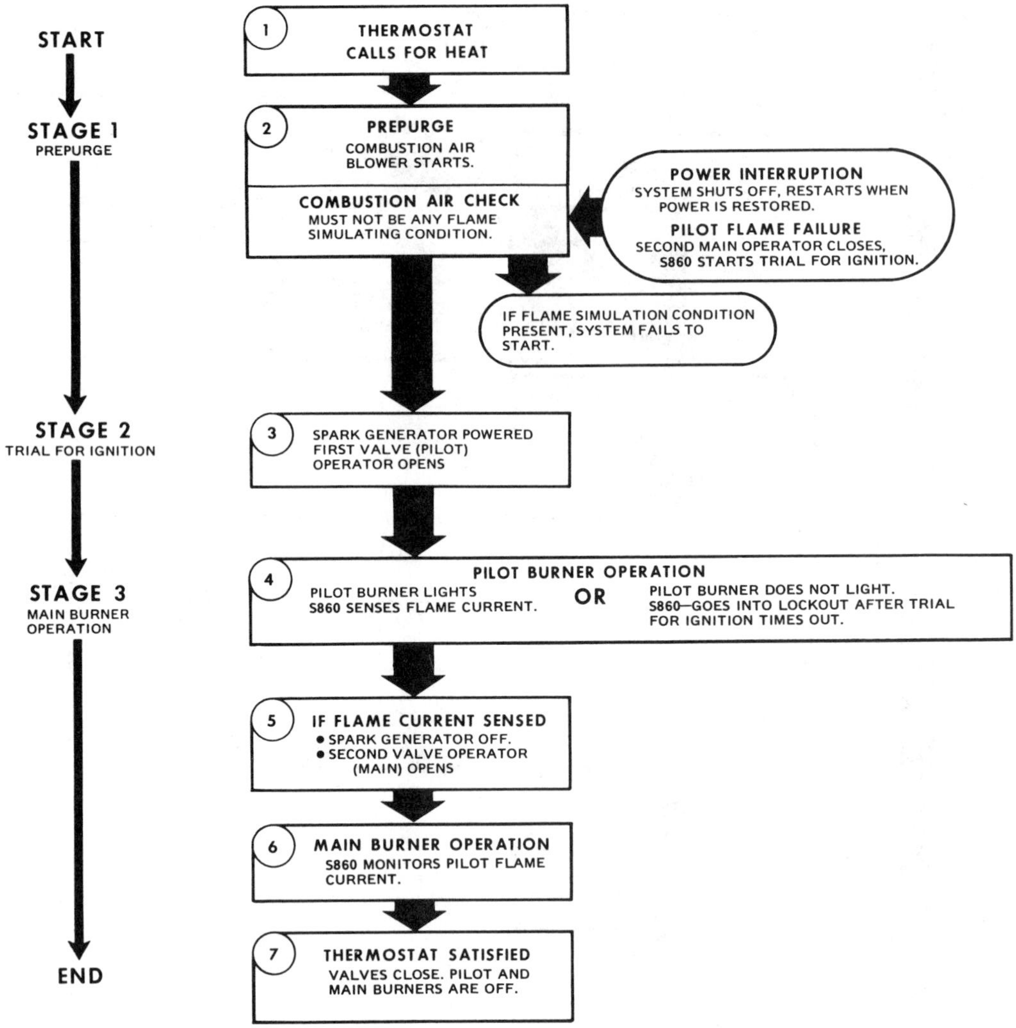
START
STAGE 1
PREPURGE
STAGE 2
TRIAL FOR IGNITION
STAGE 3
MAIN BURNER OPERATION
END
1 THERMOSTAT CALLS FOR HEAT
2 PREPURGE
COMBUSTION AIR BLOWER STARTS.
COMBUSTION AIR CHECK
MUST NOT BE ANY FLAME SIMULATING CONDITION.
POWER INTERRUPTION
SYSTEM SHUTS OFF, RESTARTS WHEN POWER IS RESTORED.
PILOT FLAME FAILURE
SECOND MAIN OPERATOR CLOSES, S860 STARTS TRIAL FOR IGNITION.
IF FLAME SIMULATION CONDITION PRESENT, SYSTEM FAILS TO START.
3 SPARK GENERATOR POWERED FIRST VALVE (PILOT) OPERATOR OPENS
4 PILOT BURNER OPERATION
PILOT BURNER LIGHTS S860 SENSES FLAME CURRENT.
OR
PILOT BURNER DOES NOT LIGHT. S860—GOES INTO LOCKOUT AFTER TRIAL FOR IGNITION TIMES OUT.
5 IF FLAME CURRENT SENSED
• SPARK GENERATOR OFF.
• SECOND VALVE OPERATOR (MAIN) OPENS
6 MAIN BURNER OPERATION
S860 MONITORS PILOT FLAME CURRENT.
7 THERMOSTAT SATISFIED
VALVES CLOSE. PILOT AND MAIN BURNERS ARE OFF.

FIGURE 14–11
S860 normal operating sequence. (*Courtesy of Honeywell.*)

Step 3: Reset the module.

1. Turn the thermostat to the lowest setting.
2. Wait one minute.

As you do steps 4 and 5, watch for points where the unit operation deviates from normal. Refer to the Troubleshooting Chart in Step 2 to correct the problem.

Step 4: Check the safety lockout operation.

TABLE 14–1 Intermittent pilot module timing survey. (*Courtesy of Honeywell.*)

Function	S86A, E	S86B, F S90A	S86C, G	S86D, H S90B	S860
Prepurge	—	—	—	—	30 sec.
Trial for ignition (Pilot gas and spark)	Continuous	Continuous	Ends with lockout	Ends with lockout[b]	Ends with lockout
Safety lockout timing	—	—	15 or 90 sec.[a]	15 or 90 sec.[a, c]	15 or 90 sec.[a]
Flame failure response[d]	0.8 sec. max.	0.8 sec. max.	0.8 sec. max.	0.8 sec. max.	0.8 sec. max.

[a]Lockout timing is stamped on module.

[b]S86H, S90B built before 6/87 continue spark after lockout until system is reset.

[c]S90B not available with 15 sec. safety lockout timing.

[d]Shutdown may be delayed several seconds if flame is lost immediately after being proved.

1. Turn off the gas supply.
2. Set the thermostat or controller above room temperature to call for heat.
3. Watch for sparking at the pilot burner either immediately or following a prepurge (see the timing summary, Table 14–1).
4. On models with timed ignition, time the length of spark operation (see the timing summary, Table 14–1).
5. On models that continue to spark after lockout, check the time before you hear a click from the gas control. A click indicates that the gas control has closed in a safety lockout.
6. Open the manual gas cock and make sure that no gas is flowing to either the pilot or the main burner.
7. Set the thermostat below room temperature and wait one minute before continuing.

Step 5: Check normal operation.

1. Set the thermostat or controller above room temperature to call for heat.
2. Make sure that the pilot lights smoothly when the gas reaches the pilot burner.
3. Make sure that the main burner lights smoothly without flashback.
4. Make sure that the burner flame operates smoothly without floating, lifting, or flame rollout to the furnace vestibule or heat buildup in the vestibule.
5. If the gas line has been disturbed, complete the gas leak test.

Gas Leak Test: Paint the gas control gasket edges and all pipe connections downstream of the gas control, including the pilot tubing connections, with a rich soap-and-water solution. Bubbles indicate gas leaks. Tighten the joints and screws or replace the component to stop the gas leak.

6. Turn the thermostat or controller below room temperature. Make sure that the main burner and the pilot flames go out.

Troubleshooting Intermittent Pilot Systems

Follow the appropriate troubleshooting guide to pinpoint the cause of the problem (Figures 14–12 and 14–13). If troubleshooting indicates an ignition problem, see the following discussion, "Ignition System Checks," to isolate and correct the problem. Following the troubleshooting procedure, perform the checkout procedure discussed earlier to be sure that the system is operating normally.

Ignition system checks. Use the following steps when making ignition system checks:

Step 1: Check the ignition cable.
Make sure that:

1. The ignition cable does not touch any metal surfaces.
2. The ignition cable is no more than 36 inches long.
3. The connections to the stud terminal and the ignitor sensor are clean and tight.
4. The ignition cable provides good electrical continuity.

Step 2: Check the ignition system grounding. Nuisance shutdowns are often caused by a poor or erratic ground.

1. A common ground, usually supplied by the pilot burner bracket, is required for the module and pilot burner/ignitor sensor.
   a. Check for good metal-to-metal contact between the pilot burner bracket and the main burner.
   b. Check the ground lead from the ground (GND) terminal on the module to the pilot burner. Make sure that the connections are clean and tight. If the wire is damaged or deteriorated, replace it with 14- to 18-gauge, moisture-resistant, thermoplastic insulated wire with 105° C minimum continuous rating.
   c. Check the temperature at the ceramic flame rod insulator. An excessive temperature will permit leakage to ground.
   d. If the flame rod or bracket is bent out of position, restore it to the correct position.
   e. Replace the pilot burner/ignitor sensor with an identical unit if the insulator is cracked.

Step 3: Check the spark ignition circuit. You will need a short jumper wire made from ignition cable or other heavily insulated wire.

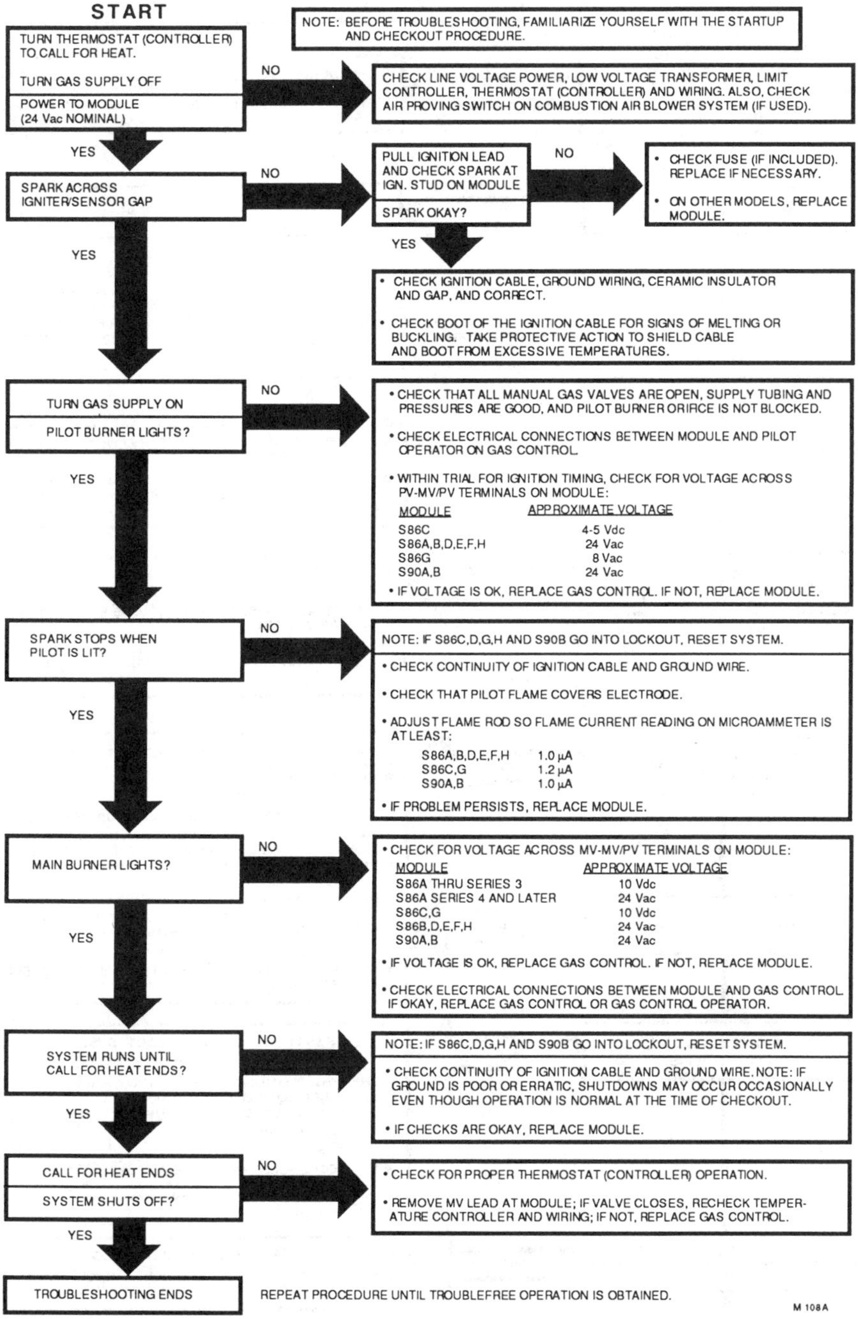
START
NOTE: BEFORE TROUBLESHOOTING, FAMILIARIZE YOURSELF WITH THE STARTUP AND CHECKOUT PROCEDURE.
TURN THERMOSTAT (CONTROLLER) TO CALL FOR HEAT.
TURN GAS SUPPLY OFF
POWER TO MODULE (24 Vac NOMINAL)
NO
CHECK LINE VOLTAGE POWER, LOW VOLTAGE TRANSFORMER, LIMIT CONTROLLER, THERMOSTAT (CONTROLLER) AND WIRING. ALSO, CHECK AIR PROVING SWITCH ON COMBUSTION AIR BLOWER SYSTEM (IF USED).
YES
SPARK ACROSS IGNITER/SENSOR GAP
NO
PULL IGNITION LEAD AND CHECK SPARK AT IGN. STUD ON MODULE
SPARK OKAY?
NO
• CHECK FUSE (IF INCLUDED). REPLACE IF NECESSARY.
• ON OTHER MODELS, REPLACE MODULE.
YES
• CHECK IGNITION CABLE, GROUND WIRING, CERAMIC INSULATOR AND GAP, AND CORRECT.
• CHECK BOOT OF THE IGNITION CABLE FOR SIGNS OF MELTING OR BUCKLING. TAKE PROTECTIVE ACTION TO SHIELD CABLE AND BOOT FROM EXCESSIVE TEMPERATURES.
YES
TURN GAS SUPPLY ON
PILOT BURNER LIGHTS?
NO
• CHECK THAT ALL MANUAL GAS VALVES ARE OPEN, SUPPLY TUBING AND PRESSURES ARE GOOD, AND PILOT BURNER ORIFICE IS NOT BLOCKED.
• CHECK ELECTRICAL CONNECTIONS BETWEEN MODULE AND PILOT OPERATOR ON GAS CONTROL.
• WITHIN TRIAL FOR IGNITION TIMING, CHECK FOR VOLTAGE ACROSS PV-MV/PV TERMINALS ON MODULE:
MODULE — APPROXIMATE VOLTAGE
S86C — 4-5 Vdc
S86A,B,D,E,F,H — 24 Vac
S86G — 8 Vac
S90A,B — 24 Vac
• IF VOLTAGE IS OK, REPLACE GAS CONTROL. IF NOT, REPLACE MODULE.
YES
SPARK STOPS WHEN PILOT IS LIT?
NO
NOTE: IF S86C,D,G,H AND S90B GO INTO LOCKOUT, RESET SYSTEM.
• CHECK CONTINUITY OF IGNITION CABLE AND GROUND WIRE.
• CHECK THAT PILOT FLAME COVERS ELECTRODE.
• ADJUST FLAME ROD SO FLAME CURRENT READING ON MICROAMMETER IS AT LEAST:
S86A,B,D,E,F,H — 1.0 μA
S86C,G — 1.2 μA
S90A,B — 1.0 μA
• IF PROBLEM PERSISTS, REPLACE MODULE.
YES
MAIN BURNER LIGHTS?
NO
• CHECK FOR VOLTAGE ACROSS MV-MV/PV TERMINALS ON MODULE:
MODULE — APPROXIMATE VOLTAGE
S86A THRU SERIES 3 — 10 Vdc
S86A SERIES 4 AND LATER — 24 Vac
S86C,G — 10 Vdc
S86B,D,E,F,H — 24 Vac
S90A,B — 24 Vac
• IF VOLTAGE IS OK, REPLACE GAS CONTROL. IF NOT, REPLACE MODULE.
• CHECK ELECTRICAL CONNECTIONS BETWEEN MODULE AND GAS CONTROL. IF OKAY, REPLACE GAS CONTROL OR GAS CONTROL OPERATOR.
YES
SYSTEM RUNS UNTIL CALL FOR HEAT ENDS?
NO
NOTE: IF S86C,D,G,H AND S90B GO INTO LOCKOUT, RESET SYSTEM.
• CHECK CONTINUITY OF IGNITION CABLE AND GROUND WIRE. NOTE: IF GROUND IS POOR OR ERRATIC, SHUTDOWNS MAY OCCUR OCCASIONALLY EVEN THOUGH OPERATION IS NORMAL AT THE TIME OF CHECKOUT.
• IF CHECKS ARE OKAY, REPLACE MODULE.
YES
CALL FOR HEAT ENDS
SYSTEM SHUTS OFF?
NO
• CHECK FOR PROPER THERMOSTAT (CONTROLLER) OPERATION.
• REMOVE MV LEAD AT MODULE; IF VALVE CLOSES, RECHECK TEMPERATURE CONTROLLER AND WIRING; IF NOT, REPLACE GAS CONTROL.
YES
TROUBLESHOOTING ENDS
REPEAT PROCEDURE UNTIL TROUBLEFREE OPERATION IS OBTAINED.
M 108A

FIGURE 14–12
S86 and S90 troubleshooting guide. (*Courtesy of Honeywell.*)

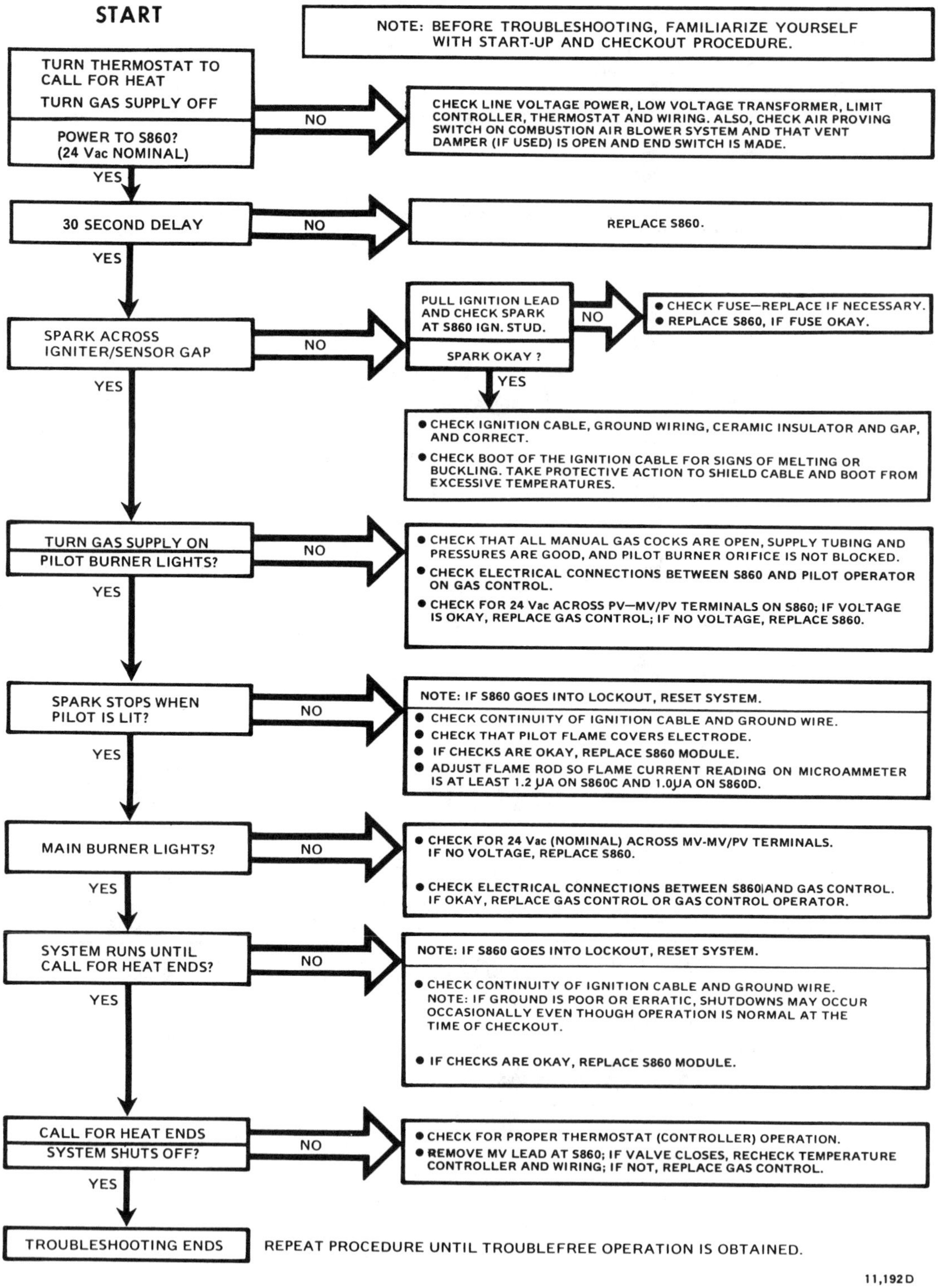
START
NOTE: BEFORE TROUBLESHOOTING, FAMILIARIZE YOURSELF WITH START-UP AND CHECKOUT PROCEDURE.
TURN THERMOSTAT TO CALL FOR HEAT
TURN GAS SUPPLY OFF
POWER TO S860? (24 Vac NOMINAL)
NO
CHECK LINE VOLTAGE POWER, LOW VOLTAGE TRANSFORMER, LIMIT CONTROLLER, THERMOSTAT AND WIRING. ALSO, CHECK AIR PROVING SWITCH ON COMBUSTION AIR BLOWER SYSTEM AND THAT VENT DAMPER (IF USED) IS OPEN AND END SWITCH IS MADE.
YES
30 SECOND DELAY
NO
REPLACE S860.
YES
SPARK ACROSS IGNITER/SENSOR GAP
NO
PULL IGNITION LEAD AND CHECK SPARK AT S860 IGN. STUD.
SPARK OKAY ?
NO
● CHECK FUSE—REPLACE IF NECESSARY.
● REPLACE S860, IF FUSE OKAY.
YES
● CHECK IGNITION CABLE, GROUND WIRING, CERAMIC INSULATOR AND GAP, AND CORRECT.
● CHECK BOOT OF THE IGNITION CABLE FOR SIGNS OF MELTING OR BUCKLING. TAKE PROTECTIVE ACTION TO SHIELD CABLE AND BOOT FROM EXCESSIVE TEMPERATURES.
YES
TURN GAS SUPPLY ON
PILOT BURNER LIGHTS?
NO
● CHECK THAT ALL MANUAL GAS COCKS ARE OPEN, SUPPLY TUBING AND PRESSURES ARE GOOD, AND PILOT BURNER ORIFICE IS NOT BLOCKED.
● CHECK ELECTRICAL CONNECTIONS BETWEEN S860 AND PILOT OPERATOR ON GAS CONTROL.
● CHECK FOR 24 Vac ACROSS PV—MV/PV TERMINALS ON S860; IF VOLTAGE IS OKAY, REPLACE GAS CONTROL; IF NO VOLTAGE, REPLACE S860.
YES
SPARK STOPS WHEN PILOT IS LIT?
NO
NOTE: IF S860 GOES INTO LOCKOUT, RESET SYSTEM.
● CHECK CONTINUITY OF IGNITION CABLE AND GROUND WIRE.
● CHECK THAT PILOT FLAME COVERS ELECTRODE.
● IF CHECKS ARE OKAY, REPLACE S860 MODULE.
● ADJUST FLAME ROD SO FLAME CURRENT READING ON MICROAMMETER IS AT LEAST 1.2 μA ON S860C AND 1.0μA ON S860D.
YES
MAIN BURNER LIGHTS?
NO
● CHECK FOR 24 Vac (NOMINAL) ACROSS MV-MV/PV TERMINALS. IF NO VOLTAGE, REPLACE S860.
● CHECK ELECTRICAL CONNECTIONS BETWEEN S860 AND GAS CONTROL. IF OKAY, REPLACE GAS CONTROL OR GAS CONTROL OPERATOR.
YES
SYSTEM RUNS UNTIL CALL FOR HEAT ENDS?
NO
NOTE: IF S860 GOES INTO LOCKOUT, RESET SYSTEM.
● CHECK CONTINUITY OF IGNITION CABLE AND GROUND WIRE. NOTE: IF GROUND IS POOR OR ERRATIC, SHUTDOWNS MAY OCCUR OCCASIONALLY EVEN THOUGH OPERATION IS NORMAL AT THE TIME OF CHECKOUT.
● IF CHECKS ARE OKAY, REPLACE S860 MODULE.
YES
CALL FOR HEAT ENDS
SYSTEM SHUTS OFF?
NO
● CHECK FOR PROPER THERMOSTAT (CONTROLLER) OPERATION.
● REMOVE MV LEAD AT S860; IF VALVE CLOSES, RECHECK TEMPERATURE CONTROLLER AND WIRING; IF NOT, REPLACE GAS CONTROL.
YES
TROUBLESHOOTING ENDS
REPEAT PROCEDURE UNTIL TROUBLEFREE OPERATION IS OBTAINED.
11,192D

FIGURE 14–13
S860 troubleshooting guide. (*Courtesy of Honeywell.*)

1. Close the manual gas shutoff valve.
2. Disconnect the ignition cable at the stud on the module. When performing the following steps, do not touch the stripped end of the jumper wire or the stud terminal. The ignition circuit generates 15 to 16 kV open circuit, and electrical shock can result.
3. Energize the module and immediately touch one end of the jumper wire firmly to the GND terminal on the module. Move the free end of the jumper slowly toward the stud terminal until a spark is established.
4. Pull the jumper wire slowly away from the stud and note the length of the gap when the sparking stops (Table 14–2).
5. Open the manual gas shutoff valve and reset the system.

TABLE 14–2 ARC length. (*Courtesy of Honeywell.*)

Arc length	Action
No arc or arc less than 1/8 in. [3 mm]	• Check external fuse, if provided. • Verify power at module input terminal • Replace module if fuse and power ok.
Arc 1/8 in. [3 mm] or longer	Voltage output is adequate.

Step 4: Check the pilot flame current. Use the following steps to check the pilot flame current.

1. Turn off the furnace at the thermostat.
2. Disconnect the ground wire from the GND terminal on the S86.
3. Disconnect the main valve wire from the thermostat (TH) terminal on the gas control.
4. Set the thermostat above room temperature to call for heat. A spark will ignite the pilot, but because the main gas valve actuator is disconnected, the main burner will not light.
5. Make sure that the pilot flame envelops $3/8$ to $1/3$ inch of the flame rod. If necessary, adjust the pilot flame by turning the pilot adjustment screw on the gas control clockwise to decrease or counterclockwise to increase the pilot flame. Always replace the pilot adjustment screw cover and tighten it firmly to prevent gas leaks.
6. Adjust the ignitor-sensor rod on the pilot until the flame current reading on the microammeter is at a maximum, and then lock it in place with the setscrew. The flame current must be at least that shown in Table 14–3. This is normally possible only on retrofit installations.
7. Remove the microammeter and correctly reconnect all the wires. Return the system to normal operation before leaving the job.

TABLE 14–3 Pilot flame current readings. (*Courtesy of Honeywell.*)

Module	Minimum flame current
S86A, B, D, E, F, H;	
S90A, B; S860D	1.0 μA
S86C, G; S860C	1.2 μA

Checking Out Direct Ignition Systems

To check out a direct ignition system, use the following steps:

Check out the gas control system.

1. At the initial installation of the appliance.
2. As a part of regular maintenance procedures. Maintenance intervals are determined by the application. See the foregoing "Applications" section for more information.
3. As a first step in troubleshooting.
4. Any time work is done on the system.

Improper installation can cause a fire or explosion and property damage, injury, or loss of life.

1. If you smell gas or suspect a gas leak, turn off the gas at the manual service valve and evacuate the house. Do not try to light any appliance or touch any electrical switch or telephone in the building until you are sure that no spilled gas remains.
2. A gas leak test, described in the following steps 1 and 5, must be done on an initial installation and any time work is done involving the gas piping.

   Step 1: Perform a visual inspection.

   1. With the electric power off, make sure that all wiring connections are clean and tight.
   2. Turn on the electric power to the appliance and the ignition module.
   3. Open the manual shutoff valves in the gas line to the appliance.
   4. Do a gas leak test ahead of the gas control if the piping has been disturbed.

   Gas Leak Test: Paint the pipe joints with a rich soap-and-water solution. Bubbles indicate a gas leak. Tighten the joints to stop the leak.

   Step 2: Review the normal operating sequence and timing summary (Figures 14–14, 14–15, and Table 14–4.)

   Step 3: Reset the module.

   1. Turn the thermostat to its lowest setting.
   2. Wait one minute.

   As you do steps 4 and 5, watch for points where the operation deviates from normal. Refer to the Troubleshooting Chart to correct the problem.

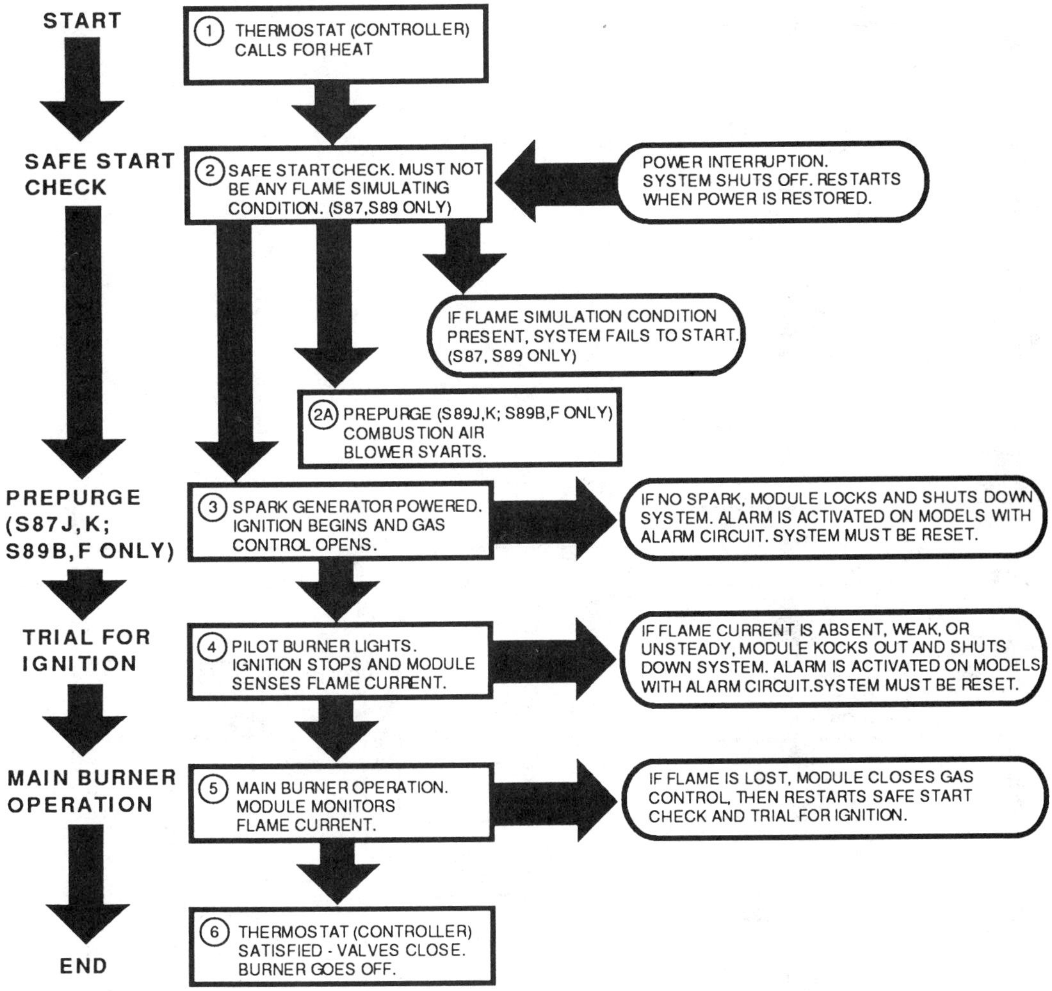
START
THERMOSTAT (CONTROLLER) CALLS FOR HEAT
SAFE START CHECK
SAFE START CHECK. MUST NOT BE ANY FLAME SIMULATING CONDITION. (S87,S89 ONLY)
POWER INTERRUPTION. SYSTEM SHUTS OFF. RESTARTS WHEN POWER IS RESTORED.
IF FLAME SIMULATION CONDITION PRESENT, SYSTEM FAILS TO START. (S87, S89 ONLY)
2A PREPURGE (S89J,K; S89B,F ONLY) COMBUSTION AIR BLOWER SYARTS.
PREPURGE (S87J,K; S89B,F ONLY)
SPARK GENERATOR POWERED. IGNITION BEGINS AND GAS CONTROL OPENS.
IF NO SPARK, MODULE LOCKS AND SHUTS DOWN SYSTEM. ALARM IS ACTIVATED ON MODELS WITH ALARM CIRCUIT. SYSTEM MUST BE RESET.
TRIAL FOR IGNITION
PILOT BURNER LIGHTS. IGNITION STOPS AND MODULE SENSES FLAME CURRENT.
IF FLAME CURRENT IS ABSENT, WEAK, OR UNSTEADY, MODULE KOCKS OUT AND SHUTS DOWN SYSTEM. ALARM IS ACTIVATED ON MODELS WITH ALARM CIRCUIT.SYSTEM MUST BE RESET.
MAIN BURNER OPERATION
MAIN BURNER OPERATION. MODULE MONITORS FLAME CURRENT.
IF FLAME IS LOST, MODULE CLOSES GAS CONTROL, THEN RESTARTS SAFE START CHECK AND TRIAL FOR IGNITION.
END
THERMOSTAT (CONTROLLER) SATISFIED - VALVES CLOSE. BURNER GOES OFF.

FIGURE 14–14
S825 and S87, S89A, B, E, F normal operating sequence. (*Courtesy of Honeywell.*)

Step 4: Check the safety lockout operation.

1. Turn off the gas supply.
2. Set the thermostat or the controller above room temperature to call for heat.
3. Watch for sparking at the ignitor immediately or following a prepurge (see the timing summary, Table 14–4).
4. After the lockout, open the manual gas cock and make sure that no gas is flowing to the main burner (or through the enrichment tube if the Q366 is used).

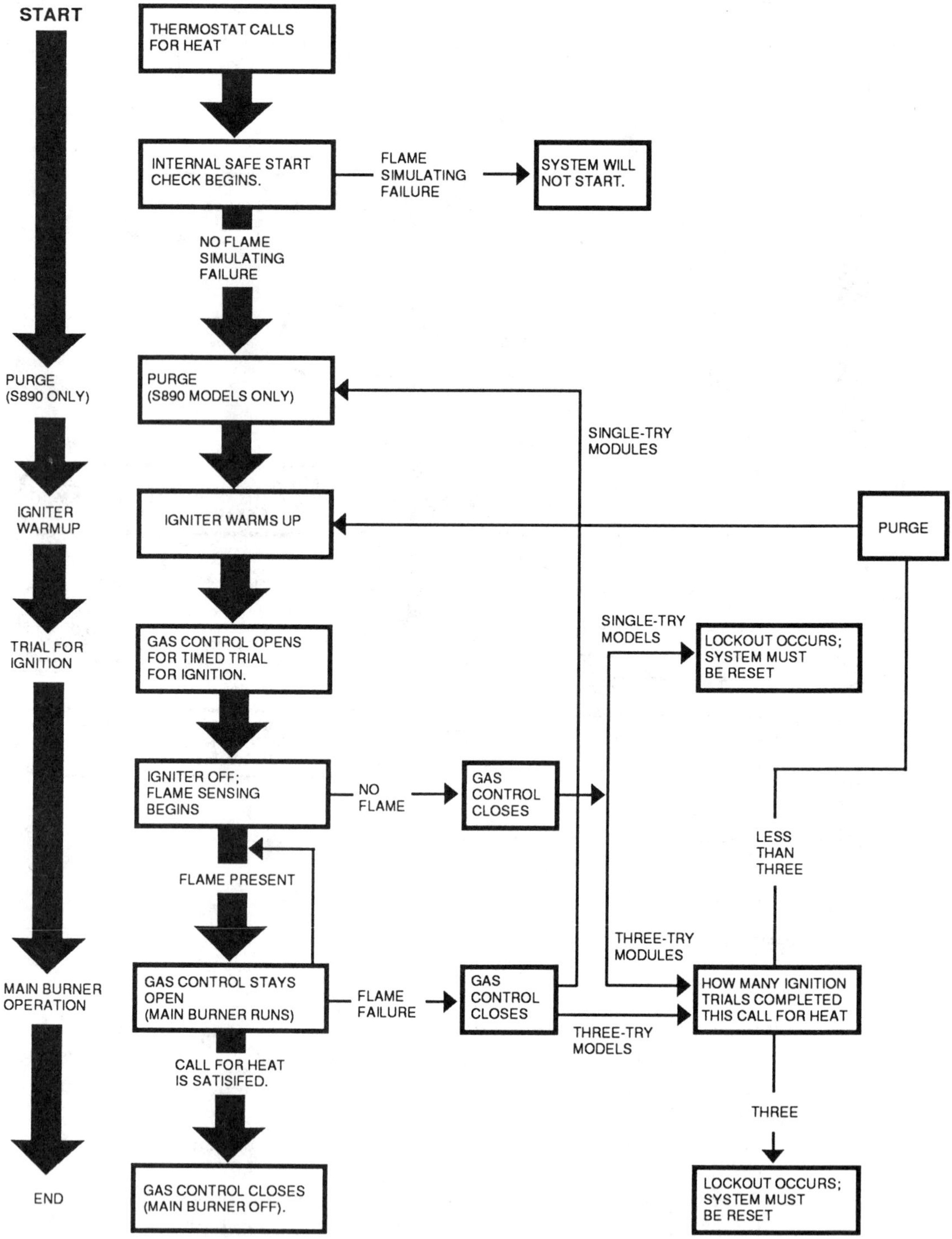
START
THERMOSTAT CALLS FOR HEAT
INTERNAL SAFE START CHECK BEGINS.
FLAME SIMULATING FAILURE
SYSTEM WILL NOT START.
NO FLAME SIMULATING FAILURE
PURGE (S890 ONLY)
PURGE (S890 MODELS ONLY)
SINGLE-TRY MODULES
IGNITER WARMUP
IGNITER WARMS UP
PURGE
TRIAL FOR IGNITION
GAS CONTROL OPENS FOR TIMED TRIAL FOR IGNITION.
SINGLE-TRY MODELS
LOCKOUT OCCURS; SYSTEM MUST BE RESET
IGNITER OFF; FLAME SENSING BEGINS
NO FLAME
GAS CONTROL CLOSES
LESS THAN THREE
FLAME PRESENT
THREE-TRY MODULES
MAIN BURNER OPERATION
GAS CONTROL STAYS OPEN (MAIN BURNER RUNS)
FLAME FAILURE
GAS CONTROL CLOSES
HOW MANY IGNITION TRIALS COMPLETED THIS CALL FOR HEAT
THREE-TRY MODELS
CALL FOR HEAT IS SATISIFED.
THREE
END
GAS CONTROL CLOSES (MAIN BURNER OFF).
LOCKOUT OCCURS; SYSTEM MUST BE RESET

FIGURE 14–15
S89 C, D, G, H normal operating sequence. (*Courtesy of Honeywell.*)

TABLE 14–4 Direct ignition timing summary. (*Courtesy of Honeywell.*)

Function	S825	S87A-D	S87J,K	S89A,E	S89B,F	S89C,D	S89G,H	S890C,D	S890G,H
Prepurge	—	—	30 sec. min.	—	30 sec. min.; 45 sec. max.	—	—	30 sec. min.; 37 sec. max.	30 sec. min.; 37 sec. max.
No. of ignition trials	1	1	1	1	1	1	3	1	3
Safety lockout timing (nom.)[a]	6, 11 or 21 sec.	4, 6, 11 or 21 sec	4, 6, 11 or 21 sec.	4, 6, 11, 15[b] or 21 sec	4, 6, 11, 15[b] or 21 sec.	4, 6, 11, 15 or 21[c] sec.	4, 6, 11 or 15 sec.	4, 6, 11 or 15 sec.	4, 6, 11 or 15 sec.
Flame failure response	0.8 sec. max.	0.8 sec. max	0.8 sec. max.	0.8 sec. max.	0.8 sec. max	2 sec. max.[d]	2 sec max.[d]	2 sec. max.[d]	2 sec. max.[d]
Igniter warmup time	—	—	—	—	—	34 sec.[e]	34 sec.	34 sec.	34 sec.

[a]Ignition continues until burner lights or system locks out.

[b]S89E,F only.

[c]S89C before 9/87 only.

[d]With 2.5 μA flame signal.

[e]S89C before 9/87 warmup period was 45 sec. with 2.5 μA flame signal.

5. Set the thermostat below room temperature and wait one minute before continuing.

Step 5: Check the normal operation.

1. Set the thermostat or controller above room temperature to call for heat.
2. Make sure that the main burner lights smoothly without flashback. Several attempts may be necessary to clear the gas line of air.
3. Make sure that the burner operates smoothly without floating, lifting, or flame rollout to the furnace vestibule or heat buildup in the vestibule.
4. If the gas line has been disturbed, complete a gas leak test.

Gas Leak Test: Paint the gas control gasket edges and all pipe connections downstream of the gas control, including the enrichment tube connections, with a rich soap-and-water solution. Bubbles indicate gas leaks. Tighten the joints and screws or replace the component to stop the gas leak.

5. Turn the thermostat or the controller below room temperature. Make sure that the main burner flame goes out.

Troubleshooting Direct Ignition Systems

Follow the appropriate troubleshooting guide to pinpoint the cause of the problem (Figures 14–16 and 14–17). If troubleshooting indicates an ignition problem, see the following discussion, "Ignition System Checks," to isolate and correct the problem.

Following the troubleshooting procedures, perform the checkout procedure discussed earlier to be sure that the system is operating normally.

Ignition system checks. Use the following procedures when making ignition system checks:

Step 1: Check the ignition cable: Make sure that the

1. Ignition cable does not touch any metal surface.
2. Ignition cable is no more than 36 inches long.
3. Connections to the stud terminal and ignitor-sensor are clean and tight.
4. Ignition cable provides good electrical continuity.

Step 2: Check the ignition system grounding. Nuisance shutdowns are often caused by a poor or erratic ground.

1. A common ground is required for the module, ignitor, flame sensor, and main burner.
   a. Check for a good metal-to-metal contact between the ignitor bracket and the main burner.
   b. Check the ground lead from the GND (burner) terminal on the module to the ignitor bracket. Make sure that the connections are clean and tight. If the wire is damaged or deteriorated, replace it with 14- to 16- gauge, moisture-resistant, thermoplastic insulated wire with 105° C minimum rating. Use a shield if necessary to protect the ground wire from radiant

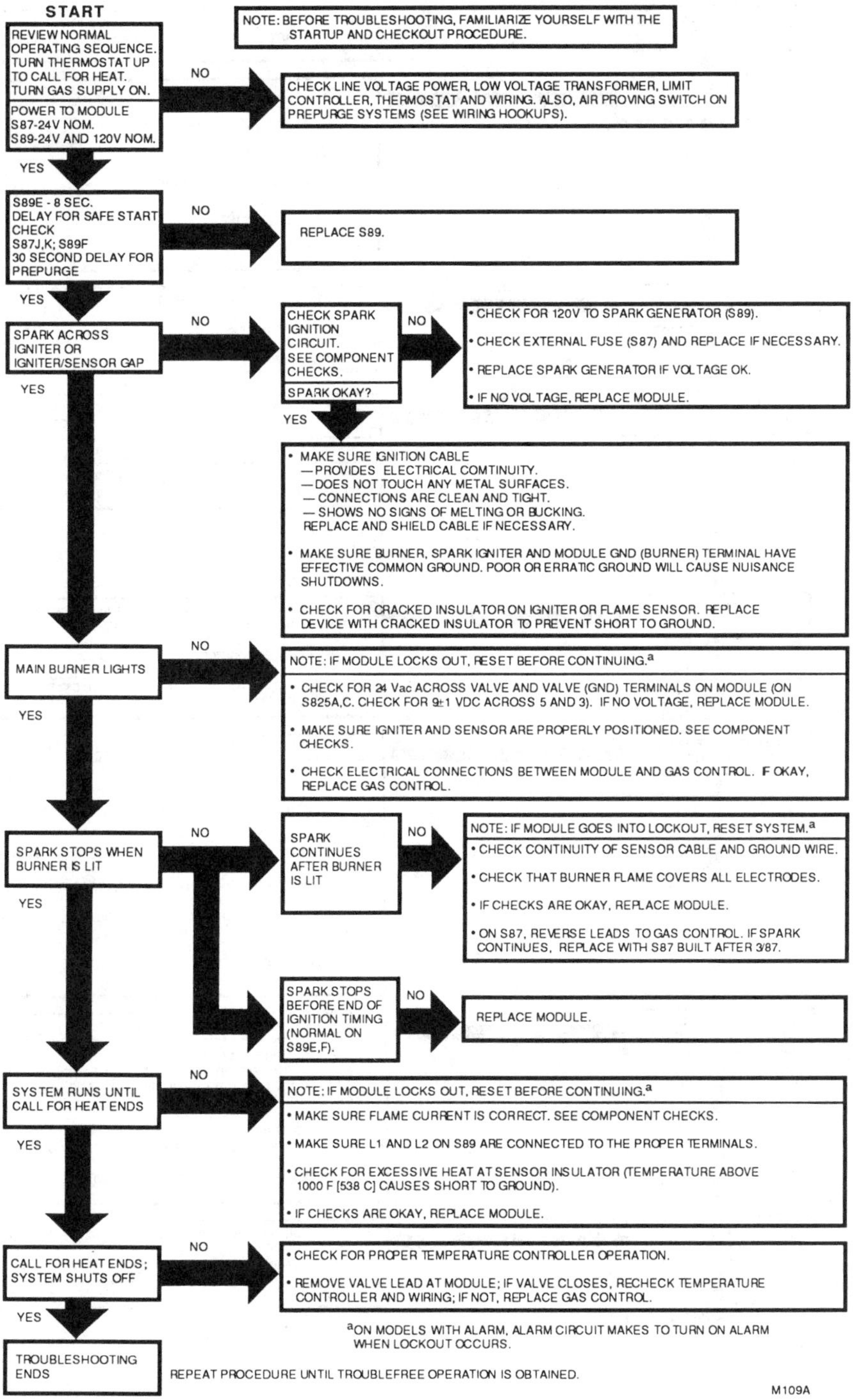
START
NOTE: BEFORE TROUBLESHOOTING, FAMILIARIZE YOURSELF WITH THE STARTUP AND CHECKOUT PROCEDURE.
REVIEW NORMAL OPERATING SEQUENCE. TURN THERMOSTAT UP TO CALL FOR HEAT. TURN GAS SUPPLY ON.
POWER TO MODULE S87-24V NOM. S89-24V AND 120V NOM.
NO
CHECK LINE VOLTAGE POWER, LOW VOLTAGE TRANSFORMER, LIMIT CONTROLLER, THERMOSTAT AND WIRING. ALSO, AIR PROVING SWITCH ON PREPURGE SYSTEMS (SEE WIRING HOOKUPS).
YES
S89E - 8 SEC. DELAY FOR SAFE START CHECK S87J,K; S89F 30 SECOND DELAY FOR PREPURGE
NO
REPLACE S89.
YES
SPARK ACROSS IGNITER OR IGNITER/SENSOR GAP
NO
CHECK SPARK IGNITION CIRCUIT. SEE COMPONENT CHECKS.
SPARK OKAY?
NO
• CHECK FOR 120V TO SPARK GENERATOR (S89).
• CHECK EXTERNAL FUSE (S87) AND REPLACE IF NECESSARY.
• REPLACE SPARK GENERATOR IF VOLTAGE OK.
• IF NO VOLTAGE, REPLACE MODULE.
YES
• MAKE SURE IGNITION CABLE
—PROVIDES ELECTRICAL COMTINUITY.
—DOES NOT TOUCH ANY METAL SURFACES.
— CONNECTIONS ARE CLEAN AND TIGHT.
— SHOWS NO SIGNS OF MELTING OR BUCKING.
REPLACE AND SHIELD CABLE IF NECESSARY.
• MAKE SURE BURNER, SPARK IGNITER AND MODULE GND (BURNER) TERMINAL HAVE EFFECTIVE COMMON GROUND. POOR OR ERRATIC GROUND WILL CAUSE NUISANCE SHUTDOWNS.
• CHECK FOR CRACKED INSULATOR ON IGNITER OR FLAME SENSOR. REPLACE DEVICE WITH CRACKED INSULATOR TO PREVENT SHORT TO GROUND.
YES
MAIN BURNER LIGHTS
NO
NOTE: IF MODULE LOCKS OUT, RESET BEFORE CONTINUING.a
• CHECK FOR 24 Vac ACROSS VALVE AND VALVE (GND) TERMINALS ON MODULE (ON S825A,C. CHECK FOR 9±1 VDC ACROSS 5 AND 3). IF NO VOLTAGE, REPLACE MODULE.
• MAKE SURE IGNITER AND SENSOR ARE PROPERLY POSITIONED. SEE COMPONENT CHECKS.
• CHECK ELECTRICAL CONNECTIONS BETWEEN MODULE AND GAS CONTROL. IF OKAY, REPLACE GAS CONTROL.
YES
SPARK STOPS WHEN BURNER IS LIT
NO
SPARK CONTINUES AFTER BURNER IS LIT
NO
NOTE: IF MODULE GOES INTO LOCKOUT, RESET SYSTEM.a
• CHECK CONTINUITY OF SENSOR CABLE AND GROUND WIRE.
• CHECK THAT BURNER FLAME COVERS ALL ELECTRODES.
• IF CHECKS ARE OKAY, REPLACE MODULE.
• ON S87, REVERSE LEADS TO GAS CONTROL. IF SPARK CONTINUES, REPLACE WITH S87 BUILT AFTER 3/87.
SPARK STOPS BEFORE END OF IGNITION TIMING (NORMAL ON S89E,F).
NO
REPLACE MODULE.
YES
SYSTEM RUNS UNTIL CALL FOR HEAT ENDS
NO
NOTE: IF MODULE LOCKS OUT, RESET BEFORE CONTINUING.a
• MAKE SURE FLAME CURRENT IS CORRECT. SEE COMPONENT CHECKS.
• MAKE SURE L1 AND L2 ON S89 ARE CONNECTED TO THE PROPER TERMINALS.
• CHECK FOR EXCESSIVE HEAT AT SENSOR INSULATOR (TEMPERATURE ABOVE 1000 F [538 C] CAUSES SHORT TO GROUND).
• IF CHECKS ARE OKAY, REPLACE MODULE.
YES
CALL FOR HEAT ENDS; SYSTEM SHUTS OFF
NO
• CHECK FOR PROPER TEMPERATURE CONTROLLER OPERATION.
• REMOVE VALVE LEAD AT MODULE; IF VALVE CLOSES, RECHECK TEMPERATURE CONTROLLER AND WIRING; IF NOT, REPLACE GAS CONTROL.
YES
TROUBLESHOOTING ENDS
aON MODELS WITH ALARM, ALARM CIRCUIT MAKES TO TURN ON ALARM WHEN LOCKOUT OCCURS.
REPEAT PROCEDURE UNTIL TROUBLEFREE OPERATION IS OBTAINED.
M109A

FIGURE 14–16
S825 and S87A, B, E, F troubleshooting guide. (*Courtesy of Honeywell.*)

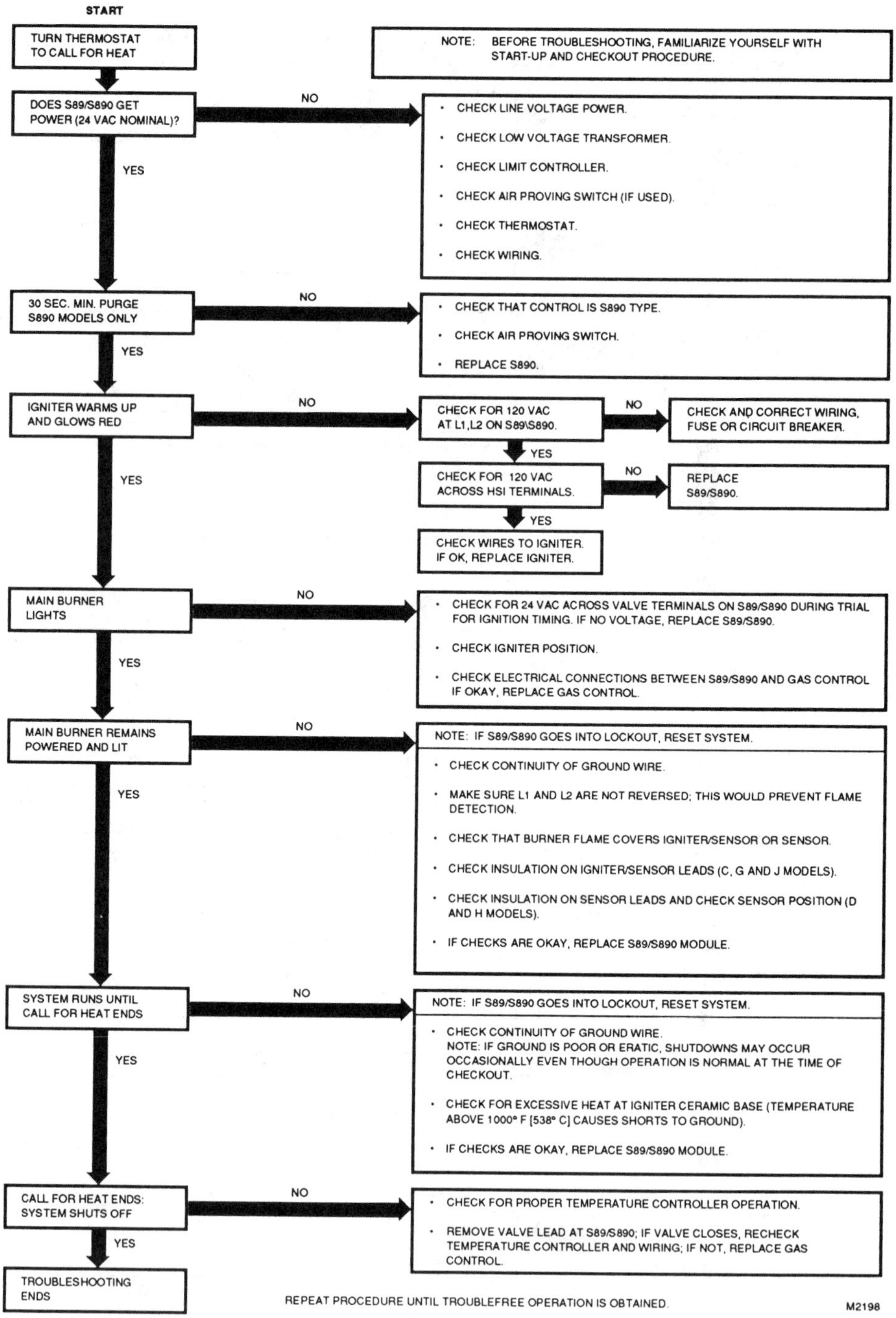
START
TURN THERMOSTAT TO CALL FOR HEAT
NOTE: BEFORE TROUBLESHOOTING, FAMILIARIZE YOURSELF WITH START-UP AND CHECKOUT PROCEDURE.
DOES S89/S890 GET POWER (24 VAC NOMINAL)?
NO
• CHECK LINE VOLTAGE POWER.
• CHECK LOW VOLTAGE TRANSFORMER.
• CHECK LIMIT CONTROLLER.
• CHECK AIR PROVING SWITCH (IF USED).
• CHECK THERMOSTAT.
• CHECK WIRING.
YES
30 SEC. MIN. PURGE S890 MODELS ONLY
NO
• CHECK THAT CONTROL IS S890 TYPE.
• CHECK AIR PROVING SWITCH.
• REPLACE S890.
YES
IGNITER WARMS UP AND GLOWS RED
NO
CHECK FOR 120 VAC AT L1,L2 ON S89\S890.
NO
CHECK AND CORRECT WIRING, FUSE OR CIRCUIT BREAKER.
YES
CHECK FOR 120 VAC ACROSS HSI TERMINALS.
NO
REPLACE S89/S890.
YES
CHECK WIRES TO IGNITER. IF OK, REPLACE IGNITER.
YES
MAIN BURNER LIGHTS
NO
• CHECK FOR 24 VAC ACROSS VALVE TERMINALS ON S89/S890 DURING TRIAL FOR IGNITION TIMING. IF NO VOLTAGE, REPLACE S89/S890.
• CHECK IGNITER POSITION.
• CHECK ELECTRICAL CONNECTIONS BETWEEN S89/S890 AND GAS CONTROL IF OKAY, REPLACE GAS CONTROL.
YES
MAIN BURNER REMAINS POWERED AND LIT
NO
NOTE: IF S89/S890 GOES INTO LOCKOUT, RESET SYSTEM.
• CHECK CONTINUITY OF GROUND WIRE.
• MAKE SURE L1 AND L2 ARE NOT REVERSED; THIS WOULD PREVENT FLAME DETECTION.
• CHECK THAT BURNER FLAME COVERS IGNITER/SENSOR OR SENSOR.
• CHECK INSULATION ON IGNITER/SENSOR LEADS (C, G AND J MODELS).
• CHECK INSULATION ON SENSOR LEADS AND CHECK SENSOR POSITION (D AND H MODELS).
• IF CHECKS ARE OKAY, REPLACE S89/S890 MODULE.
YES
SYSTEM RUNS UNTIL CALL FOR HEAT ENDS
NO
NOTE: IF S89/S890 GOES INTO LOCKOUT, RESET SYSTEM.
• CHECK CONTINUITY OF GROUND WIRE. NOTE: IF GROUND IS POOR OR ERATIC, SHUTDOWNS MAY OCCUR OCCASIONALLY EVEN THOUGH OPERATION IS NORMAL AT THE TIME OF CHECKOUT.
• CHECK FOR EXCESSIVE HEAT AT IGNITER CERAMIC BASE (TEMPERATURE ABOVE 1000° F [538° C] CAUSES SHORTS TO GROUND).
• IF CHECKS ARE OKAY, REPLACE S89/S890 MODULE.
YES
CALL FOR HEAT ENDS: SYSTEM SHUTS OFF
NO
• CHECK FOR PROPER TEMPERATURE CONTROLLER OPERATION.
• REMOVE VALVE LEAD AT S89/S890; IF VALVE CLOSES, RECHECK TEMPERATURE CONTROLLER AND WIRING; IF NOT, REPLACE GAS CONTROL.
YES
TROUBLESHOOTING ENDS
REPEAT PROCEDURE UNTIL TROUBLEFREE OPERATION IS OBTAINED.
M2198

FIGURE 14–17
S89C, D, G, H troubleshooting guide. (*Courtesy of Honeywell.*)

heat. Check the temperature at the ceramic ignitor-sensor insulator. Excessive temperature will permit leakage to ground.Provide a shield if the temperature exceeds the rating of the ignitor or sensor.

c. If the flame sensor or bracket is bent out of position, restore it to the correct position.

d. Replace the ignitor and sensor or ignitor-sensor with an identical unit if the insulator is cracked.

Step 3: Check the spark ignition circuit. You will need a short jumper wire made from ignition cable or other heavily insulated wire.

1. Close the manual shutoff valve.
2. Disconnect the ignition cable at the stud terminal on the module or the spark generator.

   When performing the following steps, do not touch the stripped end of the jumper wire or the stud terminal. The ignition circuit generates 24 kV open circuit, and an electrical shock can result.
3. Energize the module and immediately touch one end of the jumper firmly to the GND terminal on the module. Move the free end of the jumper slowly toward the stud terminal until a spark is established.
4. Pull the jumper slowly away from the stud and note the length of the gap when sparking stops (Table 14–5).
5. Open the manual shutoff valve and reset the system.

TABLE 14–5 Arc length. (*Courtesy of Honeywell.*)

Arc length	Action
No arc or arc less than 1/8 in. [3 mm] (3/16 in. [5 mm] on S825)	• Check external fuse, if provided. • Verify power at module input terminals, or at spark generator input terminals • Replace module and/or spark generator if fuse and power ok.
Arc 1/8 in. [3 mm] (3/16 in. [5 mm] on S825) or longer	Voltage output is adequate.

Step 4: Check the flame sensor circuit. These steps are used for single-rod S87 DSI only.

1. Turn off the furnace at the thermostat.
2. Connect a meter (DC microammeter scale) in series with the ground lead as shown in Figure 14–18. Use the Honeywell W136 test meter or equivalent. Connect the meter as follows.

   a. Disconnect the ground lead at the electronic control.

   b. Connect the black (negative) meter lead to the electronic control GND terminal.

   c. Connect the red (positive) meter lead to the free end of the ground lead.

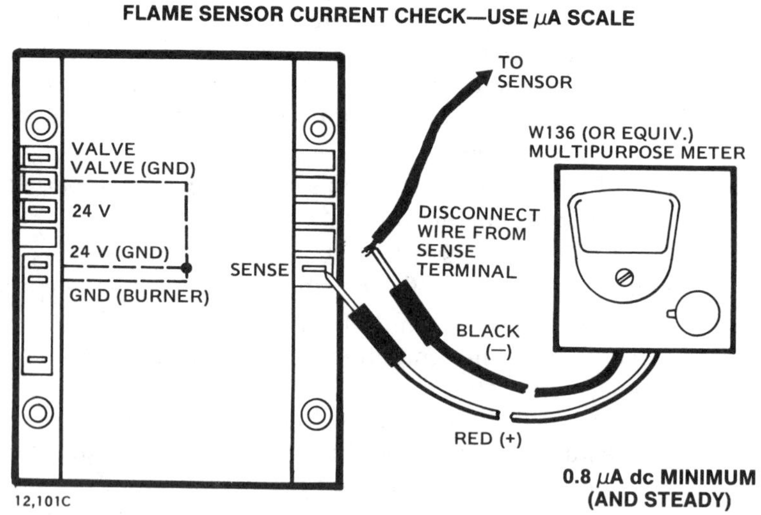
FLAME SENSOR CURRENT CHECK—USE μA SCALE
TO SENSOR
W136 (OR EQUIV.) MULTIPURPOSE METER
VALVE
VALVE (GND)
24 V
24 V (GND)
SENSE
GND (BURNER)
DISCONNECT WIRE FROM SENSE TERMINAL
BLACK (−)
RED (+)
0.8 μA dc MINIMUM (AND STEADY)
12,101C

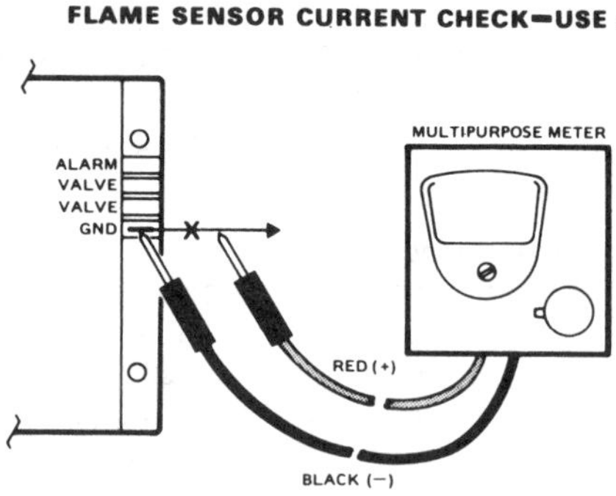
FLAME SENSOR CURRENT CHECK—USE μA SCALE
MULTIPURPOSE METER
ALARM
VALVE
VALVE
GND
RED (+)
BLACK (−)

FIGURE 14–18
Checking flame current on S87 and S89. (*Courtesy of Honeywell.*)

3. Restart the system and read the meter. The flame sensor current must be steady and at least 1.5 μA.
4. If the meter reads less than the minimum or the reading is unsteady, complete the following steps:
   a. Make sure that the burner flame is capable of providing a good rectification signal (Figure 14–19).
   b. Make sure that about 1 inch of the flame sensor or ignitor-sensor is continuously immersed in the flame for best flame signal. Turn off the system and allow it to cool; then bend the bracket or flame sensor, or relocate the sensor as necessary. Do not relocate an ignitor-sensor.
   c. Check for excessive (over 1000° F or 538° C) temperature at the ceramic insulator on the flame sensor. Excessive temperature can cause a short to ground; move the sensor to a cooler location or shield the insulator. Do not relocate an ignitor-sensor.

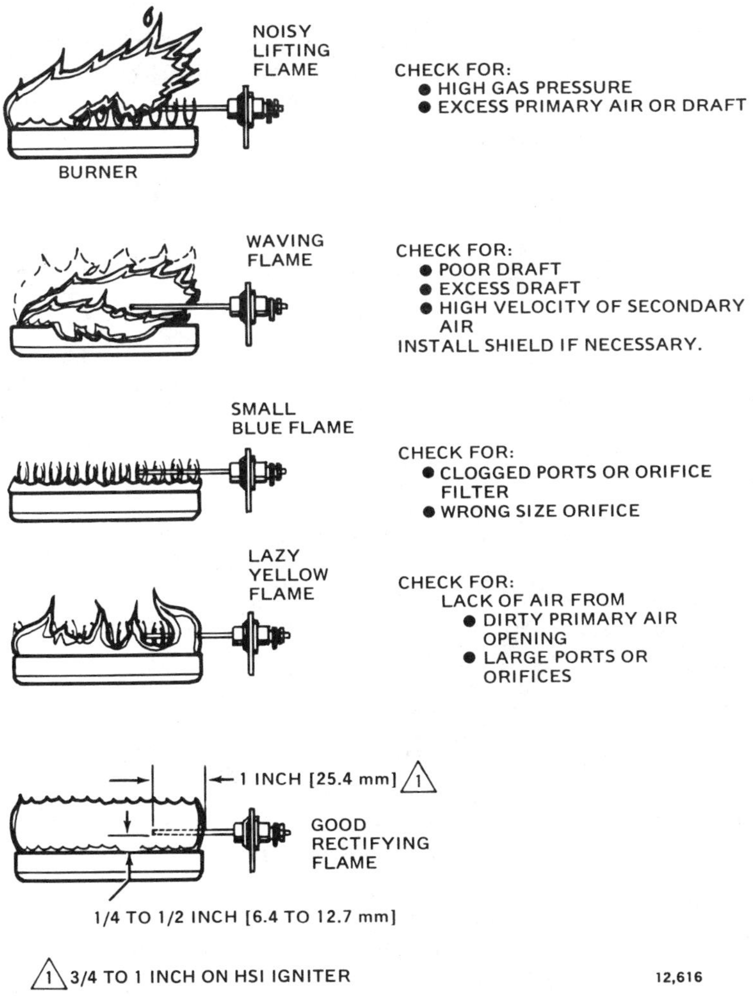
NOISY
LIFTING
FLAME
CHECK FOR:
● HIGH GAS PRESSURE
● EXCESS PRIMARY AIR OR DRAFT
BURNER
WAVING
FLAME
CHECK FOR:
● POOR DRAFT
● EXCESS DRAFT
● HIGH VELOCITY OF SECONDARY AIR
INSTALL SHIELD IF NECESSARY.
SMALL
BLUE FLAME
CHECK FOR:
● CLOGGED PORTS OR ORIFICE FILTER
● WRONG SIZE ORIFICE
LAZY
YELLOW
FLAME
CHECK FOR:
LACK OF AIR FROM
● DIRTY PRIMARY AIR OPENING
● LARGE PORTS OR ORIFICES
1 INCH [25.4 mm] 1
GOOD
RECTIFYING
FLAME
1/4 TO 1/2 INCH [6.4 TO 12.7 mm]
1 3/4 TO 1 INCH ON HSI IGNITER
12,616

FIGURE 14–19
Burner flame must provide good current path. (*Courtesy of Honeywell.*)

d. Check for a cracked ceramic insulator, which can cause a short to ground, and replace the sensor if necessary.
e. Make sure that the electrical connections are clean and tight. Replace damaged wire with moisture-resistant No. 18 wire rated for continuous duty up to 105° C.

5. Remove the microammeter and reconnect the ground wire. Return the system to normal operation.

These steps are for two-rod DSI systems only.

1. Turn off the furnace at the thermostat.
2. Connect a meter (DC microammeter) in series with the flame sensor lead (Figures 14–18 or 14–20). Use the Honeywell W136 test meter or equivalent. Connect the meter as follows.

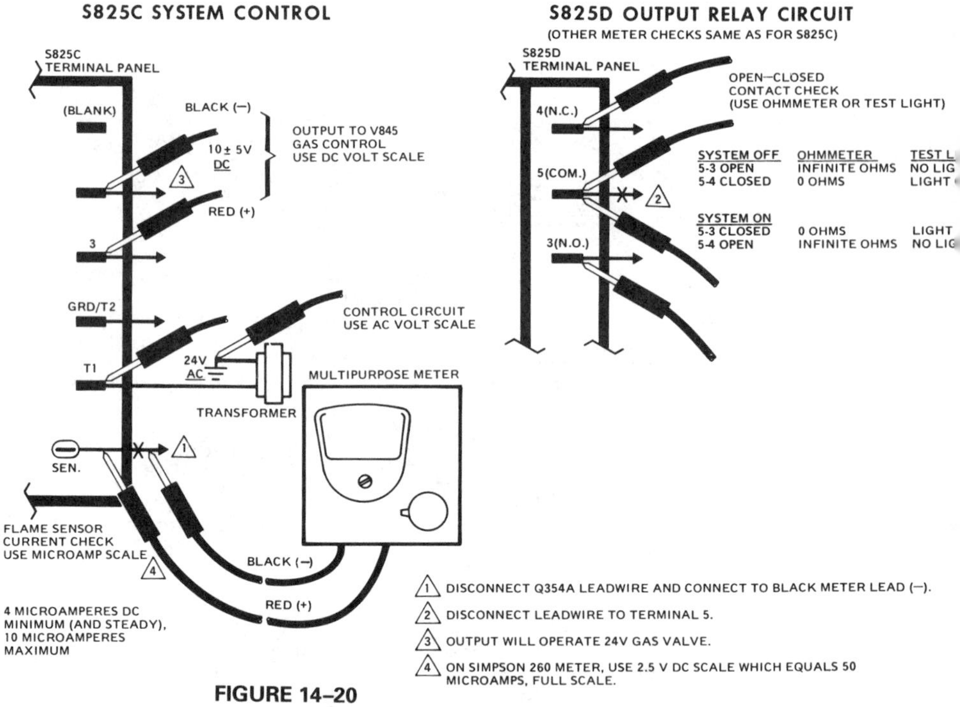
S825C SYSTEM CONTROL
S825C
TERMINAL PANEL
(BLANK)
BLACK (−)
10± 5V
DC
OUTPUT TO V845
GAS CONTROL
USE DC VOLT SCALE
RED (+)
3
GRD/T2
T1
24V
AC
CONTROL CIRCUIT
USE AC VOLT SCALE
MULTIPURPOSE METER
TRANSFORMER
SEN.
FLAME SENSOR
CURRENT CHECK
USE MICROAMP SCALE
BLACK (−)
RED (+)
4 MICROAMPERES DC
MINIMUM (AND STEADY),
10 MICROAMPERES
MAXIMUM
S825D OUTPUT RELAY CIRCUIT
(OTHER METER CHECKS SAME AS FOR S825C)
S825D
TERMINAL PANEL
OPEN—CLOSED
CONTACT CHECK
(USE OHMMETER OR TEST LIGHT)
4(N.C.)
5(COM.)
3(N.O.)
SYSTEM OFF
5-3 OPEN
5-4 CLOSED
OHMMETER
INFINITE OHMS
0 OHMS
SYSTEM ON
5-3 CLOSED
5-4 OPEN
0 OHMS
INFINITE OHMS
LIGHT
1 DISCONNECT Q354A LEADWIRE AND CONNECT TO BLACK METER LEAD (−).
2 DISCONNECT LEADWIRE TO TERMINAL 5.
3 OUTPUT WILL OPERATE 24V GAS VALVE.
4 ON SIMPSON 260 METER, USE 2.5 V DC SCALE WHICH EQUALS 50 MICROAMPS, FULL SCALE.

FIGURE 14–20
Checking S825. (*Courtesy of Honeywell.*)

a. Disconnect the sensor lead at the electronic control.
b. Connect the red (positive) meter lead to the electronic SENSE terminal.
c. Connect the black (negative) meter lead to the free end of the sensor lead.

3. Restart the system and read the meter. The flame sensor current must be steady and at least the minimum as shown in Table 14–6.
4. If the meter reads less than the minimum or the reading is unsteady, use the following procedure:

TABLE 14–6 Pilot Flame Current Readings. (*Courtesy of Honeywell.*)

Module	Minimum flame current
S87, S89A, B	1.5 μA
S89E, F	0.8 μA
S825	4.0 μA (Max. current is 10.0 μA).

a. Make sure that the burner flame is capable of providing a good rectification signal (Figure 14–19).
b. Make sure that about 1 inch of the flame sensor or ignitor to sensor is continuously immersed in the flame for the best signal (Figure 14–19). Turn off the system and allow it to cool; then bend the bracket or flame sensor, or relocate the sensor as necessary. Do not relocate an ignitor or combination ignitor-sensor.
c. Check for excessive (over 1000° F or 538° C) temperature at the ceramic insulator on the flame sensor. Excessive temperature can cause a short to ground; move the sensor to a cooler location or shield the insulator. Do not relocate an ignitor-sensor.
d. Check for a cracked ceramic insulator, which can cause a short to ground, and replace the sensor if necessary.
e. Make sure that the electrical connections are clean and tight. Replace damaged wire with moisture-resistant No. 18 wire rated for continuous duty up to 105° C.

5. On the S825, if the reading is over 10 μA, move the Q354 flame sensor so that less of the flame rod is immersed in the flame.
6. Remove the microammeter and reconnect the sensor lead. Return the system to normal operation.

These steps are for HSI (hot surface ignition) systems only.

1. Make sure that the burner flame is capable of providing a good quality rectification signal (Figure 14–19).
2. Make sure that about 3/4 to 1 inch of the flame sensor or ignitor-sensor is continuously in the flame for the best flame signal. Bend the bracket or the flame sensor, or relocate the sensor as necessary. Do not relocate an ignitor or combination ignitor-sensor.
3. Check for excessive (over 1000° F or 538° C) temperature at the ceramic insulator on the flame sensor. Excessive temperature can cause a short to ground; move the sensor to a cooler location or shield the insulator. Do not relocate an ignitor or combination ignitor-sensor.
4. Check for a cracked ceramic insulator, which can cause a short to ground, and replace if necessary.
   a. Make sure that the electrical connections are clean and tight. Replace damaged wire with moisture-resistant No. 18 wire rated for continuous duty up to 105° C (221° F).
5. If the ignitor is other than a Norton 201 or Honeywell 201071, make sure that it meets the following specifications.
   a. The ignitor must reach 1,000° C (1,832° F) within 34 seconds with 102 V AC applied.
   b. The ignitor must maintain at least 500,000Ω insulation resistance between the ignitor leadwires and the ignitor mounting bracket.
   c. The ignitor must develop an insulating layer on its surface (over time) that would prevent flame sensing.

d. The ignitor surface area immersed in the flame must not exceed one quarter of the grounded area immersed in the flame. This would prevent flame sensing.
e. The ignitor current draw at 132 V AC must not exceed 5 μA.

For troubleshooting procedures for the Honeywell spark to pilot (IID) system, see Table 14–7.

TABLE 14–7 Honeywell troubleshooting chart. (*Courtesy of Honeywell.*)

Symptoms	Possible cause	Checks & remedies
No spark	Open in ignition cable.	Check continuity of ignition cable if open, replace.
	Ignitor improperly grounded.	Check ground connections, insure good chasis ground.
	No voltage to ignition module.	Verify 25 volts AC input to ignition module.
	Ignitor improperly adjusted.	Verify correct ignitor adjustment.
	Defective control module.	Check for spark from module.
Spark, but no ignition	No gas to pilot assembly.	Verify supply pressure.
	Pilot orifice plugged.	Inspect orifice clean or replace if dirty.
	Gas supply tubing to pilot kinked.	Inspect tubing to pilot assembly correct or replace.
	No voltage to gas control.	Verify voltage across terminals PV-MV/PV of module. If no voltage, replace module, if present replace gas control.
Spark continues after ignition	Ignition cable has no continuity.	Check ignition cable for continuity.
	Poor flame impingement on sensor rod.	Check that flame covers both electrodes.
Main burner fails to light	No voltage to gas valve.	Verify voltage at terminals MV-MV/PV of modules. If no voltage, replace module.
	Defective gas valve.	Check connections between module and gas valve, if O.K., replace gas valve.
Main burner shuts down	Unit improperly grounded.	Verify ignitor assembly and module is grounded properly.
Main burner fails to shut down at end of cycle	Defective thermostat.	Check thermostat.
	Defective ignition module.	Remove MV lead from module, if gas valve closes, replace module, if not replace gas valve.

SPARK IGNITORS

These systems (Figure 14–21) may be used as direct replacement pilot lighters, or as direct spark ignition systems (Figure 14–22), on heating equipment that has been certified by an approved testing agency with the pilot lighter as part of the original equipment.

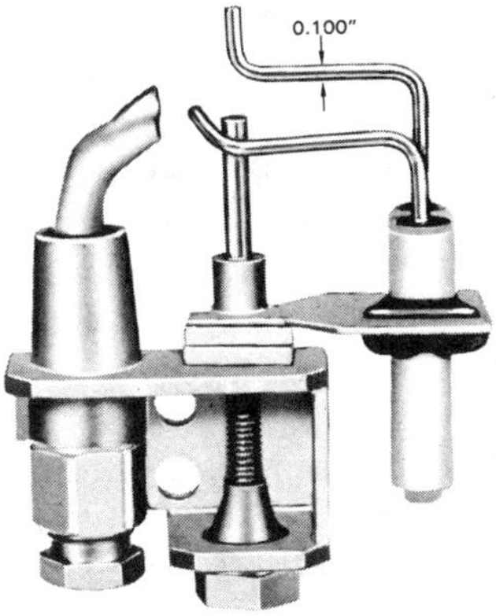
0.100"

FIGURE 14–21
Spark ignitor. (*Courtesy of Billy Langley,* Heating, Ventilating, Air Conditioning, and Refrigeration, *©1990, p. 548; reprinted by permission of Prentice-Hall, Inc., Englewood Cliffs, N.J.*)

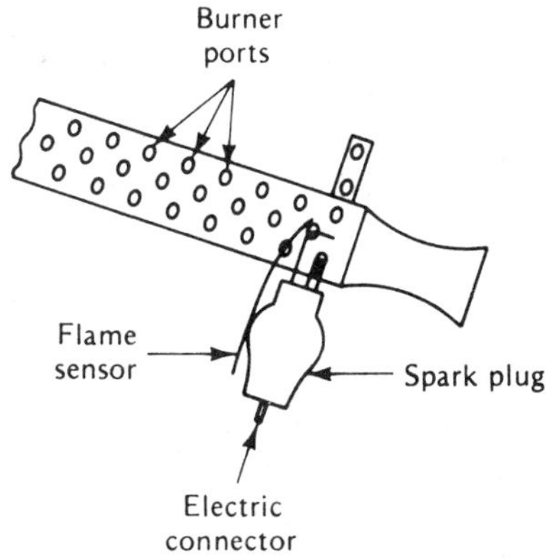
Burner ports
Flame sensor
Spark plug
Electric connector

FIGURE 14–22
Direct spark ignition system burner mounting. (*Courtesy of Billy Langley,* Heating, Ventilating, Air Conditioning, and Refrigeration, *©1990, p. 548; reprinted by permission of Prentice-Hall, Inc., Englewood Cliffs, N.J.*)

When automatic pilot relighters are not certified as an original component of the unit, their application must be limited to rooftop heating units, space heaters, open bay heaters, or on installations where the unburned gases are quickly vented.

These units have a spark frequency of approximately 100 sparks per minute. They may be used on either 24 V AC or 120 V AC. They are satisfactory for use in a temperature range of -40° F to 160° F.

Some retrofit units are not to be used on LP gas. Be sure to check the manufacturer's suggestions before making the changes.

The retrofit spark ignition system requires that some specially designed components be used to replace those in the existing control system. These control systems light the pilot flame by providing an electric spark to the pilot burner. The pilot gas is ignited and burns during each running cycle (intermittent pilot). When the pilot flame is proven, the spark ignition is discontinued.

The main gas valve is permitted to open only when the pilot flame is proven. If a loss of pilot flame occurs, the main gas valve is deenergized and the spark ignition is reactivated almost immediately. Both the main burner and the pilot burner are off during the off cycle. These units must be correctly wired to provide the desired functions (Figure 14–23).

There are models available that provide a slow opening pressure regulator to gradually increase the flow of gas to the main burner after the pilot flame has been proven.

HOT SURFACE IGNITORS (SILICON CARBIDE HSI)

There are several different types of hot surface ignition systems available; however, we will use the Johnson Series G750, 100% lockout, Microprocessor Based Hot Surface Ignition (HSI) System for this study (Figure 14–24).

Description

The G750 HSI system is designed for the direct ignition of natural, liquefied petroleum (LP), manufactured, mixed, or LP gas-air mixtures. The main gas is ignited by a commercially available hot surface ignitor, and the flame is sensed by a sensing probe (remote sensing) or by the hot surface itself (integral sensing). When the main burner has sensed the flame, the ignitor is deenergized. The main gas valve remains open to provide gas to the main burner until the room thermostat is satisfied.

If burner ignition is not proven within the allowed time trial for ignition, the main gas valve and the ignitor are deenergized. On the three-trial models a period of time known as the *interpurge* is programmed into the microprocessor to permit unburned gas to escape from the combustion chamber before another trial for ignition occurs. The number of ignition trials (one or three) will depend on the model of the unit. All G750 applications must use a redundant gas valve.

Specifications

The specifications for the G750 HSI control are listed in Table 14–8.

Operating Mode Definitions

The following are the operating mode definitions for the G750 HSI ignition system. *Note:* Some models may not use all of the operating modes listed.

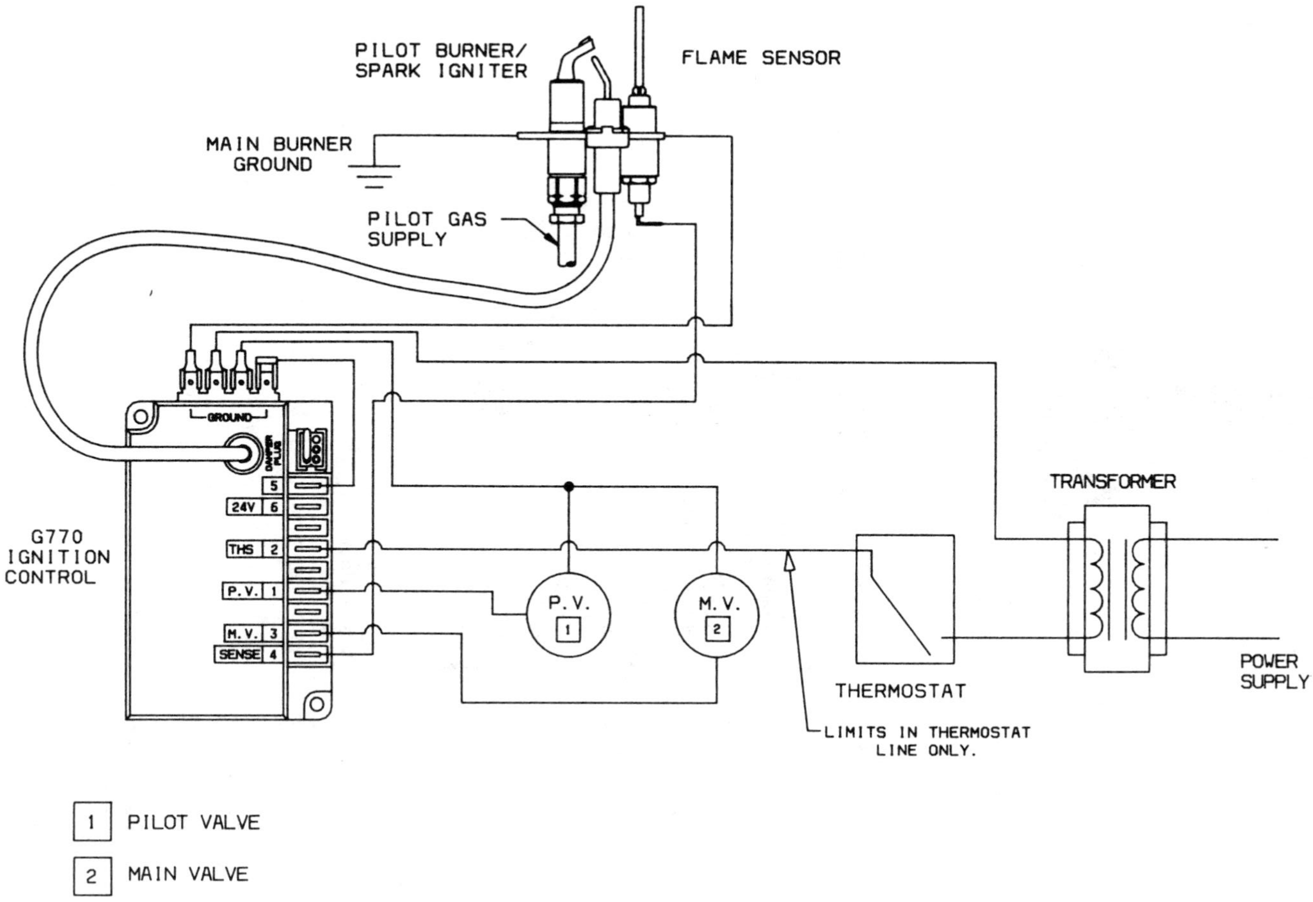
PILOT BURNER/
SPARK IGNITER
FLAME SENSOR
MAIN BURNER
GROUND
PILOT GAS
SUPPLY
GROUND
DAMPER PLUG
5
24V 6
THS 2
P.V. 1
M.V. 3
SENSE 4
G770
IGNITION
CONTROL
P.V.
1
M.V.
2
THERMOSTAT
LIMITS IN THERMOSTAT
LINE ONLY.
TRANSFORMER
POWER
SUPPLY
1 PILOT VALVE
2 MAIN VALVE

FIGURE 14–23
Wiring diagram for retrofit spark ignition system using automatic vent damper. (*Courtesy of Johnson Controls, Inc.*)

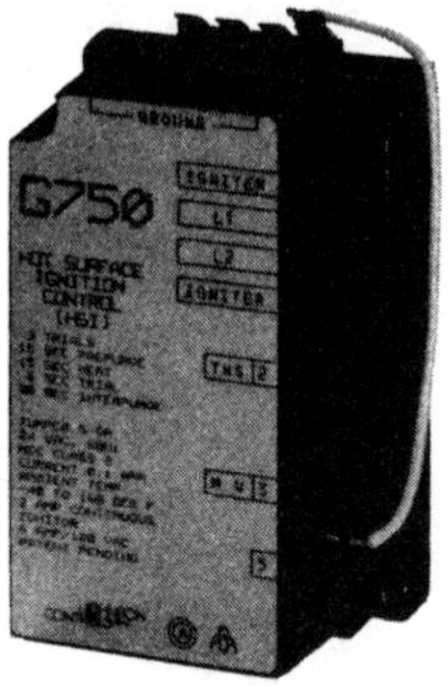
G750

FIGURE 14–24
G750 HSI control. (*Courtesy of Johnson Controls, Inc.*)

TABLE 14–8 Specifications. (*Courtesy of Johnson Controls, Inc.*)

Mounting		Surface Mount — Any Position
Electrical (Control)	**Operating Voltage**	24 Volts, 60 Hz
	Operating Current	0.1 A, 24 VAC
	Contact Rating (MV)	2.0 A Continuous / 5.0 A Inrush, 24 VAC
Electrical (Igniter)	**Operating Voltage**	120 Volts, 60 Hz
	Operating Current	6 A, 120 VAC
	Contact Rating	120 Volts (Resistive)
Ignition	**Means**	Hot Surface (HSI)
	No. of Trials	Multiple (One or Three)
	Trial Times	4, 6, 8, 10 or 15 Seconds
	Igniter Warm-Up Times	15, 30 or 45 Seconds
	Prepurge Times	0, 4, 15 or 30 Seconds
	Interpurge Time	30 Seconds
Flame Detection	**Means**	Flame Rectification
	Flame Failure Response Time	0.8 Seconds (Maximum)
	Flame Output Voltage	24 Volts RMS
	Flame Output Frequency	60 Hz
	Flame Current Signal	0.2 Microamps (Minimum)
Ambient Temp. Range		-40°F (-40°C) to 160°F (70°C)
Moisture Resistance		95% RH @ 160°F (70°C) Non-Condensing
Wiring Connections	**Igniter**	1/4″ Spade
	Control	1/4″ Spade
Case Material		Thermoplastic
Type of Gas		Natural, Liquified Petroleum (LP), Manufactured, or Mixed
Standards		ANSI Z21.20 C22.2, No. 199
Shipping Weight	**Bulk Pack of 48**	36 lbs. (19 kg)

1. Prepurge: Initial time delay of the control before ignition of the main burner gas.
2. Warm-up time: A hot surface element (ignitor) is energized for a fixed period of time to allow it to exceed the ignition temperature of the gas.
3. Trial for ignition: The main gas and the ignitor are energized for a limited period of time. (*Note:* The ignitor will be turned off before the end of the trial for ignition period to provide sensing on the integral models.)
4. Interpurge: This is the period of time between the trials for ignition when both the gas and the ignitor are deenergized to allow for any unburned gas to escape before the next trial for ignition can occur. (This step occurs only if a proper ignition did not occur during a trial for ignition period.)
5. Lockout: The main gas failed to ignite on any of the allowed trials for ignition. The thermostat contacts must be opened for at least 30 seconds to reset the control.
6. Run: The main gas is on after a successful ignition attempt.

Operation

The G750 HSI control is energized on a call from the thermostat for heat (Figure 14–25). If the control is equipped with a prepurge mode, the furnace prepurge fan or relay is also energized through the thermostat contacts. In the prepurge mode, the control will delay for the time selected (e.g., 30 seconds) before applying the electric power to hot surface ignitor. If the prepurge is not selected, the ignitor is energized within one second after the call for heat by the thermostat. The ignitor warms up and after the selected heating time (e.g., 30 seconds), and the main gas valve opens to apply gas to the main burner. A flame must be sensed within the trial for ignition time period (e.g., 4 seconds) or the electric power to the main gas valve and the ignitor is switched off by the G750 control.

When the main burner gas ignites within the chosen trial for ignition time, the ignitor is deenergized and used to sense the main burner flame current (integral sensing), signaling the G750 to keep the main gas valves open while there is a call for heat from the thermostat. (Models equipped with the remote flame sensing mode provide a similar function through the use of an individual flame sensor probe.) The control will lock out if a flame is not sensed at the end of the trial for ignition period. On multitrial (three) models, the trial for ignition will be repeated for a total of three trials after the interpurge period of 30 seconds. The control will lock out if the main burner flame is not sensed at the end of the third trial for ignition.

To reset the G750, the thermostat must be turned off for a minimum of 30 seconds. If the main burner flame goes out (flame loss with the thermostat calling for heat), the control will restart the ignition sequence.

Wiring diagrams for the G750 HSI control are shown in Figures 14–26 through 14–28.

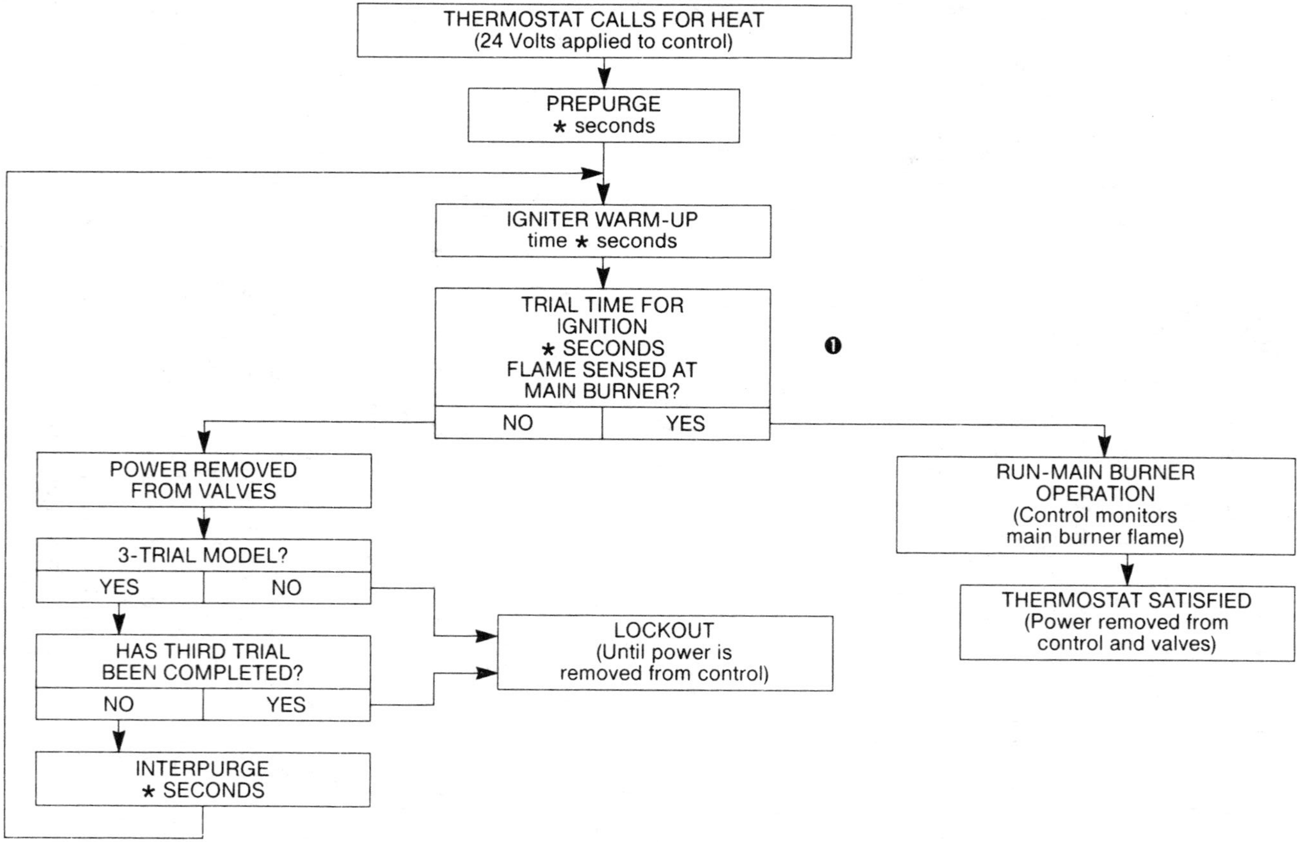
THERMOSTAT CALLS FOR HEAT
(24 Volts applied to control)
PREPURGE
★ seconds
IGNITER WARM-UP
time ★ seconds
TRIAL TIME FOR
IGNITION
★ SECONDS
FLAME SENSED AT
MAIN BURNER?
NO
YES
POWER REMOVED
FROM VALVES
3-TRIAL MODEL?
YES
NO
HAS THIRD TRIAL
BEEN COMPLETED?
NO
YES
INTERPURGE
★ SECONDS
LOCKOUT
(Until power is
removed from control)
RUN-MAIN BURNER
OPERATION
(Control monitors
main burner flame)
THERMOSTAT SATISFIED
(Power removed from
control and valves)

➊ Igniter element becomes flame sensor on ignition systems with integral sensing.
★ See control label for actual times.

FIGURE 14–25
Sequence of operation. (*Courtesy of Johnson Controls, Inc.*)

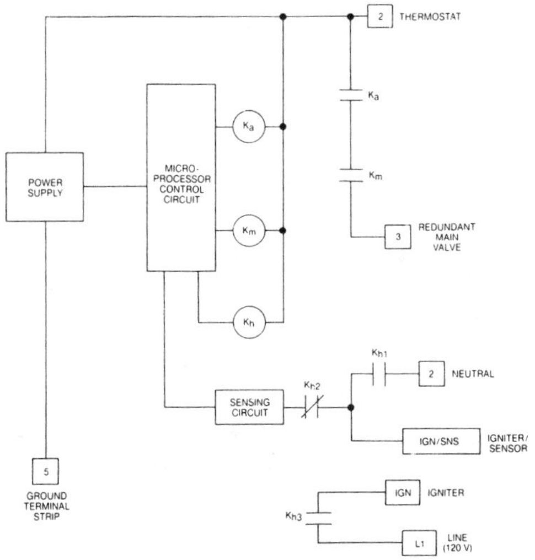

THERMOSTAT
2
Ka
Km
POWER SUPPLY
MICRO-PROCESSOR CONTROL CIRCUIT
REDUNDANT MAIN VALVE
3
Kh
Kh1
NEUTRAL
Kh2
SENSING CIRCUIT
IGN/SNS
IGNITER/ SENSOR
5
GROUND TERMINAL STRIP
IGN
IGNITER
Kh3
L1
LINE (120 V)

FIGURE 14–26
G750 wiring schematic *(Integral or Remote Sensing)*.

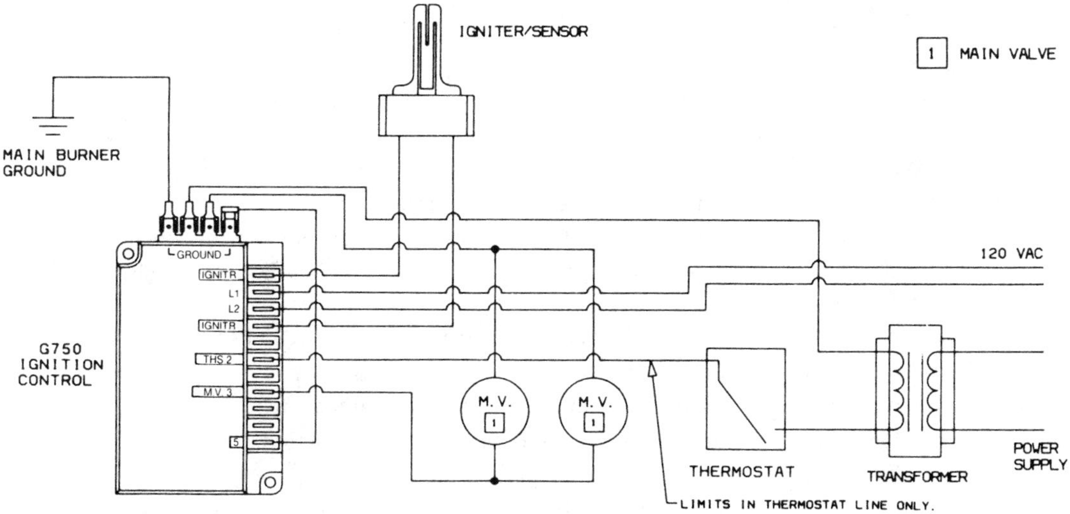

IGNITER/SENSOR
1
MAIN VALVE
MAIN BURNER GROUND
GROUND
IGNITR
L1
L2
IGNITR
THS 2
M.V. 3
5
G750 IGNITION CONTROL
120 VAC
M. V.
M. V.
THERMOSTAT
TRANSFORMER
POWER SUPPLY
LIMITS IN THERMOSTAT LINE ONLY.

Figure 14–27
Connection diagram: integral models. (*Courtesy of Johnson Controls, Inc.*)

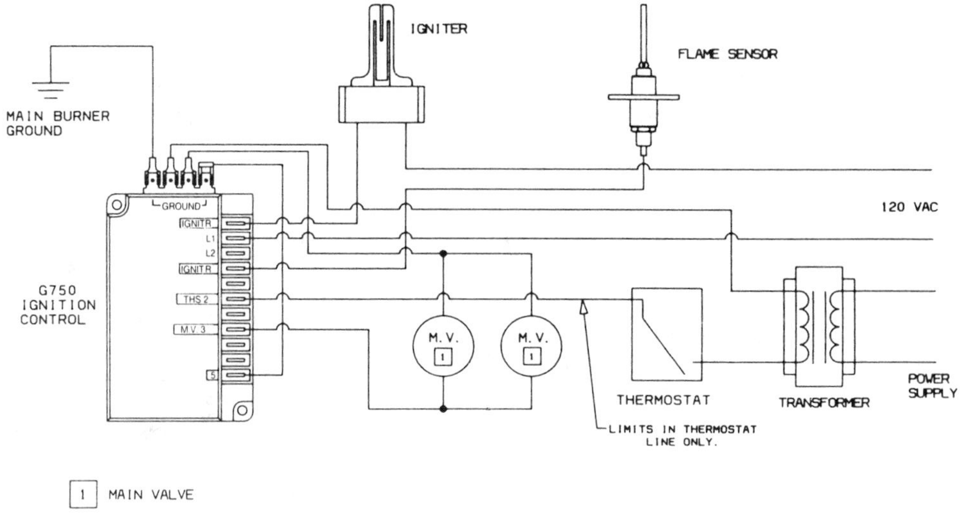
IGNITER
FLAME SENSOR
MAIN BURNER
GROUND
GROUND
IGNITR
L1
L2
IGNITR
THS 2
M.V. 3
5
G750
IGNITION
CONTROL
120 VAC
M. V.
1
M. V.
1
THERMOSTAT
TRANSFORMER
POWER
SUPPLY
LIMITS IN THERMOSTAT
LINE ONLY.
1 MAIN VALVE

FIGURE 14–28
Connection diagram: Remote sensing models. (*Courtesy of Johnson Controls, Inc.*)

REVIEW QUESTIONS

1. What are the common voltages used in electric ignition systems?
2. Why is the power interrupted to a glow coil after the pilot is lit?
3. In an electronic ignition system, what does the ignition control do when a flame is detected?
4. What is the purpose of a flame sensor in electronic ignition systems?
5. Name the two types of ignition systems.
6. What must be present before the main burner valve is energized?
7. On a DSI system, what lights the main burner?
8. When the DSI system is used, what principle applies?
9. What causes a flame to become electrically conductive?
10. Why must the flame make contact with both the pilot burner and the sensing probe?
11. In flame rectification, what causes the electric current to flow from the positive probe to the negative probe?
12. What is used to determine the amount of current flowing in an IID system?
13. What type of pilot flame is required on IID systems?
14. What is the reasoning behind eliminating standing pilot burners?
15. What is the maximum allowable length of an ignition cable?
16. Why should an ignition cable be properly supported?
17. What could an excessive temperature at the ceramic flame rod insulator cause?
18. What is the open circuit voltage that is generated by the ignition circuit?
19. What type of ground is required for the module, ignitor, sensor, and main burner?

20. What should occur between the ignitor bracket and the main burner?
21. What is the minimum flame sensor current allowed for the Honeywell DSI system?
22. When is it desirable to relocate an HSI ignitor or ignitor-sensor?
23. What should the maximum current draw be in a Honeywell HSI system operating on 132 V AC?
24. What is the initial time delay of the control before ignition of the main burner gas known as?
25. On a Johnson Controls HSI system, when will the control lock out the main burner if a flame is not sensed?

15

heating controls

The objectives of this chapter are:

- To acquaint you with the different types and operation of comfort heating thermostats.
- To acquaint you with the operation, sizing, and phasing of transformers.
- To acquaint you with the different types and operation of limit controls.
- To provide you with information on the air switch.
- To provide you with information on the discharge-air averaging thermostat.
- To acquaint you with information about water heating controls.
- To provide you with information on the outdoor reset control.
- To acquaint you with the purpose and operation of humidistats.

The residence should be heated to a comfortable temperature, with little or no variation in temperature between the floor and the ceiling.

This enormous duty falls on the control system, in conjunction with a properly designed and installed heating system. These three factors go hand in hand. Without the proper controls, however, an otherwise perfect system would be ineffective.

THE THERMOSTAT

In general, low voltage room thermostats should be used for the best temperature control (Figure 15–1). The low voltage thermostats respond much faster to temperature changes than the greater mass line-voltage devices. This has been proven on residential heating systems.

HONEYWELL

FIGURE 15–1
Low voltage heating thermostat.
(*Courtesy Honeywell, Inc.*)

From a cost standpoint, the less expensive installation of low voltage wiring more than offsets the extra cost of the transformer. Also, homeowners are accustomed to the safety of low voltage thermostats.

A room thermostat is used for sensing the room temperature and signaling the equipment either to operate or to stop operating in response to its temperature.

In its simplest form, a thermostat is a device that responds to changes in the air temperature and causes a set of electrical contacts to open or close. This is the basic function of a thermostat, but there are many different types designed to perform a variety of switching functions. They are available in heating, cooling, or heating and cooling types.

Types of Room Thermostats

There are three types of electric thermostats used in heating systems today: (1) the bimetal type, (2) the bellows-actuated type, and (3) the proportioning (modulating) type. The bimetal type is by far the most popular type in use today (Figure 15–1).

Bimetal thermostat. The bimetal thermostat gets its name from the fact that it uses a bimetal to open or close a set of electrical contacts on an increase or a decrease in room temperature (Figure 15–2). This set of contacts may be of the open type or may be enclosed in a mercury tube.

A bimetal is made of two pieces of metal which, at a given temperature, are the same length. If the temperature of the two pieces of metal is increased, one becomes longer than the other because they are different types of metal with different rates of expansion. These two metals are welded together in such a way that they become one solid piece, but they still keep their individual characteristics of different rates of expansion (Figure 15–3).

When heat is applied, one piece expands at a faster rate than the other piece. For one piece to become longer than the other, it must bend the entire bimetal into an arc (Figure 15–4).

If we anchor one end of the bimetal to something solid, the free end will move down or up with an increase in temperature. By attaching contacts to the free end and placing a stationary contact nearby, we can get different switching actions with changes in temperature (Figure 15–5).

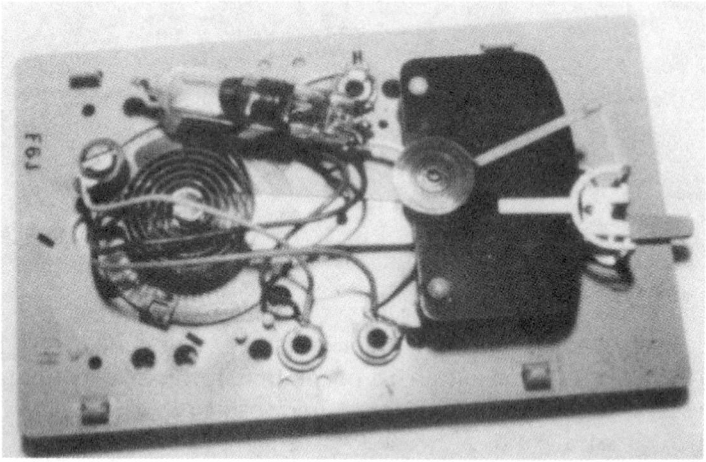

FIGURE 15–2
Bimetal thermostat with cover removed. (*Courtesy of Billy Langley,* Electric Controls for Refrigeration and Air Conditioning, *2E, ©1988, p. 90; reprinted by permission of Prentice-Hall, Inc., Englewood Cliffs, N.J.*)

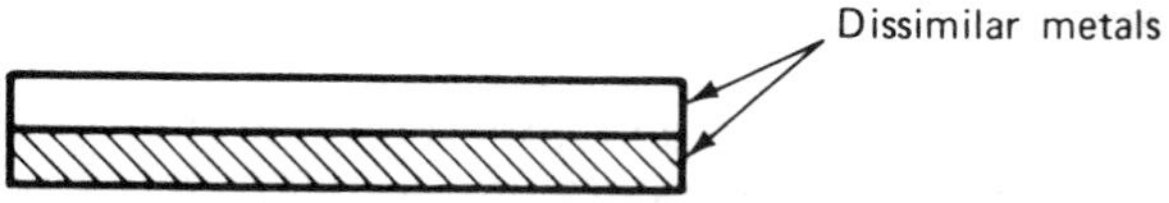
Dissimilar metals

FIGURE 15–3
Bimetal. (*Courtesy of Billy Langley,* Electric Controls for Refrigeration and Air Conditioning, *2E, ©1988, p. 90; reprinted by permission of Prentice-Hall, Inc., Englewood Cliffs, N.J.*)

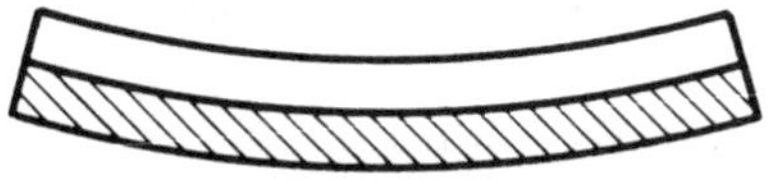

FIGURE 15–4
Bimetal in heated condition. (*Courtesy of Billy Langley,* Electric Controls for Refrigeration and Air Conditioning, *2E, ©1988, p. 90; reprinted by permission of Prentice-Hall, Inc., Englewood Cliffs, N.J.*)

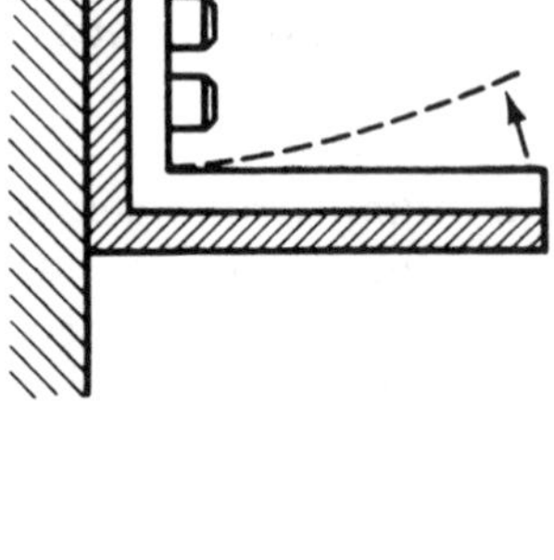

FIGURE 15–5
Anchored bimetal. (*Courtesy of Billy Langley,* Electric Controls for Refrigeration and Air Conditioning, *2E, 1988, p. 90; reprinted by permission of Prentice-Hall, Inc., Englewood Cliffs, N.J.*)

The first bimetal thermostats produced unsatisfactory results because of unstable action of the contacts. Due to the relatively small differences in room air temperature, the bimetal could not develop enough contact pressure to obtain a positive electrical connection. With the development of the permanent magnet, it was possible to obtain the convenience of a control system incorporating the best features of modern control circuits (Figure 15–6).

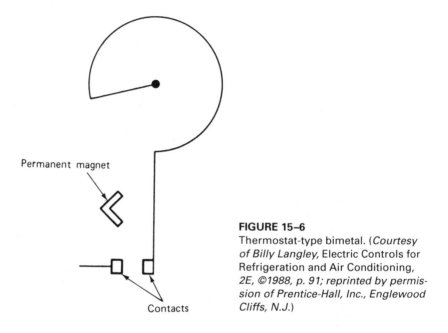

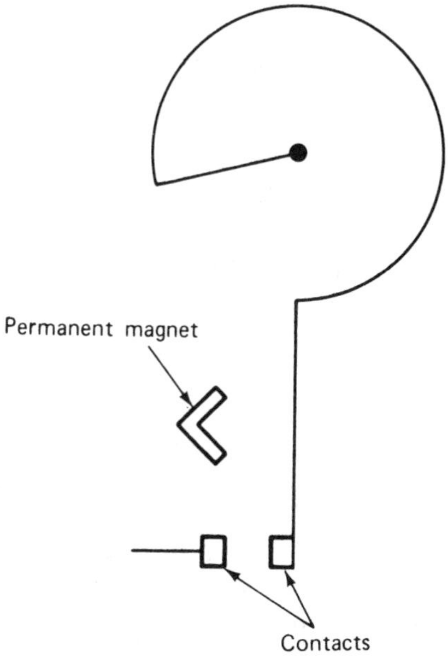
Permanent magnet
Contacts

FIGURE 15–6
Thermostat-type bimetal. (*Courtesy of Billy Langley,* Electric Controls for Refrigeration and Air Conditioning, *2E, ©1988, p. 91; reprinted by permission of Prentice-Hall, Inc., Englewood Cliffs, N.J.*)

Snap action versus mercury switch. Room thermostats are available with either snap action or mercury switches. Snap-action switches are constructed with a fixed contact securely fastened to the base of the thermostat. This contact is mounted inside a round permanent magnet, which produces a magnetic field in the area of the contact. The movable contact is attached to the bimetal, and upon a decrease in temperature (on heating models) moves slowly toward the fixed contact (Figure 15–7).

As the movable contact enters the magnetic field around the fixed contact, the magnetic field pulls the movable contact against the fixed contact with a positive snap. Because the movable contact has a floating action, it closes with a clean snap (that is, without contact bounce). This floating action also eliminates any tendency for the contacts to walk while opening. Either the walking action or a lack of positive snap will cause arcing between the contacts, which in time will burn and pit them, reducing their electrical continuity. This will eventually cause the thermostat to become less responsive to temperature changes or not make a circuit through the contacts at all.

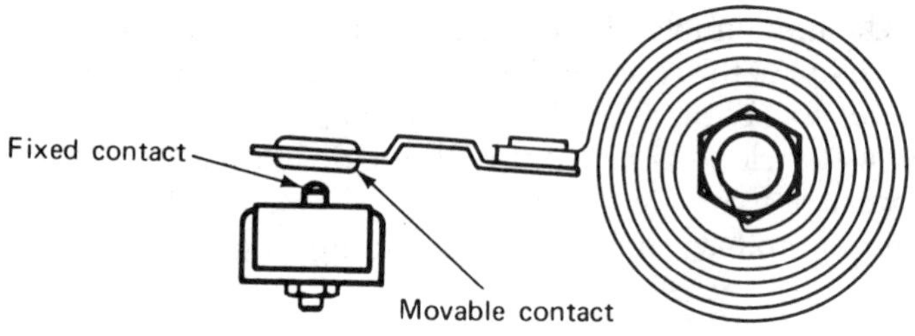
Fixed contact
Movable contact

FIGURE 15–7
Typical thermostat bimetal. (*Courtesy of Billy Langley,* Electric Controls for Refrigeration and Air Conditioning, *2E, ©1988, p. 91; reprinted by permission of Prentice-Hall, Inc., Englewood Cliffs, N.J.*)

As the bimetal becomes warmer (on a heating model), it wants to pull the movable contact away from the fixed contact. But, because the movable contact is in the magnetic field surrounding the fixed contact, the bimetal does not—at this instant—have enough force to overcome the magnetic field. As the bimetal continues to warm up and bend, is soon develops enough force to overcome the magnetic field and the movable contact breaks away with a positive snap. When an SPDT switch action is used, another fixed contact is used. This contact is located so that when one contact is broken, another is made (Figure 15–8).

All snap-action thermostats are supplied with dust covers to prevent dirt and other contaminants from getting on the contacts. Should it become necessary to clean these contacts, never use a file or sand paper. A clean business card or smooth cardboard should be inserted between the contacts. With gentle pressure on the movable contact, pull the card back and forth to clean the dirt or film from the contacts.

Mercury switches perform the same switching action as snap-action switches, but the switching action is accomplished by a globule of mercury moving between two or three fixed probes sealed inside a glass tube. Two probes are used on SPST switches (Figure 15–9). The SPDT models have three probes (Figure 15–10).

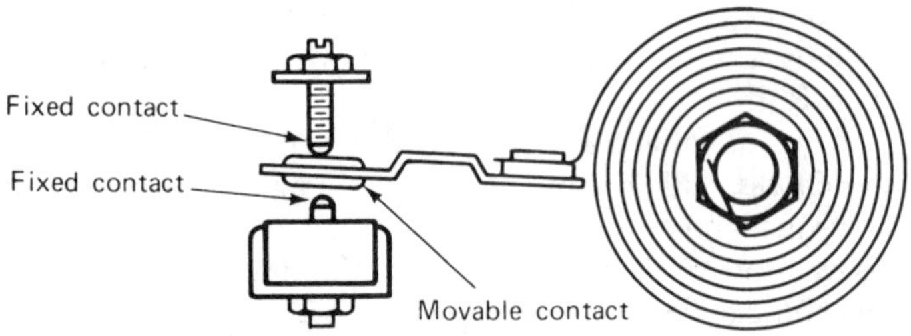
Fixed contact
Fixed contact
Movable contact

FIGURE 15–8
Single-pole-double-throw thermostat. (*Courtesy of Billy Langley,* Electric Controls for Refrigeration and Air Conditioning, *2E, ©1988, p. 92; reprinted by permission of Prentice-Hall, Inc., Englewood Cliffs, N.J.*)

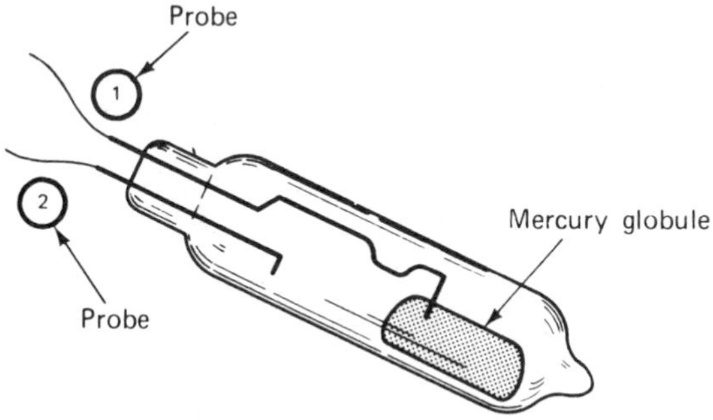
Probe
1
2
Probe
Mercury globule

FIGURE 15–9
SPST mercury switch. (*Courtesy of Billy Langley,* Electric Controls for Refrigeration and Air Conditioning, *2E, ©1988, p. 92; reprinted by permission of Prentice-Hall, Inc., Englewood Cliffs, N.J.*)

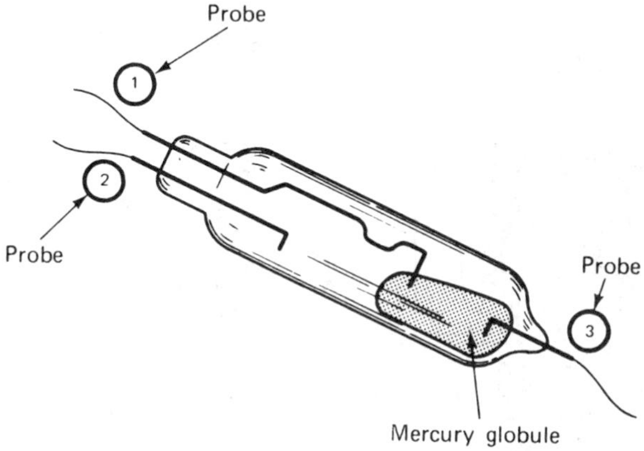
Probe
1
2
Probe
Probe
3
Mercury globule

FIGURE 15–10
SPDT mercury switch. (*Courtesy of Billy Langley,* Electric Controls for Refrigeration and Air Conditioning, *2E, ©1988, p. 92; reprinted by permission of Prentice-Hall, Inc., Englewood Cliffs, N.J.*)

These mercury switches are attached to the thermostat bimetal and perform the desired function.

Bellows-operated thermostat. The bellows-operated thermostat has the same function as the bimetal thermostat. The major difference between the two is that a bellows filled with some type of fluid that expands when heated and contracts when cooled is used instead of a bimetal element. Bellows-operated thermostats perform the same function that bimetal thermostats do.

Operation. When the room air temperature begins rising, the fluid temperature inside the bellows also begins rising. As its temperature increases, the bellows begins to expand in direct relation to the temperature. When the cut-out setting of the thermostat is reached, the contacts open and the system stops operating. As the room temperature drops, the temperature of the fluid also drops. When the cut-in temperature is reached, the thermostat contacts close the control circuit and the system starts operating.

Staging thermostat. In recent years there has been an increased demand for greater comfort and efficiency in indoor heating systems. The staging thermostat has been designed to meet these needs (Figure 15–11).

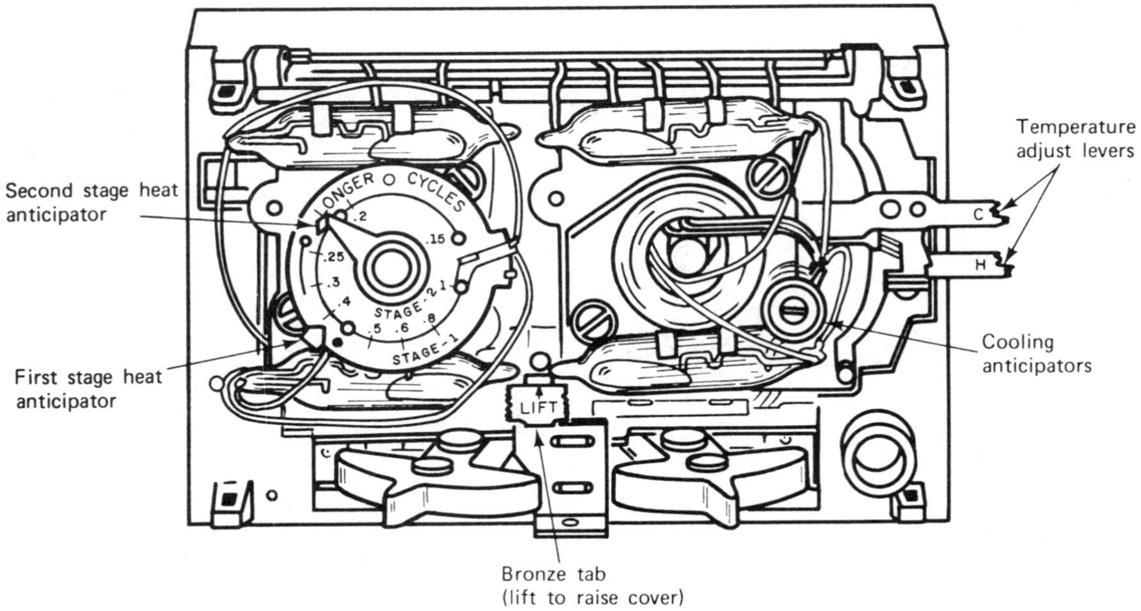
Temperature adjust levers
Second stage heat anticipator
First stage heat anticipator
Cooling anticipators
Bronze tab (lift to raise cover)
LONGER O CYCLES
.2
.15
.25
.3
.4
.5
.6
.8
1
STAGE-2
STAGE-1
LIFT
C
H

FIGURE 15–11
Multistage thermostat. (*Courtesy of Billy Langley,* Electric Controls for Refrigeration and Air Conditioning, *2E, ©1988, p. 97; reprinted by permission of Prentice-Hall, Inc., Englewood Cliffs, N.J.*)

The staging thermostat is designed to be used on systems that have more than one-stage heating or cooling or any combination of heating and cooling stages.

In a typical system using two-stage heating and two-stage cooling, when the thermostat is in the heating position the heating system operates at reduced Btu input capacity during mild weather. As the weather becomes colder, this reduced capacity is not sufficient to maintain the desired comfort level, and the thermostat automatically brings on the additional capacity of the heating system. In the cooling position the situation would be similar, except that more levels of cooling would be used rather than heating. These thermostats are also available with automatic changeover. In the automatic position all that is necessary is to set the desired level of heating and cooling, and the thermostat will automatically switch from one to the other, based on the temperature settings on the thermostat. There are two separate temperature settings on this type of thermostat, one marked C for cooling and the other marked H for heating.

Circuit wiring diagram. The circuit wiring with staging thermostats varies from installation to installation due to the number of possible stages that can be used. We shall use a two-stage heating, two-stage cooling wiring diagram to review the thermostat function (Figure 15–12). From this diagram we can see that the system switch is in the heat position and the fan switch is in the auto position. From one side of the transformer, the circuit goes to terminal RC, through the jumper wire to RH, through the internal wiring to the bar contact on heat, and on to stage 1 and stage 2 heat anticipators. there is also a circuit to terminal B, which will energize another circuit continuously. Upon a call for heat, the stage 1 heating switch closes, providing a circuit to W_1, then to the first stage of the heating system, and back to the other side of the transformer. If the temperature continues to drop, the stage 2 switch closes, giving a circuit through W_2 to the second stage of the heating system and back to the other side of the transformer.

When the selector switch is placed in the cool position, the heat bar breaks and the cool contact is made. This gives a circuit from RC to the cool contact, through the internal jumper, to the auto contact, and on to stage 1 and stage 2 contacts. When the stage 1 contacts close, there is a circuit through Y_1, stage 1 cooling, and back to the other side of the transformer. If this is not sufficient cooling, stage 2 closes and this circuit is made through Y_2, stage 2 cooling, and back to the other side of the transformer.

Fan switches. The fan switch on a thermostat has two positions, one for automatic operation and one for continuous operation.

When the fan switch is in the auto position, the fan operates only on demand from that part of the system in use at the time (heating or cooling).

If the switch is placed in the on position, the fan operates regardless of the system demand. The fan also operates in this position when the system switch (heat or cool) is turned to the off position.

Modulating thermostat. Modulating control systems are built around the Wheatstone bridge principle. Both the controller and the controlled apparatus

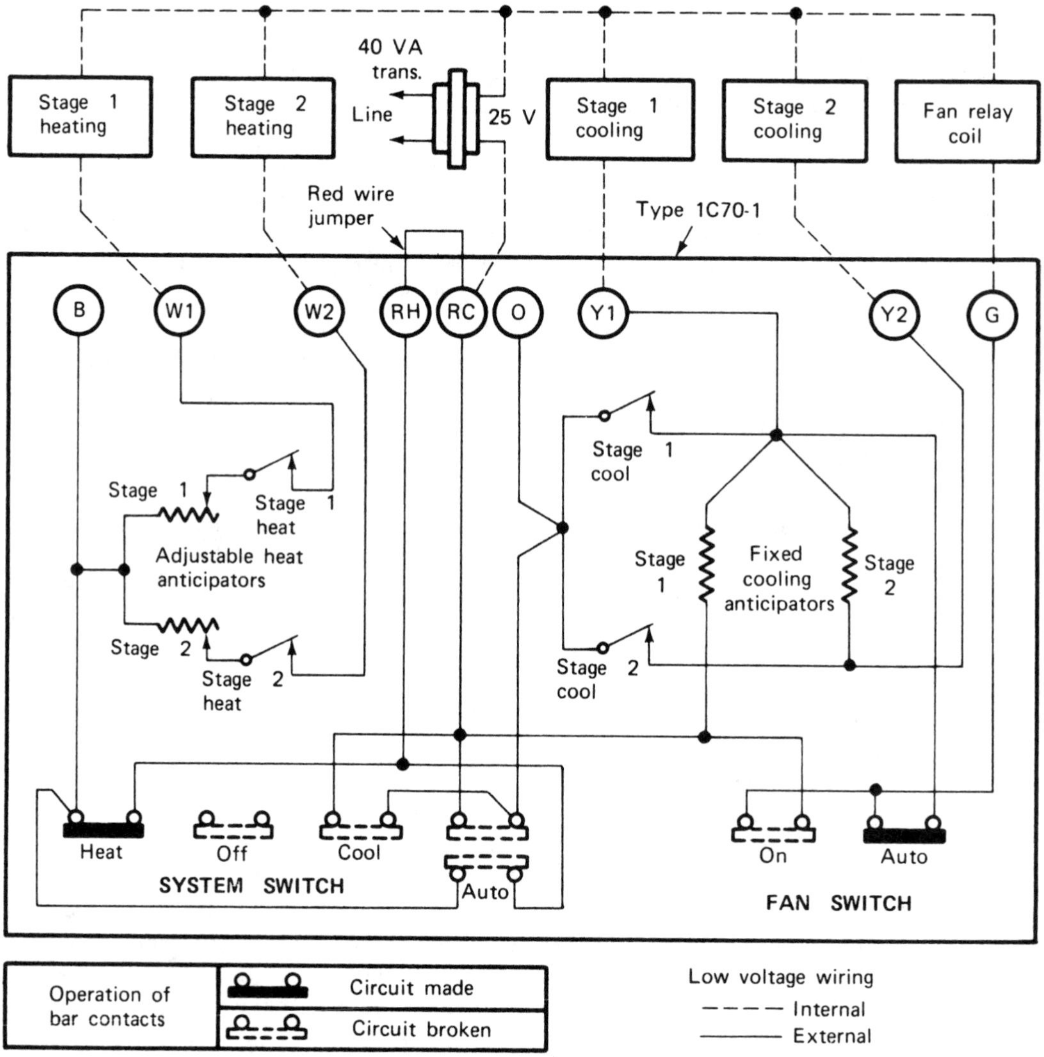
40 VA
trans.
Line
25 V
Stage 1
heating
Stage 2
heating
Stage 1
cooling
Stage 2
cooling
Fan relay
coil
Red wire
jumper
Type 1C70-1
B
W1
W2
RH
RC
O
Y1
Y2
G
Stage 1
Stage 1
heat
Adjustable heat
anticipators
Stage 2
Stage 2
heat
Stage 1
cool
Stage 1
Fixed
cooling
anticipators
Stage 2
Stage 2
cool
Heat
Off
Cool
Auto
SYSTEM SWITCH
On
Auto
FAN SWITCH
Operation of
bar contacts
Circuit made
Circuit broken
Low voltage wiring
Internal
External

FIGURE 15–12
Multistage system wiring schematic. (*Courtesy of Billy Langley,* Electric Controls for Refrigeration and Air Conditioning, *2E, ©1988, p. 98; reprinted by permission of Prentice-Hall, Inc., Englewood Cliffs, N.J.*)

(usually a modulating motor operating a damper or a water or steam valve) use this principle. Both the thermostat (controller) and the motor have potentiometers in them.

The operation of this type of circuit is as follows (Figure 15–13). If the temperature of the controller rises:

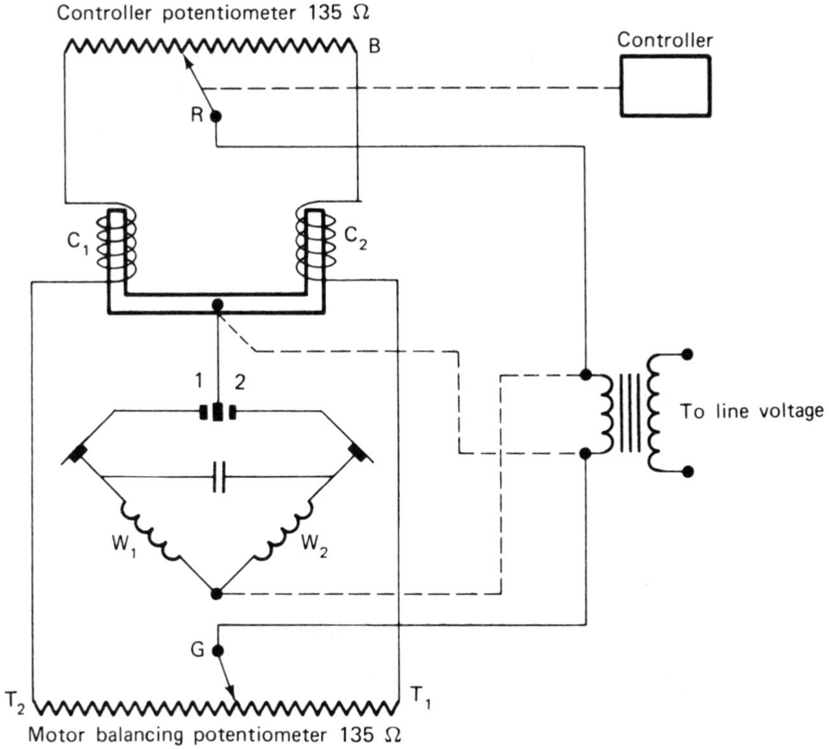
Controller potentiometer 135 Ω
B
Controller
R
C_1
C_2
1
2
To line voltage
W_1
W_2
G
T_2
T_1
Motor balancing potentiometer 135 Ω

FIGURE 15–13
Motor balancing potentiometer (135 ohms). (*Courtesy of Billy Langley, Electric Controls for Refrigeration and Air Conditioning, 2E, 1988, p. 99; reprinted by permission of Prentice-Hall, Inc., Englewood Cliffs, N.J.*)

1. The wiper of the controller potentiometer moves toward the W terminal, reducing the resistance between W and R at the controller.
2. Current flow from the transformer through the W terminal of the controller is increased, and this increased current in the corresponding relay coil pulls the DPST relay switch to contact 1.
3. Current from the transformer then flows through the relay switch 1 on the motor.
4. The motor runs counterclockwise to reposition the controlled device. As the motor runs, it moves the wiper on the potentiometer toward the G terminal.
5. When the wiper on the motor potentiometer reaches a point where the resistance between T and G on the motor potentiometer equals the resistance between R and W on the controller potentiometer, the current relay coil is equalized.

6. The relay contacts break, stopping the motor. The circuit is again in balance.

On a drop in temperature, the current flows to the other side of the relay and the motor runs in the opposite direction. The potentiometer wiper is moved through a series of levers and springs by a gas-filled bellows.

Outdoor thermostat. The outdoor thermostat provides automatic changeover from heating to cooling or cooling to heating in response to the outdoor air temperature. These thermostats may be either bimetal actuated or use a remote bulb for actuating the low voltage SPDT mercury switch, which breaks one circuit and makes another on a temperature rise or fall in the outdoor air. The operating range of this control is from about 60° F to about 90° F. These controls are used on systems where accurate temperature control is necessary. The wiring connections for this control are shown in Figure 15–14.

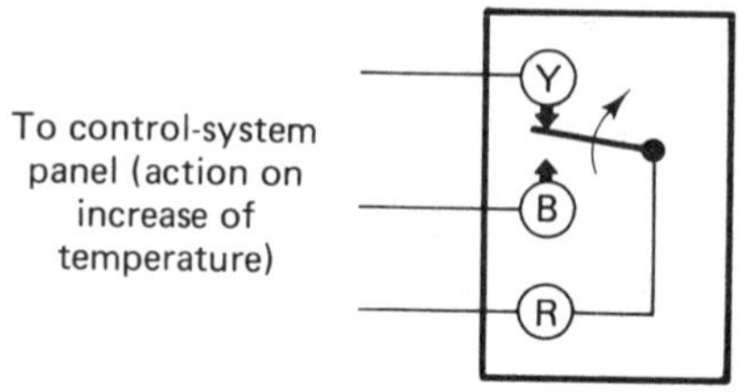
To control-system panel (action on increase of temperature)
Y
B
R

FIGURE 15–14
Outdoor thermostat wiring connections. (*Courtesy of Billy Langley,* Electric Controls for Refrigeration and Air Conditioning, *2E, ©1988, p. 102; reprinted by permission of Prentice-Hall, Inc., Englewood Cliffs, N.J.*)

Operation. In operation when the outdoor air temperature rises to the set point of the controller, the mercury bulb will dump, making one circuit and breaking another. This switches the system into the cooling mode. The switch remains in this position until the outdoor air temperature drops to the set point of the controller, stopping the cooling unit and starting the heating unit.

Fan coil thermostat. Fan coil thermostats are line-voltage controls that are designed to operate heating, cooling, or heating and cooling systems. They are used to control fan motors, relays, or water valves on fan coil units. They have fan-speed selectors as well as temperature selectors so that the most comfortable conditions can be selected. They have anticipators and bimetal-actuated snap-acting switches included for controlling the equipment. Sequenced models also have a bimetal sensing element with a dead-band (neither heat nor cold is provided) incorporated.

Operation. Fan coil thermostats operate the same as any other heating, cooling, or heating and cooling thermostat, except that the equipment cannot be operated for a certain number of degrees on the thermostat between the heating and cooling selector switches.

Outdoor heat-pump thermostat. These types of thermostats are designed to prevent operation of the auxiliary heat strips during mild weather conditions.

These thermostats are equipped with SPST normally closed (NC) switches that are mounted in the outdoor unit. The sensing element is located where it will sense the outdoor ambient temperature. The control system to the auxiliary heat strips is interrupted by the opening of the contacts on a rise in outdoor air temperature. An adjustable temperature range from 0° F to 50° F is provided on most models, which saves energy by not allowing the auxiliary heat strip to operate when the heat pump will handle the load.

Operation. During mild weather operation, the outdoor heat-pump thermostat contacts are closed until the outdoor ambient temperature has risen to the cut-out setting of the control and the auxiliary heat strips may be energized as needed. When the cut-out temperature is reached, the contacts open and heat-strip operation is not possible until the outdoor ambient temperature drops to the cut-in setting of the control.

Thermostat Location

The thermostat location is very important for successful operation of the total system.

For guidance, we offer the following suggestions for proper location of the thermostat:

1. Always locate the thermostat on an inside wall. If located on an outside wall, overheating during cold weather is likely to occur because the thermostat will always feel cold.
2. Avoid false sources of heat such as lamps, television sets, warm air ducts, or hot water pipes in the wall. Also avoid locations where heat-producing appliances like ranges, ovens, or dryers are located on the other side of the wall. Locations near windows may allow the direct sun to reach the thermostat.
3. Avoid sources of vibration like sliding doors and room doors. Always locate the thermostat at least 4 ft from sources of vibration and near a wall support if possible.

Thermostat Voltage

Thermostats are available for all common voltages and must be used with the stated voltage. If used otherwise, they will be damaged permanently or improper operation will result.

A thermostat will be damaged if it is used on a voltage which is greater than its rated voltage. If the thermostat is used on a voltage lower than the stated voltage, the anticipators will not function properly and, therefore, the temperature will not be properly maintained.

The room thermostat is provided with a heat anticipator connected in series with the rest of the control circuit (Figure 15–15). These anticipators are made of a resistance-type material that produces heat in accordance with the current drawn

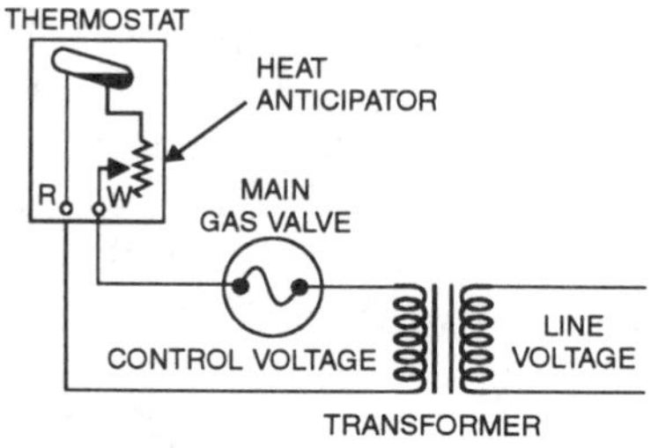
THERMOSTAT
HEAT
ANTICIPATOR
R
W
MAIN
GAS VALVE
CONTROL VOLTAGE
LINE
VOLTAGE
TRANSFORMER

FIGURE 15–15
Heat anticipator.

through them (Figure 15–16). Heat anticipators are adjustable and are normally set to correspond with the current rating of the main gas valve. The purpose of these devices is to make the room temperature more even.

In operation, when the thermostat is calling for heat, the anticipator is also producing heat to the thermostat. This heating action causes the thermostat to become satisfied before the room actually reaches the set point of the thermostat. Thus, the thermostat stops the flame and the room temperature will not overshoot, or go too high.

The heat anticipator is adjusted to match the amperage draw of the temperature control circuit. This adjustment is made by setting the heat anticipator pointer to the correct number on the scale. The correct amperage draw of the circuit is determined by actual measurement with an ammeter.

To determine the amperage draw of the control circuit, set the thermostat to demand heat, then wrap one tong of the ammeter with several turns of one of the control circuit wires, either R or W (Figure 15–17). Then divide the current reading by the number of turns. This is the best method because it measures all of the current required to operate the circuit; a part of the load may be missed when taking

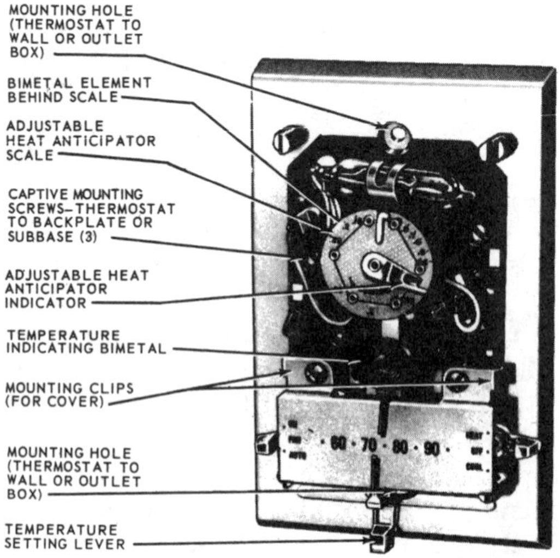
MOUNTING HOLE (THERMOSTAT TO WALL OR OUTLET BOX)
BIMETAL ELEMENT BEHIND SCALE
ADJUSTABLE HEAT ANTICIPATOR SCALE
CAPTIVE MOUNTING SCREWS–THERMOSTAT TO BACKPLATE OR SUBBASE (3)
ADJUSTABLE HEAT ANTICIPATOR INDICATOR
TEMPERATURE INDICATING BIMETAL
MOUNTING CLIPS (FOR COVER)
MOUNTING HOLE (THERMOSTAT TO WALL OR OUTLET BOX)
TEMPERATURE SETTING LEVER
60 · 70 · 80 · 90

FIGURE 15–16
Internal view of a thermostat. (*Courtesy Honeywell, Inc.*)

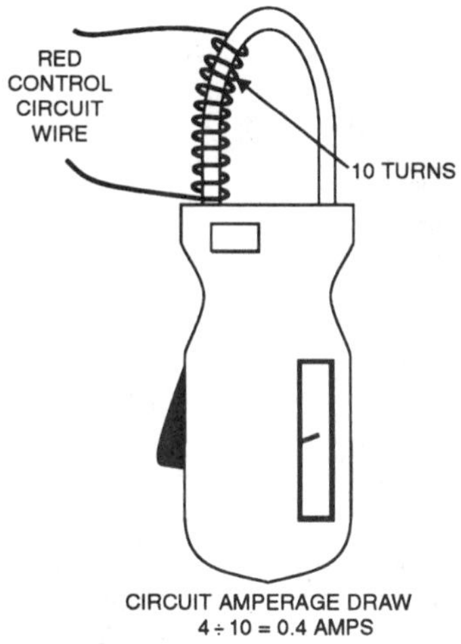
RED
CONTROL
CIRCUIT
WIRE
10 TURNS
CIRCUIT AMPERAGE DRAW
4 ÷ 10 = 0.4 AMPS

FIGURE 15–17
Checking amperage in temperature control circuit.

the rating stamped on each component. For example, if there were 10 turns of the wire on the ammeter tong, and the meter indicated that 4 amps were flowing through the circuit, the actual current flow would be 4/10 = 0.4 amps. The heat anticipator would then be adjusted to 0.4 on the scale, or a thermostat would be installed that would match this current flow.

THE TRANSFORMER

The transformer is a device used to reduce line voltage to a usable control voltage, usually 24 V. Transformers must be sized to provide sufficient power to operate the control circuit. Most are oversized sufficiently to provide enough power to operate an air conditioning control circuit also. This rating is usually 40 volt-amps (VA).

The type of transformer used in heating and air conditioning systems consists basically of two coils of insulated wire wound around an iron core (Figure 15–18). The coil connected to the line voltage is called the primary side, and the coil connected to the control circuit is called the secondary side. These are considered to be step-down transformers: they reduce the line voltage—either 120 volts or 240 volts—to 30 volts (called 24 volts).

Rating

Transformers are rated in volt-amps and are of the class 2 type. This VA rating, listed on the transformer as volt-amps, is an indication of the amount of electrical power that the transformer will provide. The transformer selected must be at least large enough to do the job required of it. This is determined by the amount of load connected to its secondary side. The rating of the transformer may be greater than needed without any problem, but if it is smaller than needed, the transformer will burn out in a short period of time.

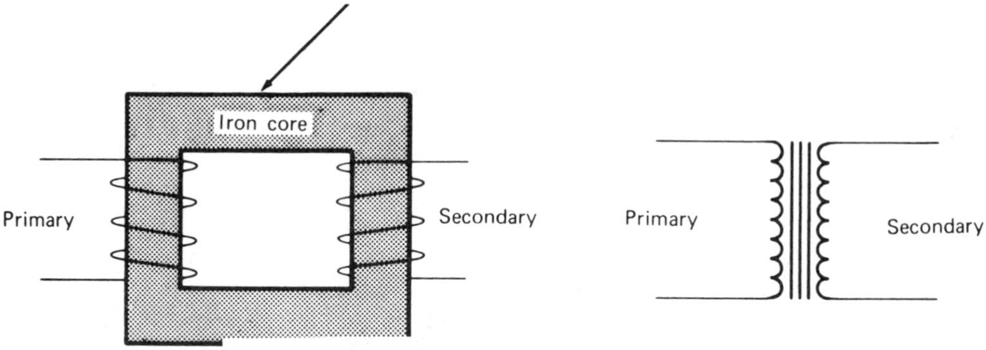
Iron core
Primary
Secondary
Primary
Secondary

FIGURE 15–18
Symbolic sketch of a transformer and schematic symbol. (*Courtesy of Billy Langley,* Electric Controls for Refrigeration and Air Conditioning, *2E, ©1988, p. 17; reprinted by permission of Prentice-Hall, Inc., Englewood Cliffs, N.J.*)

Sizing

Since most control system components are rated in amperes, it is necessary to know how to compare the VA rating to the ampere draw of the control system. The following formulas are used to make this comparison:

$$\text{amps} = \text{VA output} \div \text{secondary voltage}$$

$$\text{VA output} = \text{amps} \times \text{secondary voltage.}$$

Either one of these formulas can be used when selecting a transformer.

The only test that needs to be made is to check the ampere draw in the secondary side of the transformer with all the equipment in operation or with all the equipment in the operating mode that will draw the greatest amount of current. To check the amperage draw, start the unit and check the current draw in the red wire of the control circuit, or the wire that is common to all the loads on the secondary side of the transformer.

For example, the amperage draw of the load on the secondary side of the transformer is 0.8 amp. The VA of the transformer would need to be at least

$$\text{VA output} = 0.8 \times 24$$

$$= 19.2.$$

So we would need a transformer with a minimum output rating of 19.2 VA. Most manufacturers equip their equipment with 40-VA transformers to handle the most common types of loads to which their equipment will be connected.

Some transformers are equipped with fuses for short-circuit protection. Some fuses are of the replaceable type and others are not. When a nonreplaceable type is used, the transformer must be replaced when the control circuit has been overloaded. Therefore, the electric power to the unit should be turned off when working with the wiring connected to the secondary side of the transformer.

There are several reasons for using transformers in control circuits. First, the circuit is much safer than if line voltage were used, thus protecting the customer from electrical shock while protecting the equipment from the greater fire potential of line voltage. Second, the thermostats that can be used with low voltage circuits are made from much lighter material than the line-voltage type and are, therefore, much more responsive to changes in temperature. They are more economical to produce than line-voltage controls. Third, the low voltage wiring is greatly simplified when compared to line-voltage wiring.

Phasing Transformers

Equipment manufacturers sometimes phase transformers for various reasons, but generally phasing is done to increase the power for proper control-circuit energy. Transformer phasing assures that all the electrons are flowing in the same direction and is sometimes referred to as transformer polarizing.

To phase transformers, wire the primary sides in parallel (Figure 15–19). Connect one wire from each transformer secondary winding terminal (B and C) together (Figure 15–20). Then check the voltage across the other two leads (wires A and D) from the secondary coils. If the voltmeter indicates 48 volts, the transformers are out of phase and the wires must be reversed (Figure 15–21).

When the transformers are correctly phased, the voltmeter will indicate 0 volts between the two disconnected wires on the secondary side of the transformer. At this point, fasten the two wires (A and C) together. The transformers are now in phase. The control circuit may now be connected to the two combinations of wires (Figure 15–22).

From this example, it can be seen that there are two wrong ways to connect the transformer secondaries. One way will result in either one or both transformers being burned out. The other way will result in a voltage that is too high. This high voltage will usually burn out the thermostat immediately and will cause other controls to burn out in a short period of time. Transformers that are supplied with equipment by the manufacturer are properly phased with marked terminals and

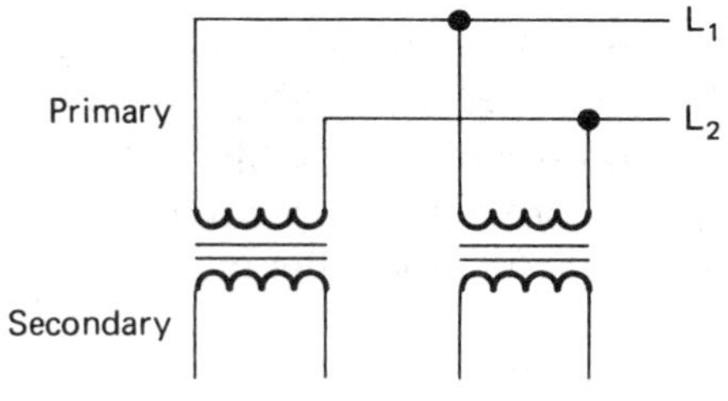
L1
L2
Primary
Secondary

FIGURE 15–19
Transformers wired in parallel. *(Courtesy of Billy Langley,* Electric Controls for Refrigeration and Air Conditioning, *2E, ©1988, p. 19; reprinted by permission of Prentice-Hall, Inc., Englewood Cliffs, N.J.)*

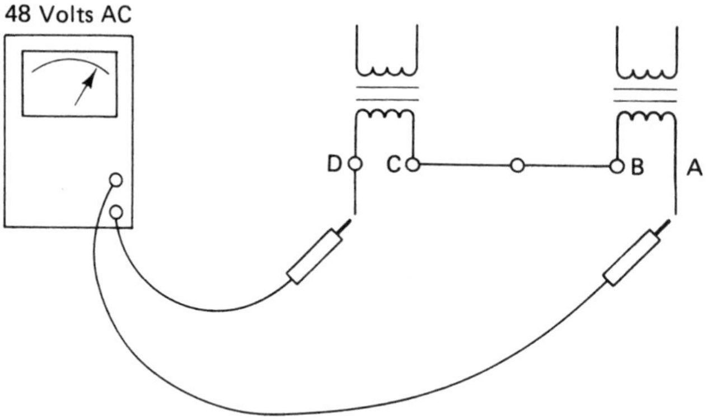
48 Volts AC
D
C
B
A

FIGURE 15–20
Checking voltage of improperly phased transformers. (*Courtesy of Billy Langley,* Electric Controls for Refrigeration and Air Conditioning, *2E, ©1988, p. 19; reprinted by permission of Prentice-Hall, Inc., Englewood Cliffs, N.J.*)

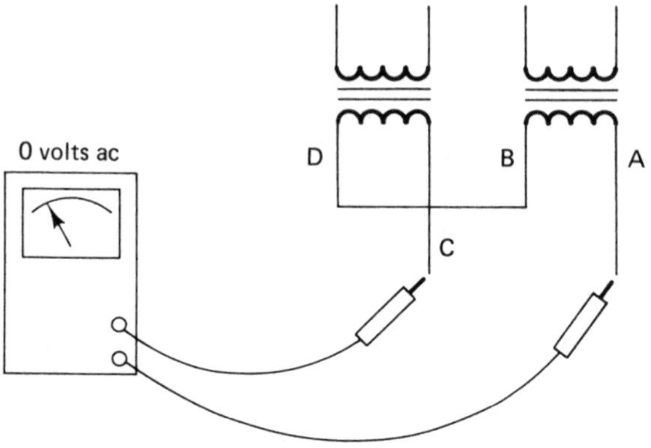
0 volts ac
D
B
A
C

FIGURE 15–21
Checking voltage of properly phased transformers. (*Courtesy of Billy Langley,* Electric Controls for Refrigeration and Air Conditioning, *2E, ©1988, p. 20; reprinted by permission of Prentice-Hall, Inc., Englewood Cliffs, N.J.*)

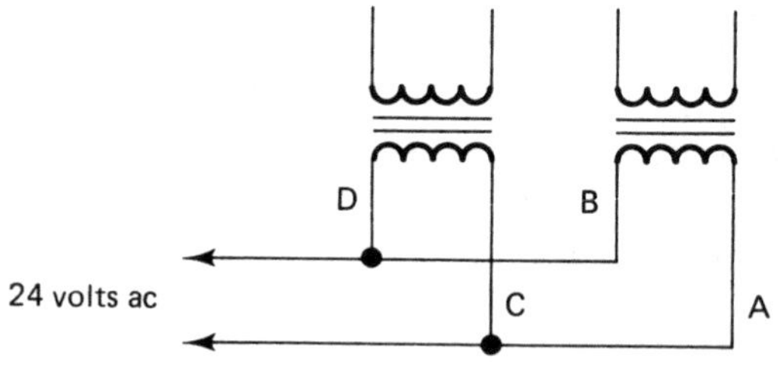
D
B
24 volts ac
C
A

FIGURE 15–22
Control circuit connected to properly phased transformers. (*Courtesy of Billy Langley,* Electric Controls for Refrigeration and Air Conditioning, *2E, ©1988, p. 20; reprinted by permission of Prentice-Hall, Inc., Englewood Cliffs, N.J.*)

are supplied with wiring diagrams. Most transformers are supplied with color-coded electrical leads to aid in proper phasing. The person who is installing the transformer can be certain that proper phasing will result by simply connecting like colors to like colors. Also, transformers that are equipped with screw terminals are color coded, as one screw is brass and the other one is either silver or nickel.

Transformer Phasing with Two Low Voltage Control Components

Phasing is required when two low voltage components with their own low voltage power supply are to be used in the same control circuit. The components may be fan centers or other such devices. There is generally only one thermostat used to control these devices (Figure 15–23). In this example, the thermostat is presumed to be open. The + and – signs indicate instantaneous voltage polarity. The voltage measured across the secondary of either transformer is indicated to be 24 volts, and the voltage across the two wires from the relays is indicated to be 48 volts.

There are two ways to phase transformers properly in this situation. The first method is as follows

1. Wire the circuit, leaving one of the wires from the thermostat loose at the relay terminal.
2. Check the voltage between the unconnected wire and the relay terminal.
3. If a voltage higher than 24 volts is indicated, reverse the two wires at the control relay.

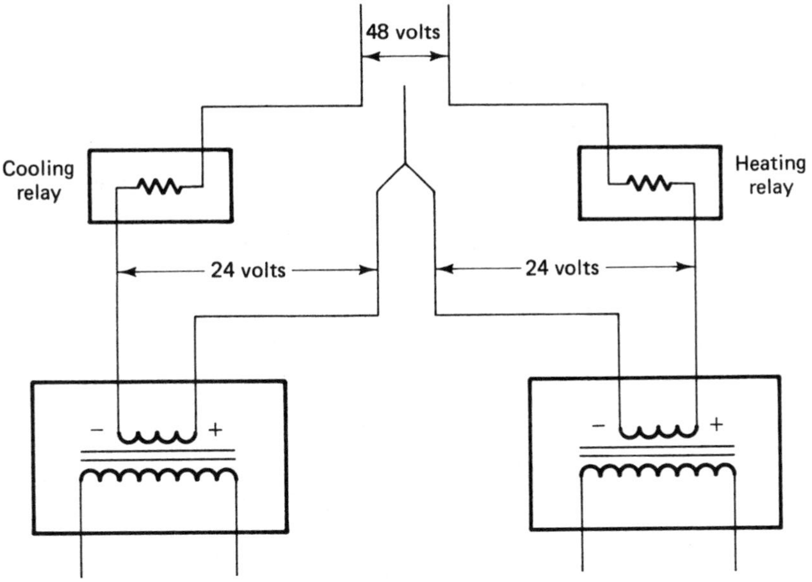
48 volts
Cooling relay
Heating relay
24 volts
24 volts
–
+
–
+

FIGURE 15–23
Phasing control components having separate power supplies. (*Courtesy of Billy Langley,* Electric Controls for Refrigeration and Air Conditioning, *2E, ©1988, p. 21; reprinted by permission of Prentice-Hall, Inc., Englewood Cliffs, N.J.*)

The second method is as follows:

1. Wire the circuit, leaving the wires disconnected at the thermostat.
2. Check the voltage between one of the wires and the other two in turn.
3. If a voltage greater than 24 volts is indicated, reverse the two thermostat wiring connections at the control relay.

FAN CONTROL

The fan control is a device used to start and stop the fan in response to the temperature inside the furnace-heat exchanger. A fan control has normally open (NO) SPST contacts, which are actuated by temperature. The control is mounted to sense the air temperature within the circulating air passages of the furnace heat exchanger.

Forced-air circulation systems have controls known as heat watchers in addition to the burning controls. These additional controls permit the circulating fan to operate only when enough heat is available to heat the conditioned space.

There are two types of fan controls: (1) temperature actuated, and (2) electrically actuated.

Temperature-Actuated Fan Control

The temperature-actuated fan control uses either a bimetal strip, a bimetal disc, a helical bimetal, or a pneumatically operated sensing element. The sensing element is inserted into the warm-air plenum or directly into the circulating air of the furnace heat exchanger. In either case these elements are mounted in such a position that they sense the temperature of the discharge air from the furnace (Figure 15–24). They are available with sensing elements of different lengths so that the temperature can be sensed in the exact place that the manufacturer has determined to be the hot spot of the unit. They are adjusted by a sliding lever on a scale that is either printed or etched on the control element dial.

Operation. In operation, when the thermostat calls for heat, the main burner gas is ignited and the furnace heat exchanger begins to warm up. As the temperature of the furnace increases, the sensing element begins to respond and

FIGURE 15–24
Fan control. (*Courtesy Honeywell, Inc.*)

move the dial toward the on setting of the control. When the on setting of the control is reached, usually between 125° F and 150° F, the snap action of the switch closes the NO contacts and completes the line voltage circuit to the fan motor (Figure 15–25).

When either the thermostat is satisfied or the limit control contacts open, the main gas valve closes, stopping the flow of gas to the main burners. As the furnace cools down, the air being delivered to the conditioned space also cools down and causes the sensing element to return to its original position. When the sensing element has returned to its original position, the switch contacts are opened and the fan motor stops. The off temperature of the fan control is usually about 100° F.

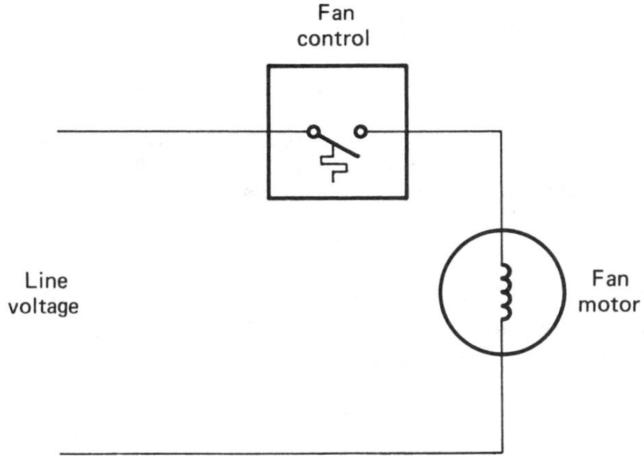
Fan
control
Line
voltage
Fan
motor

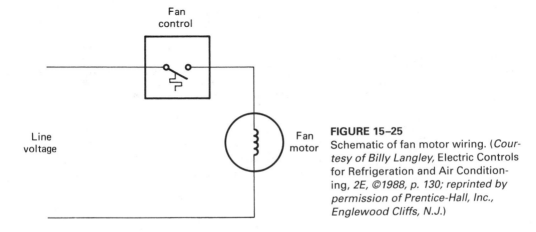

FIGURE 15–25
Schematic of fan motor wiring. (*Courtesy of Billy Langley,* Electric Controls for Refrigeration and Air Conditioning, *2E, ©1988, p. 130; reprinted by permission of Prentice-Hall, Inc., Englewood Cliffs, N.J.*)

Electrically Operated Fan Control

The electrically operated fan control provides timed fan operation for forced warm air furnaces when the heater coil is wired in parallel to the low voltage coil of the main gas valve (Figure 15–26). This control is particularly suited for counterflow and horizontal furnaces because of the peculiarity of heat buildup in the furnace after the main gas valve has closed. The heater coil uses 24 volts AC. The switch is an SPST, heater actuated, bimetal control. The fan is usually started about one minute after the thermostat calls for heat; the fan is stopped about two minutes after the thermostat is satisfied. This type of control causes the fan to operate even when there is no flame in the furnace because it is electrically actuated rather than heat actuated.

Operation. When the thermostat demands heat, the heater inside the fan control and the gas valve are energized at the same time. After the predetermined period of time has lapsed, the fan-control contacts close and start the fan motor. Theoretically, the main gas valve has also opened and the furnace has warmed up sufficiently to warm the building by this time. When the thermostat is satisfied, the

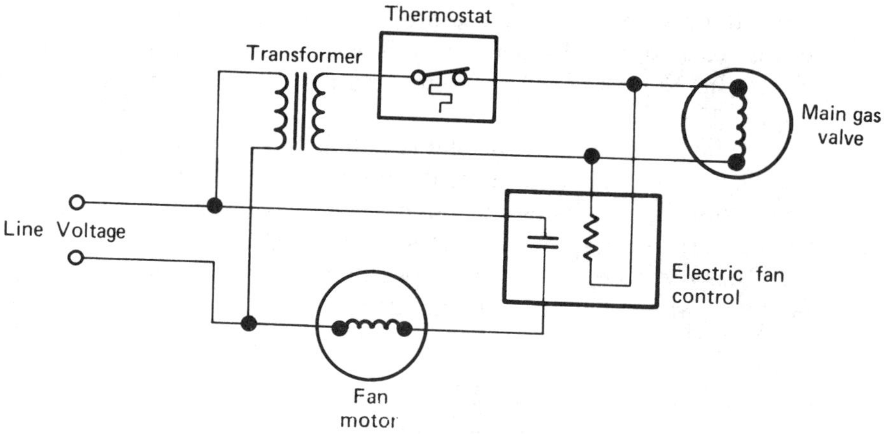
Thermostat
Transformer
Main gas
valve
Line Voltage
Electric fan
control
Fan
motor

FIGURE 15–26
Schematic diagram of an electrically operated fan control. (*Courtesy of Billy Langley,* Electric Controls for Refrigeration and Air Conditioning, *2E, ©1988, p. 130; reprinted by permission of Prentice-Hall, Inc., Englewood Cliffs, N.J.*)

main gas valve and the fan control are deenergized at the same time. The flame is extinguished immediately, and the fan continues to run and cool the furnace down. After about two minutes, the fan motor is deenergized and the furnace is ready for the next operating cycle.

LIMIT CONTROL

The limit control is a safety device used to interrupt the control circuit to the main gas valve in case of excessive temperatures inside the furnace. The limit control has NC contacts. It may be either an SPST or an SPDT switch actuated by temperature. It is mounted to sense the air temperature within the furnace.

The limit control is used on all types of warm-air furnaces to prevent excessive plenum temperatures and a possible resulting fire. They have either NC SPST switch contacts or SPDT switch contacts, depending on their specific uses. The contacts of the SPST type open on a rise in temperature and close on a fall in temperature. The SPDT contact types open the NC contacts on a rise in temperature and close the NO set at the same time. The NO set is used to energize the fan motor on horizontal and counterflow-type furnaces when excessive temperatures exist inside the furnace. The snap-action switch may be actuated by either a bimetal strip, a bimetal disc, a helical bimetal, or a pneumatically operated sensing element. The sensing element is placed so that it will sense the air temperature inside the furnace heat exchanger. The sensing elements are available in different lengths. Be sure to use the correct length so that proper operation of the equipment will be maintained (Figure 15–27).

FIGURE 15–27
Limit control. (*Courtesy Honeywell, Inc.*)

Some models are adjustable and others are nonadjustable. Those that are adjustable have a range from about 180° F to 250° F with a fixed differential of 25° F. They are adjusted by moving a sliding lever along a scale that is printed or etched on the control element dial. They usually have a stop to prevent setting the off temperature higher than 200° F. The on temperature is 25° F lower than the off temperature setting. Only the cut-out is adjusted.

Operation

In operation, when the off setting is reached, the electrical circuit to the main gas valve is interrupted. This can be done either by opening the 24-V AC control circuit directly (Figure 15–28) or by interrupting the line voltage to the primary side of the transformer (Figure 15–29). This interruption allows the main gas valve to close, stopping the flow of gas to the main burners. As the furnace cools down, the sensing element also cools. When the predetermined on setting is reached, the electrical circuit is completed to the main gas valve and operation is resumed.

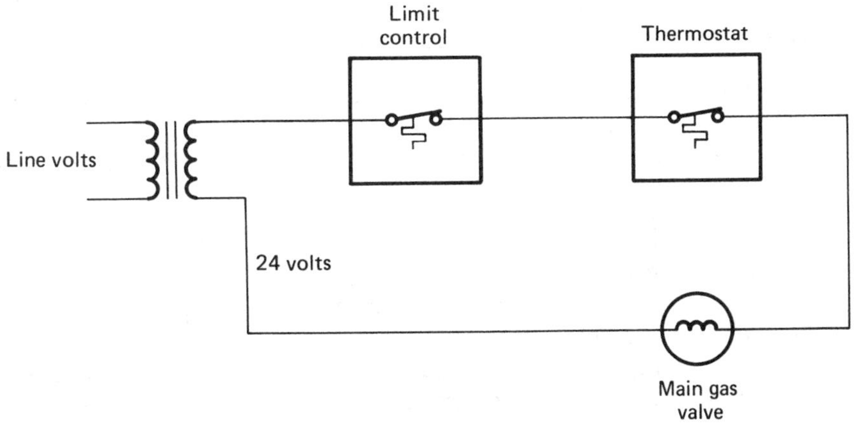
Limit control
Thermostat
Line volts
24 volts
Main gas valve

FIGURE 15–28
Limit control in 24-volt ac control circuit. (*Courtesy of Billy Langley,* Electric Controls for Refrigeration and Air Conditioning, *2E, ©1988, p. 132; reprinted by permission of Prentice-Hall, Inc., Englewood Cliffs, N.J.*)

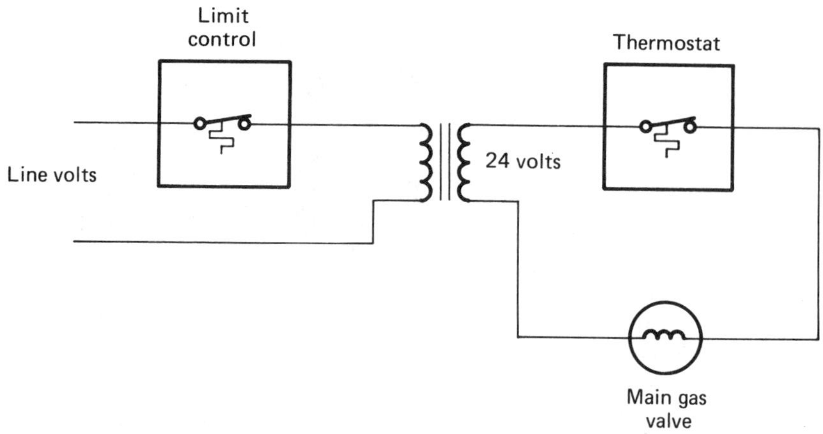
Limit
control
Thermostat
Line volts
24 volts
Main gas
valve

FIGURE 15–29
Limit control in line voltage to primary side of transformer. (*Courtesy of Billy Langley,* Electric Controls for Refrigeration and Air Conditioning, *2E, ©1988, p. 132; reprinted by permission of Prentice-Hall, Inc., Englewood Cliffs, N.J.*)

When the SPDT type is used, the operation is a little different. When the NC contacts open, the NO contacts close. At this time the fan motor is energized to blow the hot air from the furnace. When the furnace has cooled down, the contacts switch, stopping the fan motor and energizing the main gas valve.

This control is designed to function only when there is an excessive amount of heat inside the furnace. The SPST types are usually designed for pilot duty only and will not accommodate heavy current draw through the contacts. However, the SPDT type has contacts rated heavy enough to carry the fan motor current.

Combination Fan and Limit Control

Combination fan and limit controls provide fan and limit control on forced warm air furnaces and heaters. They are manufactured with a snap-acting, sealed switch. The contacts are designed for millivolt, 24-volt, 120-volt, or 240-volt control circuits.

Operation. The limit control interrupts the control circuit on a rise in temperature to stop operation of the main burner if the plenum temperature reaches the control off setting. This is the same action as that provided by the single-limit control discussed earlier.

On a temperature rise, the fan switch completes the electrical circuit to start the fan motor when the plenum temperature rises to the on setting of the control. The fan motor stops when the plenum temperature drops to the fan off setting. This is identical to the operation of the single fan control discussed earlier.

Combination fan and limit controls combine the functions of the individual fan and limit controls into a single compact unit. The scale setting is simplified. The limit action cannot be set below the fan-control setting. On some models a summer fan switch is readily accessible without removing the cover and provides for the selection of continuous fan operation or automatic fan operation.

The two switches in these controls are never wired in series. Always wire them in parallel. The limit control may be either in the low voltage circuit or it may be in the line-voltage circuit, depending on the desires of the equipment manufacturer. The fan control is always wired into the line-voltage circuit so that it can control operation of the fan directly. For some suggested wiring diagrams, see Figures 15–30 and 15–31.

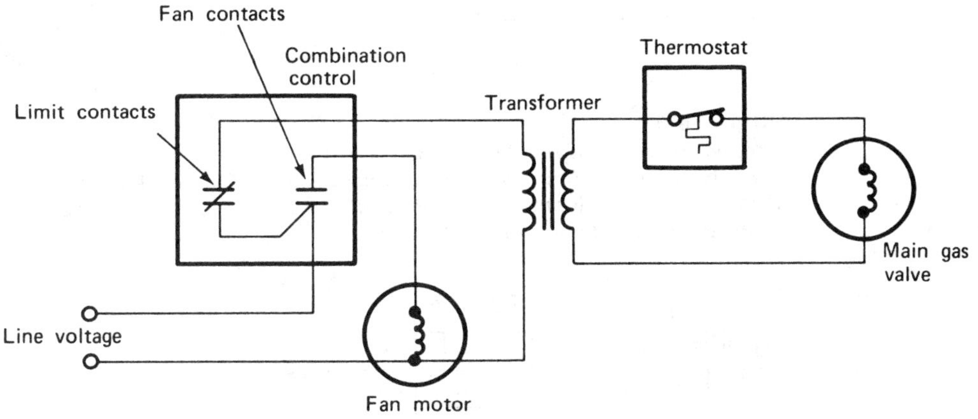
Fan contacts
Combination control
Thermostat
Limit contacts
Transformer
Main gas valve
Line voltage
Fan motor

FIGURE 15–30
Typical control circuit using combination fan and limit control. (*Courtesy of Billy Langley,* Electric Controls for Refrigeration and Air Conditioning, *2E, ©1988, p. 133; reprinted by permission of Prentice-Hall, Inc., Englewood Cliffs, N.J.*)

Auxiliary Limit Control

The auxiliary limit control is used on horizontal and counterflow furnaces. It is physically mounted in the fan compartment and senses the temperature in this area. Electrically, it is in the 24-volt temperature control circuit (Figure 15–32). The purpose of this control is to sense the air temperature and open the electrical circuit to the main gas valve if this temperature reaches a maximum of 200° F. On horizontal furnaces a reversed flow of air can be started that will prevent proper operation of the fan or the limit control, which could possibly overheat the furnace and start a fire. In counterflow furnaces the natural direction of air flow is upward. Should the limit control fail to sense a too high temperature, the auxiliary limit will open the electrical circuit to the main gas valve to prevent overheating the furnace and possibly a fire.

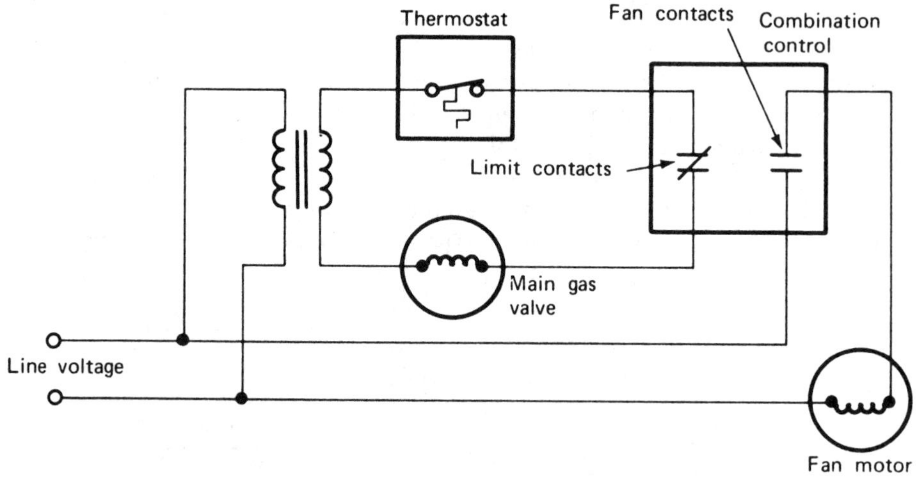
Thermostat
Fan contacts
Combination control
Limit contacts
Main gas valve
Line voltage
Fan motor

FIGURE 15–31
Typical control circuit with fan control in the line-voltage and limit control in low-voltage circuit. (*Courtesy of Billy Langley,* Electric Controls for Refrigeration and Air Conditioning, *2E, ©1988, p. 133; reprinted by permission of Prentice-Hall, Inc., Englewood Cliffs, N.J.*)

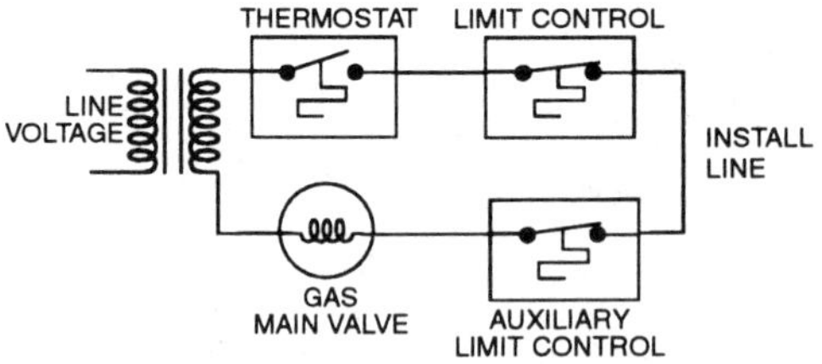
THERMOSTAT
LIMIT CONTROL
LINE VOLTAGE
INSTALL LINE
GAS MAIN VALVE
AUXILIARY LIMIT CONTROL

FIGURE 15–32
Diagram for furnace with auxiliary limit control.

Air Switch

Air switches are used to control the fan in warm-air applications to prevent reverse air circulation and excessive filter temperatures in counterflow furnaces. Two-speed fan operation is obtained through an SPDT switch. The bimetal is inserted directly into the air stream (Figure 15–33). These controls are designed for pilot duty only with a maximum current flow of 50 VA at 24 V AC.

MAIN GAS VALVE

The main gas valve is the device that acts on demand from the thermostat to either admit gas to the main burners or to stop the gas supply (Figure 15–34). This valve has many functions. It has a gas pressure regulator, a pilot safety, a main gas cock, a pilot gas cock, and the main gas solenoid all in a single unit—the combination gas control.

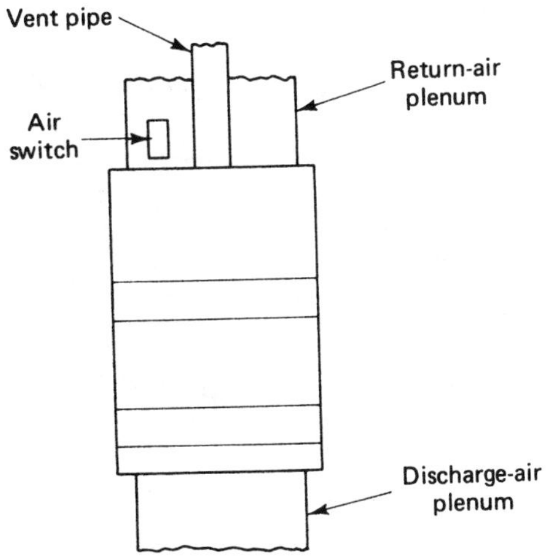
Vent pipe
Return-air plenum
Air switch
Discharge-air plenum

FIGURE 15–33
Air switch location. (*Courtesy of Billy Langley,* Electric Controls for Refrigeration and Air Conditioning, *2E, ©1988, p. 134; reprinted by permission of Prentice-Hall, Inc., Englewood Cliffs, N.J.*)

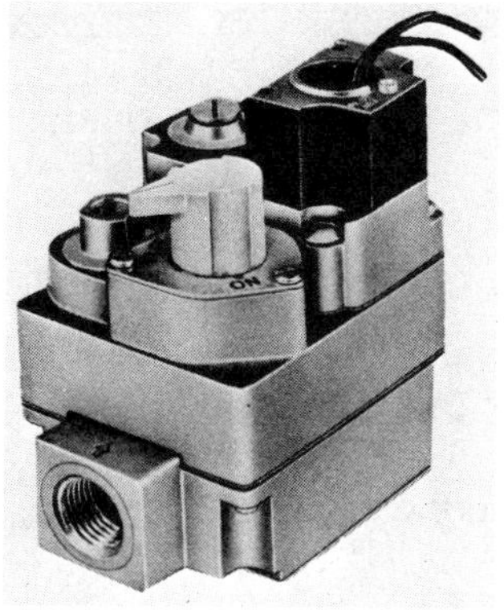

FIGURE 15–34
A main gas valve. (*Courtesy White-Rodgers Division, Emerson Electric Co.*)

Sequence of Operation

As the thermostat calls for heat, energizing the solenoid coil, the valve lever opens the cycling wave (Figure 15–35). The inlet gas now flows through the control orifice past the cycling valve. At this point, gas flow is in two directions, as follows:

1. Part of the flow is to the back of the diaphragm by means of internal passageways. The resulting increase in pressure pushes the main valve to the open position, compressing the diaphragm springs lightly.
2. Part of the flow is through the seat of the master regulator into the valve outlet by means of internal passageways. This causes the master regulator to begin its function.

The gas valve remains in this position and the master regulator continues to regulate until the relay coil is deenergized, at which time the cycling valve seals off.

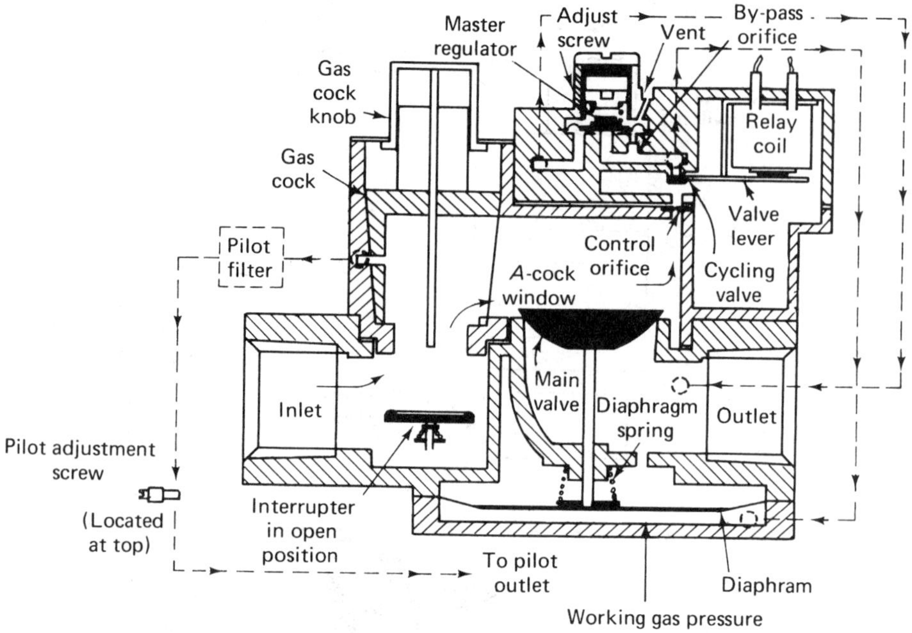
Master
regulator
Adjust
screw
Vent
By-pass
orifice
Gas
cock
knob
Relay
coil
Gas
cock
Valve
lever
Pilot
filter
Control
orifice
A-cock
window
Cycling
valve
Main
valve
Diaphragm
spring
Inlet
Outlet
Pilot adjustment
screw
(Located
at top)
Interrupter
in open
position
To pilot
outlet
Diaphram
Working gas pressure

FIGURE 15–35
Main gas valve cutaway. (*Courtesy White-Rodgers Division, Emerson Electric Co.*)

As the cycling valve closes, the regulator spring causes the seat of the master regulator to close off. The function of the bypass orifice is to permit gas in the passageways to escape into the outlet of the valve, thereby causing the main gas valve to close.

Redundant gas valves. The purpose of these valves is safety. They contain two independently operated valves to the main gas burner. If one valve should fail to close, the other will close and shut off the gas flow to the main burner (Figure 15–36). This type of valve is used on gas-fired heaters and boilers, with or without intermittent pilot ignition, in place of the regular gas valve.

The majority of problems with these valves are electrical rather than mechanical. The redundant gas valve has two electrically controlled internal shutoff valves in series with each other. Before gas can flow to the main burner, both of these valves must be open. However, only one needs to close to stop the flow of gas.

Safety, therefore, is accomplished. It is almost impossible for both of these individual valves to stick in the open position when either the thermostat or the limit control opens. The standard main gas valve does not provide this safety.

One make of redundant gas valve uses an instant-acting solenoid for controlling the flow of gas at the valve outlet and a time delay valve on the inlet to the valve. The time delay on the bimetal valve is about 10 seconds, thus providing a time delay on startup of the main burner.

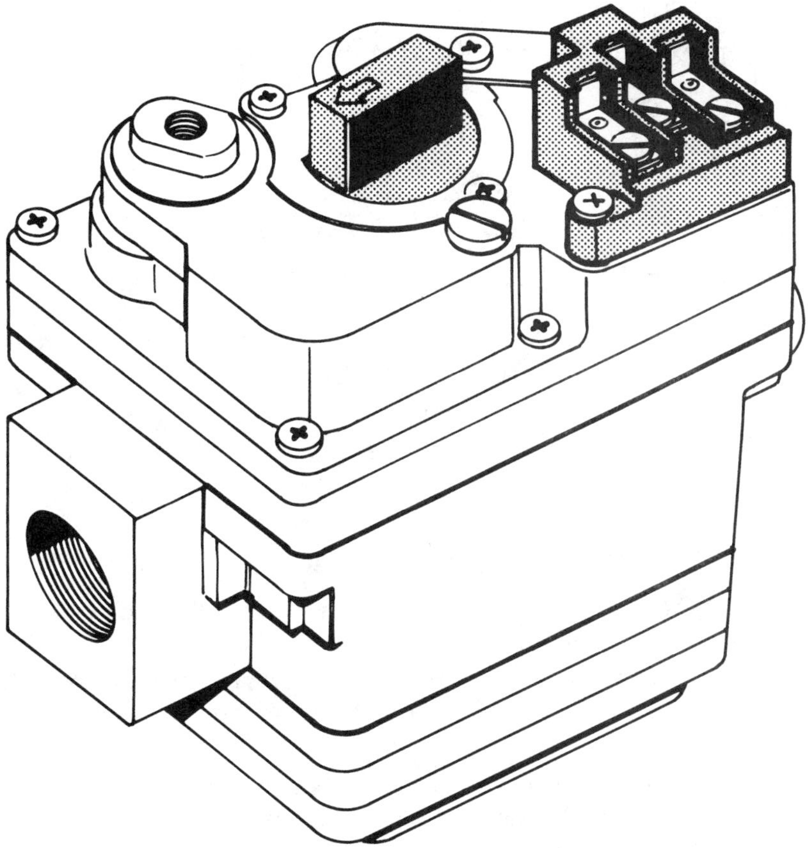

FIGURE 15–36
Redundant gas valve.

The redundant gas valve used on furnaces with standing pilots is similar to the standard type of gas valve except that an internal heat motor valve is included. There are two 24 volt coils used in these valves (Figure 15–37). One coil is shown as a solenoid valve and the other is shown as a resistor. They are parallel electrically and both are in series with the limit switch and the thermostat. The resistor is for the heat motor that operates the second or redundant gas valve.

These valves have a third internal valve which is the 100% safety shutoff valve. It is energized with the electricity (millivolts) which is produced by the thermocouple. Should the pilot flame not heat the thermocouple sufficiently, the 100% safety shutoff will close, stopping the flow of gas to the main burner.

When the contacts in either the thermostat or the limit switch open, the solenoid instantly closes. The heat motor valve, however, takes a few seconds to cool down and close the valve.

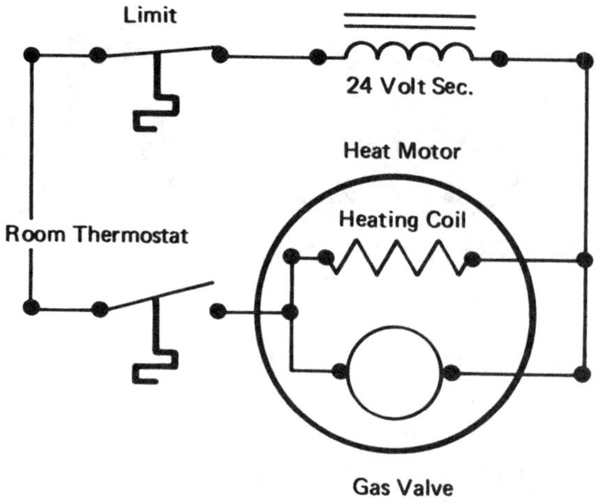
Limit
24 Volt Sec.
Heat Motor
Heating Coil
Room Thermostat
Gas Valve

FIGURE 15–37
Two 24-volt coils in one valve.

When the standing pilot valve is to be tested, use the following steps:

1. When the pilot is burning, the open circuit thermocouple test should show a minimum of 21 millivolts.
2. When the thermostat is calling for heat and the limit switch is closed, 24 volts of electricity should be at both the solenoid coil and the heating coil of the heat motor.

If proper voltage is indicated in both of these tests, the problem is not in the electrical circuit. At this point, check for gas at the valve inlet. If gas is present, the valve has a mechanical problem and should be replaced.

When an intermittent pilot ignition system is used, each time the thermostat demands heat, pilot gas will be supplied and lighted electrically. The redundant gas valve will then open and admit gas to the main burner. If the pilot gas should not be lighted within the specified amount of time, the system will shut down and will lock out on safety.

The valve used on this type of system has two solenoid valves wound on the same core (Figure 15–38). When the thermostat demands heat, an electrical circuit is completed to both coils and to the pilot gas ignitor. The pilot gas ignitor will direct a high voltage spark across the pilot, igniting the pilot gas.

The coil in the pilot safety circuit, which consists of a set of normally open and a set of normally closed contacts, is then energized by the millivolts produced by the heated thermocouple.

When the pilot flame has been established, the millivoltage generated by the thermocouple will cause the contacts in the pilot safety to change position. At this point, the number 1 solenoid coil and the pilot gas ignitor is deenergized. The coil in the number 2 solenoid coil will remain energized until the thermostat is satisfied.

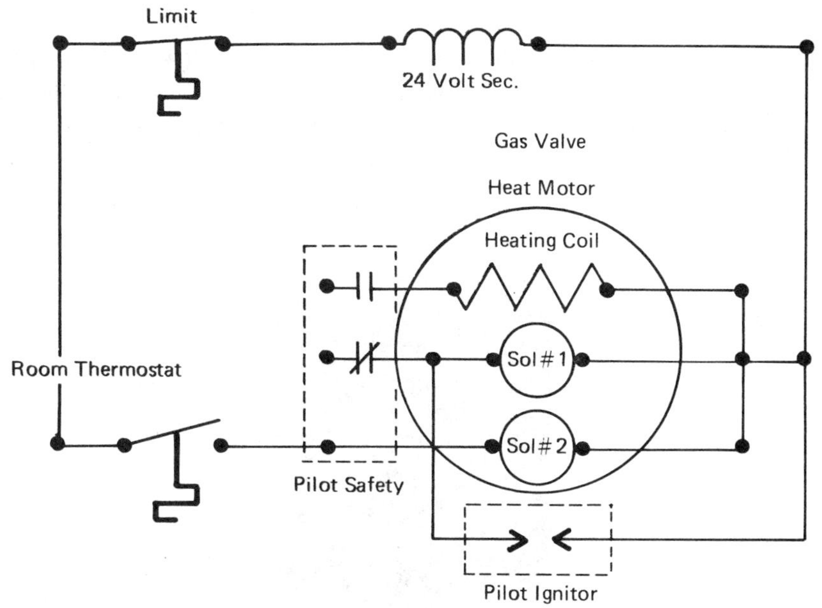
Limit
24 Volt Sec.
Gas Valve
Heat Motor
Heating Coil
Room Thermostat
Sol # 1
Sol # 2
Pilot Safety
Pilot Ignitor

FIGURE 15–38
Two solenoid coils on one core.

In this type of valve, both solenoids must be energized for the valve to open, but only one is required to keep the valve open.

As the contacts in the pilot safety change positions, an electrical circuit to the main gas valve is completed. The main gas valve is operated by the heat motor valve. It will be approximately 30 seconds before gas is admitted to the main burner.

For gas to flow to the main burner, both solenoids must be energized when the thermostat demands heat. The pilot flame must be established before the main gas valve will open.

The two solenoids being wound on the same core prevent cycling of the main burner because of sudden opening or closing of the thermostat contacts or a momentary electrical power failure. Should either of these conditions occur, solenoid number 2 will close and will not be energized again until the thermocouple is cooled down and it resets the circuit.

When troubleshooting this valve for electrical troubles, use the following steps:

1. With the pilot lit, there should be 24 volts to the main gas valve heater coil.
2. There should be 24 volts to the number 2 solenoid coil.
3. When these conditions are met and still no gas is admitted to the main burner, the valve is defective and must be replaced.

PILOT SAFETY

There are two types of pilot safety controls: non-100% safe and 100% safe. These two names refer to the amount of gas cut off when the pilot light is unsafe. The 100% safe is incorporated in the combination main gas valve. The non-100% device (Figure 15–39) incorporates a set of contacts that open the control circuit during an unsafe condition.

H
HONEYWELL

FIGURE 15–39
Non-100% safe pilot safety. (*Courtesy Honeywell, Inc.*)

These units are used in conjunction with a thermocouple to keep the control contacts closed, or the valve open, during normal operation. If at any time the control "drops out" (the contacts open), the reset button will have to be reset manually before operation of the furnace can be resumed.

THE THERMOCOUPLE

The thermocouple is a device that uses the difference in metals to provide electron flow. The hot junction of the thermocouple is put in the pilot flame where the dissimilar metals are heated (Figure 15–40). When heat is applied to the welded junction, a small voltage is produced. This small voltage, measured in millivolts (mV), is the power used to operate the pilot safety control. The output of a thermocouple is approximately 30 mV. This simple device can cause many problems if the connections are not kept clean and tight.

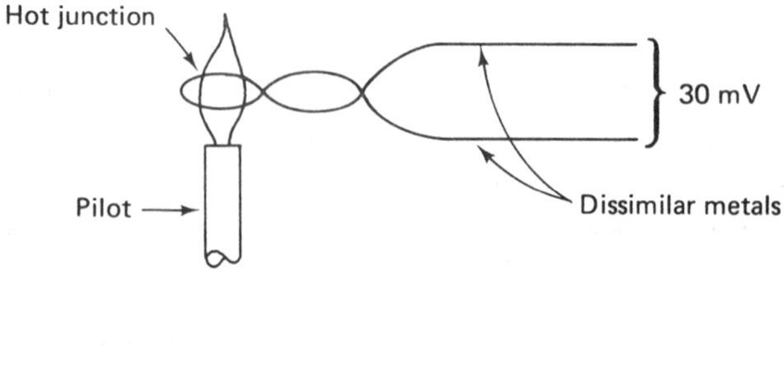
Hot junction
30 mV
Pilot
Dissimilar metals

FIGURE 15–40
Thermocouple operation.

Thermocouple Testing

There are three tests that can be made on a thermocouple to determine if it is producing enough current: (1) open circuit voltage, (2) closed circuit voltage, and (3) drop-out voltage.

Open circuit voltage. The pilot gas must be burning while making this test. Remove the end of the thermocouple that is connected to the pilot safety device. Check the voltage output at this end by touching one lead of the millivolt meter to one terminal while touching the other meter lead to the other terminal (Figure 15–41). The output of the thermocouple is DC voltage, and the meter leads may need to be reversed in order to obtain a reading. Remember that the pilot must be burning during this test. The output voltage should be 21 millivolts or more. If it is lower than this, replace the thermocouple. The thermocouple must be reconnected to the pilot safety device for normal operation.

Closed circuit voltage. To make this test, an adapter is required between the thermocouple and the pilot safety device. This adapter has points for checking the voltage on the inner contact of the thermocouple (Figure 15–42).

The output voltage is checked while the pilot gas is lit and the thermocouple is hot. The output should be a minimum of 17 millivolts. If the output is lower than this, replace the thermocouple. When this test is completed, remove the adapter and reconnect the thermocouple to the pilot safety device.

Drop-out voltage. The drop-out voltage is measured with an adapter between the thermocouple and the pilot safety device (Figure 15–42). Turn off the pilot gas and observe the voltage. When the voltage has dropped enough for the pilot safety device to close off the gas to the main burner, a low dull thud can be heard. The voltage at this point is the drop-out voltage and is usually about 5 millivolts. Remove the adapter and reconnect the thermocouple to the pilot safety device, relight the pilot gas, and put the unit back in operation. There is no recommended point for the drop-out voltage, but it should occur within approximately three minutes, and preferably in less time.

THE FIRE-STAT

The fire-stat is a safety device mounted in the fan compartment to stop the fan if the return air temperature reaches about 160° F (Figure 15–43). It is a bimetal-actuated, normally closed switch that must be manually reset before the fan can operate.

The reason for stopping the fan when the high return air temperatures exist is to prevent agitation of any open flame in the house, thus helping to prevent the spreading of any fire that may be present.

FURNACE WIRING

There are three different circuits and three different voltages in a modern furnace 24-V control system. The following diagrams will illustrate each of these:

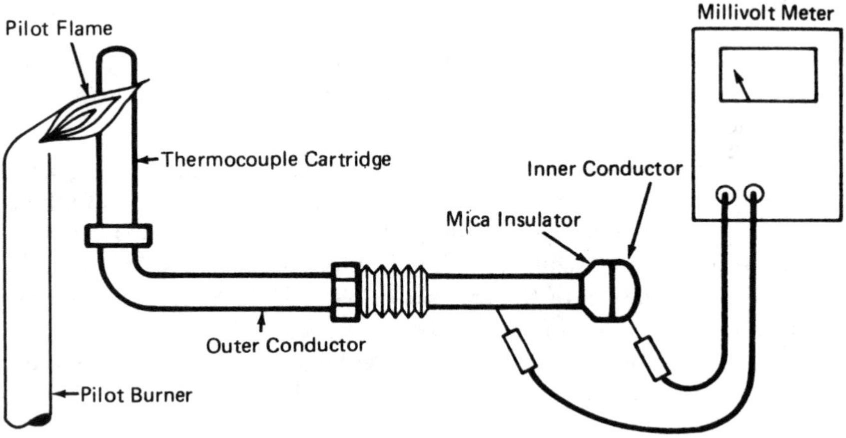
Pilot Flame
Millivolt Meter
Thermocouple Cartridge
Inner Conductor
Mica Insulator
Outer Conductor
Pilot Burner

FIGURE 15–41
Open circuit thermocouple test.

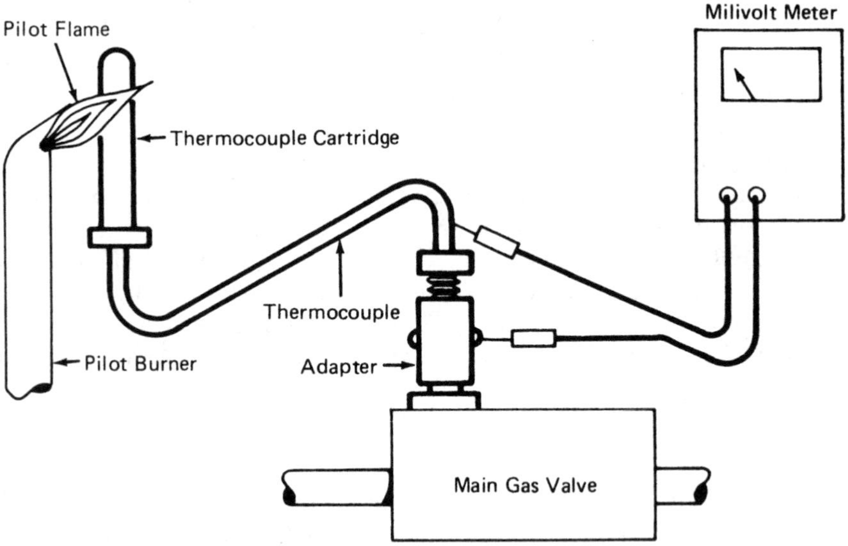
Pilot Flame
Milivolt Meter
Thermocouple Cartridge
Thermocouple
Pilot Burner
Adapter
Main Gas Valve

FIGURE 15–42
Closed circuit thermocouple test.

1. The fan or circulator circuit (Figure 15–44).
2. The temperature control circuit (Figure 15–45).
3. The pilot safety circuit, 30 mV (Figure 15–46).

PUSH
BUTTON
TO RESET
SWITCH
H
HONEYWELL

FIGURE 15–43
Fire-stat. (*Courtesy Honeywell, Inc.*)

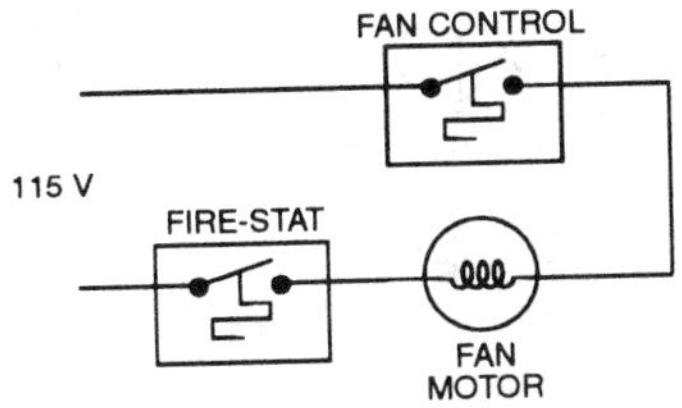
FAN CONTROL
115 V
FIRE-STAT
FAN
MOTOR

FIGURE 15–44
Fan circuit.

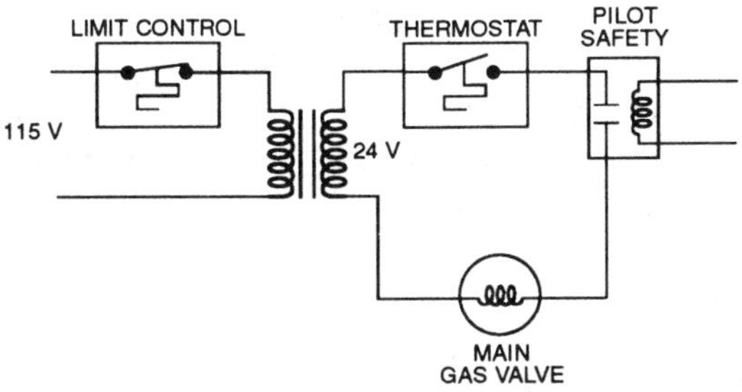
LIMIT CONTROL
THERMOSTAT
PILOT
SAFETY
115 V
24 V
MAIN
GAS VALVE

FIGURE 15–45
24-V temperature control circuit.

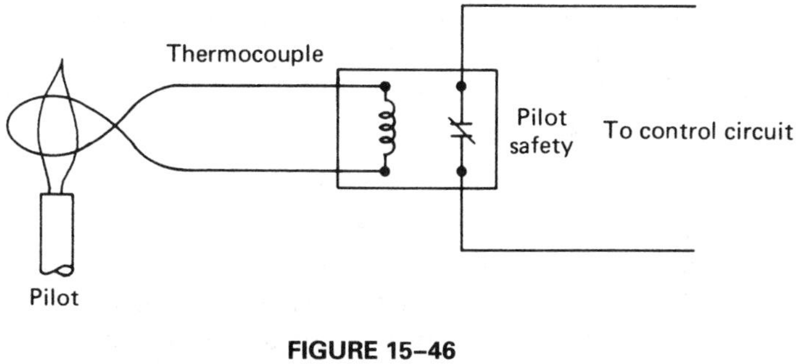
Thermocouple
Pilot
safety
To control circuit
Pilot

FIGURE 15–46
Pilot safety circuit.

When all three of these circuits are connected together (Figure 15–47), we have a modern furnace 24-V control system.

The control of electric furnaces is much the same as that just described; however, there are no pilot safety devices and the main gas valve is replaced with relays that actuate to complete the electrical circuit to the heating elements.

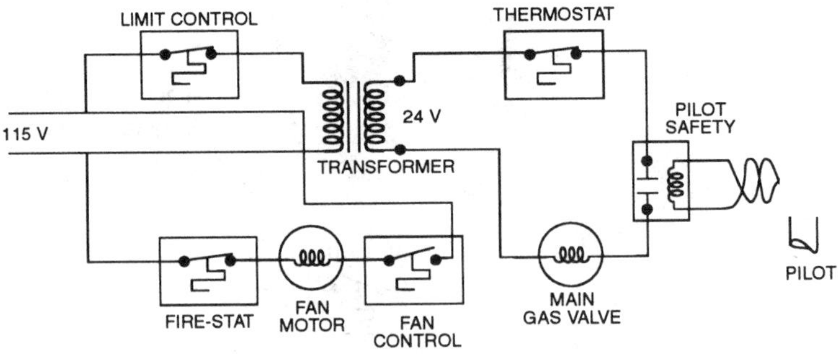
LIMIT CONTROL
THERMOSTAT
115 V
24 V
TRANSFORMER
PILOT SAFETY
PILOT
FIRE-STAT
FAN MOTOR
FAN CONTROL
MAIN GAS VALVE

FIGURE 15–47
24-V control circuit.

ELECTRIC FURNACE CONTROLS

Many of the controls used on gas furnaces are also used on electric furnaces. Some controls are slightly modified while others are a completely separate type of control. There does not seem to be any standard method of wiring the control systems on electric furnaces. The major thing, however, is to know how the control functions; then to wire it into a circuit is relatively simple.

Time Delay Relay

The time delay relay is a device used for controlling electrically operated devices at a predetermined time interval after the heater is energized (Figure 15–48). The heater timer is connected to the 24-volt temperature control circuit which is energized by the thermostat (Figure 15–49). These devices have an inherent time delay before the contacts close or open. Each relay has a given time delay for each function. Each relay will normally control a single load, such as a heating element or a fan motor (Figure 15–50). These relays may be used with a single thermostat or a multistage thermostat.

Sequencer

A more elaborate method of controlling electric heating equipment is by use of a sequencer. These multistage devices may have 10 to 12 stages depending on the size of equipment and the comfort demand (Figure 15–51). When they are used with the proper components, they will provide proportional control of multistage

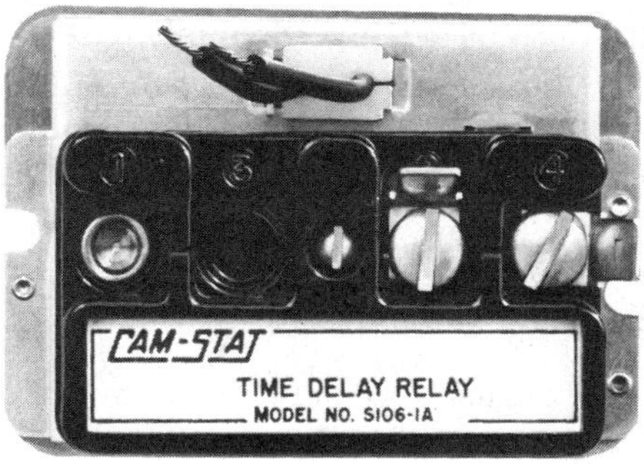
CAM-STAT
TIME DELAY RELAY
MODEL NO. S106-1A

FIGURE 15–48
Time delay relay. (*Courtesy Cam-Stat Incorporated)*

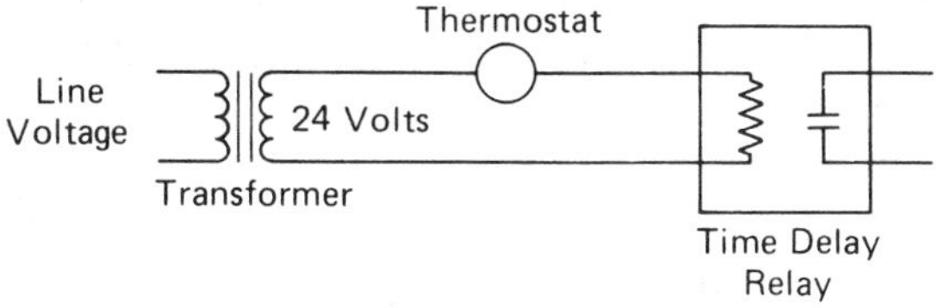
Thermostat
Line
Voltage
24 Volts
Transformer
Time Delay
Relay

FIGURE 15–49
Time delay relay wiring diagram (24 volt).

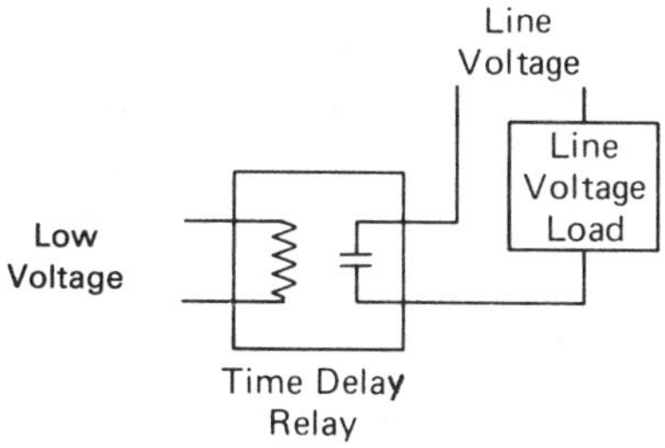
Line
Voltage
Line
Voltage
Load
Low
Voltage
Time Delay
Relay

FIGURE 15–50
Time delay relay wiring diagram (line voltage).

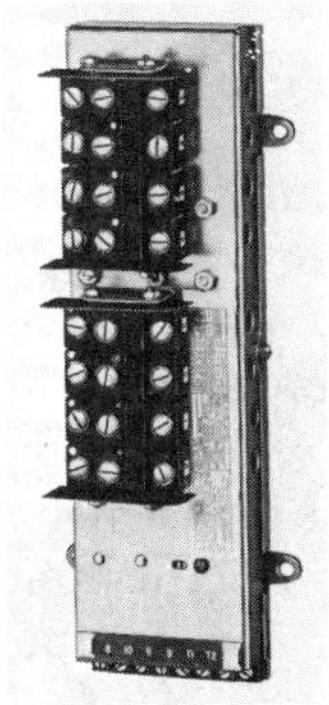

FIGURE 15–51
Electric heating sequencer. (*Courtesy Penn Controls Division*)

equipment. Normally, they are used with a modulating-type thermostat and automatically turn on or off the number of heating elements necessary to satisfy the space requirements. However, some of the newer sequencers use a thermal power element, which completely eliminates the need for motors and gears. Accurate temperature control is achieved through the use of a slide wire or solid-state sensors.

Limit Controls

Some electric furnace manufacturers use special types of limit control on their furnace. These are SPDT switches used to break one circuit and make another under high temperature conditions. When the temperature inside the furnace reaches approximately 200° F, the limit control will open the electrical circuit to the primary side of the transformer and complete an electrical circuit to the fan motor (Figure 15–52). This is done to remove heat from the furnace and help prevent damage to the heating elements.

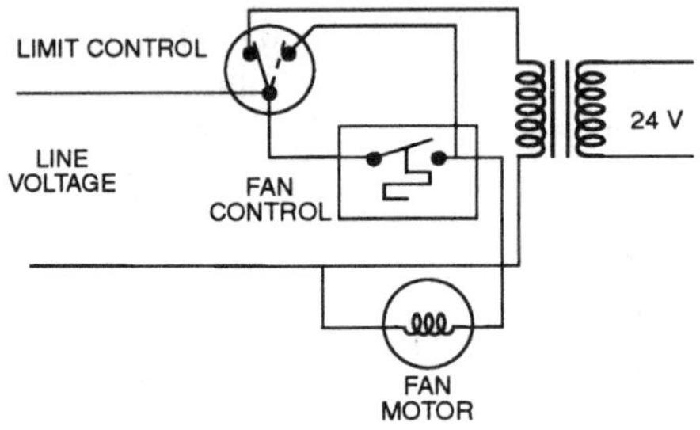
LIMIT CONTROL
LINE VOLTAGE
FAN CONTROL
24 V
FAN MOTOR

FIGURE 15–52
Electric furnace limit control wiring diagram.

Element Overcurrent Protector

These devices are installed in line with the heating element and are designed to open the electrical circuit to the element in case of an overcurrent condition (Figure 15–53). They are current sensitive and are usually very accurate. Overcurrent protectors may be automatic reset, manual reset, or replaceable type devices.

Element Overtemperature Protector

Overtemperature protectors are placed in line with the heating element and open the electrical circuit in case of an overtemperature condition of the heating element (Figure 15–54). These devices are temperature sensitive and usually require replacement if such a condition occurs. They are sometimes referred to as thermal fuses.

The variety of functions performed by a heating system is limited only by the use of controls. The more a service technician knows about controls, the easier will be his or her job in servicing such equipment. Reference to a comprehensive book

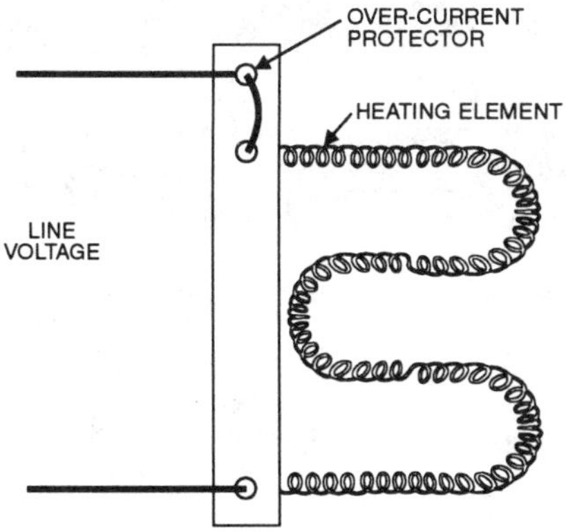
OVER-CURRENT PROTECTOR
HEATING ELEMENT
LINE VOLTAGE

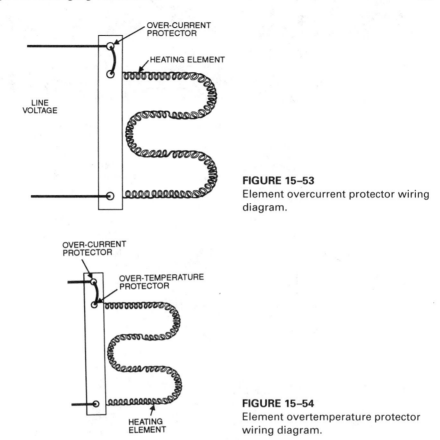

FIGURE 15–53
Element overcurrent protector wiring diagram.

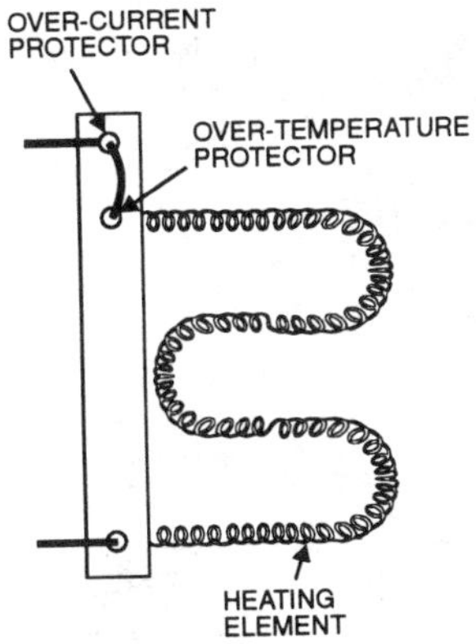
OVER-CURRENT PROTECTOR
OVER-TEMPERATURE PROTECTOR
HEATING ELEMENT

FIGURE 15–54
Element overtemperature protector wiring diagram.

on controls, such as B. C. Langley's *Electric Controls for Refrigeration and Air Conditioning* (Prentice-Hall, Inc., Englewood Cliffs, N.J., 1988), is recommended.

Electric furnace diagram: There are many different ways to wire an electric furnace for the desired operation. Figures 11–9 and 15–55 show an example of one manufacturer's diagram.

DISCHARGE-AIR AVERAGING THERMOSTAT

The discharge-air averaging thermostat is placed in the discharge-air plenum of a heating system. Its purpose is to cycle the equipment to maintain an average discharge-air temperature. Temperature averaging is a method used to save energy and, therefore, operating expenses. Discharge-air thermostats are generally SPST switches that can be used to cycle both the heating and the cooling equipment to maintain the desired average temperature.

To wire this control into the electrical system, see Figure 15–56.

WIRING DIAGRAM

MODELS 015H & 017H

⚠WARNING

Pull out furnace safety disconnect from unit control panel door before servicing the furnace.

Not suitable for use on systems exceeding 120 volts to ground.
Ne convient pas aux installations de plus de 120 volts a la terre.

NOTES:

1. **See Furnace Data Label for recommended wire sizes.**
2. **Single (1) supply circuit required for Canadian installations (see installation instructions)**
3. **Thermostat anticipator: (0.40 amps).**
4. **Use Class "K" or "RK" replacement fuses only (30 & 60 amp required).**
5. **Secure 24V pigtails connections (to thermostat) with wire nuts and tape with approved electrical tape.**
6. **24V pigtails are not used with relay control. See wiring diagram on Relay for 24V connections.**
7. **Connect Blower Plug to receptacle on Relay Control (if installed). Refer to wiring diagram on Relay.**
8. **Alternate connections for "B" & "C" Models.**
9. **Type 105C thermoplastic or equivalent must be used if replacing any original unit wiring.**

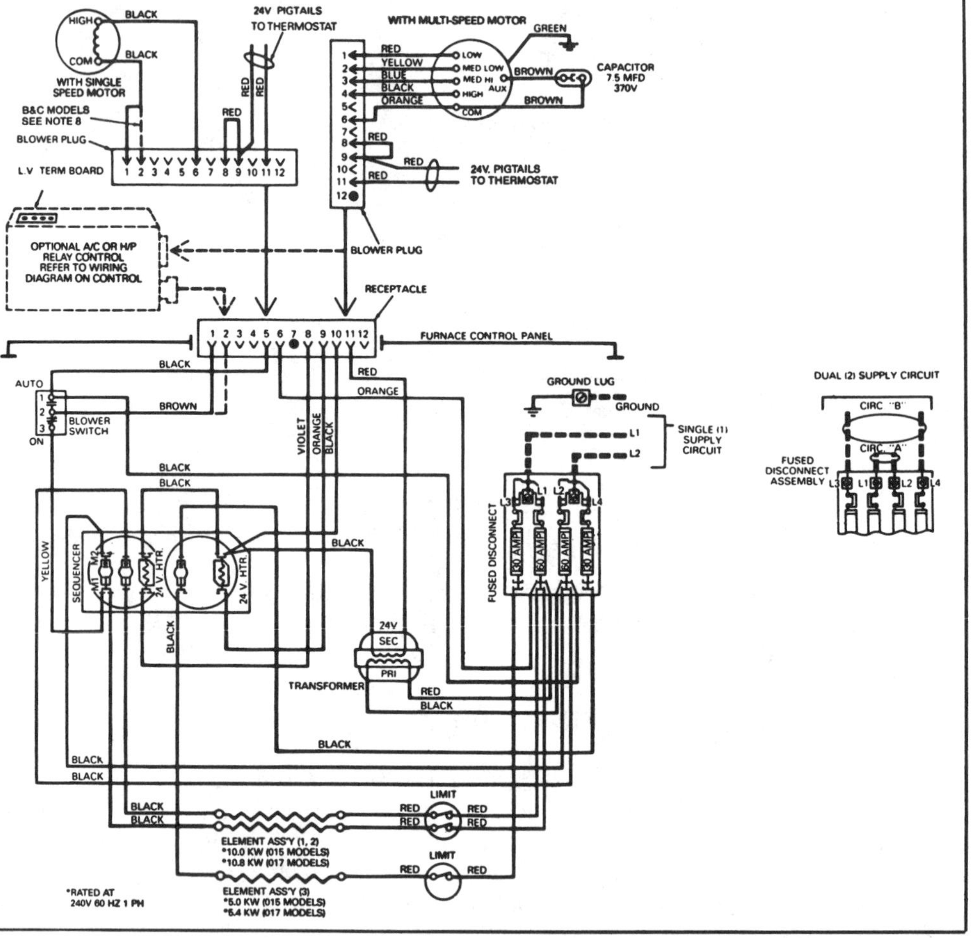

HIGH
COM
BLACK
WITH SINGLE SPEED MOTOR
B&C MODELS SEE NOTE 8
BLOWER PLUG
L.V. TERM BOARD
24V PIGTAILS TO THERMOSTAT
RED
WITH MULTI-SPEED MOTOR
GREEN
YELLOW
BLUE
ORANGE
LOW
MED LOW
MED HI
HIGH
AUX
BROWN
CAPACITOR 7.5 MFD 370V
24V. PIGTAILS TO THERMOSTAT
OPTIONAL A/C OR H/P RELAY CONTROL REFER TO WIRING DIAGRAM ON CONTROL
BLOWER PLUG
RECEPTACLE
FURNACE CONTROL PANEL
AUTO
BLOWER SWITCH
ON
VIOLET
GROUND LUG
GROUND
L1
L2
SINGLE (1) SUPPLY CIRCUIT
DUAL (2) SUPPLY CIRCUIT
CIRC "B"
CIRC "A"
FUSED DISCONNECT ASSEMBLY
FUSED DISCONNECT
30 AMP
60 AMP
YELLOW
SEQUENCER
24 V. HTR.
24V
SEC
PRI
TRANSFORMER
LIMIT
ELEMENT ASS'Y (1, 2) *10.0 KW (015 MODELS) *10.8 KW (017 MODELS)
ELEMENT ASS'Y (3) *5.0 KW (015 MODELS) *5.4 KW (017 MODELS)
*RATED AT 240V 60 HZ 1 PH

FIGURE 15–55
Electric furnace wiring diagram. (*Courtesy of Nordyne*)

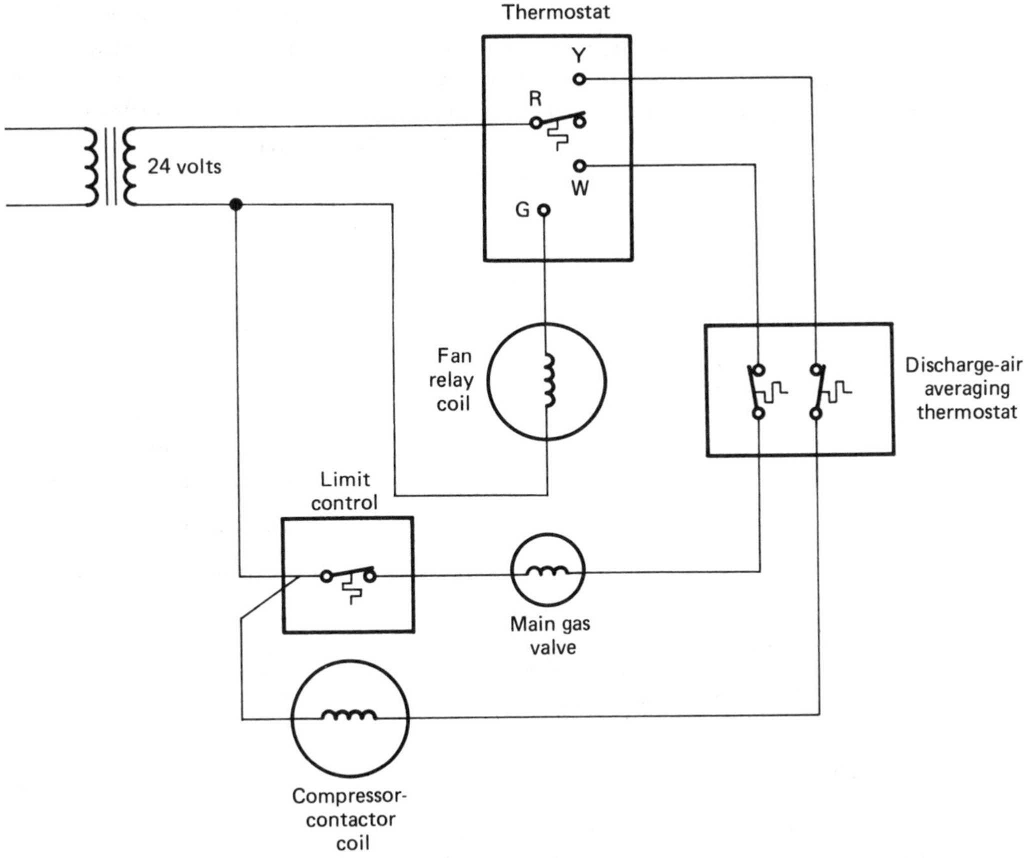
Thermostat
Y
R
24 volts
W
G
Fan
relay
coil
Discharge-air
averaging
thermostat
Limit
control
Main gas
valve
Compressor-
contactor
coil

FIGURE 15–56
Typical discharge air averaging thermostat wiring diagram. (*Courtesy of Billy Langley,* Electric Controls for Refrigeration and Air Conditioning, *2E, ©1988, p. 107; reprinted by permission of Prentice-Hall, Inc., Englewood Cliffs, N.J.*)

SAIL SWITCH

Sail switches are designed to detect air flow or the lack of air flow in ducts. They respond only to the velocity or air movement. Their purpose is to activate electronic air cleaners, humidifiers, gas valves, or other auxiliary equipment in response to air flow from the system air circulating mechanism (fan or blower). They are also used as safety devices on forced-draft burners to prevent opening of the main gas valve until sufficient combustion air is supplied to the combustion chamber. Some manufacturers use them on electric heating systems to prove a minimum air flow before the strip heaters are energized. The use of this switch allows auxiliary equipment to be wired independently of the blower motor.

FIGURE 15–55
Electric furnace wiring diagram. (*Courtesy of Nordyne*)

Such switches are made with a sail, which is mounted on an SPDT, micro, snap-acting switch. Some of them have a maximum operating temperature of 180° F. Therefore, when replacing a sail switch, make certain that one for the correct ambient temperature is being used. Some of them are positionally mounted; use caution to install the replacement correctly. A typical wiring diagram for an electronic air cleaner is shown in Figure 15–57.

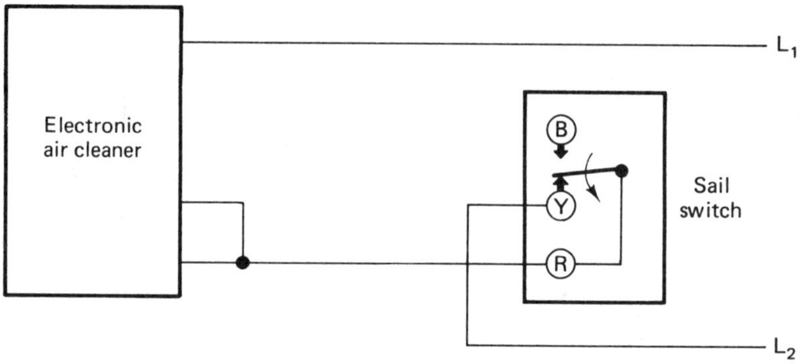
Electronic
air cleaner
B
Y
R
Sail
switch
L1
L2

FIGURE 15–57
Typical sail switch schematic for an electronic air cleaner. (*Courtesy of Billy Langley,* Electric Controls for Refrigeration and Air Conditioning, *2E, ©1988, p. 124; reprinted by permission of Prentice-Hall, Inc., Englewood Cliffs, N.J.*)

Operation

The thermostat generally controls the operation of the fan motor, which in turn moves the air through the system as required. The operating contacts of a sail switch are normally open. As the fan starts blowing air through the area being measured, the sail is caused to move. When sufficient air is flowing through the area, the sail moves far enough to close the contacts in the switch. At this time the controlled equipment is energized, and the system functions as designed.

If for any reason the air flow is reduced or is not sufficient to maintain the closed condition of the NO contacts, the equipment is automatically shut down.

When the thermostat is fully satisfied, the fan motor is deenergized, stopping the air flow through the system. As the air flow is reduced, the sail switch moves to open the operating contacts which, in turn, deenergize all the equipment under its command.

WATER HEATING CONTROLS

Heating a structure with water is common in large commercial, small commercial, and residential systems and when energy conservation is of importance. Control of the boiler requires a special type and application of controls.

Boiler Control

A boiler control may be defined as any control that provides safe, economical, and automatic operation of a boiler.

Water-Level Control

The purpose of a water-level control is to stop the burners in the event of low water in a boiler in order to prevent damage to the boiler. The water-level control is a float-operated control, which senses the water level inside the boiler and will cut off the fuel to the main burner when a predetermined level is reached. (Figure 15–58). This control is designed to operate on boilers with pressures up to approximately 30 pounds per square inch gauge. It is mounted on the boiler so that the switch will interrupt the control circuit to the main fuel valve when the boiler water level drops to a dangerously low level.

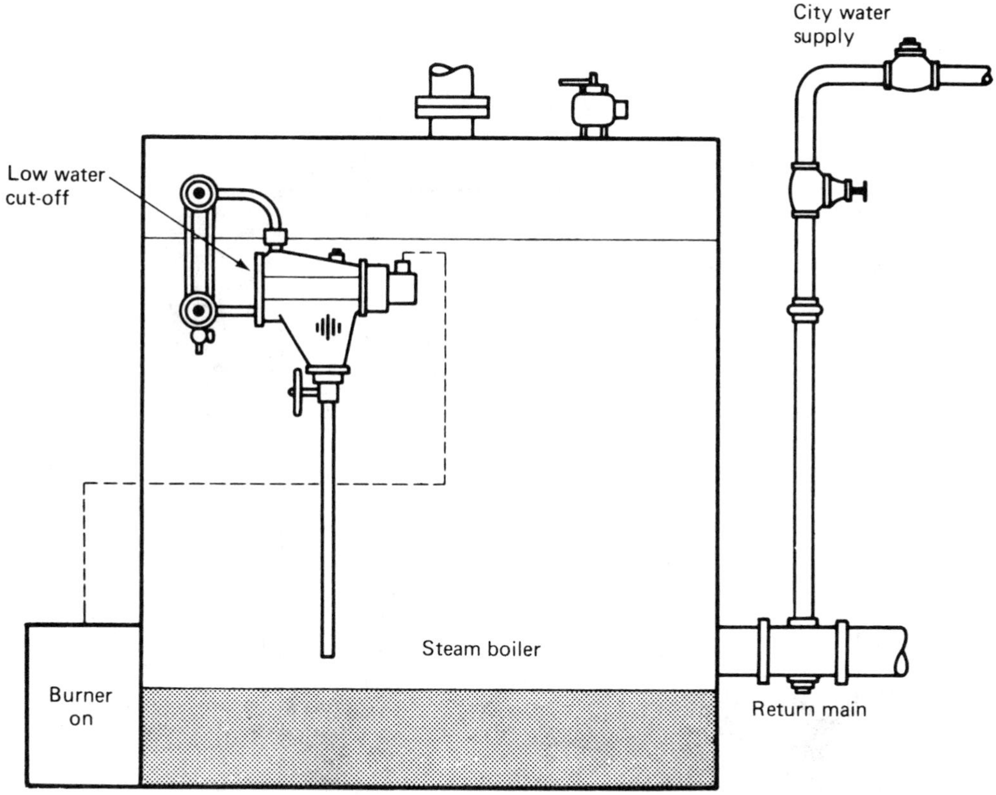
City water supply
Low water cut-off
Steam boiler
Burner on
Return main

FIGURE 15–58
Float-operated low-water cutoff. (*Courtesy of Billy Langley,* Electric Controls for Refrigeration and Air Conditioning, *2E, ©1988, p. 153; reprinted by permission of Prentice-Hall, Inc., Englewood Cliffs, N.J.*)

The low water cut-off control is mounted on the boiler at the desired water level (Figure 15–59). There are several locations and piping arrangements available for this type of installation. Because there is no normal water line to be maintained in a hot-water boiler, any location of the control above the lowest permissible water level is satisfactory. A steam boiler does, however, have a specific water level that must be maintained, and the recommendations of the boiler manufacturer should be followed.

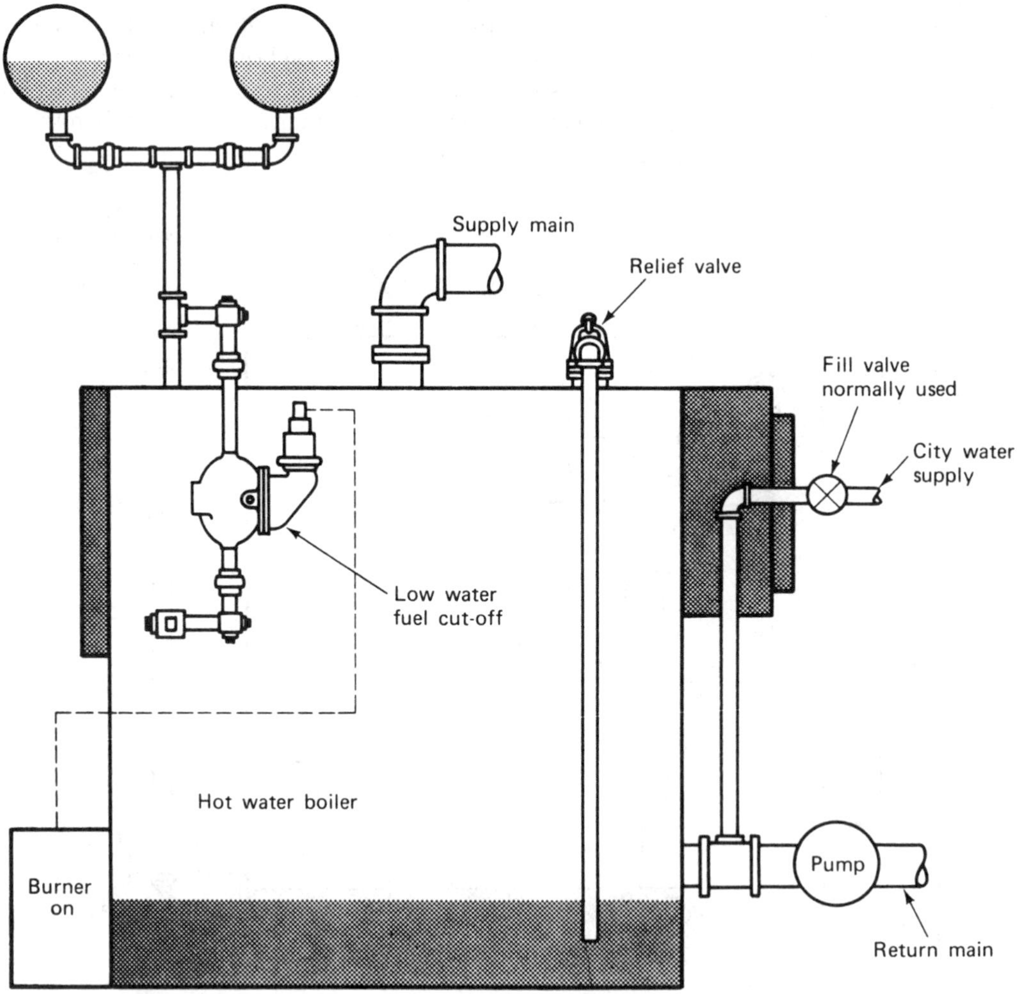
Supply main
Relief valve
Fill valve
normally used
City water
supply
Low water
fuel cut-off
Hot water boiler
Burner
on
Pump
Return main

FIGURE 15–59
Installation of low-water cutoff. (*Courtesy of Billy Langley,* Electric Controls for Refrigeration and Air Conditioning, *2E, ©1988, p. 155; reprinted by permission of Prentice-Hall, Inc., Englewood Cliffs, N.J.*)

The construction of a hot-water boiler is essentially the same as a steam boiler. The major difference is in the way the steam boiler is operated. Most of the conditions causing low water to occur in a steam boiler will also hold true for a hot-water boiler.

Operation. The water line in the boiler and the water line in the control drop at the same time and are at the same level. This lowering of the water line in the control-float chamber causes the float to drop. The action of the float causes the control switch to open and interrupt the control circuit to the automatic burner (Figure 15–60).

This action provides a basic safety control for boilers. It is a means of stopping the automatic firing device if the water level should drop below the minimum safe level.

Combination Feeder and Low Water Cut-Off

If the low water cut-off could be absolutely relied on to stop the automatic burner each time a low water condition occurred, then the problem would be solved. However, experience proves that under certain circumstances, the low water cut-off cannot fulfill its duties.

The combination feeder and low water cut-off offers much more safety than the low water cut-off alone can. It covers almost all installations and provides the most complete measure of safety (Figure 15–61). This control provides:

1. The mechanical operation of feeding water to the boiler as fast as it is discharged through the relief valve.
2. The electrical operation of stopping the burner when a low water situation occurs.

This combination of water feeding and control of the electrical circuit offers the best safeguards, which are the best and most complete recommendations for boilers.

Operation. In operation, the combination feeder and low water cut-off control admits water to the boiler to maintain the desired water level, while at the same time completing the electrical power to the control circuit that operates the burner in response to the requirements of the boiler. Should the water leave the boiler faster than the feeder can admit it, the low water cut-off stops the main burner before the boiler becomes overheated and perhaps ruined.

Switches

The low water cut-off and the combination feeder and low water cut-off control have electrical contacts of various configurations (Figure 15–62). These contacts are directly connected to and are operated by the position of the water float in the control-float chamber.

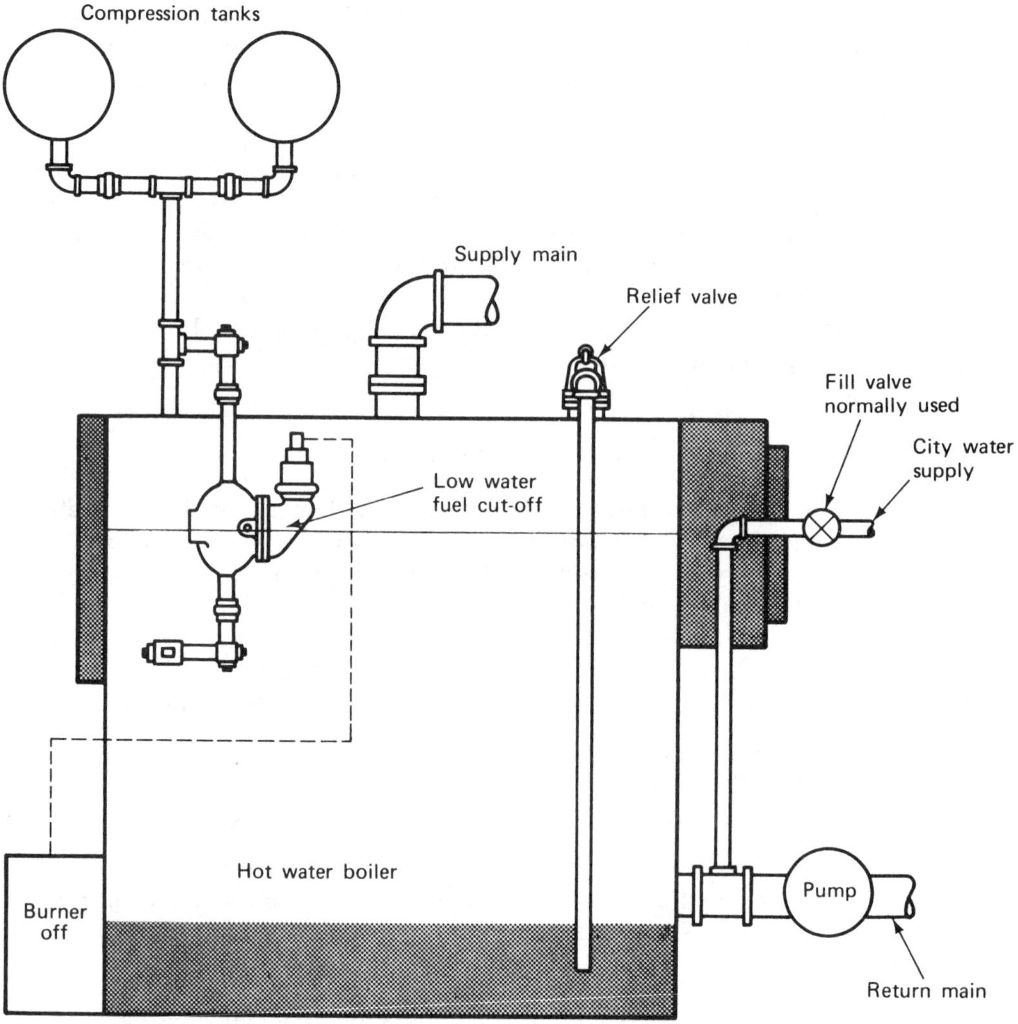
Compression tanks
Supply main
Relief valve
Fill valve
normally used
City water
supply
Low water
fuel cut-off
Hot water boiler
Burner
off
Pump
Return main

FIGURE 15–60
Low-water line-stops burner. (*Courtesy of Billy Langley,* Electric Controls for Refrigeration and Air Conditioning, *2E, ©1988, p. 155; reprinted by permission of Prentice-Hall, Inc., Englewood Cliffs, N.J.*)

In Figure 15–62(b), the NC contacts complete the control circuit and allow the burner to operate on demand from the temperature or pressure control.

Figure 15–62(c) shows that when a low water level condition occurs, the control circuit is opened and the NO alarm contacts are made, sending a danger signal to the user.

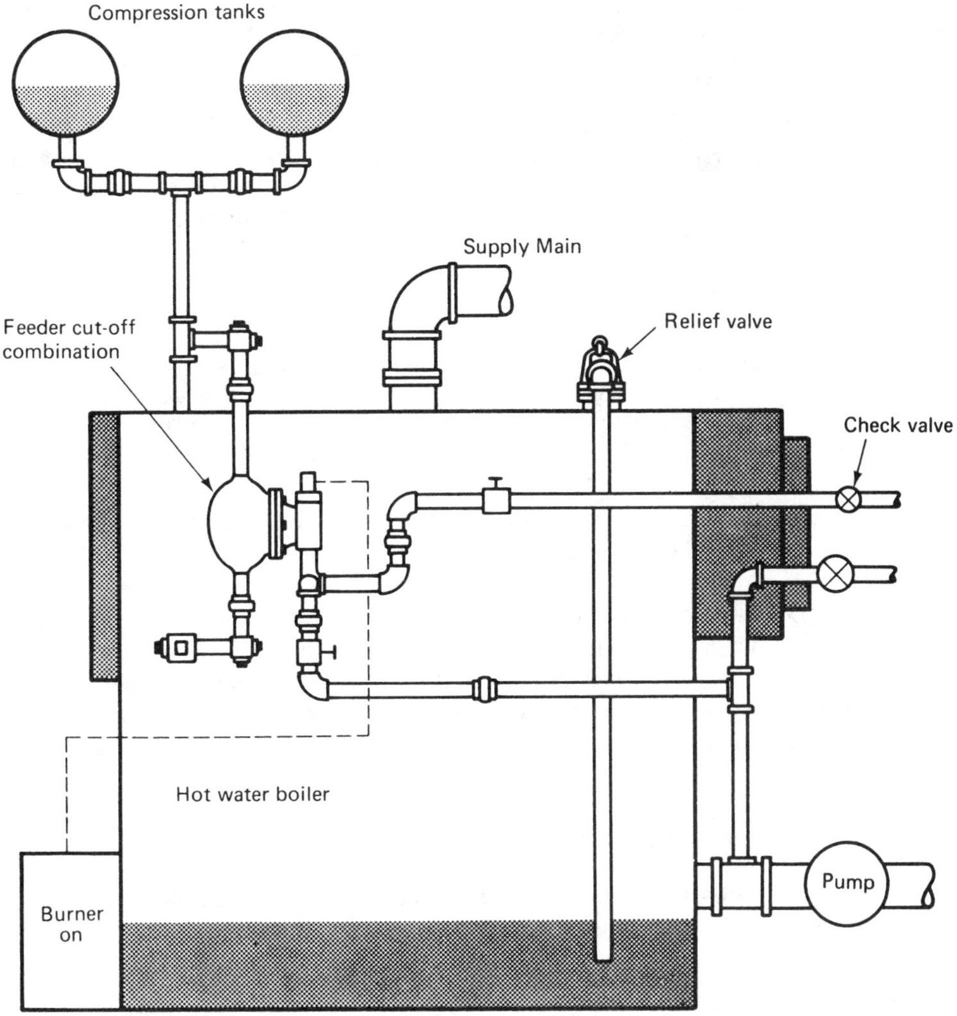
Compression tanks
Supply Main
Feeder cut-off combination
Relief valve
Check valve
Hot water boiler
Burner on
Pump

FIGURE 15–61
Combination feeder and low-water cutoff. (*Courtesy of Billy Langley,* Electric Controls for Refrigeration and Air Conditioning, *2E, ©1988, p. 156; reprinted by permission of Prentice-Hall, Inc., Englewood Cliffs, N.J.*)

Boiler Temperature Control (Aquastat)

In normal operation, the boiler temperature control regulates the operation of the main burner by interrupting the control circuit. The sensing element is installed in the boiler through openings provided by the boiler manufacturer. The switch may be either an SPST snap-acting switch or a mercury tube. Both of these switches are

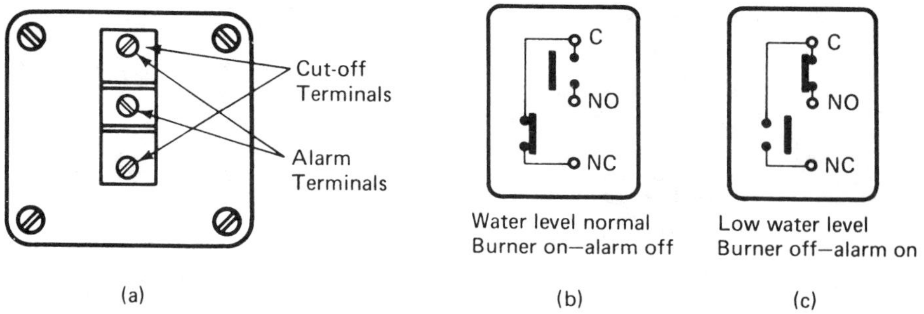
Cut-off
Terminals
Alarm
Terminals
(a)
C
NO
NC
Water level normal
Burner on—alarm off
(b)
C
NO
NC
Low water level
Burner off—alarm on
(c)

FIGURE 15–62
(a) Switch terminal locations, (b) Water level normal, burner on–alarm off, (c) Low-water level, burner off–alarm on. *(Courtesy of Billy Langley,* Electric Controls for Refrigeration and Air Conditioning, *2E, ©1988, p. 155; reprinted by permission of Prentice-Hall, Inc., Englewood Cliffs, N.J.)*

actuated by a helically wound bimetal. These controls are adjustable so that the proper operating temperature of the unit can be selected.

Operation. As the water temperature inside the boiler drops to the cut-in temperature, the sensing element moves to a position that causes the switch to close and complete the control circuit to the main burner gas valve (Figure 15–63). As the water is heated to the cut-out temperature setting of the aquastat, the switch contacts open and interrupt the control circuit, deenergizing the main burner gas valve.

Boiler Pressure Control

The boiler pressure control is used on steam boilers and is mounted above the boiler proper to sense the most critical point in the system (Figure 16–64). The switch is

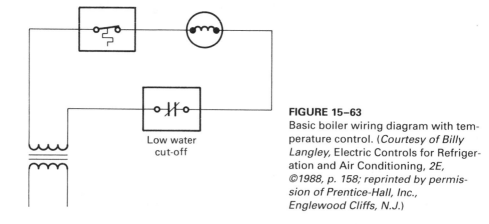

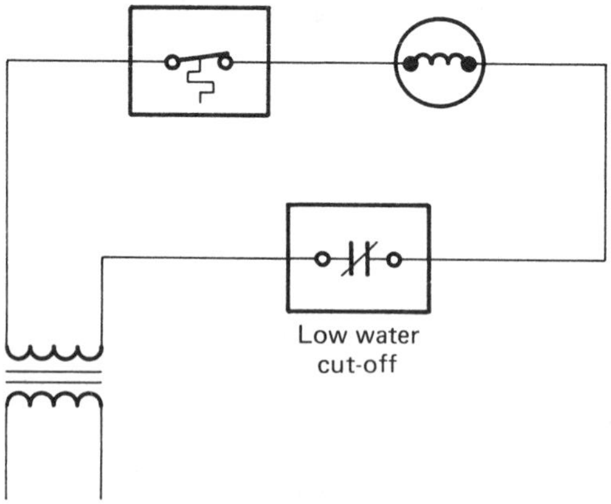
Low water
cut-off

FIGURE 15–63
Basic boiler wiring diagram with temperature control. (*Courtesy of Billy Langley,* Electric Controls for Refrigeration and Air Conditioning, *2E, ©1988, p. 158; reprinted by permission of Prentice-Hall, Inc., Englewood Cliffs, N.J.*)

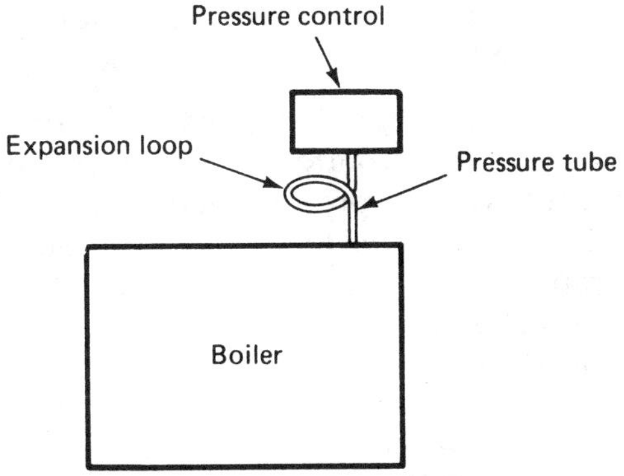
Pressure control
Expansion loop
Pressure tube
Boiler

FIGURE 15–64
Boiler pressure control installation. (*Courtesy of Billy Langley,* Electric Controls for Refrigeration and Air Conditioning, *2E, ©1988, p. 158; reprinted by permission of Prentice-Hall, Inc., Englewood Cliffs, N.J.*)

actuated by a diaphragm and bellows, which expand with an increase in pressure and contract with a decrease in pressure. The contacts in these controls are wired in series in the control circuit and either make or break the control circuit in response to the pressure inside the system (Figure 15–65). Boiler pressure controls have an adjustable scale so that the proper operating pressure can be maintained.

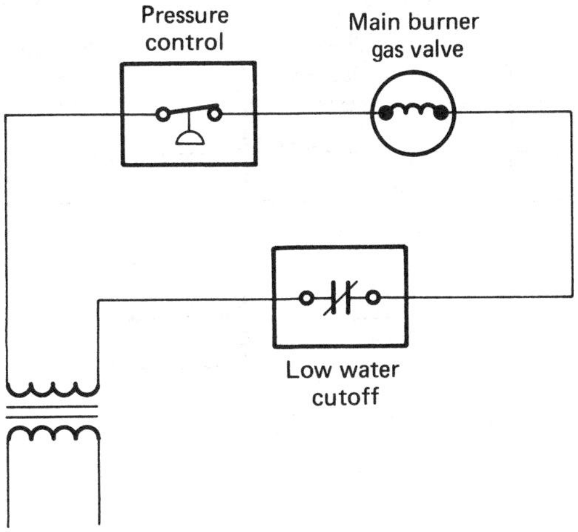
Pressure control
Main burner gas valve
Low water cutoff

FIGURE 15–65
Basic boiler wiring diagram with pressure control. (*Courtesy of Billy Langley,* Electric Controls for Refrigeration and Air Conditioning, *2E, ©1988, p. 158; reprinted by permission of Prentice-Hall, Inc., Englewood Cliffs, N.J.*)

High Limit Control

The high limit control is a safety device that is wired into the control circuit to prevent overheating the boiler in case the temperature or pressure control fails to function properly. These controls may use either SPST or SPDT switches. The SPST type only interrupts the control circuit and stops operation of the main-burner gas valve. The SPDT models may be used for several different functions, such as energizing a circulator pump or energizing an alarm circuit to alert the user that a problem has occurred (Figure 15–66). The high limit control is wired into the control circuit in series with the other controls (Figure 15–67).

Operation. If the boiler temperature rises above the setting of the pressure or temperature controller, the high limit control will interrupt the control circuit, deenergizing the main-burner gas valve to prevent overheating the boiler. The secondary circuit, if used, will then be energized to perform the desired function, such as starting the circulator pump or energizing the alarm circuit. As the temperature or pressure of the boiler drops to the cut-in setting of the control, the contacts close, completing the control circuit. The boiler is again started for normal operation.

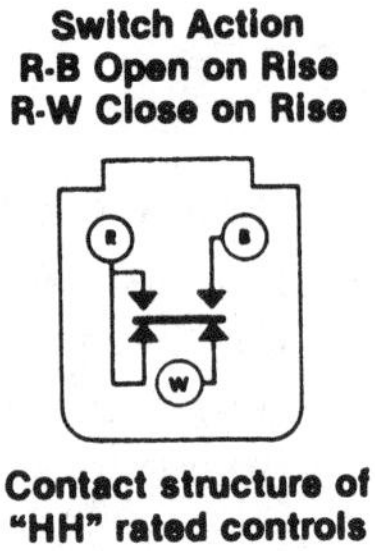
Switch Action
R-B Open on Rise
R-W Close on Rise
Contact structure of
"HH" rated controls

FIGURE 15–66
Contact configuration of a high-limit control. (*Courtesy of Billy Langley,* Electric Controls for Refrigeration and Air Conditioning, *2E, ©1988, p. 159; reprinted by permission of Prentice-Hall, Inc., Englewood Cliffs, N.J.*)

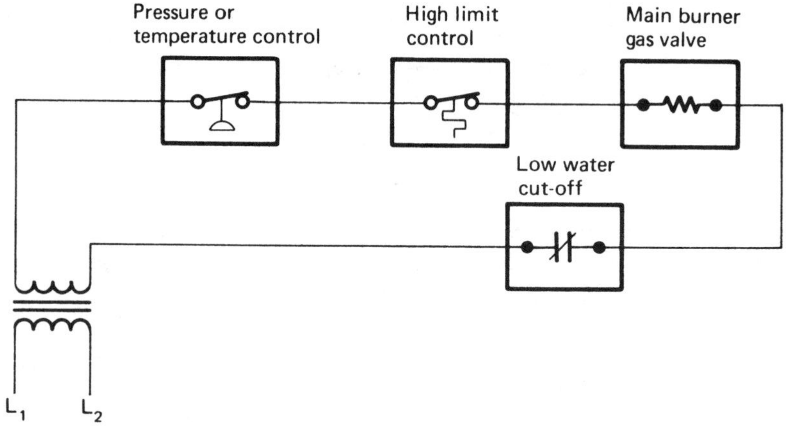
Pressure or
temperature control
High limit
control
Main burner
gas valve
Low water
cut-off
L_1
L_2

FIGURE 15–67
Basic boiler wiring diagram. (*Courtesy of Billy Langley,* Electric Controls for Refrigeration and Air Conditioning, *2E, ©1988, p. 160; reprinted by permission of Prentice-Hall, Inc., Englewood Cliffs, N.J.*)

OUTDOOR RESET CONTROL

Outdoor reset controls are used to maintain a proper balance between the heating medium and the outdoor ambient temperature. They automatically raise or lower the temperature of the heating medium (water, steam, or warm air) control point as the outdoor air temperature changes. They use two sensing bulbs. One sensing bulb is mounted to sense the heating medium, and the other is set to sense the outdoor air temperature. Outdoor reset controls are equipped with a set of SPDT contacts that change position on a change in temperature. They may be used on line-voltage, low voltage (24 volts), or millivoltage systems. This control is not designed to replace the high limit control. The differential adjustment of this control is adjustable within the limits of the control.

Operation

Because both the outdoor air temperature and the temperature of the heating medium are measured, a combination of these two temperatures affects the operation of this control. When the outdoor air temperature is high and the system capacity demand is low, the temperature of the heating medium is also low, reducing the amount of energy used to heat the medium. When the outdoor air temperature is low, the temperature of the heating medium is higher to compensate for the additional heat loss from the building.

As the temperature of the outdoor air begins to fall, the outdoor sensing bulb signals the controller that additional heat is needed to maintain the temperature inside the building. The temperature of the heating medium is raised just enough to compensate for the additional heat loss. Likewise, as the outdoor air temperature rises, the outdoor sensing bulb signals the controller that a lower heating medium temperature will be enough. This lowers the heating cost for the building while maintaining the indoor temperature at the desired level.

HUMIDISTATS

Humidistats are used to control humidification equipment on heating systems. It is desirable to add humidity to some buildings because the air becomes drier when heated by the heating equipment. Today, there are many different types of humidity-sensitive materials available. For the most part they are organic and include such materials as nylon, wood, human hair, and, in some instances, animal membranes. In addition, there are other materials that are sensitive to humidity changes. These materials change in electrical resistance with a change in humidity. Only a few of these are successfully adaptable for use in humidity controllers.

The following are the most commonly used types of humidity controllers and the sensors used in them.

Mechanical-type controls. Mechanical sensors are dependent on a change in the length of the size of the sensing element in direct relation to a change in relative humidity. The sensors used in modern controllers are commonly made from human hair or some type of synthetic polymer. They are normally attached to linkage in the controller, which in turn controls the mechanical, electrical, or pneumatic switching element in a valve or motor.

In general, they are designed to control at some set point that is adjustable by the user or operator of the equipment. Also available are controllers which automatically change the set point within the building in response to the outdoor air temperature. This reduces condensation on windows and building walls.

Electronic-type controllers. With a change in humidity, the sensors used in electronic humidity controllers also undergo a change in their electrical resistance. These sensors are made of two electrically conductive materials that are separated by a humidity-sensitive pyroscopic insulating material, such as polyvinyl acetate, polyvinyl alcohol, or a solution of certain types of salt.

When air is passed over the sensing element, any small changes in humidity are detected. The electrical resistance between the two conductors varies in response to the humidity present in the air. In most applications the sensor is placed in one leg of a Wheatstone bridge, where the output signal can be used to control the equipment or as a readout. Generally, a signal amplification is required.

Electronic humidity controllers are most popular in applications where very close control of the humidity is required. In some applications they may be used for controlling the speed of the fan so that a more accurate humidity control can be achieved.

There are several ways to wire these controls into a system. However, they should be wired so that the humidifier operates only when the fan is operating to prevent moisture from collecting on the heat exchanger during the off cycle. One suggested diagram is shown in Figure 15–68. This diagram provides a safety

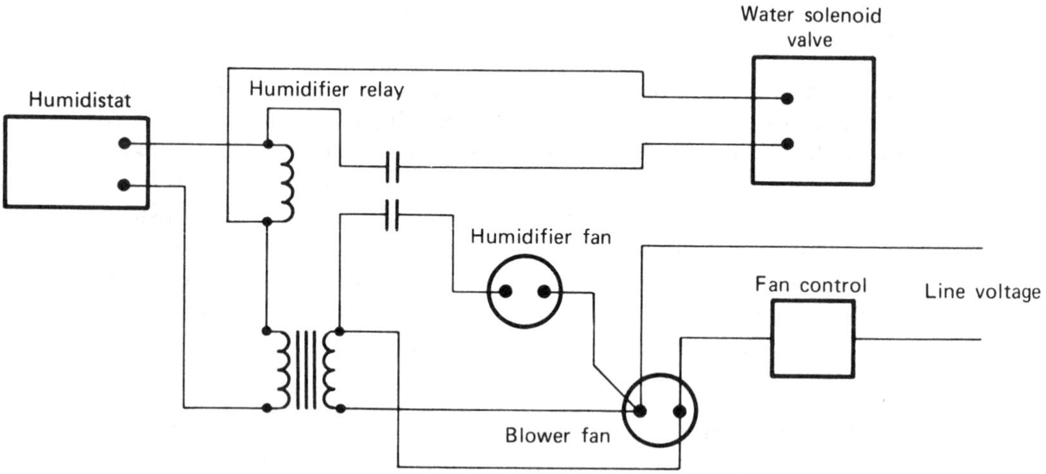
Water solenoid valve
Humidistat
Humidifier relay
Humidifier fan
Fan control
Line voltage
Blower fan

FIGURE 15–68
Typical humidifier wiring diagram. (*Courtesy of Billy Langley,* Electric Controls for Refrigeration and Air Conditioning, *2E, ©1988, p. 106; reprinted by permission of Prentice-Hall, Inc., Englewood Cliffs, N.J.*)

feature—the water solenoid is not allowed to be energized unless both sources of electricity are available.

Most humidistats may be used on either line-voltage or low voltage control systems. They have a moisture-sensitive nylon ribbon used to actuate the contacts, which open on a rise in humidity. Positive on-off settings are sometimes provided for manual operation. The switch is an SPST snap-acting type.

REVIEW QUESTIONS

1. What is the purpose of a room thermostat?
2. Name the three types of electric room thermostats.
3. What types of switches are available on room thermostats?
4. What are the two positions of a fan switch on a thermostat?
5. What principle is used in the modulating-type thermostat?
6. What is the approximate operating range of the outdoor thermostat?
7. Why is a low voltage thermostat desired over a line-voltage thermostat?
8. Which is more economical to install, the low voltage or the line-voltage thermostat?
9. How does a heat anticipator operate?
10. To what is the heat anticipator adjusted?
11. What is the purpose of the transformer?
12. How are transformers rated?
13. Write the formula used to determine the VA rating of a transformer.
14. Why is transformer phasing desirable?
15. What is the purpose of the fan control in a furnace?
16. What is the insert made of on a fan control?
17. What does the sensing element of a fan control measure?
18. What type of fan operation does an electrically operated fan control provide?
19. Are the fan control contacts normally open or closed?
20. In what circuit is the fan control located?
21. What is the purpose of a limit control?
22. In what circuit is the limit control?
23. At what temperature does the limit control function?
24. Are the limit control contacts normally open or normally closed?
25. What is the purpose of the combination fan and limit control?
26. What causes the main gas valve to open or close?
27. In what circuit is the main gas valve?
28. Name the six components incorporated in the combination gas control.
29. Name the two types of pilot safety devices.
30. What provides electricity for the pilot safety device?
31. Of what is the thermocouple made?
32. How many circuits are in a modern gas furnace?
33. How many voltages are in a modern furnace control circuit?

34. What is the purpose of a time delay relay in an electric furnace?
35. Where is the overcurrent protector installed?
36. What type of special limit controls are sometimes used on electric furnaces?
37. What are overtemperature protectors sometimes called?
38. What type of thermostat is normally used with electric furnace sequencers?
39. What type of control is used to control the fan in warm-air applications to prevent circulation and excessive filter temperatures in counterflow furnaces?
40. What control is used to cycle heating equipment to maintain an average discharge-air temperature?
41. What control is designed to detect air flow?
42. Define a boiler control.
43. What control is used to stop the burners in a boiler when low water conditions occur?
44. Under what pressures are water level controls designed to operate?
45. What control provides a safer boiler operation than a low water cut-off?
46. On a normal hot-water boiler, what regulates the operation of the main burners?
47. On what type of system is a boiler pressure control used?
48. What control prevents overheating of a boiler?
49. What is the purpose of an outdoor reset control?
50. In what applications are electronic humidity controls popular?

16

heat load calculation and operating cost estimating

This chapter uses the T U Electric heat load calculation chart, and information applicable to the T U Electric service area, but the general steps are the same for any area. The electric utility company in your area probably has the same, or similar, charts, tables, and information. Check with the utility company for assistance. Some of the information provided in this chapter was taken from the *ARI Energyguide Directory,* which may be obtained from ARI (1501 Wilson Blvd., Arlington VA 22209).

INTRODUCTION

We will use the T U Electric Weather Chart, the Infiltration Construction Definition Chart, an example, and a floor plan for the house to which the example applies. The example is of a typical house that has been constructed to meet energy efficiency standards. There will also be instructions for both ARI and the degree day methods to estimate the operating cost for cooling and heating, followed by a heating example using the Gas Appliance Manufacturers Association (GAMA) method to estimate gas heating cost.

Use the following procedure:

Step 1. Refer to the T U Electric Form 219-R (Figure 16–10) and the T U Electric Weather Chart (Chart 16–2). The cooling design conditions are usually selected for 100° F outside temperature and 75° F inside temperature. Place this information in the proper space at the top of the form. Subtract the two numbers to obtain the temperature difference between the two numbers (25° F) (Figure 16–1).

DESIGN CONDITIONS	
COOLING D.B. °F.	
Outside Temp.	100
Inside Temp.	75
Temp. Diff.	25

FIGURE 16–1
Design conditions on Form 219-R. (*Courtesy of T U Electric*)

Step 2. The heating design is usually calculated on a 75° F inside temperature. From the T U Electric Weather Chart (Chart 16–2), the outside winter design temperature is selected for the geographical location involved. For this example, Irving, Texas was chosen; it has a 20° F outside design temperature (Chart 16–1).

CHART 16–1
T U Electric weather conditions. (*Courtesy of T U Electric*)

	Outside Temperature Design (°F)*		Heating Degree Days (65°Base)**	Annual Heating Consumption Factor***
	Summer	Winter		
DALLAS DIV., FT WORTH & CENTRAL REGION:	100	20	2,400	310
Arlington, Dallas, Duncanville, Euless, Ft. Worth, Garland, Grand Prairie, Grapevine, Farmers Branch, Irving, Lancaster, Mesquite, Plano, Richardson				

CHART 16–2
Method of reducing infiltration of unconditioned air. (*Courtesy of T U Electric*)

(Ranges from 0.67 to 1.75 Air Changes Per Hour)

INFILTRATION REDUCED BY:	AIR CHANGES SAVED PER HOUR
A. Soleplate Sealed (33%)	0.3564
B. Wiring & Plumbing Holes and Furrdowns Sealed (26%)	0.2808
C. Exterior Doors & Windows Weather Stripped (7%)	0.0756
D. Exterior Door & Window Rough Openings Caulked (17%)	0.1836
E. Attic Access Outside Conditioned Space or Weather Stripped (2%)	0.0216
F. Outside Sheathing Holes Sealed and Polyethylene Film Installed (7%)	0.0756
G. Ventless or Dampered Range Hood Installed or No Range Vent (8%)	0.0864
TOTAL	1.0800

DESIGN CONDITIONS

COOLING D.B. °F.		HEATING D.B. °F.	
Outside Temp.	100	Inside Temp.	75
Inside Temp.	75	Outside Temp.	20
Temp. Diff.	25	Temp. Diff.	55

FIGURE 16–2
Design conditions in Form 219-R. (*Courtesy of T U Electric*)

Step 3. Insert the inside and outside design temperatures on the form and subtract them to find the temperature differential (Figure 16–2).

Step 4. Refer to the set of house plans to find additional design criteria (Figure 16–3). In addition to the floor plans, refer to other parts of the plans, such as the roof plan, foundation plan, cross sections, and elevations. If some of the needed information cannot be found on the plans, it can be obtained by consultation with the architect, builder, or owner.

Much of the design information can be found on Figure 16–3, but more is needed. The following also apply:

1. The house is for a family of four people.
2. It will be built on a slab without perimeter insulation.
3. The house will have a hip roof with a 2-ft overhang.
4. The windows will be single glazed glass. The doors will be made of wood.
5. Inside shading will be used over the glass areas.
6. The walls will have R-16 insulation; the ceilings will have R-30 insulation.
7. The attic will be vented. The duct insulation will be 2 inches thick.
8. The ceilings will be 8 ft high.
9. A heat pump unit with a Seasonal Energy Efficiency ratio (SEER) of 9 will be used for both heating and cooling the house.
10. As seen from the arrow pointing to the north on the floor plan (Figure 16–3) the front of the house will face the west.

Step 5. Use the data that apply to the house construction to fill in the remainder of the data block at the top of the form (Figure 16–4).

Step 6. Item 1 (Figure 16–5)—Glass Solar Gain: Locate the correct column under the cooling factor heading. In this example, the overhang is 2 ft and there is inside shading. Circle the 2′ under the Inside Shading and Overhang column. The factors directly below the 2′ figure are the number of Btu of radiant heat gain per square foot of glass for each direction of exposure (Figure 16–5).

Step 7. Circle the solar gain factors under the 2′ column which apply to each direction of glass exposure (Figure 16–6). The glass is exposed on the north, east, south, and west elevations. The solar gain factors which apply are circled directly under the 2′ overhang column opposite the direction of exposure.

Step 8. Item 1: Under the quantity column heading, insert the square feet of glass for the corresponding direction of exposure after determining the glass area

WINDOW SCHEDULE

MARK	TYPE	SIZE	MFG.#	MAT'L.
A	RANCH WALL	9'–0" X 6'–1"	9061	ALUMN.
B	RANCH WALL	9'–0" X 6'–1"	9061	ALUMN.
C	SINGLE HUNG	9'–0" X 6'–1"	9061	ALUMN.
D	SINGLE HUNG	9'–0" X 6'–1"	9061	ALUMN.
E	SINGLE HUNG	9'–0" X 6'–1"	9061	ALUMN.
F	HORIZ. SLIDER	9'–0" X 6'–1"	9061	ALUMN.

DOORS

MARK	SIZE
1	3'–0" X 6'–8"
2	2'–8" X 6'–8"
12	6'–0" X 6'–8"

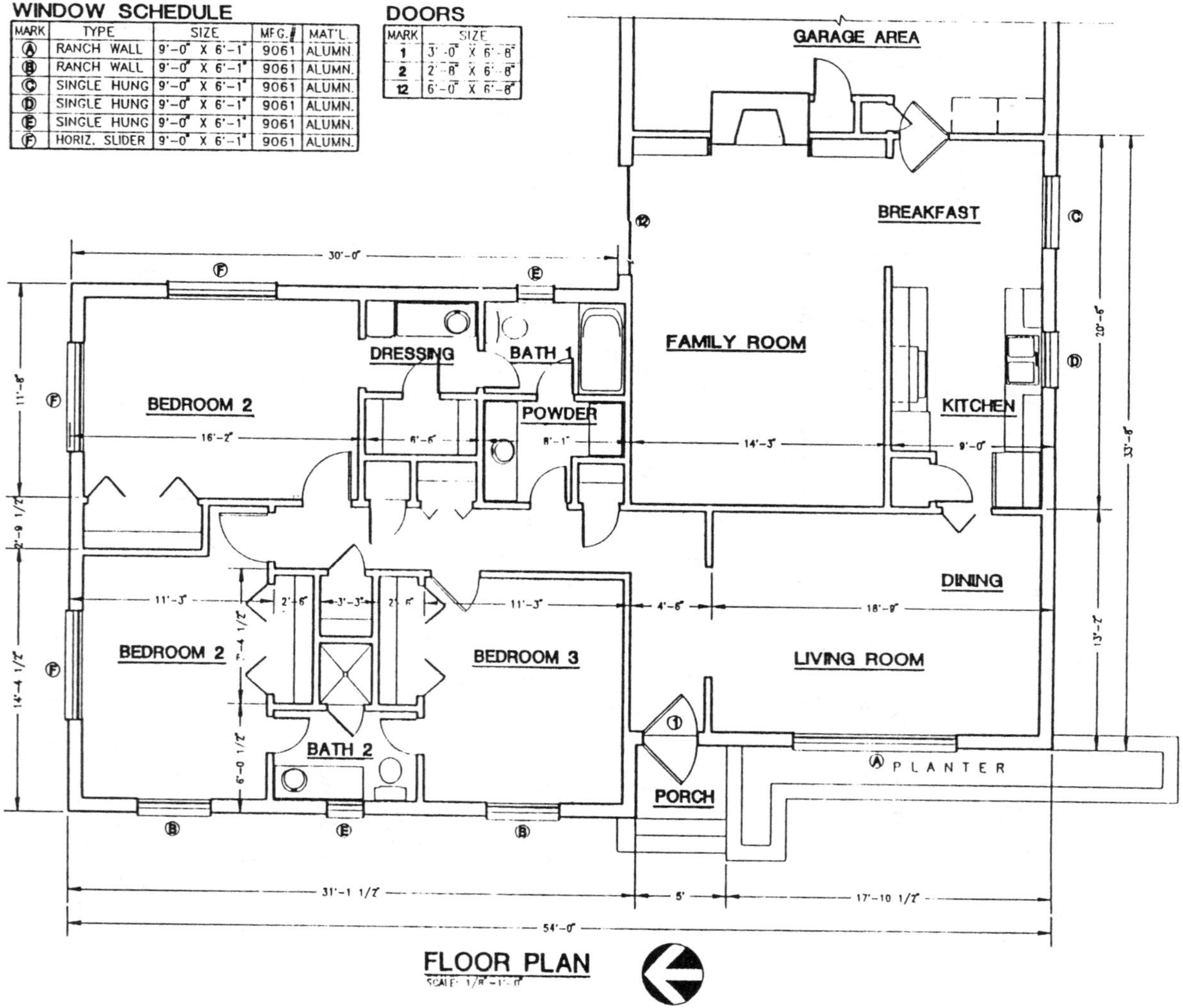

GARAGE AREA
BREAKFAST
FAMILY ROOM
KITCHEN
DINING
LIVING ROOM
PORCH
PLANTER
BEDROOM 2
DRESSING
BATH 1
POWDER
BEDROOM 2
BATH 2
BEDROOM 3
30'–0"
16'–2"
6'–6"
8'–1"
14'–3"
9'–0"
11'–3"
2'–6"
3'–3"
4'–6"
18'–9"
11'–8"
2'–9 1/2"
14'–4 1/2"
6'–0 1/2"
20'–6"
13'–2"
33'–8"
31'–1 1/2"
5'
17'–10 1/2"
54'–0"

FLOOR PLAN

FIGURE 16–3 House plans. (*Courtesy of T U Electric*)

INSULATION		HOUSE FACES	OVERHANG OR SHADING
Ceiling	R-30	☐ North	2'
Wall	R-16	☐ East	2'
Floor	SLAB	☐ South	2'
Window	STORM W-TB	☒ West	2'

FIGURE 16–4
Construction data. (*Courtesy of T U Electric*)

CONSTRUCTION		COOLING FACTOR						
1. **GLASS—SOLAR GAIN Windows-Doors-Skylights** (Sq. Ft.) (*Note A)	SC	No Shading	Inside Shading & Overhang 0'	1'	(2')	3'	4'	Outside Shades
North		32	14	0	0	0	0	0
Northeast		50	23	21	19	19	17	17
East		50	23	20	19	17	14	13
Southeast		80	36	26	19	8	0	21
South		100	45	33	10	0	0	26
Southwest		155	70	51	39	17	0	42
West		190	86	82	80	67	57	57
Northwest		128	58	54	51	51	49	37
Horizontal		185						

FIGURE 16–5
Locating shading factors. (*Courtesy of T U Electric*)

CONSTRUCTION		COOLING FACTOR						
1. **GLASS—SOLAR GAIN Windows-Doors-Skylights** (Sq. Ft.) (*Note A)	SC	No Shading	Inside Shading & Overhang 0'	1'	(2')	3'	4'	Outside Shades
North		32	14	0	(0)	0	0	0
Northeast		50	23	21	19	19	17	17
East		50	23	20	(19)	17	14	13
Southeast		80	36	26	19	8	0	21
South		100	45	33	(10)	0	0	26
Southwest		155	70	51	39	17	0	42
West		190	86	82	(80)	67	57	57
Northwest		128	58	54	51	51	49	37
Horizontal		185						

FIGURE 16–6
Marking the correct shading factors. (*Courtesy of T U Electric*)

for each exposure. Refer to Figure 16–3, the floor plan and window schedule. (Remember that the house is to face west.)

The glass area is found for each window by multiplying the window width in feet by the window height in feet. The glass door area is found in a similar manner.

The total area of glass exposure for each direction is the sum of the individual window areas and glass door areas facing in that direction.

Example:

(Refer to the window and door schedule on the floor plan in Figure 16–3.)

Glass Facing North

2-F	2(6 × 2) = 24
1-Sliding door	(6 × 7) = 42
	66 sq. ft.

Glass Facing East

1-F	(6 × 2) = 12
1-E	(2 × 3) = 6
	18 sq. ft.

Glass Facing South

1-D	(3 × 3) = 9
1-C	(3 × 4 ½) = 13 ½
	22 ½ or 23 sq. ft.

Glass Facing West

2-B	2(4 × 6) = 48
1-E	(2 × 3) = 6
1-A	(9 × 6) = 54
	108 sq. ft.

Fill in the square feet of glass in the quantity column opposite the direction of exposure and under the appropriate shading category (Figure 16–7). Insert a value for SC (shading coefficient) to compensate for plateglass or multiple layers, as shown in Figure 16–10.

CONSTRUCTION		COOLING FACTOR							HEATING FACTOR	QUANTITY	COOLING Btuh	HEATING Btuh
1. GLASS—SOLAR GAIN Windows-Doors-Skylights (Sq. Ft.) (*Note A)	SC	No Shading	Inside Shading & Overhang 0′	1′	2′ (circled)	3′	4′	Outside Shades				
North		32	14	0	0 (circled)	0	0	0		66		
Northeast		50	23	21	19	19	17	17				
East		50	23	20	19 (circled)	17	14	13		18		
Southeast		80	36	26	19	8	0	21				
South		100	45	33	10 (circled)	0	0	26		23		
Southwest		155	70	51	39	17	0	42				
West		190	86	82	80 (circled)	67	57	57		108		
Northwest		128	58	54	51	51	49	37				
Horizontal		185										

FIGURE 16–7

Filling in the glass factor values. (*Courtesy of T U Electric*)

Step 9. Item 1: Multiply the cooling factor by the quantity of glass for each exposure and insert these figures in the cooling Btu/h column. Note A gives the shading factors to modify the cooling Btu/h should something other than standard single-pane window glass be used. For instance, the Btu/h for double glass would be the product of 0.85 times the cooling factor times the area of glass in square feet

(Figure 16–10). For all practical purposes, the sun can impose a maximum solar gain from only one vertical wall or exposure at any given time. Therefore, only the one largest vertical calculated Btu/h solar gain will be included in the total cooling load. Identify the largest cooling Btu/h figure by marking it in some manner, such as with an asterisk or by crossing out the other amounts (Figure 16–8).

CONSTRUCTION		COOLING FACTOR							HEATING FACTOR	QUANTITY	COOLING Btuh	HEATING Btuh
1. **GLASS—SOLAR GAIN Windows-Doors-Skylights** (Sq. Ft.) (*Note A)	SC	No Shading	Inside Shading & Overhang 0′	1′	2′	3′	4′	Outside Shades				
North		32	14	0	0	0	0	0				
Northeast		50	23	21	19	19	17	17		66	~~0~~	
East		50	23	20	19	17	14	13				
Southeast		80	36	26	19	8	0	21		18	~~342~~	
South		100	45	33	10	0	0	26				
Southwest		155	70	51	39	17	0	42		23	~~230~~	
West		190	86	82	80	67	57	57				
Northwest		128	58	54	51	51	49	37		108	8640	
Horizontal		185										

FIGURE 16–8
Identifying largest cooling factors. (*Courtesy of T U Electric*)

A skylight was not used in the house in this example, so the "horizontal" glass area category listed under Glass—Solar Gain is ignored. If horizontal glass (i.e., skylight) was used, then the total square feet (area) of the horizontal quantity should be multiplied by the solar gain cooling factor to obtain a cooling Btu/h for solar gain from the horizontal glass. This should be marked with an asterisk and be added along with the largest vertical area solar gain Btu/h to obtain the total cooling Btu/h load when calculating the subtotal for Item 9.

Step 10. Item 1: From the established design conditions (steps 1 and 3), determine the temperature differentials for heating and cooling. In the last line of Item 1, circle the heating and cooling temperature differentials (Figure 16–9) for a cooling temperature differential of 25° F and a heating temperature differential of 55° F.

CONSTRUCTION	U	COOLING FACTOR (DESIGN D.B. TEMP. DIFFERENTIAL)				HEATING FACTOR (DESIGN D.B. TEMP. DIFFERENTIAL)				QUANTITY	COOLING Btuh	HEATING Btuh
		20	22	25	30	65	60	55	50			

FIGURE 16–9
Marking design temperature differentials. (*Courtesy of T U Electric*)

Step 11. Items 2 through 7: Circle all heating and cooling factors which apply and are to be utilized in items 2 through 6. These values are located directly under the circled design dry bulb (DB) temperature differential for both heating and cooling across from the type of construction involved. Item 7 has only one heating and one cooling factor for each design dry bulb temperature differential. See Step 12 for additional instructions on Item 7 (Figure 16–10).

FORM [illegible]

TU ELECTRIC

HEATING AND COOLING ESTIMATE
RESIDENTIAL

Customer ______________________

Address ______________________

Town ______________ District ______________

Prepared by ______________ Date ______________

Accepted by (*Note E) ______________ Organization ______________

DESIGN CONDITIONS

COOLING D.B. °F.
Outside Temp. ________
Inside Temp. ________
Temp. Diff. ________

HEATING D.B. °F.
Inside Temp. ________
Outside Temp. ________
Temp. Diff. ________

INSULATION
Ceiling ________
Wall ________
Floor ________
Window ________

HOUSE FACES
☐ North
☐ East
☐ South
☐ West

OVERHANG OR SHADING
North ________
East ________
South ________
West ________

CONSTRUCTION	SC	COOLING FACTOR: No Shading	Inside Shading & Overhang 0′	1′	2′	3′	4′	Outside Shades	HEATING FACTOR	QUANTITY	COOLING Btuh	HEATING Btuh
1. GLASS—SOLAR GAIN Windows-Doors-Skylights (Sq. Ft.) (*Note A)												
North		32	14	0	0	0	0	0				
Northeast		50	23	21	19	19	17	17				
East		50	23	20	19	17	14	13				
Southeast		80	36	26	19	8	0	21				
South		100	45	33	10	0	0	26				
Southwest		155	70	51	39	17	0	42				
West		190	86	82	80	67	57	57				
Northwest		128	58	54	51	51	49	37				
Horizontal		185										

	U	DESIGN D.B. TEMP. DIFFERENTIAL 20	22	25	30	65	60	55	50	QUANTITY	COOLING Btuh	HEATING Btuh
2. GLASS TRANSMISSION (Sq. Ft.)		U x TD				U x TD						
Standard—Single Glazing	1.13	22.6	24.9	28.2	33.9	73.4	67.8	62.2	56.5			
Insulating—Double Glazing	.78	15.6	17.2	19.5	23.4	50.7	46.8	42.9	39.0			
Storm Window	.67	13.4	14.7	16.8	20.1	43.6	40.2	36.8	33.5			
Storm or Insulating Glass with Thermal Break	.56	11.2	12.3	14.0	16.8	36.4	33.6	30.8	28.0			
3. DOORS—(Sq. Ft.)		U x TD				U x TD						
Solid Wood or Hollow Core	.55	11.0	12.1	13.8	16.5	35.8	33.0	30.3	27.5			
Wood with Storm Door	.34	6.8	7.5	8.5	10.2	22.1	20.4	18.7	17.0			
Metal with 1½″ Urethane	.11	2.2	2.4	2.8	3.3	7.2	6.7	6.1	5.6			
4a. FLOOR: SLAB (Linear Ft. Exposed Edge)		U x zero				U x TD						
No Edge Insulation	.81					52.6	48.6	44.6	40.5			
R-4 Edge Insulation	.68					44.2	40.8	37.4	34.0			
R-7 Edge Insulation	.55					35.8	33.0	30.2	27.5			
4b. FLOORS: ENCLOSED CRAWL SPACE (Sq. Ft.)		U x zero				U x (TD-20)						
No Insulation	.270					12.2	10.8	9.4	8.1			
R-7 Insulation	.093					4.2	3.7	3.2	2.8			
R-11 Insulation	.073					3.3	2.9	2.6	2.2			
R-13 Insulation	.060					2.7	2.4	2.1	1.8			
R-19 Insulation	.046					2.1	1.8	1.6	1.4			
R-22 Insulation	.039					1.8	1.6	1.4	1.2			
4c. FLOORS: OPEN CRAWL SPACE (Sq. Ft.)		U x (TD-5)				U x TD						
No Insulation	.374	5.6	6.4	7.5	9.4	24.3	22.4	20.6	18.7			
R-7 Insulation	.103	1.5	1.8	2.1	2.6	6.7	6.2	5.7	5.2			
R-11 Insulation	.073	1.1	1.2	1.5	1.8	4.7	4.4	4.0	3.6			
R-13 Insulation	.064	1.0	1.1	1.3	1.6	4.2	3.8	3.5	3.2			
R-19 Insulation	.046	0.7	0.8	0.9	1.2	3.0	2.8	2.5	2.3			
R-22 Insulation	.041	0.6	0.7	0.8	1.0	2.7	2.5	2.3	2.0			
5. WALLS—(Sq. Ft.) NET		U x TD				U x TD						
No Insulation—Solid Masonry	.389	7.8	8.6	9.7	11.7	25.3	23.3	21.4	19.5			
No Insulation—Wood Siding	.320	6.4	7.0	8.0	9.6	20.8	19.2	17.6	16.0			
No Insulation—Brick Veneer	.240	4.8	5.3	6.0	7.2	15.6	14.4	13.2	12.0			
R-5 Insulation	.128	2.6	2.8	3.2	3.8	8.3	7.7	7.0	6.4			
R-7 Insulation	.109	2.2	2.4	2.7	3.3	7.1	6.5	6.0	5.5			
R-11 Insulation	.075	1.5	1.7	1.9	2.3	4.9	4.5	4.1	3.8			
R-13 Insulation	.065	1.3	1.4	1.6	2.0	4.2	3.9	3.6	3.3			
R-16 Insulation	.054	1.1	1.2	1.4	1.6	3.5	3.2	3.0	2.7			
R-19 Insulation	.047	0.9	1.0	1.2	1.4	3.1	2.8	2.6	2.4			
R-24 Insulation	.038	0.8	0.8	1.0	1.1	2.5	2.3	2.1	1.9			

FIGURE 16–10
Heating and cooling estimate form (residential). (*Courtesy of T U Electric*)

CONSTRUCTION	U	COOLING FACTOR				HEATING FACTOR				QUANTITY	COOLING Btuh	HEATING Btuh
		DESIGN D.B. TEMP. DIFFERENTIAL										
		20	22	25	30	65	60	55	50			
6a. **CEILINGS—WITH ATTIC** (Sq. Ft.) (*Note B)		U x (TD+40)				U x TD						
No Insulation	.598	35.9	37.1	38.9	41.9	38.9	35.9	32.9	29.9			
R-4 Insulation	.176	10.6	10.9	11.4	12.3	11.4	10.6	9.7	8.8			
R-7 Insulation	.114	6.8	7.1	7.4	8.0	7.4	6.8	6.3	5.7			
R-11 Insulation	.079	4.7	4.9	5.1	5.5	5.1	4.7	4.3	4.0			
R-19 Insulation	.048	2.9	3.0	3.1	3.4	3.1	2.9	2.6	2.4			
R-22 Insulation	.042	2.5	2.6	2.7	2.9	2.7	2.5	2.3	2.1			
R-26 Insulation	.036	2.2	2.2	2.3	2.5	2.3	2.2	2.0	1.8			
R-30 Insulation	.032	1.9	2.0	2.1	2.2	2.1	1.9	1.8	1.6			
R-33 Insulation	.029	1.7	1.8	1.9	2.0	1.9	1.7	1.6	1.5			
R-38 Insulation	.025	1.5	1.6	1.6	1.8	1.6	1.5	1.4	1.2			
6b. **CEILINGS—NO ATTIC** (Sq. Ft.) (*Note B)		U x (TD+45)				U x TD						
No Insulation	.470	30.6	31.5	32.9	35.3	30.6	28.2	25.9	23.5			
R-4 Insulation	.160	10.4	10.7	11.2	12.0	10.4	9.6	8.8	8.0			
R-5 Insulation	.130	8.5	8.7	9.1	9.8	8.5	7.8	7.2	6.5			
R-7 Insulation	.109	7.1	7.3	7.6	8.2	7.1	6.5	6.0	5.5			
R-11 Insulation	.076	4.9	5.1	5.3	5.7	4.9	4.6	4.2	3.8			
R-19 Insulation	.047	3.1	3.1	3.3	3.5	3.1	2.8	2.6	2.4			
R-26 Insulation	.035	2.3	2.3	2.4	2.6	2.3	2.1	1.9	1.8			
R-30 Insulation	.031	2.0	2.1	2.2	2.3	2.0	1.9	1.7	1.6			
7. **INFILTRATION:VOL METHOD** (Cu. Ft.) - **SENSIBLE**	$q_s = .018$	$q_s \times w_c \times TD$ (w_c @ 7.5 m/h wind= 0.72)				$q_s \times w_c \times TD$ (w_c @ 15.0 m/h = 1.00)						
Air Changes Per Hour Saved — **Improvement to Structure**												
0.3564 — **Soleplate Sealed**												
0.2808 — **Wiring & Plumbing Holes and Furrdowns Sealed**												
0.0756 — **Exterior Doors & Windows Weather Stripped**												
0.1836 — **Exterior Doors & Windows Rough Opening Caulked**												
0.0216 — **Attic Access outside Conditioned Space or Weather Stripped**												
0.0756 — **Outside Sheathing Holes Sealed and Polyethylene Film Installed**												
0.0864 — **Ventless or Dampered Range Hood Installed (or no Range Vent)**												
1.75 – ______ (*Note C) = ______ Air Changes/hr X		0.26	0.29	0.32	0.39	1.17	1.08	0.99	0.90			
8. **PEOPLE—SENSIBLE HEAT** (Avg. No.)		250										
9. **SUB TOTAL—SENSIBLE HEAT** (Total 1 through 8) Btuh												
10. **TOTAL SENSIBLE—INCLUDING DUCT LOSS** (Multiply Line 9 Sub Total by Duct Factor)						DUCT FACTOR (From Table below)					X ______	X ______
11. **UNITARY COOLING EQUIPMENT REQUIREMENT** (Btuh): Line 10 x 1.25												
12. **PEOPLE—LATENT HEAT** (Avg. No.)		200										
13. **TOTAL LOAD Btuh** (Total 11, 12)												
										(*Note D.)		
											Tons	kW

DUCTWORK LOCATION	INSUL. THICKNESS	COOLING FACTOR	HEATING FACTOR
Attic—vented	1″	1.15	1.25
—vented	2″	1.10	1.15
—unvented	1″	1.20	1.20
—unvented	2″	1.15	1.10
Crawl space—vented	2″	1.10	1.15
—unvented	1″	1.05	1.10
Within conditioned area	1"	1.00	1.00
within slab	0"	1.20	1.25

*Note A. 1. Use only the one largest solar gain.
2. Multiply the factors by shading coeffient (SC) of .90 for plate glass, .85 for double glass.
3. No solar gain is used if overhang exceeds 4', or if glass is shaded by permanent structure.

*Note B. For light colored roof, Multiply COOLING Factor by .75

*Note C. Sum of air changes per hour saved

*Note D. 12,000 Btuh/ton; 3413 Btuh/kW

*Note E. See Ft. Worth H.U.D. Circular Letter No. 87-2. The warrantor of the HVAC equipment accepts the calculated loads as his/her own calculations

FIGURE 16–10a
Heating and cooling estimate form (residential), continued. (*Courtesy of T U Electric*)

The following characteristics apply to the structure. (Factors which apply are circled under the 25° Cooling and 55° Heating Differential columns for items 2, 3, 4a, 5, 6a, and 7 in Figure 16–10.

- Standard glass—single glazing (Item 2)
- Wooden doors (Item 3)
- Slab floor with no insulation (Item 4a)
- Wall insulation of R-16 (Item 5)
- Ceiling with R-30 insulation and attic space above (Item 6a)
- Doors and windows are weatherstripped and the attic access is in the garage (unconditioned space). The range has a dampered range hood.

Step 12. For Item 7, determine the construction features of the house structure that have resulted in reduced infiltration. Identify these on the form by an arrow or by circling the reduction factor. (Most electric houses of the 1960–1975 period had the windows and doors weatherstripped and the attic access located in the garage, outside the conditioned area. A few also had the furrdowns sealed. Since there are no separate data on furrdown infiltration, any improvement to furrdowns is ignored in this particular instance.) After making the air change reduction determinations, add the marked numbers to obtain a total and place it in the blank provided beneath the "Air Change per Hour Saved" column. This figure is subtracted from the 1.75 (the number of air changes per hour in an unimproved structure) to find the total number of air changes per hour. Chart 16–2 indicates the amount of infiltration of unconditioned air and the air changes saved per hour with each method.

Step 13. Determine the quantity of items 2 through 7. Items 2 and 3 are in square feet. Item 4a is in linear feet. Item 7 is in cubic feet. (The units for each quantity are indicated.) Insert these figures on the form under the quantity column and directly across from the item to which they apply (Figure 16–10).

The following calculations were made after reference to the floor plan and to additional information supplied in the example in Step 4.

Item 2, Glass Transmission: This is the total of the glass area found under quantity of solar gain, Item 1 (Figure 16–11).

$$66 + 18 + 23 + 108 = 215 \text{ square feet}$$

Item 3, Doors: The area of the exterior doors is found by totaling the areas of all exterior doors. The area of each door is the length in feet times the width in feet of each door.

$$2\,^1/_2 \times 7 = 18$$
$$3 \times 7 = 21$$
$$39 \text{ square feet}$$

Item 4a, Slab: The quantity is in linear feet of the exposed edge. It is the number of feet measured around the house, but does not include the garage in this example.

A. The Building

1. Location ____*T U ELECTRIC*____ Identification ____*JOHN BROWN*____
2. Building Heat Gain *27574* Btuh Address *IRVING*
3. Building Heat Loss *34920* Btuh Date *11-14-91*
4. Local Power Rates Prepared by *B.J.*

   a. Summer *.075* $/kWh b. Winter *.046* $/kWh

B. System Data

1. Equipment Manufacturer *SEE ARI DIRECTORY*
2. Model Number *SEE ARI DIRECTORY*
3. Unit Cooling Capacity (Btuh) *29000* SEER *11*
4. Unit Heating Capacity at 47° F (Btuh) *30000* HSPF *7.5*
5. Unit Average National Annual Operating Cost (from ARI Directory)

   a. Cooling $ *208* b. Heating $ *505*

ARI METHOD

C. Operating Hours for Cost Factors

1. Operating Hours for Cooling and Heating (see Map Figs. 1 and 2 in ARI Directory)
   a. Summer Cooling Load Hours from ARI Fig. 1 *1600*
   b. Winter Heating Load Hours from ARI Fig. 2 *1400*
2. Climatic Region, Heating from Fig. 2: *II*
3. Select Cooling Cost Factor from Table 1: *0.92*
4. Select Heating Cost Factor from Table 2: *1.036*

D. Estimated Cooling Cost Calculation

Cooling Cost = ((A.4.a) (B.5.a) (C.1.a) (C.3)) / ((1,000 hrs) (*.0804*))*
= ((*.075*) (*208*) (*1600*) (.92)) / ((1,000 hrs) (*.0804*))* = $ *285*

E. Estimated Heating Cost Calculation

Heating Cost = ((A.4.b) (B.5.b) (C.1.b) (C.4)) / ((2,080 hrs) (*.0804*))*
= ((*.046*) (*505*) (*1400*) (*1.030*)) / ((2,080 hrs) (*.0804*))* = $ *201*

* See National Average kWh cost in ARI manual in EnergyGuide Section. Total $ *486*/yr.

DEGREE DAY METHOD

F. Cooling: *27.574* MBtuh / *11* SEER = *2.5* kW x *1600* hours/season = *4010* KWh/season @
(219R line 13)
$*.075*/kWh = $*300*/season

Use 1600 hrs. for 75° maintained

G. HEATING: *34.920* MBtuh x *310* kWh/MBtuh/season / *7.5* HSPF x 3.413 = *4926* kWh/season @
(219R line 13)
$*.046*/kWh = $ *226*/season

If electric furnace, use SPF = 1.0, HSPF = 3.413

If heat pump, use data on the specific unit to be used or, SPF = 2.1, HSPF = 7.167

Total $ *526*/yr

FIGURE 16–11
Annual operating cost estimate of an air conditioner, electric furnace or heat pump. (*Courtesy of T U Electric*)

$$2(37) + 2(54) = 182 \text{ ft.}$$

Item 5, Walls: The net wall area = (total length of the outside wall in feet × the ceiling height in feet) – (door and window area in square feet).
total area = 182 × 8 = 1456 square feet

$$\text{net area} = 1{,}456 - 39 - 215 = 1{,}202 \text{ square feet}$$

Item 6a, Ceiling: The length of the structure in feet times the width of the structure in feet [Figure 16–10(a)].

$$31 \times 29 = 899$$

$$33 \times 23 = 759$$

$$1{,}658 \text{ square feet}$$

Item 7, Infiltration Volume: The area of the structure in square feet times the ceiling height in feet = volume.

$$1{,}658 \times 8 = 13{,}264 \text{ cubic feet}$$

Insert these calculated quantities on the form opposite the corresponding description of the construction. Quantity for Item 2 is shown in Figure 16–10(a). Refer to Figure 16–10 for additional quantities, items 3 through 7.

Step 14. Determine the cooling Btu/h for each Item (2 through 6) by multiplying the circled factor times the quantity and inserting this figure in the cooling Btu/h column. Determine the heating Btu/h for each item (2 through 6) by multiplying the circled heating factor times the quantity and inserting this figure in the heating Btu/h column. Do the same for Item 7, except in addition multiply by the air changes figure. Btu/h calculations are shown in Item 2 and Item 7. Others can be seen in Figure 16–10.

Step 15. Item 8: Determine the average number of persons to occupy the structure, and insert this figure in the quantity column. Multiply 250 times the estimated number of occupants and insert this figure in the cooling Btu/h column.

Step 16. Item 9: Obtain a subtotal for the cooling requirement by totaling items 1 through 8, which appear under the cooling Btu/h column. Be sure to use only the single largest vertical solar gain plus any horizontal solar gain in Item 1. Make a subtotal for the heating Btu/h by totaling items 2 through 7 under the heating Btu/h column.

Step 17. Item 10: From the ductwork location chart in the lower left section on the back of the form [Figure 16–10(a)], determine the heating and cooling duct loss factors for the appropriate location and insulation level of the duct system. Circle the duct loss factors. In the example we are referring to the conditioned air supply ductwork, not the return air ductwork. The factors are 1.10 and 1.15, respectively, for a vented attic and duct insulation of 2-inch thickness.

Step 18. Item 10: Place the duct loss factors in the appropriate blanks in line 10. Determine the total sensible cooling Btu/h of the structure by multiplying line

9, cooling Btu/h, by the cooling duct loss factor. Determine the total sensible heating Btu/h of the structure by multiplying line 9, heating Btu/h, by the heating duct loss factor.

Step 19. Item 11: Multiply the sensible cooling Btu/h in Item 10 by 1.25, the latent factor, to obtain the structure cooling Btu/h.

Step 20. Item 12: Insert the average number of people under the quantity column. It is the same as in Item 8. Multiply this figure by 200 to obtain the latent cooling Btu/h.

Step 21. Item 13: The total cooling Btu/h is obtained by adding the cooling Btu/h in items 11 and 12. The total heating Btu/h is the same as that shown in Item 10.

Step 22. Tons of cooling is found by dividing the cooling Btu/h in Item 13 by 12,000 (see Note D).

Step 23. The kilowatt capacity of electric heating is found by dividing the heating Btu/h in Item 13 by 3,413 (see Note G).

DESIGN DRY BULB TEMPERATURES OTHER THAN THOSE SHOWN

It is possible to make a heating and cooling cost estimate for design dry bulb temperatures other than those shown on the form (Figure 16–10). The procedure is to calculate the heating and cooling factors from the formulas which apply to the construction materials involved. An overall conductance factor (U) is shown for each of these construction sections, and it is used along with the desired design temperature differentials (TD) in the appropriate formula. For example, the U for single glazed glass is 1.13, and the formula is $U \times TD$. For a cooling design temperature difference of 20° F, the cooling factor would be 1.13×20, or 22.6. The heating factor for a heating design temperature difference of 64° F would be 1.13×64, or 72.3. Factors for other materials may be found in a similar manner. Heating and cooling estimate calculations otherwise are performed in the manner previously described.

Figure 16–10 shows the cooling and heating costs which result from using energy-efficient methods which met the more energy-efficient construction standards. A heat pump was used for the heating and cooling.

Since insulated glass was used rather than standard window glass, it is necessary to use a shading coefficient multiplying factor of 0.85 for the solar gain calculations in Item 1 on the form. [See Note A in Figure 16–10(a).] For the west solar gain calculations, $80 \times 108 \times 0.85 = 7{,}344$.

OPERATING COST ESTIMATE INSTRUCTIONS FOR USING THE 219-OC DEGREE DAY METHOD

Formulas are provided for making cooling and heating cost estimates on the Form 219-OC (Figure 16–12). In the top formula (for cooling calculations), the MBtu/h is the cooling Btu/h from Item 13 on the Form 219R (Figure 16–10) divided by 1,000. (Move the decimal point in the Btu/h figure three places to the left.)

The SEER is determined from the design or nameplate information for the equipment selected. (Refer to the example in Step 4 for the design conditions. Equipment with a SEER of 11 was selected.)

Divide the MBtu/h by the SEER to obtain the kW requirement, and place it in the next blank.

The cooling full load equivalent hours of operation are determined by the temperature maintained (1,600 hours for 75° F indoor temperature).

A. The Building

1. Location TU ELECTRIC — Identification JOHN BROWN
2. Building Heat Gain 42502 Btuh — Address IRVING
3. Building Heat Loss 604049 Btuh — Date 11-14-91
4. Local Power Rates — Prepared by B.J.

 a. Summer .075 $/kWh b. Winter .046 $/kWh

B. System Data

1. Equipment Manufacturer SEE ARI DIRECTORY
2. Model Number SEE ARI DIRECTORY
3. Unit Cooling Capacity (Btuh) 43500 SEER 11
4. Unit Heating Capacity at 47° F (Btuh) 40000 HSPF 7.7
5. Unit Average National Annual Operating Cost (from ARI Directory)

 a. Cooling $ 310 b. Heating $ 656

ARI METHOD

C. Operating Hours for Cost Factors

1. Operating Hours for Cooing and Heating (see Map Figs. 1 and 2 in ARI Directory)
 a. Summer Cooling Load Hours from ARI Fig. 1 1600
 b. Winter Heating Load from ARI Fig. 2 1400
2. Climatic Region, Heating from Fig. 2: II
3. Select Cooling Cost Factor from Table 1: 1.0
4. Select Heating Cost Factor from Table 2: 1.38

D. Estimated Cooling Cost Calculation

Cooling Cost = ((A.4.a) (B.5.a) (C.1.a) (C.3)) / ((1,000 hrs) (.0804))*
= ((.075) (310) (1600) (1.0)) / ((1,000 hrs) (.0804))* = $ 462

E. Estimated Heating Cost Calculation

Heating Cost = ((A.4.b) (B.5.b) (C.1.b) (C.4)) / ((2,080 hrs) (.0804))*
= ((.046) (656) (1400) (1.38)) / ((2,080 hrs) (.0804))* = $ 348

* See National Average kWh Cost in ARI manual in EnergyGuide Section. Total $ 810 /yr.

FIGURE 16–12
Annual operating cost estimate of an air conditioner, electric furnace or heat pump. (*Courtesy of T U Electric*)

The kW figure times the hours of operation gives the cooling kWh/season. This time the cost per kWh gives the cost per season.

The cost of seasonal heating is found by making calculations using the bottom formula on Form 219-OC (Figure 16–12).

The heating Btu/h shown on Form 219-R (Figure 16–10) is divided by 1,000 to obtain the MBtu/h figure. The next figure (kWh/MBtu/h) is found in the T U Electric Weather Chart (Chart 16–3). This figure is found under the Annual Heating

	Outside Temperature Design (°F)*		Heating Degree Days	Annual Heating Consumption
	Summer	Winter	(65° Base)**	Factor***
DALLAS DIV. FT. WORTH & CENTRAL REGION: Arlingon, Dallas, Duncanville, Euless, Ft. Worth, Garland, Grand Prairie, Grapevine, Farmers Branch, Irving, Lancaster, Mesquite, Plano, Richardson	100	20	2,400	310
EASTERN REGION: Athens, Corsicana, Crockett, Lufkin Nacogdoches, Palestine, Terrell, Tyler	100	25	2,300	330
NORTHERN REGION: Decatur, Denison, Gainesville, McKinney, Mineral Wells, Paris, Sherman, Sulphur Springs	100	20	2,750	344
NORTHWEST REGION: Archer City, Breckenridge, Burkburnett, DeLeon, Eastland, Electra, Graham, Henrietta, Iowa Park, Wichita Falls	100	20	2,875	355
SOUTHERN REGION: Brownwood, Cleburne, Hillsboro, Killeen, Round Rock, Stephenville, Taylor, Temple, Waco, Waxahachie	100	25	2,225	321
WESTERN REGION: Andrews, Big Spring, Colorado City, Crane, Lamesa, Midland, Monahans, Odessa, Snyder, Sweetwater	100	20	2,750	344

* Generalized from ASHRAE Fundamentals Handbook 1985, Chapter 24

** Accuracy ±10%

*** Factors for maintaining 75° indoors at design conditions. Units = kWh/MBtuh

CHART 16–3
T U Electric weather chart. (*Courtesy of T U Electric*)

Consumption column. (Irving, for example, has a heating factor of 310 for a room temperature of 75° F.)

The HSPF or Heating Seasonal Performance Factor is 3.413 (100% efficiency) for an electric furnace or about 7.17 for a heat pump. Use the actual HSPF data for the unit if it is known. If an SPF is given, multiply it by 3.413 to obtain the HSPF.

A sample calculation using the formula is shown on Form 219-OC (Figure 16–13).

DEGREE DAY METHOD

F. Cooling: 27.574 MBtuh / 11 SEER = 2.5 kW x 1600 hours/season = 4010 kWh/season @
(219R line 13)
$.075/kWh = $300/season

Use 1600 hrs. for 75° maintained

G. HEATING: 34.920 MBtuh x 310 kWh/MBtuh/season / 7.5 HSPF x 3.413 = 4926kWh/season @
(219R line 13)
$.046/kWh = $ 226/season

If electric furnace, use SPF = 1.0, HSPF = 3.413

If heat pump, use data on the specific unit to be used or, SPF = 2.1, HSPF = 7.167

Total $ 526/yr

FIGURE 16.13 Operating cost estimate using Degree Day Method. (Courtesy of T U Electric.)

ARI GUIDE FOR ESTIMATING ANNUAL OPERATING COST OF A CENTRAL AIR CONDITIONER OR HEAT PUMP

This guide is designed to assist customers in estimating their annual operating costs of central air conditioners and heat pumps covered by the U.S. Federal Trade Commission (FTC) appliance labeling rules. Along with the energy efficiency information, this guide is designed to assist customers in making purchasing decisions. The guide contains step-by-step instructions on how to perform the operating cost estimates. A sample worksheet is also provided. This directory is mailed free on a limited basis upon written request on business letterhead to ARI from an individual representing the air conditioning or allied industries. A directory will be sent automatically, including a single copy of a monthly supplement, unless ARI is notified to stop or the post office advises that the new address is no longer deliverable. For all other requests, the price per copy is $10.00; request an order form for a discounted price schedule for multiple copies.

In the following text, energy efficiency is expressed in three ways:

SEER: Seasonal Energy Efficiency Ratio (for cooling).

HSPF: Heating Seasonal Performance Factor (for heating).

EER: Energy Efficiency Rating (a term used by the Federal Trade Commission to mean either SEER or HSPF, whichever is applicable).

To assist customers in making informed decisions regarding equipment selection, the FTC has determined that the minimum and maximum product energy efficiency ratings available are as follows:

Description		Energy Efficiency (EER) Ranges			
		Packaged Units		Split Systems	
		Minimum	Maximum	Minimum	Maximum
AIR CONDITIONER	SEER	5.60	10.20	5.85	15.00
HEAT PUMP	SEER	6.50	10.50	7.25	13.05
HEAT PUMP	HSPF	5.05	7.80	5.30	8.90

Performance data for central air conditioners are shown in Section AC of the directory. The cooling capacity and the SEER are listed for each single-package air conditioner and for each split-system combination of condensing (outdoor) unit and indoor coil.

Performance data for heat pumps are shown in Section HP of the directory. Capacities and efficiencies for cooling and heating are listed for each single-package heat pump and for each split-system combination of outdoor and indoor coil. Cooling and heating efficiencies are expressed in SEER and HSPF, respectively.

The directory lists the average national annual operating cost for each air conditioner in Section AC. It also lists the average national annual operating cost for cooling and heating for each heat pump in Section HP.

Estimates of operating costs may be either higher or lower than your average operating costs. They are affected by many factors, which can vary widely. For example, since no two heating or cooling seasons are identical, operating costs will vary from year to year. Operating costs are also affected by the temperature that is to be maintained (thermostat setting), with higher settings costing more in winter and lower settings costing more in summer. Other factors that affect system operation include the number of occupants, location within a region, activities that generate or release heat within the structure, and living habits such as the opening of windows, etc. Nevertheless, the estimates will be helpful in determining approximately how much a system will cost to operate and in comparing the performance of different systems.

The average annual operating costs listed in the *Directory* are based on U.S. government standard tests and national averages of 1,000 cooling load hours, 2,080 heat load hours, and national average electric rates of 8.04 cents per kilowatt hour. The step-by-step procedure calls for your local heating and cooling load hours and electric rates.

Step-by-Step Procedure

The following is a step-by-step procedure for estimating your annual operating costs for cooling with a central air conditioner or heat pump and for heating when using a heat pump. Enter the required information on the worksheet in Figure 16–12. It will be helpful to look at the sample worksheet while you read the procedure.

A. The building (home). Enter the identified information on the worksheet line indicated.

1. Location: The city and state in which the building is located.
2. Building heat gain: The calculated rate of heat flow from the environment, including heat gain from the sun, into the building at summer design conditions (sometimes called cooling load), expressed in Btu per hour (Btu/h). Building heat gain should be calculated by an engineer, contractor, or dealer.
3. Building heat loss: The calculated rate of heat flow from the building at winter design conditions, expressed in Btu per hour (Btu/h). Building heat loss should be calculated by an engineer, contractor, or dealer.
4. Local power costs: Dollars per kilowatt hour ($/kWh)
   a. Summer rate
   b. Winter rate

Local power costs are available from your local power company or from your electric bill.

B. System data: Locate the information requested in Section AC of this directory for air conditioners and Section HP for heat pumps.

1. Equipment manufacturer's name.
2. Model number. Include both condensing unit and coil for split-system air conditioners, or outdoor unit and indoor unit for split-system heat pumps.
3. Unit cooling capacity, Btu/h. Btu/h = directory cooling capacity in MBtu/h × 1,000.
4. Unit heating capacity at 47° F for heat pump. Btu/h = directory heating capacity in MBtu/h × 1,000.
5. Unit average national operating costs, dollars.
   a. Cooling
   b. Heating

C. Operating hours and cost factors: Locate the required information in the figures and tables in this Guide Section of the *Directory*.

1. Operating hours for your geographic location for cooling and heating.
   a. Summer cooling load hours from map, Figure 16–14.
   b. Winter heating load hours from map, Figure 16–15.
2. Climatic region: Heating from map, Figure 16–15 (Roman numerals).
3. Select cooling cost factor from Table 16–1, as follows:
   a. In the Unit Capacity column, locate the row with the capacity nearest your air conditioner or heat pump cooling capacity (line B.3 of worksheet). In

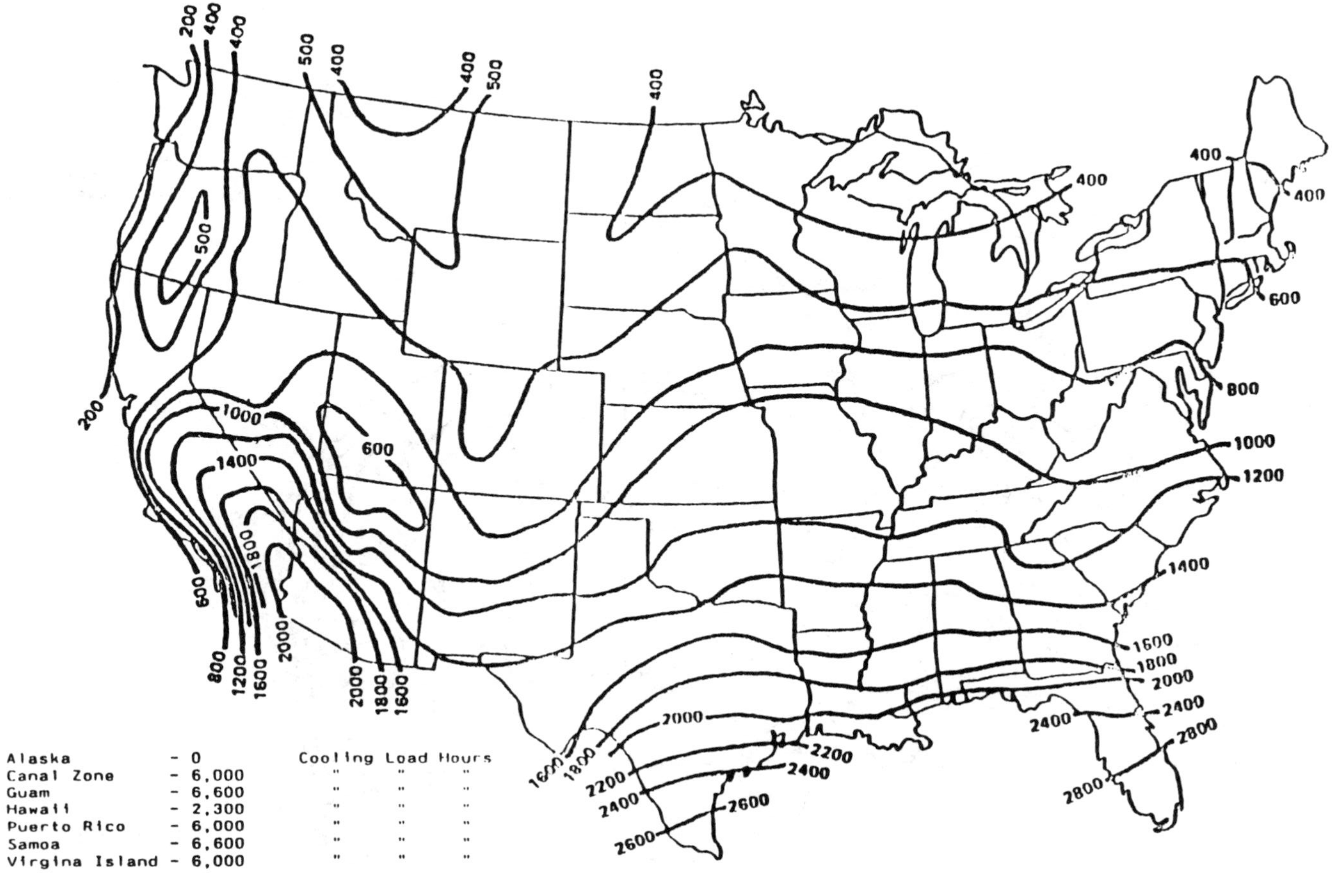
Alaska - 0
Canal Zone - 6,000
Guam - 6,600
Hawaii - 2,300
Puerto Rico - 6,000
Samoa - 6,600
Virgina Island - 6,000
Cooling Load Hours

FIGURE 16–14
Summer cooling load hours. (*Courtesy of Air Conditioning and Refrigeration Institute*)

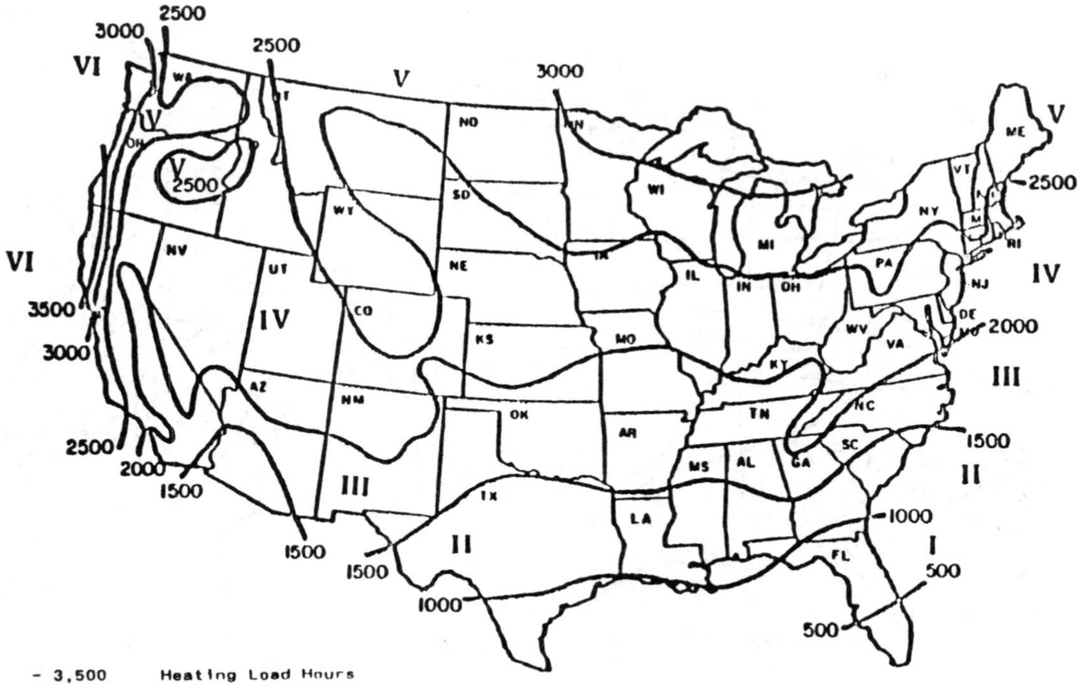

VI
V
IV
III
II
I
3500
3000
2500
2000
1500
1000
500
WA
OR
ID
MT
ND
SD
NE
KS
OK
TX
WY
CO
NM
UT
NV
AZ
MN
IA
MO
AR
LA
WI
IL
MI
IN
OH
KY
TN
MS
AL
GA
FL
SC
NC
VA
WV
PA
NY
VT
ME
NJ
DE
RI

Alaska - 3,500 Heating Load Hours
Hawaii and Territories - 0 " " "

FIGURE 16–15
Winter heating load hours. (*Courtesy of Air Conditioning and Refrigeration Institute*)

TABLE 16–1 Cooling cost factor. (Courtesy of Air Conditioning and Refrigeration Institute)

	COOLING COST FACTOR			
RANGE OF COMPARABILITY Btu/hr	UNIT CAPACITY Btu/hr	BUILDING HEAT GAIN Btu/hr		
7.500-12.499		7.500	10.000	12.500
	7.500	1.00	1.33	1.67
	10.000	0.75	1.00	1.25
	12.499	0.50	0.80	1.00
12.500-17.499		12.500	15.000	17.500
	12.500	1.00	1.20	1.40
	15.000	0.83	1.00	1.17
	17.499	0.71	0.86	1.00
17.500-22.499		17.500	20.000	22.500
	17.500	1.00	1.14	1.29
	22.000	0.86	1.00	1.13
	27.499	0.78	0.89	1.00
22.500-27.499		22.500	25.000	27.500
	22.500	1.00	1.11	1.22
	25.000	0.90	1.00	1.10
	27.499	0.82	0.91	1.00
27.500-32.499		27.500	30.000	32.500
	27.500	1.00	1.09	1.18
	30.000	0.92	1.00	1.08
	32.499	0.85	0.92	1.00
32.500-37.499		32.500	35.000	37.500
	32.500	1.00	1.08	1.15
	35.000	0.83	1.00	1.07
	37.500	0.87	0.93	1.00
37.500-44.999		37.500	41.250	45.000
	37.500	1.00	1.10	1.20
	41.250	0.91	1.00	1.09
	45.000	0.83	0.92	1.00
45.000-49.499		45.000	47.500	50.000
	45.000	1.00	1.06	1.11
	47.500	0.85	1.00	1.05
	49.999	0.90	0.95	1.00
50.000-54.999		50.000	52.500	55.000
	50.000	1.00	1.05	1.10
	52.500	0.95	1.00	1.05
	54.999	0.90	0.95	1.00
55.000-59.999		55.000	57.500	60.000
	55.000	1.00	1.05	1.09
	57.500	0.96	1.00	1.04
	59.999	0.92	0.96	1.00
60.000-64.999		60.000	62.500	65.000
	60.000	1.00	1.04	1.08
	62.500	0.96	1.00	1.04
	65.000	0.82	0.96	1.00

the same row, select the cooling cost factor in the column with the heat gain nearest your building heat gain (line A.2 of worksheet). Enter the cooling cost factor on line C.3.a of worksheet.

4. Select heating cost factor from Table 16–2 as follows:

a. Identify range of comparability, first column, that includes heat pump heating capacity (line B.4 of worksheet). Within that range, identify climatic region (line C.2 of worksheet). Select heating cost factor in column headed by the value nearest your building heat loss (line A.3 of worksheet). Enter on line C.4.a of worksheet.

D. Calculate estimated operating cost for cooling using the following equation:

$$\text{cooling cost} = [(\text{A.4.a})\ (\text{B.5.a})\ (\text{C.3.a})] \div [(1000)\ (0.0804)]$$

where

A.4.a = line A.4.a = local power cost, summer \$/kWh
B.5.a = line B.5.a = unit average national annual cost, cooling, \$
C.1.a = line C.1.a = summer cooling load hours
C.3.a = line C.3.a = cooling cost factor
1,000 = average national cooling load hours
0.0804 = national average power cost, \$/kWh.

E. Calculate estimated operating cost for heating using the following equation:

$$\text{heating cost} = [(\text{A.4.b})\ (\text{B.5.a})\ (\text{C.1.a})\ (\text{C.4.a})] \div [(2{,}080)\ (0.0804)]$$

where

A.4.b = line A.4.b = local power cost, winter \$/kWh
B.5.b = line B.5.b = unit average national annual cost heating, \$
C.1.b = line C.1.b = average national heating load hours
C.4.a = line C.4.a = heating cost factor
2,080 = average national heating load hours, climate Region IV
0.0804 = national average power cost, \$/kWh.

F. Calculate estimated total operating cost for cooling and heating.

$$\text{total operating cost} = \text{cooling cost} + \text{heating cost}$$

GAMA PROCEDURE FOR ESTIMATING ANNUAL HEATING REQUIREMENTS AND COMPARING THE COSTS OF OPERATION

To estimate the amount of fuel a home will use to provide comfort heating for a year, it is necessary to use both (a) information specific to the particular home's need for heat, and (b) the size and efficiency information on models listed in the manual.

This section presents a process (based on the Department of Energy efficiency test procedure for furnaces and boilers) for first estimating the annual amount of fuel used for heating a specific installation and then for making a comparison of the estimated operating costs of operation of models of essentially the same size but different efficiencies. Because of the number of variables involved in this procedure, the process shown here is only for the purpose of (1) estimating the amount of fuel that will be consumed in one year for the specific installation, and (2) comparing models of various Annual Fuel Utilization Efficiencies (AFUEs) that can assist in making a purchasing decision. The method can be used for comparing forced-air furnaces or boilers using the same fuel (gas to gas) or different fuels (gas to oil).

The procedure is based on the Department of Energy test procedure for estimating the annual operating cost of gas- or oil-fired furnaces or boilers. It includes the cost of annual fuel usage (natural gas, propane gas, or Number 2 heating oil), in addition to the annual electrical cost to operate furnace circulating blowers, pumps, and power burners.

When selecting models for comparison, make sure the units are the same configurations needed for the particular installation (i.e., upflow models, etc.). A worksheet (Figure 16–16) is provided for ease in compiling and calculating the information needed. The steps required to complete the worksheet are as follows:

Step 1. Determine the heating load hours (HLHs) for your area. To find an approximate number of heating load hours for your specific area, use the map illustrated in Figure 16–15.

heating load hours, HLH____________hours

Step 2. Estimate the design heating requirement (DHR) of the specific installation. Because the design heating requirement is dependent on a number of variables (such as size of the house, building materials, insulation, architectural features, special climatic conditions, etc.), this determination must usually be performed by a knowledgeable local heating contractor or other qualified source.

design heating requirement, (DHR) ________ Btu/h

Step 3. Identify the size of model needed to satisfy the design heating requirements determined above. Models listed in the *Directory* of essentially the same input can now be evaluated for an estimate of an annual operating cost. Models of the same fuel type (gas to gas) as well as different fuel types (gas to oil) can be compared using this procedure.

input ________ Btu/h

heating capacity ________ Btu/h

Annual Fuel Utilization Efficiency (AFUE) ________ %

Average annual fuel consumption, *Ef* ________ (Average annual fuel energy consumption) *MMBTu* (millions of Btu per year)

Average annual electrical consumption, *Eae* ________ (Average annual auxiliary electrical consumption) kWh

General instructions: After determining size of unit needed, refer to the listing of models of essentially the same size and select models for comparison. Fill in blocks with the listing information and perform determination and calculation requirements to obtain the Estimated Cost in Step 6.

	MODEL 1	MODEL 2
Brand: Model Number: Input Rating: AFUE:	GAMA GAMA 60,000 78.0	GAMA GAMA 60,000 85.0
STEP 1. Heating Load Hours (from Figure 1)	1500	1500
STEP 2. Design Heating Requirements (DHR) divided by 1000 (Btu/hr)	34.920	34.920
STEP 3. Input Rating Btu/hr (from Directory)	60,000	60,000
STEP 4. Calculate Operating Hours of models (OH) OH = .77 x Heating Load Hours (HLH) x Design Heating Requirements (DHR) x A where .77 = adjustment factor HLH = Heating Load Hours (Step 1) DHR = Design Heating Requirement (Step 2) $A = \frac{100{,}000}{\text{Input (Btu/hr)} \times (\text{AFUE})}$ Input (Btu/hr) = Model Input from Chapter 1 AFUE = Annual Fuel Utilization Efficiency from Chapter 1 (use exact number as published)	OH = 0.77 x 1500 x 35 x .0214 = 865.1 $A = \frac{100{,}000}{60{,}000 \times 78}$ = .0214	OH = 0.77 x 1500 x 35 x .0196 = 792.3 $A = \frac{100{,}000}{60{,}000 \times 85}$ = .0196
STEP 5. Estimated Annual Energy Consumption (EAEC) Btu EAEC = Input (Btu/hr) (From Step 3) x Operating Hours (From Step 4)	EAEC = 60,000 x 865.1 = 51,906,000	EAEC = 60,000 x 792.3 = 47,538,000
STEP 6. Estimated Annual Operating Cost (EAOC) EAOC = Estimated Annual Energy Consumption x $\frac{1}{\text{Btu Content of Fuel *}}$ x Fuel Cost where EAEC = In Btu's from Step 5. Btu Content of Fuel * (see footnote) Fuel Cost = Natural Gas - in ¢ per therm, Propane Gas - in ¢ per gallon, and Heating Oil (No. 2) - in $ per gallon	$\frac{51{,}906{,}000}{100{,}000}$ x 52.6 = 273 $273	$\frac{47{,}538{,}000}{100{,}000}$ x 52.6 = 250 $250

*Btu Content of Fuel is as follows:

Natural Gas - 1 therm	= 100,000 Btu's
Propane Gas - 1 gallon	= 91,000 Btu's
Heating Oil (No. 2) - 1gallon	= 138,700 Btu's

FIGURE 16.16 Worksheet. (Courtesy of Appliance Manufacturer's Association)

Step 4. Determine the rated design heating requirement, RDHR, for the selected model. The RDHR for an appliance is the DHR value which was used to calculate the Ef and Eae ratings included in the directory. Using the heating capacity from Step 3, read the RDHR from Table 16–2.

RDHR (rated design heating requirement) ________ *Btu/h*

Step 5. Calculate the adjustment factor, AF, which is required to correct the energy usage figures from the *Directory* for your specific installation.

$$AF = (HLH \times DHR) \div (2{,}080 \times RDHR)$$

where

HLH = heating load hours from Step 1

DHR = design heating requirements from Step 2

$2{,}080$ = average annual heating load hours

$RDHR$ = rated design heating requirement from Step 4.

Step 6. Calculate the estimated annual fuel usage, EAFU, for your specific installation.

$$EAFU = AF \times Ef$$

where

AF = adjustment factor from Step 5

Ef = average annual fuel consumption from Step 3

EAFU ________ *MMBtu.*

Step 7. Calculate the estimated annual electrical usage, EAEU, for your specific installation.

$$\text{EAEU} = \text{AF} \times \text{Eae}$$

where

AF = adjustment factor from Step 5

Eae = average electrical consumption from Step 3

EAEU ________ *kWh.*

Step 8. Calculate the estimated annual operating cost, EAOC, for your specific installation and selected model.

	U	DESIGN D.B. TEMP. DIFFERENTIAL										
		20	22	25	30	65	60	55	50			
2. **GLASS TRANSMISSION** (Sq. Ft.)		U x TD				U x TD						
Standard—Single Glazing	1.13	22.6	24.9	28.2	33.9	73.4	67.8	62.2	56.5			
Insulating—Double Glazing	.78	15.6	17.2	19.5	23.4	50.7	46.8	42.9	39.0	215		
Storm Window	.67	13.4	14.7	16.8	20.1	43.6	40.2	36.8	33.5			
Storm or Insulating Glass with Thermal Break	.56	11.2	12.3	14.0	16.8.	36.4	33.6	30.8	28.0			

FIGURE 16–17
Entering total glass values on form. (*Courtesy of T U Electric*)

TABLE 16–2 Heating cost factor. (Courtesy of Air Conditioning and Refrigeration Institute.)

Range Of Comparability Btu/hr		HEATING COST FACTOR Building Heat Loss M5TU/HR (MBtu/hr Times 1000 * BTU/HR) 5	10	15	20	25	30	35	40	50	60	70	80	90	100	110	130
UP TO 12.499	REGION I	0.417	0.791	1.708													
	REGION II	0.444	0.836	1.286	1.785												
	REGION III	0.474	0.893	1.330	1.856	2.452											
	REGION IV		1.000	1.501	2.086	2.768	3.511										
	REGION V		1.166	1.747	2.389	3.098	3.862										
	REGION VI	0.432	0.814	1.203	1.679												
12.500-17.499	REGION I	0.283	0.558	0.820	1.113												
	REGION II	0.311	0.590	0.859	1.158	1.506											
	REGION III		0.624	0.907	1.210	1.557	1.853										
	REGION IV		0.682	1.000	1.343	1.729	2.163	2.636	3.137								
	REGION V		0.780	1.152	1.545	1.971	2.431	2.924	3.438								
	REGION VI	0.303	0.577	0.836	1.111	1.428	1.785	2.172									
17.500-22.499	REGION I		0.425	0.617	0.817	1.039											
	REGION II		0.450	0.652	0.856	1.086	1.350										
	REGION III			0.690	0.905	1.136	1.398	1.693	2.023								
	REGION IV			0.755	1.000	1.266	1.561	1.888	2.249	3.037							
	REGION V			0.869	1.159	1.467	1.797	2.151	2.529	3.342	4.205						
	REGION VI		0.440	0.637	0.831	1.038	1.274	1.543	1.846								
22.500-27.499	REGION I		0.347	0.505	0.661	0.825	1.006										
	REGION II			0.534	0.695	0.862	1.048	1.256	1.486								
	REGION III			0.564	0.736	0.909	1.095	1.300	1.527	2.046							
	REGION IV				0.803	1.000	1.211	1.441	1.690	2.257	2.883						
	REGION V				0.921	1.152	1.394	1.650	1.923	2.514	3.157	3.834	4.532				
	REGION VI		0.523	0.680	0.837	1.005	1.193	1.400	1.879	2.374							
27.500-32.499	REGION I			0.435	0.567	0.701	0.842										
	REGION II			0.461	0.599	0.735	0.878	1.036	1.207								
	REGION III			0.484	0.631	0.775	0.922	1.078	1.246	1.635							
	REGION IV				0.677	0.836	1.000	1.175	1.363	1.788	2.276						
	REGION V				0.761	0.946	1.138	1.338	1.549	2.007	2.512	3.056	3.623				
	REGION VI			0.451	0.587	0.719	0.853	0.997	1.156	1.512	1.919						

32.500-37.499	REGION I	0.371	0.485	0.595	0.708	0.827									
	REGION II		0.515	0.630	0.746	0.868	0.998	1.291							
	REGION III		0.544	0.667	0.790	0.915	1.045	1.335	1.670						
	REGION IV			0.723	0.859	1.000	1.147	1.470	1.835	2.244	2.675	3.133			
	REGION V			0.822	0.981	1.145	1.314	1.672	2.062	2.485	2.929	3.394			
	REGION VI		0.503	0.617	0.728	0.841	0.962	1.230	1.532						
37.500-44.999	REGION I	0.328	0.429	0.528	0.625	0.725	0.831								
	REGION II			0.560	0.661	0.763	0.870	1.107	1.380						
	REGION III				0.699	0.807	0.917	1.153	1.421						
	REGION IV					0.876	1.000	1.266	1.563	1.898	2.269	2.662	3.077		
	REGION V					0.995	1.144	1.450	1.779	2.136	2.521	2.824	3.349	3.784	
	REGION VI			0.548	0.647	0.746	0.845	1.061	1.310	1.590	1.902				
45.000-49.999	REGION I		0.349	0.429	0.508	0.587	0.669								
	REGION II			0.456	0.538	0.620	0.703	0.881	1.086						
	REGION III				0.568	0.654	0.741	0.821	1.122	1.348	1.600				
	REGION IV						0.799	1.000	1.222	1.470	1.745	2.045	2.360	2.691	
	REGION V						0.907	1.143	1.382	1.661	1.852	2.261	2.586	2.925	3.627
	REGION VI			0.446	0.527	0.607	0.687	0.853	1.038	1.248	1.477	1.726	1.978	2.232	2.808
50.000-54.999	REGION I		0.353	0.435	0.515	0.594	0.673	0.840							
	REGION II				0.539	0.621	0.703	0.872	1.059	1.267					
	REGION III					0.652	0.739	0.915	1.102	1.307	1.543	1.782			
	REGION IV							1.000	2.215	1.449	1.709	1.984	2.292	2.607	
	REGION V							1.156	1.408	1.671	1.957	2.254	2.573	2.900	3.587
	REGION VI			0.530	0.610	0.690	0.649	1.021	1.209	1.419	1.646	1.903	2.162	2.682	
55.000-59.999	REGION I			0.366	0.433	0.499	0.565	0.699							
	REGION II				0.526	0.594	0.731	0.877	1.038						
	REGION III					0.626	0.770	0.917	1.O76	1.254	1.447				
	REGION IV						0.634	1.000	1.179	1.374	1.585	1.817	2.060		
	REGION V						0.950	1.145	1.348	1.565	1.794	2.037	2.291	2.833	
	REGION VI				0.516	0.583	0.715	0.851	0.999	1.161	1.344	1.542			
60.000-64.999	REGION I			0.370		0.438	0.505	0.572	0.706	0.849					
	REGION II					0.533	0.602	0.739	0.884	1.042	1.219				
	REGION III							0.777	0.926	1.087	1.254	1.447	1.656		
	REGION IV								1.000	1.177	1.371	1.581	1.808	2.055	2.590
	REGION V								1.140	1.345	1.563	1.792	2.037	2.293	2.841
	REGION VI					0.523	0.591	0.725	0.859	1.005	1.164	1.341	1.534		

$$EAOC = (EAFU \times 1{,}000{,}000 \times fuel\ cost) \div Btu\ content + (EAEU \times electrical\ cost)$$

where

EAFU = estimated annual fuel usage in MMBtu from Step 6

1,000,000 = conversion factor for MMBtu to Btu

fuel cost = cost of fuel in your area in:

$ per therm (100,000 Btu) for natural gas

$ per gallon for propane gas

$ per gallon for heating oil

Btu content = 100,000 Btu per therm for natural gas

91,000 Btu per gallon for propane gas

138,000 Btu per gallon for heating oil

EAEU = estimated annual electrical usage in kWh from Step 7

electrical cost = cost of electricity in your area in $ per kWh

EAOC, $________ per year.

This procedure can be repeated for other models of essentially the same size with different efficiencies to compare their estimated annual operating costs. Models with higher efficiency (AFUE) ratings will consume less fuel and cost less to operate, but generally have a higher purchase price and may have a higher installation cost. Therefore, there is a period of time referred to as the *payback period*, before the savings that result from the lower operating costs of a more efficient model make up the difference in price of that furnace or boiler as compared to a less efficient model.

Example:

Assume you are intending to buy a new gas furnace and have calculated the estimated annual operating cost (AEOC) using the foregoing procedure for two models of essentially the same size to meet your heating requirements but with different efficiency (AFUE) ratings.

Price of Furnace	AFUE	Estimated Annual Operating Cost
Model A, $600 installed	78%	$425
Model B, $775 installed	91%	$365

The additional cost of the more efficient model (Model B) is $775 – 600 = $175 higher installation cost.

The estimated annual savings in operating costs for Model B as compared to Model A are

$$\$425 - 365 = \$60 \text{ lower operating cost per year.}$$

The payback period is the ratio of the higher installed cost to the lower annual operating cost:

$$\text{payback period} = \$175 \div \$60 = 2.9 \text{ years.}$$

Definitions. The following are some definitions of the terms used in this procedure:

1. Input, MBtu/h: This figure represents the amount of fuel that the model consumes in one hour. It defines the rate of energy supplied in a fuel to a furnace or boiler when operating under continuous burning (steady-state) conditions.

a. MBtu/h: Stands for thousands of British thermal units used in one hour and is a term used to measure energy; for example, a few hundred Btu provide enough energy to make a pot of coffee. Some larger homes may require as much as 80,000 to 90,000 Btu an hour for heating on very cold days.

b. Steady-State Conditions: Conditions of continuous burner operation during which fuel consumption of the furnace or boiler is measured, somewhat like measuring your car's gasoline mileage under steady highway driving conditions.

2. Heating Capacity, MBtu/h: This figure tells how much heat the model can produce in one hour operating under steady-state conditions expressed in thousands of Btu per hour. For isolated combustion or outdoor units, the heating capacity is determined by multiplying the specified input by the steady-state efficiency, as tested, and subtracting an additional therm to account for jacket loss which would go into an unheated environment.
3. PE Watts: This is the electrical energy input rate supplied to the power burner (combustion air blower, fuel pump, damper motor) of a furnace or boiler operating under continuous burning (steady-state) conditions.
4. Eae, Kwh/yr: This is the average annual auxiliary electrical energy consumption for a gas furnace or boiler in kilowatt hours per year. It is a measure of the total electrical energy supplied to a furnace or boiler during a one-year period.
5. Ef, MMBtu/yr: This is the average annual fuel energy consumption for a gas furnace or boiler. It is shown in millions of Btu per year.
6. AFUE, %: AFUE stands for annual fuel utilization efficiency and is the efficiency rating of the model shown. Unlike steady-state conditions, this rating is based on average usage, including on and off cycling, as set out in the

standardized Department of Energy test procedures. Remember, the higher the AFUE rating, the more efficient the model will be.

7. Primary heat exchanger material (boilers only).

Note: Items 2 and 6 are certified values, while items 3, 4, and 5 are provided for application use. Eae and Ef are based on national averages. For the purpose of comparison, use the calculation procedure in this section.

17

troubleshooting charts

ELECTRIC HEAT

Condition	Possible Cause	Corrective Action
Unit will not run	1. Blown fuse 2. Burned transformer 3. Thermostat not calling for heat 4. Defective thermostat 5. Defective heating relay	1. Replace fuse and correct cause 2. Replace transformer and correct cause 3. Set thermostat 4. Replace thermostat 5. Replace relay
Fan will not run	1. Burned fan motor 2. Broken fan belt 3. Burned contacts in fan relay 4. Defective fan control 5. Defective wiring connections	1. Repair or replace fan motor 2. Replace fan belt 3. Replace fan relay 4. Replace fan control 5. Repair wiring or connections
Fan motor hums but will not start	1. Defective fan motor bearings 2. Defective fan motor starting switch 3. Defective starting capacitor 4. Burned start windings in motor 5. Defective blower bearings 6. Loose wiring connections in motor starting circuit	1. Replace bearing or fan motor 2. Repair starting switch or replace motor 3. Replace capacitor 4. Repair or replace motor 5. Replace bearings 6. Replace motor

Fan motor cycles	1. Defective motor bearings 2. Defective blower bearings 3. Defective run capacitor 4. Defective fan control control 5. Defective fan relay 6. Defective motor windings	1. Replace bearings or motor 2. Replace blower bearings 3. Replace capacitor 4. Replace fan control 5. Replace relay 6. Repair or replace motor
Fan blows cold air	1. Defective heat sequencing relays 2. Burned heating elements 3. Loose wiring connections 4. Defective thermostat 5. Fan set to on position 6. Defective fan control	1. Replace relays 2. Replace elements 3. Repair wiring 4. Replace thermostat 5. Set to auto position 6. Replace fan control
Not enough heat	1. Dirty air filters 2. Unit too small 3. Too little air flow through furnace 4. Thermostat heat anticipator not properly set 5. Defective fan motor 6. Air conditioning evaporator coil dirty 7. Thermostat not properly located 8. Thermostat set too low 9. Thermostat out of calibration 10. Low voltage 11. Air ducts not insulated 12. Burned heating elements 13. Defective heat sequencing relays 14. Defective thermosat 15. Defective element 16. Outdoor thermostat set too low	1. Clean or replace filters 2. Install more elements 3. Increase air flow; remove restrictions 4. Set heat anticipator 5. Repair or replace fan motor 6. Clean evaporator 7. Relocate thermostat 8. Set thermostat 9. Calibrate thermostat 10. Correct cause 11. Insulate ducts 12. Replace elements 13. Replace relays 14. Replace thermostat 15. Replace limits 16. Reset thermostat
Too much heat	1. Unit too large 2. Thermostat heat anticipator not properly set	1. Reduce Btu input 2. Set heat anticipator 3. Relocate thermostat 4. Set thermostat

	3. Thermostat not properly located 4. Thermostat set too high 5. Thermostat out of calibration	5. Calibrate thermostat
High humidity in building	1. Humidity due to cooking 2. Humidity due to bathing 3. Humidity due to rain	1. Vent cook stove 2. Vent bathroom 3. Increase temperature rise through furnace
Blown element limits	1. Shorted heating element 2. Dirty filers 3. Dirty blower 4. Broken or slipping belt 5. High or low voltage 6. Defective blower motor 7. Not enough air through furnace 8. Loose electrical connections	1. Replace element and correct cause 2. Clean or replace filters 3. Clean blower 4. Adjust or replace belt 5. Notify power company 6. Replace motor 7. Remove restriction 8. Repair connections
High operating costs	1. Unit too small 2. Dirty air filters 3. Dirty air conditioning evaporator 4. Air ducts not properly insulated 5. Thermostat in wrong location 6. Dirty blower 7. Defective thermostat 8. Fan belt slipping 9. Low or high voltage 10. Thermostat setting too high	1. Increase number of elements 2. Clean or replace filters 3. Clean evaporator 4. Insulate ducts 5. Relocate thermostat 6. Clean blower 7. Replace thermostat 8. Replace or adjust fan belt 9. Notify power company 10. Set to lower setting

GAS HEAT

Condition	Possible Cause	Corrective Action
Unit will not run	1. Blown fuse 2. Burned transformer 3. Pilot out 4. Thermostat not calling for heat 5. Defective wiring or connections	1. Replace fuse and correct cause 2. Replace transformer and correct cause 3. See entry "Pilot not burning properly or out" 4. Set thermostat 5. Repair wiring or connections

Fan will not run	1. Burned fan motor 2. Broken fan belt 3. Burned contacts in fan relay 4. Defective fan control 5. Defective wiring connections	1. Repair or replace motor 2. Replace fan belt 3. Replace fan relay 4. Replace fan control 5. Repair wiring or connections
Fan motor hums, but will not start	1. Defective bearings in fan motor 2. Defective starting switch in fan motor 3. Defective starting capacitor capacitor 4. Burned start winding in motor 5. Defective blower bearings 6. Loose wiring connections in motor starting circuit	1. Repair or replace motor 2. Repair starting switch or replace motor 3. Replace capacitor 4. Repair or replace motor 5. Replace bearings 6. Repair wiring
Fan motor cycles	1. Defective motor bearings 2. Defective blower bearings 3. Defective run capacitor 4. Defective fan control 5. Return air too cold 6. Fan control differential too close 7. Fan control off setting too high 8. Too much air flow through furnace 9. Defective motor windings 10. Fan control "on" setting too low	1. Replace bearings or fan motor 2. Replace blower bearings 3. Replace run capacitor 4. Replace fan control 5. Allow air to warm 6. Adjust fan control 7. Adjust fan control 8. Reduce air flow 9. Repair or replace motor 10. Adjust control
Pilot not burning properly or out	1. Faulty thermocouple 2. Dirty or corroded thermocouple connection 3. Gas supply turned off 4. Pilot burner orifice dirty 5. Thermocouple not installed in flame properly 6. Drafts affecting pilot flame 7. Defective pilot safety	1. Replace thermocouple 2. Clean connection 3. Restore gas supply 4. Clean orifice 5. Properly install thermocouple 6. Shield pilot from drafts 7. Replace pilot safety device safety device

Fan cycles while main burner stays on	1. Wrong size orifices in burners 2. Low manifold gas pressure 3. Too much air flowing through furnace 4. Too cold return air 5. Electrical or motor problems	1. Replace orifices with proper size 2. Increase gas pressure 3. Reduce air flow 4. Allow air to warm 5. See previous entry "Fan motor cycles"
Main burner cycles while blower stays on	1. Dirty air filters 2. Wrong size orifices in burners 3. High mainfold gas pressure 4. Faulty limit control 5. Too little air flow through furnace	1. Clean or replace filters 2. Replace orifices with proper size 3. Reduce gas pressure 4. Replace limit control 5. Increase air flow; clear obstructions
Not enough heat	1. Dirty air filters 2. Wrong size orifices in burners 3. Low manifold gas pressure 4. Too little air flow through furnace 5. Thermostat heat anticipator not properly set 6. Defective fan motor 7. Unit too small 8. Air conditioning evaporator dirty 9. Thermostat not properly located 10. Thermostat set too low 11. Thermostat out of calibration	1. Clean or replace filters 2. Replace orifices with proper size 3. Increase gas pressure 4. Increase air flow; clear obstructions 5. Set heat anticipator 6. Replace or repair fan motor 7. Replace unit with proper size 8. Clean evaporator 9. Move thermostat 10. Set thermostat 11. Calibrate thermostat
Too much heat	1. Unit too large 2. Thermostat heat anticipator not properly set 3. Thermostat not properly located 4. Thermostat set too high 5. Thermostat out of calibration	1. Reduce Btu input; replace unit 2. Set heat anticipator 3. Move thermostat 4. Set thermostat 5. Calibrate thermostat

Pilot burning, main gas valve will not operate	1. Blown fuse 2. Defective gas valve 3. Burned transformer 4. Burned thermostat heat anticipator 5. Bad thermostat 6. Bad electrical connections 7. Broken thermostat wire	1. Replace fuse and check for cause 2. Replace gas valve 3. Replace transformer and check for cause 4. Replace thermostat 5. Replace thermostat 6. Repair connections 7. Repair broken wire
Delayed ignition of main burner	1. Poor flame travel to the burner 2. Poor flame distribution over the burner 3. Low manifold gas pressure 4. Defective step-opening regulator	1. Correct flame travel 2. Correct flame distribution 3. Adjust gas pressure 4. Adjust or replace regulator
Roll-out on main burner ignition	1. Restricted heat exchanger 2. Quick opening main gas valve	1. Clear restrictions 2. Install surge arrestor
Flame flashback	1. Low manifold gas pressure 2. Extremely small main burner flame 3. Distorted burner or carry-over wing slots 4. Defective main burner orifice 5. Orifice misaligned 6. Erratic gas valve operation 7. Dirty burner 8. Improper gas-air mixture 9. Unstable gas supply pressure	1. Adjust manifold gas pressure 2. Adjust primary air 3. Repair burner or carry-over wing slots 4. Replace orifices 5. Replace orifices 6. Replace gas valve 7. Clean burners 8. Be sure that proper gas is being used 9. Install a two-stage pressure regulator
Resonance (loud rumbling noise)	1. Excess primary air to main burner 2. Defective orifice spud 3. Dirty orifice spuds	1. Adjust primary air 2. Replace orifice spuds 3. Clean orifices
Yellow flame	1. Too little primary air 2. Dirty orifice spuds 3. Orifice spuds misaligned 4. Restricted heat exchanger 5. Poor vent operation	1. Adjust primary air 2. Clean spuds 3. Align orifice spuds 4. Clean heat exchanger 5. Correct venting

Floating main burner flame	1. Air blowing into heat exchanger 2. Restricted heat exchanger 3. Negative pressure in furnace room	1. Check for defective heat exchanger 2. Clean heat exchanger 3. Increase air supply
Main burner flame too large	1. Orifices too large 2. Excessive manifold gas pressure 3. Defective gas pressure regulator 4. Wrong type of gas being used	1. Replace orifices 2. Adjust pressure regulator 3. Replace regulator 4. Install changeover kit
Main burner flame too small	1. Dirty orifice spuds 2. Low manifold gas pressure 3. Orifices too small 4. Wrong type of gas being used	1. Clean orifice spuds 2. Adjust pressure regulator 3. Replace orifices 4. Install changeover kit
Odor in	1. Vent not operating properly 2. Poor ventilation	1. Correct venting problem 2. Check flame conditions and correct
High operating costs	1. Unit too small 2. Dirty air filters 3. Dirty air conditioning evaporator 4. Air ducts not insulated properly 5. Thermostat in wrong location 6. Dirty blower	1. Install proper size unit 2. Clean or replace filters 3. Clean evaporator 4. Insulate ducts 5. Relocate thermostat 6. Clean blower

HEATING (OIL)

Condition	Possible Cause	Corrective Action
Burner will not start	1. Thermostat off or set too low 2. Burner motor overload tripped 3. Primary control off on safety switch 4. Dirty thermostat contacts 5. Bad thermostat circuit breaker 7. Disconnect switch open 8. Shorted flame detector circuit	1. Turn thermostat on or set to higher temperature 2. Push motor overload reset button 3. Reset safety switch lever 4. Clean thermostat contacts 5. Replace thermostat 6. Replace fuse or set circuit breaker 7. Close switch 8. Replace flame detector 9. Separate and insulate leads 10. Protect detector from light 11. Replace element or control 12. Replace element or control

	9. Shorted flame detector leads 10. Flame detector exposed to direct light 11. Faulty friction clutch 12. Hot contacts stuck 13. Dirty cold contacts 14. Flame detector carboned 15. Loose connection or broken wire on flame detector 16. Low line voltage or power failure 17. Limit control open 18. Open electric circuit to limit control 19. Defective internal primary control circuit 20. Dirty burner relay contacts in primary control 21. Defective burner motor 22. Binding burner blower wheel 23. Seized fuel pump	13. Clean contacts 14. Clean bimetal 15. Repair connection or replace wire 16. Notify power company 17. Set limit control to 200° F, then jumper control terminals; if burner starts, replace the control 18. Repair or replace wiring 19. Replace control 20. Clean contacts 21. Replace burner motor 22. Turn off power and rotate blower by hand; if binding, free it 23. Turn power off and rotate blower by hand; if binding, replace fuel pump
Burner starts and fires but short cycles	1. Thermostat in warm draft 2. Heat anticipator set wrong set wrong 3. Furnace blower running too slow 4. Limit control set too low 5. Dirty air filter 6. Return air restriction 7. Low or fluctuating voltage 8. Loose wiring connection	1. Relocate thermostat 2. Correct anticipator setting 3. Speed up blower to obtain an 85° F to 95° F temperature 4. Reset limit to 200° F 5. Clean or replace filter 6. Clear restriction 7. Notify power company 8. Repair connection
Burner starts and fires but does not heat enough (short cycles)	1. Vibration at thermostat 2. Thermostat in warm location 3. Heat anticipator set wrong 4. Furnace blower running too slow 5. Dirty air filter 6. Defective blower motor bearings 7. Defective blower	1. Correct vibration or relocate thermostat 2. Relocate thermostat 3. Correct anticipator setting 4. Speed up blower to obtain 85° F to 95° F temperature rise 5. Clean or replace filter 6. Replace motor 7. Replace bearings 8. Clean blower wheel 9. Change rotation or replace motor 10. Clear restriction

	bearings 8. Dirty furnace blower rotation 9. Wrong blower motor rotation 10. Return air restricted 11. Limit control set low 12. Low or fluctuation voltage 13. Loose wiring connection	11. Reset limit to 200° F 12. Notify the power company 13. Repair connection
Burner starts and fires; then locks out on safety	1. Too little primary air; long dirty flame 2. Too much primary air; short lean flame 3. Unbalanced flame 4. Too little or restricted draft 5. Excessive draft 6. Dirty flame detector bimetal element 7. Faulty flame detector friction clutch 8. Welded or shorted cold contacts in flame detector 9. Air leaking into flue pipe around flame detector mount 10. Dirty flame detector cad cell face 11. Loose or defective 12. Faulty flame detector cad cell; resistance exceeds 1,500 ohms 13. Defective primary control circuit	1. Increase combustion air 2. Reduce combustion air 3. Replace nozzle 4. Correct draft or remove restriction 5. Adjust barometric damper 6. Clean element 7. Replace flame detector control 8. Replace flame detector control 9. Seal air leaks 10. Clean cad cell face 11. Repair or replace cad cell holder and wires 12. Replace cad cell 13. Replace primary control
Burner starts, but no flame is established	1. Oil tank empty 2. Oil tank shutoff valve closed 3. Water in oil tank 4. Air leak in oil supply line 5. Oil filter plugged 6. Oil pump strainer plugged 7. Restricted oil line 8. Excessive combustion air 9. Excessive vent draft 10. Off-center spray	1. Contact oil distributor 2. Open valve 3. Remove water 4. Repair leak 5. Install new filter 6. Clean strainer 7. Repair or replace line 8. Adjust air supply 9. Adjust barometric damper to between 0.030 and 0.035 inches water column 10. Replace nozzle 11. Replace nozzle

Problem	Possible cause	Remedy
	from nozzle 11. Nozzle strainer plugged 12. Nozzle orifice plugged 13. Faulty oil pump 14. Low fuel pressure 15. Faulty pump coupling 16. Low line voltage to transformer primary 17. Faulty transformer 18. No or weak ignition spark 19. Dirty or shorted ignition electrodes 20. Improper position or gap of ignition electrodes 21. Cracked or burned lead insulation 22. Loose or disconnected electrode leads 23. Defective electrode lead insulators 24. Oil pump or blower overloading motor 25. Faulty oil pump motor 26. Low voltage	12. Replace nozzle 13. Replace pump 14. Adjust pressure to desired pressure 15. Replace coupling 16. Notify power company 17. Replace transformer 18. Properly ground transformer case 19. Clean electrodes 20. Correctly position and reset electrode gap 21. Replace electrode leads 22. Repair or replace leads 23. Replace electrodes 24. Remove overload condition 25. Replace motor 26. Notify power company
Burner starts and fires but loses flame and locks out on safety	1. Dirty face and cad cell 2. Faulty cad cell; resistance exceeds 1,500 ohms 3. Loose or defective cad cell wires 4. Stack control bimetal dirty 5. Faulty friction clutch in stack control 6. Air leaking into vent pipe around stack control mount 7. Defective stack control cold contacts 8. Too much combustion air 9. Too little combustion air 10. Unbalanced flame 11. Excessive vent draft 12. Too little vent draft 13. Vent restricted 14. Oil pump looses prime 15. Air leak in oil supply line 16. Partially plugged nozzle	1. Clean cad cell face 2. Replace cad cell 3. Repair or replace wires 4. Clean bimetal element 5. Replace stack control 6. Seal air leaks 7. Replace stack control 8. Adjust combustion air damper 9. Adjust combustion air damper 10. Replace nozzle 11. Adjust barometric damper 12. Adjust barometric damper 13. Clear restriction 14. Prime pump at bleed port 15. Repair leaks 16. Replace nozzle 17. Replace nozzle 18. Remove water from tank 19. Change to Number 1 oil 20. Clean strainer or replace pump 21. Clear restriction

	17. Partially plugged nozzle strainer 18. Water in oil storage tank 19. Oil too heavy 20. Plugged fuel pump strainer 21. Restricted oil line	
Too much heat; burner runs continuously	1. Defective thermostat 2. Shorted thermostat 3. Thermostat in cold location 4. Thermostat not level 5. Defective primary control	1. Repair or replace thermostat 2. Repair or replace wires 3. Relocate thermostat 4. Level thermostat 5. Replace control
Too little heat; burner runs continuously	1. Too much combustion air 2. Air leaking into heat exchanger 3. Excessive vent draft 4. Wrong burner head adjustment 5. Plugged heat exchanger 6. Too little combustion air 7. Insufficient vent draft 8. Insufficient indoor air flow 9. Dirty indoor blower 10. Dirty furnace filter 11. Partially plugged nozzle 12. Nozzle too small 13. Low oil pressure	1. Reduce combustion air 2. Repair leaks 3. Adjust barometric damper 4. Correct burner head setting 5. Clean heat exchanger and adjust burner 6. Increase combustion air 7. Adjust barometric damper 8. Speed blower to obtain 85° F to 95° F temperature rise 9. Clean blower 10. Clean or replace filter 11. Replace nozzle 12. Replace nozzle with larger nozzle 13. Increase to proper pressure

HEAT PUMP (COOLING CYCLE)

Condition	Possible Cause	Corrective Action
No cooling, but compressor runs continuously	1. Defective compressor valves 2. Shortage of refrigerant 3. Defective reversing valve 4. Air or noncondensables in system 5. Wrong superheat setting on indoor expansion valve 6. Loose thermal bulb on indoor expansion valve 7. Dirty indoor coil 8. Dirty indoor filters 9. Indoor blower belt slipping 10. Restriction in refrigerant system	1. Replace valves and valve plate or compressor 2. Repair leak and recharge 3. Replace reversing valve 4. Purge noncondensables 5. Adjust superheat setting 6. Tighten thermal bulb 7. Clean coil 8. Clean or replace filters 9. Replace or adjust belt 10. Locate and remove restriction

Too much cooling; compressor run continuously	1. Faulty wiring 2. Faulty thermostat 3. Wrong thermostat location	1. Locate and repair wiring 2. Replace thermostat 3. Relocate thermostat
Liquid refrigerant flooding compressor (thermostatic expansion valve–TXV system)	1. Wrong superheat setting on indoor expansion valve 2. Loose thermal bulb on indoor expansion valve 3. Faulty indoor expansion valve 4. Defective indoor check valve 5. Refrigerant overcharge	1. Adjust superheat 2. Tighten thermal bulb 3. Replace expansion valve 4. Replace check valve 5. Purge refrigerant overcharge
Liquid refrigerant flooding compressor (capillary tube system)	1. Refrigerant overcharge 2. High head pressure 3. Dirty indoor filter 4. Dirty indoor coil 5. Indoor blower belt slipping 6. Indoor check valve defective	1. Purge overcharge 2. See entry "High head pressure" 3. Clean or replace filter 4. Clean coil 5. Replace or adjust belt 6. Replace check valve

HEAT PUMP (HEATING CYCLE)

Condition	Possible Cause	Corrective Action
No heating, but compressor runs continuously	1. Refrigerant shortage 2. Compressor valves defective 3. Leaking reversing valve 4. Defective defrost control, time clock, or relay	1. Repair leak and recharge 2. Replace valves and valve plate or compressor 3. Replace reversing valve 4. Replace defrost control, time clock, or relay
Too much heat; compressor runs continuously	1. Faulty wiring 2. Faulty thermostat 3. Wrong thermostat location	1. Repair wiring 2. Replace thermostat 3. Relocate thermostat
Compressor cycles on low pressure control at end of defrost cycle	1. Defective reversing valve 2. Defective power element on indoor expansion valve 3. Shortage of refrigerant	1. Replace reversing valved 2. Replace power element 3. Repair leak and recharge
Unit runs in cooling cycle but pumps down in heating cycle	1. Faulty outdoor expansion valve 2. Defective power element on outdoor expansion valve	1. Clean or replace expansion valve 2. Replace power element 3. Replace reversing valve 4. Clean coil

	3. Defective reversing valve 4. Dirty outdoor coil 5. Belt slipping on outdoor blower 6. Defective indoor check valve 7. Restriction in refrigerant circuit	5. Replace or adjust belt 6. Replace check valve 7. Locate and remove restriction
Defrost cycle will not terminate	1. Shortage of refrigerant 2. Defrost control out of adjustment 3. Defective defrost control, time clock or relay 4. Defective reversing valve 5. Defective compressor valves 6. Faulty electrical wiring	1. Repair leak and recharge 2. Adjust control 3. Replace defrost control, time clock, or relay 4. Replace reversing valve 5. Replace valves and valve plate or compressor 6. Repair wiring

HEAT PUMP (HEATING OR COOLING CYCLE)

Condition	Possible Cause	Corrective Action
Compressor hums but will not start	1. Faulty fuse 2. Faulty wiring 3. Loose electrical terminals 4. Compressor overload 5. Faulty starting capacitor 6. Faulty starting relay 7. Burned compressor motor 8. Defective compressor bearings 9. Stuck compressor	1. Replace fuse and correct cause 2. Repair wiring 3. Repair loose connections 4. Locate and remove overload 5. Replace capacitor 6. Replace relay 7. Replace compressor 8. Replace bearings or compressor 9. Replace compressor
Compressor cycling on overload	1. Low voltage 2. Loose electrical terminals 3. Single phasing of phase power 4. Defective contactor contacts 5. Defective compressor overload 6. Compressor overloaded 7. Defective start capacitor 8. Defective run capacitor 9. Defective starting relay	1. Determine reason and repair 2. Repair terminals 3. Replace fuse or repair wiring; or notify power company 4. Replace contacts or contactors 5. Replace overload 6. Locate and remove overload 7. Replace capacitor 8. Replace capacitor 9. Replace starting relay 10. Purge overcharge 11. Replace bearings or compressor

	10. Refrigerant overcharge 11. Defective compressor bearings 12. Air or noncondensables in system (high head pressure)	12. Purge noncondensables from system
Compressor off on high pressure control	1. Refrigerant overcharge 2. Control out of adjustment 3. Defective indoor fan motor 4. Defective outdoor fan motor 5. Defective fan relay on either outdoor or indoor section 6. Too long defrost cycle 7. Defective reversing valve 8. Blower belt slipping on indoor or outdoor coil 9. Indoor or outdoor coil dirty 10. Dirty indoor air filters 11. Air bypassing indoor or outdoor coil 12. Air volume too low over indoor or outdoor coil 13. Auxiliary heat strips ahead of indoor coil	1. Purge overcharge 2. Adjust control 3. Repair or replace motor 4. Repair or replace motor 5. Repair or replace relay 6. Replace time clock, defrost relay, or termination thermostat 7. Replace reversing valve 8. Adjust or replace belt 9. Clean proper coil 10. Replace or clean filters 11. Prevent air bypass 12. Increase ductwork or remove restriction from coils 13. Locate heat strips downstream of indoor coil
Compressor cycles on low pressure control	1. Refrigerant shortage 2. Low suction pressure 3. Defective expansion valve 4. Dirty indoor or outdoor coil 5. Slipping blower belt 6. Dirty indoor air filter 7. Ductwork restriction 8. Liquid drier or suction strainer restricted 9. Defrost thermostat element loose or making poor contact 10. Air temperature too low for evaporation 11. Defrost cycle too long 12. Defective fan motor	1. Repair leak and recharge 2. Increase load (See heading "Suction pressure low") 3. Repair or replace expansion valve 4. Clean coil 5. Replace or adjust blower belt 6. Clean or replace filter 7. Increase ductwork 8. Replace drier or strainer 9. Tighten or increase contact 10. Relocate unit or provide adequate air temperature 11. Replace time clock, defrost relay, or termination thermostat 12. Repair or replace fan motor

Outdoor fan runs, but compressor will not	1. Faulty electrical wiring or loose connections 2. Defective starting capacitor 3. Defective starting relay 4. Defective run capacitor 5. Shorted or grounded compressor motor 6. Stuck compressor 7. Compressor overloaded 8. Defective contactor contacts 9. Single phasing of three-phase power 10. Low voltage	1. Repair wiring or connections 2. Replace starting capacitor 3. Replace starting relay 4. Replace run capacitor 5. Replace compressor 6. Replace compressor 7. Determine and remove overload 8. Replace contactor or contacts 9. Locate problem and repair or contact power company 10. Locate and correct cause
Outdoor fan motor will not start	1. Faulty electrical wiring or loose connections 2. Defective outdoor fan motor 3. Defective outdoor fan relay 4. Defective defrost control, timer, or relay	1. Repair wiring or connections 2. Repair or replace motor 3. Replace fan relay 4. Replace control timer or relay
Outdoor section does not run	1. No electrical power 2. Blown fuse 3. Faulty electrical wiring or loose terminals 4. Compressor overloaded 5. Defective transformer 6. Burned contactor coil 7. Compressor overload open 8. High pressure control open 9. Low pressure control open 10. Thermostat off	1. Inform power company 2. Replace fuse and correct fault 3. Repair wiring or terminals 4. Determine overload and correct 5. Replace transformer 6. Replace contactor coil 7. Determine cause and correct 8. Determine cause and correct 9. Determine cause and correct 10. Turn thermostat on and set
Indoor blower will not run	1. Blown fuse 2. Faulty electrical wiring or loose connections 3. Burned transformer 4. Indoor fan relay defective 5. Faulty indoor fan motor 6. Faulty thermostat	1. Replace fuse and correct cause 2. Repair wiring or connections 3. Replace transformer 4. Replace fan relay 5. Repair or replace motor 6. Replace thermostat
Indoor coil iced over	1. Dirty filters 2. Dirty coil	1. Clean or replace filters 2. Clean coil

	3. Blower fan belt slipping 4. Outdoor check valve sticking closed 5. Defective indoor expansion valve 6. Low indoor air temperature 7. Shortage of refrigerant	3. Replace or adjust belt 4. Replace check valve 5. Clean or replace expansion valve 6. Increase temperature 7. Repair leak and recharge
Noisy compressor	1. Low oil level in compressor 2. Defective suction and/or discharge valves 3. Loose hold-down bolts 4. Broken internal springs 5. Inoperative check valves 6. Loose thermal bulb on indoor expansion valve 7. Improper superheat setting on indoor expansion valve 8. Stuck open indoor expansion valve	1. Determine reason for loss of oil and correct. Replace oil. 2. Replace valves and valve plate or compressor 3. Tighten 4. Replace compressor 5. Repair or replace check valve 6. Tighten thermal bulb 7. Adjust superheat 8. Clean or replace valve
Compressor loses oil	1. Refrigerant shortage 2. Low suction pressure 3. Restriction in refrigerant circuit 4. Indoor expansion valve stuck open	1. Repair leak and recharge 2. Increase load on evaporator 3. Remove restriction 4. Clean or replace expansion valve
Unit operates normally in one cycle, but high suction pressure on other cycle	1. Leaking check valve 2. Loose thermal bulb on outdoor or indoor expansion valve 3. Leaking reversing valve 4. Expansion valve stuck open on indoor or outdoor coil	1. Replace checkvalve 2. Tighten thermal bulb 3. Replace reversing valve 4. Repair or replace expansion valve
Unit pumps down in cool or defrost cycle but operates normally in heat cycle	1. Defective reversing valve 2. Defective power element on indoor expansion valve 3. Restriction in refrigerant circuit	1. Replace reversing valve 2. Replace power element 3. Locate and remove restriction 4. Clean or replace expansion valve 5. Replace check valve

	4. Clogged indoor expansion valve 5. Check valve in outdoor section sticking closed	
High head pressure	1. Overcharge of refrigerant 2. Air or noncondensables in system 3. High air temperature supplied to condensing coil 4. Dirty indoor or outdoor coil 5. Dirty indoor air filters 6. Indoor or outdoor blower belt slipping 7. Air bypassing indoor or outdoor coil	1. Remove overcharge 2. Purge noncondensables 3. Reduce air temperature 4. Clean coil 5. Clean or replace filters 6. Replace or adjust blower belt 7. Prevent air bypassing
High suction pressure	1. Defective compressor suction valves 2. High head pressure 3. Excessive load on cooling 4. Leaking reversing valve 5. Leaking check valves 6. Indoor or outdoor expansion valve stuck open 7. Loose thermal bulb on indoor or outdoor expansion valve	1. Replace valves and valve plate or compressor 2. See previous entry, "High head pressure" 3. Determine cause and correct 4. Replace reversing valve 5. Replace check valve 6. Clean or replace expansion valve 7. Tighten bulb
Low suction pressure	1. Shortage or refrigerant 2. Blower belt slipping on indoor or outdoor blower 3. Dirty indoor air filters 4. Defective check valves 5. Restriction in refrigerant circuit 6. Ductwork small or restricted 7. Defective expansion valve 8. Clogged indoor or outdoor expansion valve 9. Wrong superheat setting on indoor or outdoor expansion valve 10. Dirty indoor or outdoor coil 11. Bad contactor contacts	1. Repair leak and recharge 2. Replace belt or adjust 3. Clean or replace 4. Replace check valves 5. Locate and remove restriction 6. Repair or replace ductwork 7. Replace power element 8. Clean or replace valve 9. Adjust superheat setting 10. Clean coil 11. Replace contactor or contacts

THERMOSTAT

Condition	Possible Cause	Corrective Action
Thermostat jumpered; system will not operate	1. Thermostat not at fault 2. Limit control set too low 3. No low voltage 4. Open control circuit 5. Main gas valve inoperative 6. Bad, loose, or corroded connection in control circuit	1. Check for another problem 2. Check and adjust limit control setting 3. Check transformer and replace 4. Check for cause and repair 5. Check and replace gas valve 6. Repair connections
Thermostat jumpered; system operates	1. Dirty thermostat contacts 2. Damaged thermostat	1. Clean contacts or replace thermostat 2. Replace thermostat
Room temperature overshoots thermostat setting	1. Thermostat located in cold location 2. Wiring hole in wall not sealed 3. Thermostat exposed to cold drafts 4. Thermostat not sensing circulating air 5. Mercury switch thermostat not installed level 6. Thermostat out of calibration 7. Anticipator set wrong 8. Heating unit too large 9. Nonanticipated thermostat	1. Relocate thermostat 2. Plug hole 3. Relocate thermostat 4. Relocate thermostat 5. Level thermostat 6. Calibrate or replace thermostat 7. Set anticipator 8. Replace unit or reduce Btu input 9. Replace thermostat with one having an anticipator
Room temperature does not reach thermostat setting	1. Thermostat not mounted level 2. Thermostat not properly calibrated 3. Heating unit too small 4. Limit control set too low 5. Thermostat sensing direct sun rays 6. Thermostat affected by external heat source 7. Thermostat mounted on warm wall 8. Dirty thermostat contacts	1. Level thermostat 2. Calibrate thermostat 3. Install larger unit 4. Adjust limit control 5. Relocate thermostat 6. Relocate thermostat 7. Relocate thermostat 8. Clean contacts or replace thermostat 9. Repair connections 10. Clean or replace filter

	9. Bad, loose, or corroded connections in control circuit 10. Dirty indoor air filter	
Thermostat seems out of calibration	1. Thermostat not properly mounted 2. Thermostat not properly calibrated 3. Bad, loose, or corroded connections in control circuit 4. Dirty indoor air filter	1. Remount thermostat 2. Recalibrate thermostat 3. Repair connections 4. Clean or replace filter
Thermostat short cycles	1. Heat anticipator improperly set	1. Set anticipator
Thermostat stays on too long	1. Thermostat not exposed to circulating air 2. Heating unit too small 3. Heat anticipator set too high 4. Nonanticipated thermostat 5. Dirty thermostat contacts	1. Relocate thermostat 2. Install new unit or increase Btu input 3. Set anticipator 4. Replace with anticipated thermostat 5. Clean contacts or replace thermostat
Excessive room temperature variations	1. Thermostat not exposed to circulating air 2. Anticipator set too high 3. Heating unit too large 4. Nonanticipated thermostat	1. Relocate thermostat 2. Set anticipator 3. Replace unit or reduce Btu input 4. Replace with anticipated thermostat

PILOT BURNER

Condition	Possible Cause	Corrective Action
Cannot light pilot	1. Pilot gas turned off 2. Air in pilot gas line 3. Burner orifice clogged 4. Lighting knob not depressed 5. Reset button not depressed 6. Lighting knob not at pilot position	1. Restore gas supply 2. Purge air from line 3. Clean or replace orifice 4. Depress knob and light 5. Depress button and light 6. Move knob to pilot position and depress 7. Open pilot adjustment

	7. Pilot adjustment closed off	
Pilot goes out when reset knob is released	1. Lighting knob released too soon 2. Reset button released too soon 3. Bad thermocouple or thermopile 4. Bad pilotstat power unit 5. Loose, corroded, or dirty power unit connections 6. Pilot flame too small 7. Bad powerpile terminals 8. Plugged pilot filter	1. Depress knob and hold down about one minute after flame is burning 2. Depress button and hold down about one minute after flame is burning 3. Check output and replace 4. Replace pilotstat 5. Repair and/or clean connections 6. Adjust pilot gas pressure or clean or replace orifice 7. Repair terminals 8. Replace filter
Pilot burning but unit on pilot safety shutdown	1. Low pilot gas supply pressure 2. Pilot affected by excessive draft 3. Bad thermocouple or thermopile 4. Bad pilotstat power unit 5. Bad pilotstat power unit connections 6. Too small pilot flame 7. Improper furnace venting 8. Gas line restricted or too small	1. Adjust pilot gas supply pressure 2. Shield pilot from excessive draft 3. Check output and replace 4. Replace pilotstat 5. Repair connections 6. Adjust pilot flame 7. Install proper venting system 8. Clear restriction or install new line
Lazy, yellow pilot flame	1. Plugged pilot burner lint screen 2. Pilot primary air opening clogged 3. Pilot burner orifice too large 4. Excessive ambient temperature 5. Improper venting	1. Clean screen 2. Clean pilot 3. Install proper size orifice 4. Relocate pilot to another satisfactory position 5. Install proper venting system
Waving, blue pilot flame	1. Too low gas supply pressure 2. Combustion products affecting pilot	1. Increase gas suply pressure 2. Shield pilot from combustion products
Small, blue pilot flame	1. Clogged pilot burner orifice 2. Pilot gas adjustment	1. Clean or replace orifice 2. Open adjustment to proper level 3. Replace with pilot proper size orifice

	closed off 3. Too small pilot burner orfice 4. Clogged pilot filter	4. Replace filter
Noisy, lifting, blowing pilot flame	1. Pilot gas pressure too high	1. Adjust pilot gas pressure
Hard, sharp pilot flame	1. Pilot burner orifice too small 2. Typical of manufactured butane-air, and propane-air pilot burners	1. Replace with proper size orifice 2. Normal, no problem

glossary

The definitions given here apply only to heating. They are defined particularly as used in this book.

Absolute Humidity: The weight of water vapor in grains actually contained in 1 ft^3 of the mixture of air and moisture.

Air: An elastic gas. It is a mechanical mixture of oxygen and nitrogen and slight traces of other gases. It may also contain moisture known as humidity. Dry air weighs 0.075 lb/ft^3. One Btu will raise the temperature of 55 ft^3 of air 1° F.

Air Change: The number of times in an hour the air in a room is changed by mechanical means or by infiltration of outside air leaking into the room through cracks around doors and windows, etc.

Air Cleaner: A device used for the purpose of removing airborne impurities, such as dust, fumes, and smoke. (Air cleaners include air washers and air filters.)

Air Conditioning: The simultaneous control of the temperature, humidity, air motion, and air distribution within an enclosure. When human comfort and health are involved, a reasonable air purity with regard to dust, bacteria, and odors is also included. The primary requirement of a good air conditioning system is a good heating system.

Air Infiltration: The leakage of air into a house through cracks and crevices and through doors, windows, and other openings, caused by wind pressure and/or temperature difference.

Air Valve: See *vent valve.*

Atmospheric Pressure: The weight of a column of air 1 in^2 in a cross section and extending from the earth to the upper level of the blanket of air surrounding the earth. This air exerts a pressure of 14.7 psi at sea level,where water will boil at 212° F. High altitudes have a lower atmospheric pressure with correspondingly lower boiling point temperatures.

Automatic Flue Damper: A device between the vent system and the draft diverter on a heating furnace to prevent heat escaping through the vent system.

Boiler: A closed vessel in which steam is generated or in which water is heated by some heat source.

Boiler Heating Surface: The area of heat transmitting surfaces in contact with the water (or steam) in the boiler on one side and the fire or hot gases on the other.

Boilout: The evaporation of water which occurs when the storage water is heated to the boiling point and solar heating continues.

British Thermal Unit (Btu): The quantity of heat required to raise the temperature of 1 pound of water 1° F. This definition is approximate, but sufficiently accurate for any work discussed in this book.

Bucket Trap (Inverted): A float trap with an open float. The float or bucket is open at the bottom. When the air or steam in the bucket has been replaced by condensation, the bucket loses its buoyancy, and when it sinks it opens a valve to permit the condensate to be pushed into the return line.

Bucket Trap (Open): The bucket (float) is open at the top. Water surrounding the bucket keeps it floating and the pin is pressed against the seat. Condensate from the system drains into the bucket. When enough has drained into it so that the bucket loses its buoyancy, it sinks and pulls the pin off its seat and steam pressure forces the condensate out of the trap.

Carbon Dioxide (CO_2): CO_2 is one of the products of complete combustion of carbon formed by the union of one atom of carbon with two atoms of oxygen. It is an inert gas which in itself is harmless to human beings. An excessive amount of it is harmful, however, as it displaces the oxygen in the air which is vital to human life.

Carbon Monoxide (CO): Carbon monoxide is a colorless, odorless, and tasteless gas which cannot be detected by any of the senses. It is so noxious that continued exposure to two tenths of 1% (0.2%) of it will cause unconsciousness and death, and its effect is so insidious that a person breathing it may have little or no warning before collapsing.

Central Fan System: A mechanical indirect system of heating, ventilating, or air conditioning consisting of a central plant where the air is heated and/or conditioned and then by fans or blowers circulated through a system of distributing ducts.

Chimney: A vertical shaft enclosing one or more flues conveying the flue gases to the outdoor atmosphere.

Chimney Effect: The tendency in a duct or other vertical air passage for air to rise when heated due to its decrease in density.

Circulating Pipe (Hot water system): The pipe and orifice in a Hoffman Panelmatic® Hot Water Control System through which the return water bypasses the boiler until the temperature of the circulating stream is too low, at which time part of it is replaced by the correct quantity of hot water to restore its temperature.

Coefficient of Heat Transmission (Overall) *U*: The amount of heat (Btu) transmitted from air to air in 1 h/ft^2 of the wall, floor, roof, or ceiling for a difference in temperature of 1° F between the air on the inside and the outside of the wall, floor, roof, or ceiling.

Combustion: The rapid oxidation of fuel accompanied by the production of heat and some light. The complete combustion of a fuel is possible only in the presence of an adequate supply of oxygen and heat.

Combustion Area or Zone: The space within a furnace in which combustion takes place.

Combustion Products: The products that result from the burning of a fuel. These include the nitrogen and other inert materials present in the air which are used in the combustion process, but do not include the excess air that may be present.

Comfort Line: The effective temperature at which the largest percentage of adults feel comfortable.

Comfort Zone (Average): The range of effective temperatures at which the majority of adults feel comfortable.

Condensate: In steam heating, the water formed by cooling steam as in a radiator. The capacity of traps, pumps, etc., is sometimes expressed in pounds of condensate they will handle per hour. For instance, 1 pound of condensate per hour is equal to approximately 4 ft^2 of steam heating surface (240 Btu/h/ft^2).

Conductance (Thermal) C: The amount of heat (Btu) transmitted from surface to surface in one hour through 1 ft^2 of a material or construction for the thickness or type under construction for a difference in temperature of 1° F between the two sides.

Conduction (Thermal): The transmission of heat through and by means of matter.

Conductivity (Thermal) K: The amount of heat (Btu) transmitted in one hour through 1 ft^2 of a homogenous material 1 inch thick for a difference of 1° F between the two surfaces of the material.

Conductor (Thermal): A material capable of readily transmitting heat by means of conduction.

Convection: The transmission of heat by circulation (either natural or forced) of a liquid or a gas such as air. If natural, it is caused by the difference in weight of a hotter and colder fluid.

Convector: A concealed radiator. An enclosed heating unit located (with enclosure) either within, adjacent to, or exterior to the room or space to be heated, but transferring heat to the room or space mainly by the process of convection. A shielded heating unit is also termed a convector. If the heating unit is located exterior to the room or space to be heated, the heat is transferred through one or more ducts or pipes.

Cooling Leg: A length of uninsulated pipe through which the condensate flows to a trap and which has sufficient surface to permit the condensate to dissipate enough heat to prevent flashing when the trap opens. In the case of a thermostatic trap, a cooling leg may be necessary to permit the condensate to drop a sufficient amount in temperature to permit the trap to open.

Degree Day (Standard): A unit which is the difference between 65° F and the daily average temperature when the latter is below 65° F. The number of degree days in any one day is equal to the number of degrees F that the average temperature for that day is below 65° F.

Dew Point Temperature: The air temperature corresponding to saturation (100% relative humidity) for a given moisture content. It is the lowest temperature at which air can retain the water vapor it contains.

Differential Thermostat: A device which provides the control functions for domestic hot water systems by measuring the boiler leaving water temperature and comparing this to the outdoor ambient temperature.

Dilution: Dilution refers to the mixing of air with the flue gases which takes place at the draft hood. Dilution lowers the dew point temperature and helps prevent condensation.

Direct-Return System (Hot Water): A two-pipe hot water system in which the water, after it has passed through a heating unit, is returned to the boiler along a direct path so that the total distance traveled by the water from each radiator is the shortest feasible. There is, therefore, a considerable difference in the lengths of the several circuits composing the system.

Downfeed One-Pipe Riser (Steam): A pipe that carries steam downward to the heating units and into which the condensation from the heating unit drains.

Downfeed (Steam): A steam heating system in which the supply mains are above the level of the heating units which they serve.

Draft: A gravitational force or pressure difference that causes the vent gases to move to the outdoors. Draft is present in a flue when the vent gases are warmer (and lighter) than the outside air.

Draft Hood: A device that is built into a furnace, or made a part of the vent connector from a furnace, which is designed to (1) assure the ready escape of the flue gases in the event of no draft, back draft, or stoppage beyond the draft hood; (2) prevent a back draft from entering the furnace; and (3) neutralize the effect of stack action of the chimney or gas vent upon the operation of the furnace.

Dry Bulb Temperature: The temperature of air as determined by an ordinary thermometer.

Dry Return (Steam): A return pipe in a steam heating system that carries both air and water from condensation.

Dry Saturated Steam: Saturated steam containing no water in suspension.

Excess Air: The air supplied to a flame and not used in the combustion process. Usually 50% of the air required for complete combustion of the fuel.

Exhaust System: A system incorporating a hood, ductwork, and a blower by means of which contaminated air is moved to the outdoors.

Extended Heating Surface: Heating surface consisting of fins or ribs that receive heat by conduction from the prime surface.

Extended Surface Heating Unit: A heating unit having a relatively large amount of extended surface that may be integral with the core containing the heating medium or assembled over such a core, making good thermal contact by pressure or by being soldered to the core or by both pressure and soldering. (An extended surface heating unit is usually placed within an enclosure and therefore functions as a convector.)

Fahrenheit: A thermometer scale at which the freezing point of water is at 32° F and its boiling point is 212° F above 0° F. It is generally used in the United States for expressing temperature.

Flash (Steam): The rapid passing of water into steam at a high temperature when the pressure it is under is reduced so that its temperature is above that of its boiling point for the reduced pressure. For example, if hot condensate is discharged by a trap into a low pressure steam return or into the atmosphere, a certain percentage of the water will immediately be transformed into steam. It is also called reevaporation.

Float and Thermostatic Trap: A float trap with a thermostatic element for permitting the escape of air into the return line.

Float Trap: A steam trap operated by a float. When enough condensate has drained (by gravity) into the trap body, the float is lifted, which in turn lifts the pin off its seat and permits the condensate to flow into the return line until the float has been sufficiently lowered to close the port. Temperature does not affect the operation of a float trap.

Flue: The passageway that is provided for the flow of the products of combustion from the furnace to the outdoors.

Flue Collar: That part of a furnace which is designed to connect onto the draft hood or vent connector.

Flue Gases: The gases that are expelled from the combustion chamber of a furnace, the safe removal of which is the function of the venting system. These include the products of combustion and excess air, but not the dilution air which is brought in through the draft hood. These are the gases that are analyzed when determining the unit efficiency.

Forced Draft: In a forced-draft combustion system the combustion air is forced into the combustion chamber under a pressure. The entire combustion zone and the vent system is under a slight positive pressure.

Furnace: That part of a boiler or warm air heating plant in which combustion takes place. Sometimes also the complete heating unit of a warm air heating system.

Greenhouse Principle: A method of converting and trapping radiation from the sun in the form of heat.

Grille: A perforated covering for an air inlet or outlet usually made of wire screen, pressed steel, or other material.

Head: Unit of pressure usually expressed in feet of water.

Heat: That form of energy into which all other forms may be changed. Heat always flows from a body of higher temperature to a body of lower temperature. See also *latent heat, sensible heat, specific heat, total heat, heat of the liquid.*

Heat Exchanger: A device used to transfer heat from one medium to another.

Heat of the Liquid: The heat (Btu) contained in a liquid due to its temperature. The heat of the liquid for water is 0 at 32° F and increases 1 Btu approximately for every degree rise in temperature.

Heating Medium: A substance such as water, steam, or air used to convey heat from the boiler, furnace, or other source of heat to the heating units from which the heat is dissipated.

Heating Surface: The exterior surface of a heating unit. See also *extended heating surface.*

Heating Unit: Radiators, convectors, baseboards, finned tubing, coils embedded in the floor, wall, or ceiling, or any device which transmits the heat from the heating system to the room and its occupants.

Hot Water Heating System: A heating system in which water is used as the medium by which heat is carried through pipes from the boiler to the heating units.

Humidistat: An instrument that controls the relative humidity of the air in a room.

Humidity: The water vapor mixed with air.

Hysteresis Circuit: A special circuit of the differential thermostat which prevents the pump from being unnecessarily turned on and off immediately following an initial pump turn-on or turn-off event.

Induced Draft: An induced-draft combustion system pulls the air through the combustion zone and blows it out the vent. The combustion zone will be under a slightly negative pressure. The venting system may have either a slightly negative or positive pressure, depending on the design.

Insulation (Thermal): A material having a high resistance to heat flow.

Latent Heat: Hidden heat that cannot be felt or measured with a thermometer. It also causes the change of state of a substance.

Latent Heat of Vaporization: The heat (Btu/lb) necessary to change 1 pound of liquid into vapor without raising its temperature. In round numbers this is equal to 960 Btu/lb of water.

Mechanical Equivalent of Heat: The mechanical energy equivalent of 1 Btu, which is equal to 778 foot-pounds.

One-Pipe Supply Riser (Steam): A pipe which carries steam to a heating unit and which also carries the condensation from the heating unit. In an upfeed riser, steam travels upward and the condensate travels downward, while in a downfeed both steam and condensate travel down.

One-Pipe System (Hot Water): A hot water heating system in which one pipe serves as a supply main and also a return main. The heating units have separate supply and return pipes, but both are connected to the same main.

One-Pipe System (Steam): A steam heating system consisting of a main circuit in which the steam and condensate flow in the same pipe. There is but one connection to each heating unit, which must serve as both the supply and return.

Overhead System: Any steam or hot water system in which the supply main is above the heating units. With a steam system the return must be below the heating units; with water, the return may be above the heating units.

Panel Heating: A method of heating involving the installation of the heating units (pipe coils) within the wall, floor, or ceiling of the room.

Panel Radiator: A heating unit placed on, or flush with, a flat wall surface and intended to function essentially as a radiator. Do not confuse with panel heating system.

Plenum Chamber: An air compartment maintained under pressure and connected to one or more distribution ducts.

Pressure: Force per unit area such as pound per square inch. See *static, velocity, total gauge,* and *absolute pressures.* Unless otherwise qualified, it refers to unit static gauge pressure.

Pressure-Reducing Valve: A piece of equipment for changing the pressure of a gas or liquid from a higher to a lower one.

Prime Surface: A heating surface having the heating medium on one side and air (or extended surface) on the other.

Radiant Heating: A heating system in which the heating is by radiation only. Sometimes applied to panel heating systems.

Radiation: The transmission of heat in a straight line through space.

Radiator: A heating unit located within the room to be heated and exposed to view. A radiator transfers heat by radiation to objects it can "see" and by conduction to the surrounding air, which in turn is circulated by natural convection.

Relative Humidity: The amount of moisture in a given quantity of air compared with the maximum amount of moisture the same quantity of air could hold at the same temperature. It is expressed as a percentage.

Return Mains: The pipes that return the heating medium from the heating units to the source of heat supply.

Sealed Combustion System: A self-contained furnace which is designed so that all of the air supplied for the combustion process is isolated from the space that it is heating.

Sensible Heat: Heat which only increases the temperature of objects as opposed to latent heat.

Specific Heat: In the foot-pound-second system, the amount of heat (Btu) required to raise 1 pound of a substance 1° F. In the centimeter-gram-second system, the amount of heat (cal) required to raise 1 pound of a substance 1° C. The specific heat of water is 1.

Split System: A system in which the heating system utilizes radiators or convectors and ventilation by a separate apparatus.

Square Foot of Heating Surface: Equivalent direct radiation (EDR). By definition, that amount of heating surface which will give off 340 Btu/h when filled with a heating medium at 215° F and surrounded by air at 70° F. The equivalent ft^2 of heating surface may have no direct relation to the actual surface area.

Static Pressure: The pressure which tends to burst a pipe. It is used to overcome the frictional resistance to flow through the pipe. Expressed as a unit of pressure, it may be given in either absolute or gauge pressure. It is frequently expressed in feet of water column.

Steam: Simply stated, the term steam means water in the vapor phase. The vapor formed when water has been heated to its boiling point, corresponding to the pressure it is under. See also *dry saturated steam, wet saturated steam, superheated steam.*

Steam Heating System: A heating system in which the heating units give up their heat to the room by condensing the steam furnished to them by a boiler or other source.

Steam Quality: The quality of steam is expressed in terms of percent. For instance, if a quantity of steam consists of 90% steam and 10% moisture, the quality of the steam is 90%. The vapor formed when water has been heated to its boiling point, corresponding to the pressure it is under. See also *dry saturated steam, wet saturated steam, superheated steam.*

Strainer: A device such as a screen or filter to retain solid particles while the liquid passes through.

Superheat: Temperature of a vapor above the boiling temperature of its liquid at that pressure.

Superheated Steam: Steam that is at a temperature above its saturation temperature at that pressure.

Temperature: The measure of heat intensity.

Temperature, Ambient: The temperature of the air around the object under consideration.

Temperature, Final: The temperature of a substance as it leaves a piece of apparatus.

Temperature, Saturation: The boiling point of a substance at a given pressure.

Temperature, Wet Bulb: The temperature of the air measured with a thermometer having a bulb covered with a moistened wick.

Thermocouple: A device which generates electricity using the principle that if two dissimilar metals are welded together and the junction is heated, a voltage will develop across the open ends.

Thermometer: A device for measuring temperatures.

Thermostat: A device which is responsive to ambient temperatures.

Valve, Reversing: A valve used to reverse the flow of refrigerant in a heat pump depending on whether heating or cooling is desired.

Vapor: A word used to denote a vaporized fluid rather than the word *gas*.

Vapor Barrier: A thin plastic or metal foil used in air conditioned structures to prevent water vapor from entering the insulating material.

Vapor, Saturated: A vapor condition which will result in the condensation of droplets of liquid as the vapor temperature is reduced.

Vent Cap: A fitting that is designed to keep rain and other foreign matter from entering the vent system and to deflect air currents that may cause downdrafts. It is required that type B and type BW gas vents terminate in an approved cap or roof assembly.

Vent Connector: The pipe which connects a fuel-burning furnace to a vent system.

Vent Gases: The products of combustion from gas furnaces plus the excess air, plus the dilution air in the vent system above the draft hood or regulator.

Vent Manifold: A breeching or lateral section of a common vent by means of which two or more furnaces can be connected to a vertical venting system.

Venting System: A continuous open passageway from the flue collar or draft hood of a fuel-burning furnace to the outdoor atmosphere for the purpose of removing the products of combustion.

Wet Bulb: A device used in the measurement of relative humidity. Evaporation of moisture lowers the temperature of a wet bulb compared to a dry bulb temperature in the same area.

Wet Saturated Steam: Steam at the saturation temperature corresponding to a given pressure; it contains water particles in suspension.

index